W9-CEF-584

Ecological Studies, Vol. 172

Analysis and Synthesis

Ecological Studies

Volumes published since 1998 are listed at the end of this book.

Springer

Berlin
Heidelberg
New York
Hong Kong
London
Milan
Paris
Tokyo

E. Matzner (Ed.)

Biogeochemistry of Forested Catchments in a Changing Environment

A German Case Study

With 172 Figures, 4 in Color, and 51 Tables

Professor Dr. Egbert Matzner
Lehrstuhl für Bodenökologie
Universität Bayreuth
95440 Bayreuth
Germany

ISSN 0070-8356
ISBN 3-540-20973-5 Springer-Verlag Berlin Heidelberg New York

Library of Congress Control Number: 2004104100

Springer-Verlag Berlin Heidelberg New York
Springer-Verlag is a part of Springer Science+Business Media

springeronline.com

Printed in Germany

Production: Friedmut Kröner, 69115 Heidelberg, Germany
Cover design: *design & production* GmbH, 69126 Heidelberg
Typesetting: Kröner, 69115 Heidelberg, Germany

31/3150 YK – 5 4 3 2 1 0 – Printed on acid free paper

Preface

Forest ecosystems represent a major type of land use in Germany and in Europe. They provide a number of functions, or ecosystem services, beneficial to humans, namely biomass production, regulation of the water- and energy cycle, C and N sequestration, erosion control, recreation, and they act as habitat for numerous species. The stability of forest ecosystems in Europe as influenced by the deposition of air pollutants has been a matter of debate for more than 20 years. Besides atmospheric deposition, other environmental conditions affecting forest ecosystems, such as temperature, CO_2 content of the atmosphere and precipitation, have significantly changed in the past and continue to change in the future. Quantifying and predicting the effects of these changes on ecosystem functioning are a challenge to ecosystem research and also a requirement to establish sustainable use of forest ecosystems in the future.

This book summarizes results of long-term, interdisciplinary ecosystem research conducted in two forested catchments and coordinated at the Bayreuth Institute of Terrestrial Ecosystem Research (BITÖK), University of Bayreuth, Germany. It does not aim to summarize all the research of BITÖK in the past decade, which would go far beyond the studies in these two catchments. Instead, we concentrate here on the long-term developments in the biogeochemistry of carbon and mineral elements and on the water cycle, at both the plot and the catchment scale.

The results presented in this book should be valuable to researchers and students from many disciplines who work in terrestrial ecosystem- and landscape ecology, and, furthermore, to hydrologists, foresters and those making decisions relevant to environmental conditions.

The book is based on research that was made possible by the efforts of a large number of persons, not explicitly mentioned in the various chapters, to whom we are deeply grateful.

We would like to thank the Federal Ministry of Research and Technology (BMBF) for the generous funding of BITÖK from 1990–2004. We also thank the Bavarian State Ministry of Sciences Research and the Arts which made BITÖK possible by assuring the continuation of the Departments and Chairs in the University structure after the funding of the Federal Ministry.

Furthermore, our thanks go to the Bavarian State Office for Water Management and to the Bavarian State Forest authorities, who supported BITÖK through intensive cooperation and by providing the research sites.

Thanks are due to the University of Bayreuth for continuous support of the Institute.

BITÖK would not have been established without the pioneering work of Prof. Dr. E.-D. Schulze, who was the initiator of BITÖK and mentor in the first years. We are grateful for his efforts dedicated to BITÖK and also for his constructive comments which helped to improve the book.

Our thanks are also extended to Dr. Thomas Gollan, Scientific Secretary of BITÖK, who solved an incredible number of financial, technical and administrative problems over the years, making the life of scientists at BITÖK productive and enjoyable. Last, but not least, we would like to thank Ingeborg Vogler for her excellent help during the editing of the manuscripts.

Bayreuth, April 2004

Prof. Dr. Egbert Matzner
Director of BITÖK

Contents

Contributors

(bold letters indicate principal authors of chapters and principal investigators of project)

Alewell C.

Environmental Geosciences, University of Basel, Bernoullistr. 32, 4056 Basel, Switzerland

Anthoni P.

Max-Planck-Institute for Biogeochemistry, 07745 Jena, Germany

Berg B.

Department of Soil Ecology, BITÖK, University of Bayreuth, 95440 Bayreuth, Germany, e-mail: bjoern.berg@bitoek.uni-bayreuth.de, Tel.: +49-921-555762, Fax:+49-921-555799

Berger M.

Grundbachtal 6, 01737 Tharandt/Kurort Hartha, Germany, e-mail: martina.berger@epost.de, Tel.: +49-35203-31917, Fax: +49-35203-31918

Bittersohl J.

Bayerisches Landesamt für Wasserwirtschaft, Lazarettstr. 67, 80636 Munich, Germany

Buchmann N.

ETH-Zürich, Institute of Plant Sciences, LFW C56, 8092 Zürich, Switzerland

Büttcher H.

Institute for Water and Soil Dr. Uhlmann, Langobardenstr. 48, 01239 Dresden, Germany

Delany A.C.

National Centre of Atmospheric Research, Boulder, Colorado 80307-3000, USA

Dittmar C.

Department of Forests and Forestry, Weihenstephan University of Applied Sciences, Am Hochanger 5, 85354 Freising, Germany, e-mail: christoph.dittmar@freenet.de, Tel.: +49-9201-799181, Fax: +49-9201-799182

Dlugi R.

Gernotstr. 11, 80804 Munich, Germany

Elling W.

Department of Forests and Forestry, Weihenstephan University of Applied Sciences, Am Hochanger 5, 85354 Freising, Germany

Falge E.

Department of Plant Ecology, University of Bayreuth, 95440 Bayreuth, Germany

Faltin W.

Department of Plant Ecology, University of Bayreuth, 95440 Bayreuth, Germany

Fleck S.

Laboratory for Forest, Nature, and Landscape Research, University of Leuven, Vital Decosterstraat 102, 3000 Leuven, Belgium, e-mail: stefan.fleck@gmx.de, Tel.: +49-5504-937483

Foken T.

Department of Micrometeorology, University of Bayreuth, 95440 Bayreuth, Germany, e-mail: thomas.foken@uni-bayreuth.de, Tel.: +49-921-552293, Fax: +49-921-552366

Gerstberger P.

BITÖK, University of Bayreuth, 95440 Bayreuth, Germany, e-mail: pedro.gerstberger@bitoek.uni-bayreuth.de, Tel.: +49-921-555703, Fax: +49-921-555799

Göckede M.

Department of Micrometeorology, University of Bayreuth, 95440 Bayreuth, Germany

Graus M.

Institute of Ion Physics, Leopold-Franzens-University, 6020 Innsbruck, Austria

Hansel A.

Institute of Ion Physics, Leopold-Franzens-University, 6020 Innsbruck, Austria

Held A.

Institute for Landscape Ecology, University of Münster, Robert-Koch-Str. 26, 48149 Münster, Germany

Kalbitz K.

Department of Soil Ecology, BITÖK, University of Bayreuth, 95440 Bayreuth, Germany, e-mail: karsten.kalbitz@bitoek.uni-bayreuth.de, Tel.: +49-921-555624, Fax: +49-921-555799

Klemm O.

Institute for Landscape Ecology, University of Münster, Robert-Koch-Str. 26, 48149 Münster, Germany, e-mail: otto.klemm@uni-muenster.de, Tel.: +49-251-8333921, Fax: +49-251-8338352

Köstner B.

Meteorology, Institute of Hydrology and Meteorology, Technical University of Dresden, 01737 Tharandt, Germany, e-mail: koestner@forest.tu-dresden.de, Tel.: +49-351-46339100, Fax: +49-351-46339103

Küsel K.

Research Group for Limnology, Institute of Ecology, University of Jena, Carl-Zeiss-Promenade 10, 07745 Jena, Germany, e-mail: kirsten.kuesel@uni-jena.de, Tel.: +49-3641-643257, Fax: +49-3641-6433325

Lange H.

Norwegian Forest Research Institute (Skogforsk), 1432 Ås, Norway

Lindenmair J.

Institute Phytosphere, ICG-III, Research Centre Jülich, 52425 Jülich, Germany, e-mail: j.lindenmair@fz-juelich.de, Tel.: +49-2461-618682, Fax: +49-2461-612492

Lischeid G.

Hydrogeology, BITÖK, University of Bayreuth, 95440 Bayreuth, Germany, e-mail: gunnar.lischeid@bitoek.uni-bayreuth.de, Tel.: +49-921-555632, Fax: +49-921-555799

Mangold A.

Research Centre Jülich, Institute for Chemistry and Dynamics of the Geosphere I, 52425 Jülich, Germany

Matzner E.

Department of Soil Ecology, BITÖK, University of Bayreuth, 95440 Bayreuth, Germany, e-mail: egbert.matzner@bitoek.uni-bayreuth.de, Tel.: +49-921-555610, Fax: +49-921-555799

Michalzik B.

Geographic Institute, University of Göttingen, Section of Landscape Ecology, Junior Professor in Soil Geography and Soil Ecology, Goldschmidtstr. 5, 37077 Göttingen, Germany

Moritz K.

Bayerisches Landesamt für Wasserwirtschaft, Lazarettstr. 67, 80636 Munich, Germany

Nowak A.

Institute for Tropospheric Research, 04318 Leipzig, Germany

Park J.-H.

School of Life Sciences, Arizona State University, P.O. Box 874501, Tempe, Arizona 85287, USA

Rappenglück B.

Institute of Meteorology and Climate Research (IMK-IFU), Forschungszentrum Karlsruhe GmbH, 82467 Garmisch-Partenkirchen, Germany

Rebmann C.

Max-Planck-Institute for Biogeochemistry, Postbox 100164, 07701 Jena, Germany, email: crebmann@bgc-jena.mpg.de, Tel.: +49-3641-576141, Fax: +49-3641-577863

Ruppert J.

Department of Micrometeorology and Department of Climatology, BITÖK, University of Bayreuth, 95440 Bayreuth, Germany

Scheer C.

BITÖK, University of Bayreuth, 95440 Bayreuth, Germany

Schmidt M.

Department of Plant Ecology, University of Bayreuth, 95440 Bayreuth, Germany

Schulze E.-D.

Max-Planck-Institute for Biogeochemistry, 07745 Jena, Germany

Stadler B.

Department of Animal Ecology, BITÖK, University of Bayreuth, 95440 Bayreuth, Germany, e-mail: bernhard.stadler@bitoek.uni-bayreuth.de, Tel.: +49-921-555760, Fax: +49-921-555799

Steinbrecher R.

Institute of Meteorology and Climate Research (IMK-IFU), Forschungszentrum Karlsruhe GmbH, 82467 Garmisch-Partenkirchen, Germany, e-mail: rainer.steinbrecher@imk.fzk.de, Tel.: +49-8821-183217, Fax: +49-8821-183217

Subke J.-A.

Second University of Naples, Department of Environmental Sciences, Via Vivaldi 43, 81100 Caserta, Italy, e-mail: jens.subke@unina2.it, Tel.: +39-0823-274656, Fax: +39-0823-274605

Tenhunen J.D.

Department of Plant Ecology, University of Bayreuth, 95440 Bayreuth, Germany

THOMAS C.

Department of Micrometeorology and Department of Climatology, BITÖK, University of Bayreuth, 95440 Bayreuth, Germany

WICHURA B.

German Meteorological Service, Michendorfer Chaussee 23, 14473 Potsdam, Germany, email: bodo.wichura@dwd.de, Tel.: +49-331-316360, Fax: +49-331-316299

WIEDENSOHLER A.

Institute for Tropospheric Research, 04318 Leipzig, Germany

WRZESINSKY T.

Institute for Landscape Ecology, University of Münster, Robert-Koch Str. 26, 48149 Münster, Germany, e-mail: wrzesinsky@uni-muenster.de, Tel.: +49-251-8339771, Fax: +49-251-8338352

ZIMMERMANN R.

Max-Planck-Institute for Biogeochemistry, 07745 Jena, Germany

ZUBER T.

Department of Soil Ecology, BITÖK, University of Bayreuth, 95440 Bayreuth, Germany

I Introduction

1 Introduction

E. Matzner

1.1 General Introduction

Terrestrial ecosystems can be defined as functionally or structurally discrete units of the landscape. Primary and secondary produces are necessarily parts of terrestrial ecosystems as well as abiotic compounds, namely those that are involved in feedbacks to organisms and to the environment. The latter is the case for various soil components, like organic matter, exchangeable ions, some secondary minerals and the soil solution, which are all considered as part of the ecosystem. Furthermore, the atmospheric surface layer (Prandtl layer) which is influenced by the structure of the vegetation can be considered as part of the ecosystem. On the other hand, the primary soil minerals are considered as part of the environment of the ecosystem, since the mineral composition is mostly geogene and only to a minor extent affected by the organisms of the ecosystem at time scales relevant to ecosystem research. The element release by irreversible weathering of minerals is then seen as an external input (Ulrich 1992).

Tansley (1935), who introduced the term 'ecosystem' in 1935, views the ecosystem as "the basic unit of nature on the earth surface". The integration of abiotic and biotic compounds and their interaction through feedbacks differentiate the ecosystem from other levels of biological organization.

Forest ecosystems fulfil various functions that are important for humans: biomass production, habitat, immobilization of air pollutants, regulation of water and elemental cycling, erosion control and recreation. These functions are used by society to a different degree depending on local conditions and priorities. Besides the use of timber, in Germany, forest ecosystems have become more and more important for ground- and surface water as well as for recreation. While society benefits from forest ecosystems, their functioning has changed in the past and continues to change in the future due to alterations in environmental conditions and management practices. These changes may sometimes be seen as beneficial, as in the case of increasing growth rates of timber (Kauppi et al. 1992; Spieker et al. 1996; Mund et al.

Ecological Studies, Vol. 172
E. Matzner (Ed.), Biogeochemistry of Forested Catchments in a Changing Environment

2002) or C-sequestration (Goodale et al. 2002), but sometimes they may also be detrimental to ecosystem functioning, as in the case of acidification and eutrophication (Fenn et al. 1998; Alewell et al. 2000a).

Ecosystem research is relatively young in comparison to other ecological disciplines. Milestone projects of forest ecosystem research started in the 1960s with the well-known Hubbard Brook catchment study in the United States (Likens et al. 1977) and the German Solling project (Ellenberg et al. 1986). In the Solling project, the detrimental effects of atmospheric pollution on forest ecosystems were first highlighted and the widespread decline of forest ecosystems forecasted (Ulrich et al. 1979). In fact, forest decline phenomena were recognized at the beginning of the 1980s in many regions of Europe and have initiated a large number of research projects in the following years on the functioning of forest ecosystems and the potential reasons for the decline. A synthesis of these is given in Schulze et al. (1989a), Ulrich (1989a) and in Kratz and Lohner (1997).

By the end of the 1980s, German forest ecosystem research was concentrated in 'ecosystem research centers', namely the Forest Ecosystems Research Center at the University of Göttingen, the Bayreuth Institute of Terrestrial Ecosystem Research (BITÖK) at the University of Bayreuth and the Höglwald-Project at the Ludwig Maximilian University in Munich, all working on more basic research questions related to environmental conditions affecting forest ecosystem functions and structures. Recently, forest ecosystem research has focused on more applied questions related to forest management. New coordinated projects were established in addition to the existing research centers at the University of Freiburg, the University of Dresden and at the Forest Research Institute in Eberswalde.

1.2. Goals and Approaches of BITÖK Research

The general task of ecosystem research, and also at the Bayreuth Institute of Terrestrial Ecosystem Research, is to investigate relationships between structure and function in ecosystems and their interactions with the environment and management practices (Fig. 1.1). The evaluation of ongoing changes, the prediction of their effects, mitigation measures as well as the development of criteria for the sustainable use of forest ecosystems should be based on understanding these relationships.

The research at BITÖK is concentrating on forested catchments in the heterogeneous landscape of NE Bavaria, Germany (Chap. 2), but the activities go far beyond the studies at these sites. The approaches used to address the various questions include measurements of fluxes and structural properties in intensive sites, both at the plot and catchment scale, assessments of gradients and patterns at different scales, experiments in the field to test hypotheses at

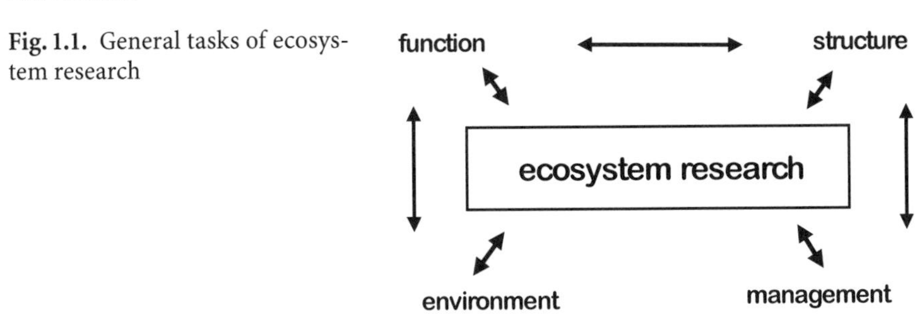

Fig. 1.1. General tasks of ecosystem research

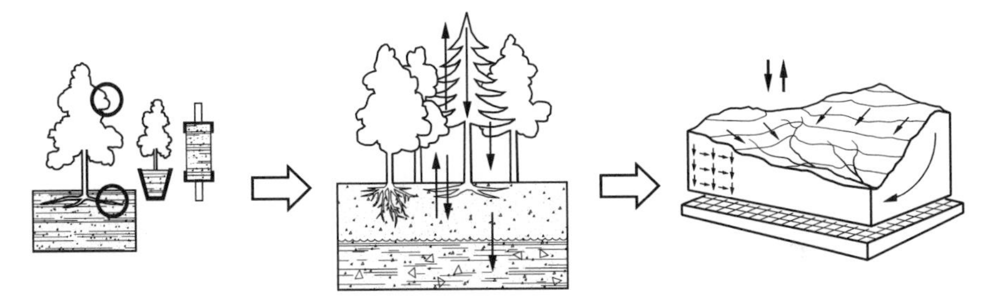

Fig. 1.2. Ecosystem research at different scales

the ecosystem scale, experiments in the laboratory to gain detailed process understanding, model application at different scales and regionalization using empirical indicators of ecosystem functioning (Fig. 1.2).

1.3. Scope of the Synthesis and Problems Addressed

This volume is a synthesis of up to 15 years of biogeochemical research at the two major study sites of BITÖK, the forested Lehstenbach and the Steinkreuz catchments in NE Bavaria.

These catchments were chosen because they each represent typical landscape units of NE Bavaria: The Lehstenbach catchment stands for higher-elevation sites on poor magmatic and metamorphic rocks, often covered by Norway spruce stands. These stands are highly affected by the deposition of air pollutants (Schulze et al. 1989a). The Steinkreuz catchment represents lower-elevation sites on acidic sedimentary rocks often covered by deciduous species, mainly European beech and sessile oak. These sites generally are supposed to be less affected by air pollutant deposition.

As will be shown in Chapter 3, the environmental conditions of forest ecosystems in Germany have changed in the last decades. Thus, special emphasis will be given to the effects of these changes, especially to the response of water- and CO_2 exchange and to the cycling of mineral elements and dissolved organic matter. Furthermore, specific structure–function relations and management effects are addressed.

The **acidification of soils and waters** has been a subject of ecosystem research for more than two decades. Depletion of nutrient cations from forest soils and the release of potentially toxic Al-ions from soil minerals as a result of acidic deposition were often seen as a major cause of forest decline in areas subjected to acidic deposition (Ulrich 1989a, 1989b; Cronan and Grigal 1995; Matzner and Murach 1995; Shortle et al. 2000). There is a general consensus that acidic deposition, namely the deposition of sulfuric acid, is the cause of surface and groundwater acidification in central and northern Europe as well as in North America (Sullivan 2000). Today, under conditions of decreasing SO_4 and H^+ depositions in Europe and North America, the question of the recovery of acidified soils and waters is under debate (Stoddard et al. 1999; Alewell et al. 2000b). The time needed for recovery and the processes involved seem to differ in various regions mainly in relation to soil conditions (Jenkins et al. 2001), with soil S pools and N behaviour as the main driving variables.

In order to address this subject, we present long-term trends on atmospheric deposition of acidifying substances, on soil solution and runoff chemistry and on element budgets. We will investigate the role of soil S, deposited Ca, Mg and N for the recovery process. The role of riparian zones and of NO_3 and SO_4 reduction for the water quality in runoff will be addressed, as well as the dynamics of dissolved organic matter (DOM) and its relevance for the acidity and Al concentration of the runoff.

Central European forests have been subjected to high deposition rates of nitrogen for several decades. The **fate of deposited N and N-saturation** of the ecosystems are major questions presently addressed in ecosystem sciences. While the actual rates of N deposition in forests cannot be exactly quantified and the processes involved in the deposition of N are today not fully understood, the rates for central European forests are estimated at a range of 20–45 kg ha^{-1} $year^{-1}$ (Harrison et al. 2000; de Vries et al. 2002). These rates by far exceed the N demand for the timber accumulation in a growing forest. Since the high rates of deposition have lasted for several decades, substantial amounts of N (several hundreds of kilograms per hectare) were accumulated in the ecosystems, transported to ground- and surface waters in the form of nitrate or emitted to the atmosphere as nitrose gases (Papen and Butterbach-Bahl 1999). The fate of deposited N is related to the C turnover (Schulze 2000) and a major N sink seems to be the soil.

The long-term shifts in ecosystem functioning as a consequence of N deposition are often summarized under the term 'N-saturation'. According to Aber et al. (1989), N saturation is indicated if the storing capacity of the ecosystem is exhausted and nitrate losses with seepage and runoff exceed the natural background losses.

Considering the fate of deposited N and the time needed to reach N saturation, the soil N pool and soil N turnover are of major importance since the N pool of soils by far (often ten times) exceeds the N pool of the vegetation. The mechanisms of N accumulation in forest soils, their kinetics and limitations are yet unresolved and predictions of the N cycle in response to changing deposition and environmental conditions are thus very uncertain.

Here, we present results on the long-term development of nitrate concentrations and fluxes in soil solution and runoff in highly N-polluted forest ecosystems. Actual rates of N deposition and the role of fog deposition will be estimated. We will contribute to the fate of N by estimating the N accumulation in the forest floor during litter decomposition and we will address the role of dissolved organic N for the N turnover in the ecosystem. The effect of high leaf N concentrations on gas-exchange will be elucidated. The phyllosphere organisms and their influence on temporal and spatial patterns of throughfall N fluxes will be part of the synthesis as well as the hydrological conditions related to nitrate in runoff.

Knowledge on climatic controls of **water-vapor exchange between vegetation and atmosphere** is important to predict evapotranspiration, a significant component of the catchment water balance. The relevant controlling variables are available energy, saturation vapor pressure deficit of the air, the aerodynamic (g_a) and canopy conductance (g_c, average stomatal conductance) of the vegetation (Monteith 1965). Within these parameters, the estimation of g_c is most difficult because it represents the highly sensitive reaction of approximately 10^{11} stomata per tree (Larcher 2001) to their highly variable micrometeorological situation. This makes g_c dependent on plant internal structural and physiological variations and micrometeorological changes, as well as plant external variations in genetic properties and vegetation structure. The latter dependencies may be retrieved in spatially integrating measures of vegetation structure, e.g., the temporal and spatial variability of plant age, size, height, stand density and leaf area index (LAI) (Shuttleworth 1989; Oliver 1997). Thus, knowledge on vegetation controls must still be based on empirical estimates of g_c derived and up-scaled from different levels of integration (leaf, tree, forest canopy), and the investigation of effects of vegetation change on the water balance has been confined to studies at the catchment scale (Bosch and Hewlett 1982). Structural dynamics of vegetation may be strongly influenced by human activities, e.g. increased N input or forest management practices. Therefore, predicting and valuing potential effects of environmental change requires understanding and quantification of dependencies between vegetation structure itself and atmospheric exchange processes. Up to now, plant parameters in land surface–atmosphere exchange models are mostly confined to LAI, vegetation height and/or roughness length. However, the importance of LAI and vegetation height for stand-level transpiration and atmospheric coupling may change significantly with species, stand dynamics during aging or spatial distribution of plant individuals or patches (e.g. Magnani et al. 2000; Köstner et al. 2001; Binkley et al. 2002). Especially in mixed broad-leaved forests, the vertical distribution of leaf and branch structures shows high adaptability to light environment and space capture (e.g. Küppers 1994; Niinemets et al. 1998). However, relatively little information is available on mixed-species stands and predictions at the whole-stand level are difficult (Kelty 1992). Forest management practices in central Europe are currently converting coniferous monocultures to more natural, mixed-species stands. It is therefore important to address potential functional changes related to structural change.

Here, we present studies on mixed stands of European beech (Fagus sylvatica) and pedunculate oak (Quercus petraea) with varying abundance of beech and oak. Structural properties and exchange processes with the atmosphere are presented at the leaf, branch and crown internal level as well as at the tree and forest canopy level. The relevance of crown-internal gradients for upscaling of conductances and transpiration is investigated in a detailed study at the branch level. Effects of gradual structural changes on canopy transpiration and carbon gain are simulated for the stand-level under comparable climatic conditions. The results are compared with coniferous monocultures as well as with natural mixed broad-leaved stands.

In the context of the global C budget and the increase in atmospheric CO_2, the **terrestrial C sinks and the C sequestration in forest ecosystem** are key issues (Kauppi et al. 1992; IPCC 2001; Goodale et al. 2002). The soil C pool in temperate forest ecosystems often exceeds the C pool of the vegetation and the C sequestration of soils may lead to a long-lasting C sink if the C is allocated to the stable soil C fractions. The processes determining the sink or source function of soils for C in relation to vegetation, climate, atmospheric CO_2 and N deposition are only qualitatively understood. In addition, changes in soil C pools are difficult to measure in comparison to changes in the vegetational pools. Thus, quantifications of soil and ecosystem C pool development at all scales in non-disturbed forests are subject to large uncertainties. Various methods are presently used to establish C budgets of terrestrial ecosystems, namely C inventories of soil and vegetation, chronosequences (Schulze 2000) and microclimatic C flux measurements above the forest canopy (Valentini et al. 2000). All these methods have different advantages and shortcomings and we still have not reached a common conclusion on the magnitude and location of terrestrial C sinks and their development.

We address these questions by presenting the C budget of a forest ecosystem based on micrometeorological and C-isotopic measurements, investigating the growth development of the trees and estimating the C sequestration of the soil. Furthermore, the role of dissolved organic carbon for the C turnover in the ecosystem and especially in the soil will be assessed.

Predicting the effects of global climatic change on the function of terrestrial ecosystem is a big challenge to ecosystem research (Walker et al. 1999). Increasing average temperatures up to several degrees are expected for the upcoming decades. Changes in precipitation intensity, amount and distribution are also likely to occur, but their prognosis is rather uncertain (IPCC 2001). Evaluating and predicting future ecosystem functioning require the quantitative understanding of the **climatic controls on processes and fluxes.** These are addressed today in several ways: by experimental manipulations in the field, like soil warming (Bergh and Linder 1999; Rustad et al. 2000) or precipitation manipulation (Park and Matzner 2003), by laboratory experiments (Kirschbaum 1995), by use of regional gradients (Glardina and Ryan 2000) and by empirical analysis of flux measurement (Köstner et al. 2001; Lischeid 2001). The biggest problem involved in these approaches is to use the observations normally made at relatively small temporal and spatial scales for extrapolation to larger scales. For example, the temperature dependence of soil processes resulting from laboratory experiments has been recently criticized for its use in large-scale models (Glardina and Ryan 2000; Schlesinger et al. 2000) because the short-term effects significantly differed from the long-term effects of temperature.

In relation to climatic controls of fluxes and processes in forest ecosystems we present here results on the water and C exchange between vegetation and atmosphere, on deposition processes, the regulation of soil respiration, the dynamics of dissolved organic matter and mineral elements in soil solution and on the generation of runoff.

Developing sustainable use of natural resources is a major political goal worldwide. Sustainable use can be defined as one that optimizes the needs of the present generation without disturbing the potential of future generations to satisfy their needs (Hauff 1987). While there is international agreement that ecosystems should be managed to allow sustainable use, no general agreement exists on the **criteria used to identify sustainable use** of ecosystems (Cocklin 1989). In relation to terrestrial ecosystem functions, one may use the cycling of elements and water as criteria for sustainability. These cycles should be managed to sustainably fulfil the ecosystem functions without polluting the atmosphere and hydrosphere. Furthermore, the protection of species and their functions may be taken as a criterion (Holling 1986; Chapin et al. 1996; Christensen et al. 1996).

In relation to the effects of atmospheric deposition on forested ecosystems and their sustainability, the 'critical load' concept has been introduced (Nilsson and Grennfelt 1988; Schulze et al. 1989b). The critical load is defined as "a quantitative estimate of an exposure to one or more pollutants below which significant harmful effects on specific sensitive elements of the environment do not occur according to the present knowledge". While the concept has several shortcomings (Cresser 2000), critical loads have been estimated for European forests in the case of S, acidity and N deposition and have found access to European legislation on air pollution. Their calculation is mostly based on soil processes and requires, first, the definition of target parameters ('critical criteria'), like, for example, soil parameters, runoff chemistry and tree nutritional parameters which are supposed to be reactive to changing deposition rates (van der Salm and de Vries 2001). Second, the quantification of ecosystem processes and fluxes like N immobilization, weathering of minerals and biomass accumulation is required.

Our results will be of importance in this context by addressing temporal developments of parameters relevant to critical loads in response to changing deposition and by presenting long-term element budgets for N and nutrient cations. The estimates of N sequestration rates in soil organic matter are highly relevant for the calculation of critical loads for N.

The contributions to this book are organized in four major parts, dealing (1) with the documentation of environmental changes, (2) the response of vegetation, (3) soil response and (4) catchment response. With respect to the questions addressed above, the results will be integrated in the synthesis chapter.

References

Aber JD, Nadelhoffer KJ, Steuder P, Melillo JM (1989) Nitrogen saturation in northern forest ecosystems – hypotheses and implications. BioScience 39:378–386

Alewell C, Manderscheid B, Gerstberger P, Matzner E (2000a) Effects of reduced atmospheric deposition on soil solution chemistry and elemental contents of spruce needles in NE-Bavaria, Germany. J Plant Nutr Soil Sci 163:509–516

Alewell C, Manderscheid B, Meesenburg H, Bittersohl J (2000b) Is acidification still an ecological threat? Nature 407:856–857

Bergh J, Linder S (1999) Effects of soil warming during spring on photosynthetic recovery in boreal Norway spruce stands. Global Change Biol 5:245–253

Binkley D, Stape JL, Ryan MG, Barnard HR, Fownes J (2002) Age-related decline in forest ecosystem growth: an individual-tree, stand-structure hypothesis. Ecosystems 5:58–67

Bosch JM, Hewlett JD (1982) A review of catchment experiments to determine the effect of vegetation change on water yield and evapotranspiration. J Hydrol 55:3–23

Chapin FG, Torn MS, Tateno M (1996) Principles of ecosystem sustainability. Am Nat 148:1016–1034

Christensen NL, Bartuska AM, Brown JH, Carpenter S, D'Antonio C, Francis R, Franklin JF, MacMahon JA, Noss RF, Parsons DJ, Peterson CH, Turner MG, Woodmansee RG (1996) The report of the Ecological Society of America Committee on the scientific bases for ecosystem management. Ecol Appl 6:665–691

Cocklin CR (1989) Methodological problems in evaluating sustainability. Environ Conserv 16:343–351

Cresser MS (2000) The critical loads concept: milestone or millstone for the new millennium? Sci Total Environ 249:51–62

Cronan CS, Grigal DF (1995) Use of calcium/aluminium ratios as indicators of stress in forest ecosystems. J Environ Qual 24:209–226

De Vries W, Reinds GJ, van Dobben H, de Zwart D, Aamlid D, Neville P, Posche M, Auée J, Voogd JCH, Vel EM (2002) Intensive monitoring of forest ecosystems in Europe. European Commission, Brussels

Ellenberg H, Meyer R, Schauermann J (1986) Ökosystemforschung im Solling: Ergebnisse des Sollingprojektes 1966–1986. Ulmer Verlag, Stuttgart

Fenn ME, Poth MA, Aber JD, Baron JS, Bormann BT, Johnson DW, Lemley AD, McNulty SG, Ryan DF, Stottlemeyer R (1998) Nitrogen excess in North American ecosystems: predisposing factors, ecosystem responses, and management strategies. Ecol Appl 8:706–733

Glardina CP, Ryan MG (2000) Evidence that decomposition rates of organic carbon in mineral soil do not vary with temperature. Nature 404:858–861

Goodale CL, Apps MJ, Birdsey RA, Field CB, Heath LS, Houghton RA, Jenkins JC, Kohlmaier GH, Kurz W, Liu S, Nabuurs GJ, Nilsson S, Shvidenko AZ (2002) Forest carbon sinks in the northern hemisphere. Ecol Appl 12:891–899

Harrison AF, Schulze ED, Gebauer G, Bruckner G (2000) Canopy uptake and utilization of atmospheric nitrogen. Ecological studies 142. Springer, Berlin Heidelberg New York, pp 171–188

Hauff V (1987) Unsere gemeinsame Zukunft. Der Brundtland-Bericht der Weltkommission für Umwelt und Entwicklung. Eggenkamp Verlag, Greven

Holling CS (1986) Resilience of ecosystems: local surprise and global change. In: Clark WC, Munn RE (eds) Sustainable development and the biosphere. Cambridge University Press, Cambridge, pp 292–317

IPCC (2001) Technical summary of the working group I report. http://www.ipcc.ch

Jenkins A, Ferrier RC, Wright RF (2001) Assessment of recovery of European surface waters from acidification 1970–2000. Hydrol Earth Syst Sci (Spec Issue) 5:273–541

Kauppi PE, Mielikäinen K, Kuusela K (1992) Biomass and carbon budget of European forests, 1971–1990. Science 256:70–74

Kelty MJ (1992) The ecology and silviculture of mixed-species forests. Kluwer, Dordrecht

Kirschbaum MUF (1995) The temperature dependence of soil organic matter decomposition, and the effect of global warming on soil organic storage. Soil Biol Biochem 27:753–760

Köstner B, Tenhunen JD, Alsheimer M, Wedler M, Scharfenberg HJ, Zimmermann R, Falge E, Joss U (2001) Controls on evapotranspiration in a spruce forest catchment of the Fichtelgebirge. In: Tenhunen JD, Lenz R, Hantschel R (eds) Ecosystem approaches to landscape management in central Europe. Ecological studies, vol 147. Springer, Berlin Heidelberg New York, pp 377–415

Kratz W, Lohner H (1997) Evaluation of the results of forest damage research (1982–1992) to explain complex cause–effect relationships by means of systems analytical methods. UBA-Texte 6/97. Umweltbundesamt, Berlin

Küppers M (1994) Carbon gaps: competitive light interception and economic space filling – a matter of whole-plant allocation. In: Caldwell MM, Pearcy RW (eds) Exploitation of environment heterogeneity by plants. Ecophysiological processes above- and belowground. Academic Press, San Diego, pp 111–114

Larcher W (2001) Ökophysiologie der Pflanzen. Ulmer Verlag, Stuttgart

Likens GE, Borman FH, Pierce RS, Eaton JS, Johnson NM (1977) Biogeochemistry of a forested ecosystem. Springer, Berlin Heidelberg New York

Lischeid G (2001) Investigating short-term dynamics and long-term trends of SO_4 in the runoff of a forested catchment using artificial neural networks. J Hydrol 243:31–42

Magnani F, Mencuccini M, Grace J (2000) Age-related decline in stand productivity: the role of structural acclimation under hydraulic constraints. Plant Cell Environ 23:251–263

Matzner E, Murach D (1995) Soil changes induced by air pollutant deposition and their implication for forests in central Europe. Water Air Soil Pollut 85:63–76

Monteith JL (1965) Evaporation and environment. In: Fogg GE (ed) The state and movement of water in living organisms. Symp Soc Exp Biol 19. Academic Press, New York, pp 205–234

Mund M, Kummets E, Hein M, Bauer GA, Schulze ED (2002) Growth and carbon stocks of a spruce forest chronosequence in central Europe. For Ecol Manage 171:275–296

Niinemets Ü, Kull O, Tenhunen JD (1998) An analysis of light effects on foliar morphology, physiology, and light interception in temperate deciduous woody species of contrasting shade tolerance. Tree Physiol 18:681–696

Nilsson J, Grenfelt P (1988) Critical loads for sulphur and nitrogen: report. Nordic Council of Ministers, Copenhagen

Oliver CD (1997) Similarities of stand structures and stand development processes throughout the world – some evidence and applications to silviculture through adaptive management. In: Kelty MJ (ed) The ecology and silviculture of mixed-species forests. Kluwer, Dordrecht, pp 11–26

Papen H, Butterbach-Bahl K (1999) A 3-year continuous record of nitrogen trace gas fluxes from untreated and limed soil of a N-saturated spruce and beech forest ecosystem in Germany, 1. N_2O emission. J Geophys Res 104:18487–18503

Park JH, Matzner E (2003) Controls on the release of dissolved organic carbon and nitrogen from a deciduous forest floor investigated by manipulations of aboveground litter inputs and water flux. Biogeochemistry 66:265–286

Rustad LE, Melillo JM, Mitchell MJ, Fernandez IJ, Steudler PA, McHale PJ (2000) Effects of soil warming on carbon and nitrogen cycling. Responses of northern U.S. In: Mickler RA, Birdsey RA, Hom J (eds) Forests to environmental change. Ecological studies, vol 139. Springer, Berlin Heidelberg New York, pp 357–381

Schlesinger WH, Winkler JP, Megonigal JP (2000) Soils and the global carbon cycle. Wiley, New York

Schulze E-D (2000) Carbon and nitrogen cycling in European forest ecosystems. Ecological studies, vol 142. Springer, Berlin Heidelberg New York

Schulze E-D, Lange OL, Oren R (1989a) Forest decline and air pollution. Ecological studies, vol 77. Springer, Berlin Heidelberg New York

Schulze E-D, de Vries W, Hauhs M, Rosen K, Rasmussen L, Tamm O, Nilsson J (1989b) Critical loads for nitrogen deposition on forest ecosystems. Water Air Soil Pollut 48:451–456

Shortle WC, Smith KT, Minocha R, Minocha S, Wargo PM, Vogt KA (2000) Tree health and physiology in a changing environment. Ecological studies 139. Springer, Berlin Heidelberg New York, pp 229–274

Shuttleworth WJ (1989) Micrometeorology of temperate and tropical forest. Philos Trans R Soc Lond B 324:299–334

Spieker KH, Mielikäinen K, Köhl M, Skovsgaard JP (1996) Growth trends in European forests. Springer, Berlin Heidelberg New York

Stoddard JL, Jeffries DS, Lükewille A, Clair TA, Dillon PJ, Driscoll CT, Forsius M, Johannessen M, Kahl JS, Kellogg JH, Kemp A, Mannio J, Monteith DT, Murdoch PS, Patrick S, Rebsdorf A, Skjelkvale BL, Stainton MP, Traaen T, van Dam H, Webster KE, Wieting J, Wilander A (1999) Regional trends in aquatic recovery from acidification in North America and Europe. Nature 401:575–578

Sullivan TJ (2000) Aquatic effects of acidic deposition. Lewis, London
Tansley AG (1935) The use and abuse of vegetational concepts and terms. Ecology 16:284–307
Ulrich B (1989a) Effects of acid deposition on forest ecosystems in Europe. In: Adriano AC, Havas M (eds) Advances in environmental science. Acidic precipitation, vol 2. Springer, Berlin Heidelberg New York, pp 169–272
Ulrich B (1989b) Forest decline in ecosystem perspective. In: Ulrich B (ed) Proc Int Congr on Forest Decline Research: State of Knowledge and Perspectives. Forschungszentrum Karlsruhe, Karlsruhe
Ulrich B (1992) Forest ecosystem theory based on material balance. Ecol Modell 63:163–183
Ulrich B, Mayer R, Khanna PK (1979) Deposition von Luftveruntreinigungen und ihre Auswirkungen in Waldökosystemen im Solling. Schr Forstl Fak Univ Göttingen 58:21–42
Valentini R, Matteucci G, Dolman AJ, Schulze E-D, Rebmann C, Moors EJ, Granier A, Gross P, Jensen NO, Pilegaard K, Lindroth A, Grelle A, Bernhofer C, Grünwald T, Aubinet M, Ceulemans R, Kowalski AS, Vesala T, Rannik Ü, Berbigier P, Loustau D, Gudmundsson J, Thorgelrsson H, Ibrom A, Morgenstern K, Clement R, Moncrieff J, Montagnani L, Minerbi S, Jarvis PG (2000) Respiration as the main determinant of carbon balance in European forests. Nature 404:861–865
Van der Salm C, de Vries W (2001) A review of the calculation procedure for critical acid loads for terrestrial ecosystems. Sci Total Environ 271:11–25
Walker B, Steffen W, Canadell J, Ingram J (1999) The terrestrial biosphere and global change. Implications for natural and managed ecosystems. International geosphere-biosphere programme book series 4. Cambridge University Press, Cambridge

2 The Lehstenbach and Steinkreuz Catchments in NE Bavaria, Germany

P. Gerstberger, T. Foken, and K. Kalbitz

2.1 The Lehstenbach Catchment in the Fichtelgebirge Mountains

2.1.1 Brief Overview of the Region

The arched, densely forested Fichtelgebirge (ca. 1,000 km^2) lies in the northeastern part of Bavaria (district of Oberfranken; near the frontier to the Czech Republic; Fig. 2.1) at the confluence of three Palaeozoic, Variscian mountain ridges of central Europe: the Franconian forest, Ore Mountains, and Bohemian forest. The Fichtelgebirge comprises a large granitic pluton surrounded by metamorphic rock series as gneiss, mica schists, and phyllites. The summits of the Fichtelgebirge, Schneeberg (1,053 m), and Ochsenkopf (1,023 m) do not reach the timberline. The mountain ridge acts as a main European catchment between the Atlantic North Sea and the Danubian Black Sea.

In the Pleistocene the Fichtelgebirge was not glaciated, but erosion and solifluction occurred, leaving typical 'woolsack' rock formations of granite (Waldstein, Rudolfstein, Nußhardt) on the summits and forming large boulder streams (Haberstein, Platte, Schneeberg).

The region was, for several decades, heavily influenced by air pollution from a power plant in Arzberg (now being shut down) and from industrial agglomerations in the Czech Republic with strong emissions of SO_2. The long-lasting pollution of the region resulted in severe acidification of the soils, and forest decline at higher altitudes of the Fichtelgebirge occurred in the 1980s (Schulze et al. 1989) mainly in the upper regions between 800 and 1,053 m a.s.l. The SO_2 concentration of the air has dropped dramatically from values of about 60–80 ppb in 1987 to 1–4 ppb today, due to smoke cleaning, filtering, and the ceasing of lignite use for electric power production. However, the input of N is still high and shows no decreasing trend (Matzner et al., this Vol.).

The main experimental plots are located in the Lehstenbach catchment in the so-called Waldstein hillsides, a mountainous ridge of the northwestern

Ecological Studies, Vol. 172
E. Matzner (Ed.), Biogeochemistry of Forested Catchments in a Changing Environment

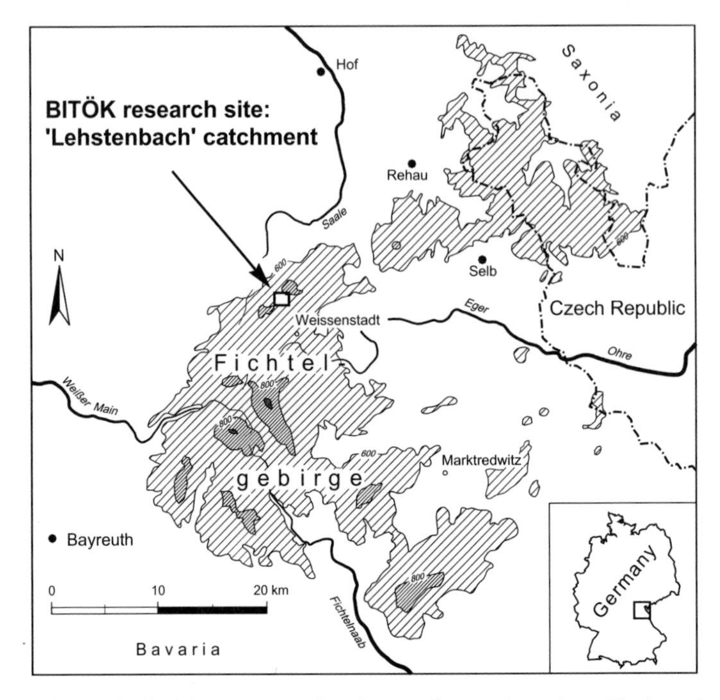

Fig. 2.1. The Fichtelgebirge mountains in northeast Bavaria with location of the Lehstenbach catchment

Fichtelgebirge mountains which reaches up to 877 m a.s.l. (Grosser Waldstein). The Lehstenbach brook flows into the Eger, which runs through the Czech Republic as the Ohre and flows into the Elbe which leads into the North Sea. General characteristics of the Lehstenbach catchment are given in Table 2.1.

2.1.2 Natural Vegetation of the Region

The natural climax vegetation of the Fichtelgebirge (according to investigations of peat bogs; Firbas and von Rochow 1956) would be a mixed broad-leaved forest dominated by beech (*Fagus sylvatica*) with a high portion of fir (*Abies alba*, locally up to 40 %), sycamore (*Acer pseudo-platanus*), and elm (*Ulmus glabra*). On exposed sites and on infertile or shallow soils, light-demanding pioneer species such as downy birch (*Betula pubescens* subsp. *carpatica*), aspen (*Populus tremula*), rowan (*Sorbus aucuparia*), and Scots pine (*Pinus sylvestris*) occur (Reif 1989). Scots pine becomes more dominant on the well-drained and shallow sites on metamorphic phyllites at the margins of the granitic pluton. On local peat deposits on saddles and flat areas

Table 2.1. Properties of the catchments Lehstenbach in the Fichtelgebirge and Steinkreuz in the Steigerwald

	Lehstenbach catchment	Steinkreuz catchment
Catchment size	4.5 km^2	0.5 km^2
Location	50°8′35″N, 11°52′8″E (tower of Weidenbrunnen)	49°52′21″N, 10°27′45″E
Elevation	877–695 m a.s.l. (tower of Weidenbrunnen: 775 m a.s.l.)	400–460 m a.s.l.
Annual precipitation	1,156,5 mm (1971–2000)	approx. 700–800 mm
Mean annual temperature	5.3 °C (1971–2000)	7.9 °C
Geology	Porphyritic granites, phyllites, and quartzite of Ordovician age; weathered up to a depth of 30 m	Upper Keuper (Blasensandstein, Coburger Sandstein, Lower Burgsandstein)
Soil type	Haplic Podzols, Dystric Cambisols, Hostosols	Dominant: Dystric Cambisols; further: Gleyic and Stagnic Cambisols
Tree species	Norway spruce (*Picea abies*)	European beech (*Fagus sylvatica*), sessile oak (*Quercus petraea*)
Age of stand	50 years (Weidenbrunnen), 130 years (Coulissenhieb)	80–130 years
Main understorey species	*Calamagrostis villosa*, *Deschampsia flexuosa*, *Vaccinium myrtillus*, *Dryopteris dilatata*, *Oxalis acetosella*, *Dicranum scoparium*	Sparse understorey vegetation: *Luzula albida*, *Oxalis acetosella*, *Anemone nemorosa*, *Maianthemum bifolium*, *Carex montana*, *C. brizoides*, *Deschampsia flexuosa*
Understorey cover	60–80 %	10–25 %

with poor drainage (today nearly destroyed completely), the endangered *Pinus mugo* subsp. *rotundata* formed nearly pure stands (Geiger 1994). At the highest altitudes (Schneeberg), around widely distributed bogs, in frost hollows, and around boulder fields, Norway spruce (*Picea abies*) occurred naturally as a displaced species after beech invasion between ca. 700 B.C. and 1200 A.D. After the vast exploitation of the forests over the last 400 years in order to supply the mining industry with construction material and charcoal, by litter raking (used for padding the stables of livestock in winter), potash burning (a component in the making of glass), the porcelain industry, etc., the region was mostly free of a closed forest cover and was partly changed to heathland and birch bush. Most of the area has been afforested since the 19th century with Norway spruce, irrespective of the habitat. Unaware of the importance of seed origin, spruce seeds of lowlands (from Poland, etc.) were used for the montane reforestations, resulting in unadapted spruce forest with a high percentage of crown break, due to heavy snow cower and severe hoarfrost cover in winter.

Today, the Fichtelgebirge is dominated by spruce forests (ca. 93 %), with many plots of the same age class. Spruce forests with the dominant understorey grass species *Calamagrostis villosa* (*Calamagrostis villosa–Picea abies* community) are the most frequent community at higher altitudes; at altitudes between 600 and 800 m a.s.l. the understorey grass species *Deschampsia* (=*Avenella*) *flexuosa* and the blueberry (*Vaccinium myrtillus*) dominate the spruce forests. Few other species comprise the low ground cover: *Oxalis acetosella*, *Carex pilulifera*, *Trientalis europaea*, and *Rumex acetosella* (Reif 1989; Reif and Leonhardt 1991).

2.1.3 Experimental Sites and Vegetation

In the forested Lehstenbach catchment (size: 4.5 km^2; 695–877 m a.s.l.) up to six investigation plots were established: Coulissenhieb, Weidenbrunnen, Schlöppner Brunnen I and II, Gemös, LFW01, and other small, short-term plots (Table 2.2, Fig. 2.2).

After 7 years of intensive ecosystem research in a mature, 130-year-old spruce stand, an area of 0.6 ha of the Coulissenhieb stand was felled in 1999–2000 (site 'clearcut'; see Table 2.2), the whole aboveground biomass of the trees was removed, and one half of the area was replanted with tree species of the natural vegetation (beech: *Fagus sylvatica*, white fir: *Abies alba*, and maple: *Acer pseudoplatanus*). Other species such as rowan (*Sorbus aucuparia*), downy and common birch (*Betula pubescens* subsp. *carpatica* and *Betula pendula*), and willow (*Salix caprea*) established themselves on the clearcut, whereas all seedlings of spruce were removed. On the other half of the area, seedlings of spruce were kept undisturbed, so that they can build up another new spruce stand. Thus, investigations on the impacts of the two veg-

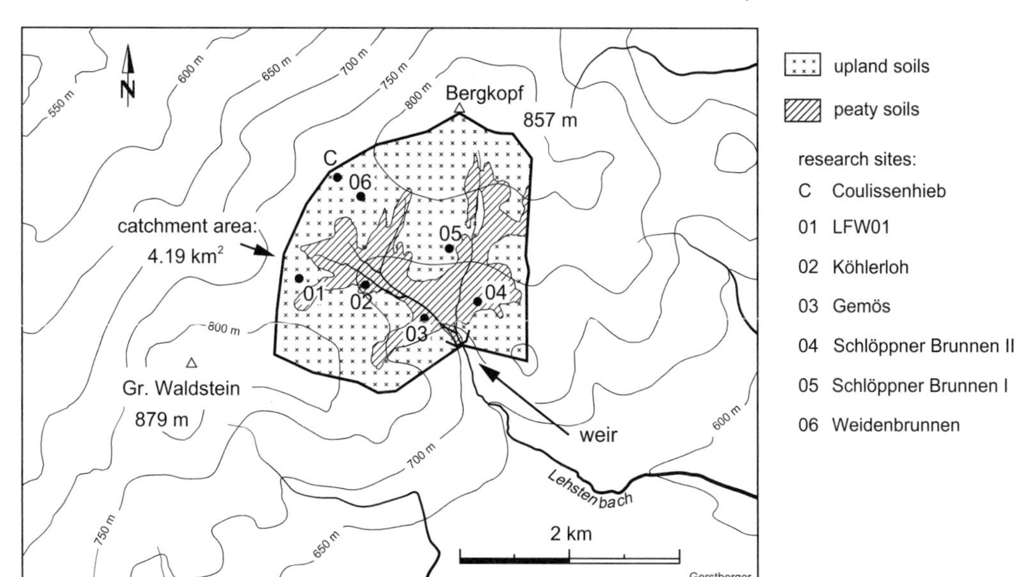

Fig. 2.2. Lehstenbach catchment with contour lines and BITÖK research sites

etation types or tree species on the soil and nutrient cycling started in 2000 at the Coulissenhieb site.

The main focus of the ecosystem research of BITÖK lies in the tree vegetation as the main actor and regulator in exchange processes between the atmosphere and vegetation, element cycling, turnover, and soil and groundwater development. Understorey vegetation was thought to be of minor importance, but as Wedler et al. (1997) showed, the transpiration of the understorey of central European spruce forests comprises about one quarter that of the tree layer.

Inside the plot Coulissenhieb (area: 2.3 ha), the location of every spruce tree (*Picea abies*) and every sensor was mapped by tachymetry. The understorey vegetation was mapped on a 85 × 32-m plot in summer 1996. There the vegetation is composed of about a dozen acidophilic plant species (mainly *Deschampsia flexuosa*, *Vaccinium myrtillus*, *Calamagrostis villosa*, *Galium harcynicum*, and the fern *Dryopteris dilatata*). Different moss species are common, covering about 45–60 % of the soil (*Dicranum scoparium*, *Brachythecium curvifolium*, *Plagiothecium undulatum*, *Polytrichum formosum*, *Pleurozium schreberi*, *Pohlia nutans*, *Lophocolea bidentata*, and *Ptilidium ciliare*). Plant indicators of high nitrogen input from the atmosphere to the soil ('nitrophytes') are still not prevalent on the Coulissenhieb site, presumably because of (1) the very low pH value of the upper soil horizon between 2.6 and

Table 2.2. Properties of the main investigation sites in the Lehstenbach catchment. *LAI* Leaf area index; *n.d.* not determined

Site	Investigated since	Area (ha)	Altitude (m a.s.l.)	Exposition	Slope (°)	Soil	Mean age of trees (2003)	Number of trees (ha^{-1})	Mean height (m)	Basal area ($m^2 ha^{-1}$)	Sapwood area (m^2)	LAI trees ($m^2 m^{-2}$)	LAI understorey ($m^2 m^{-2}$)
Coulissenhieb	1992	2.3	765–785	W	4	Haplic Podzol	130	322	26.7	39.4	13.6	6.5	2.06
Coulissenhieb – ‘clearcut’	1997	0.6	770–785	W	4	Haplic Podzol	5						
Weidenbrunnen	1993	0.95	770–780	SW	2	Haplic Podzol	50	1,007	16.1	30.9	20.3	5.3	0.31
Schlöppner Brunnen	1993	0.4	750	S	5	Histosol	45	1,678	14.7	39.5	23.1	6.3	0.16
Gemös	1995	0.24	740	S	7	Histosol	115	446	25.7	42.4	18.8	7.6	1.01
LFW01	1987	0.13	770	NO	4	Haplic Podzol	108	522	26.5	43.7	n.d.	n.d.	n.d.

2.9 pH ($CaCl_2$) and (2) the dense tree canopy. However, *Urtica dioica*, *Senecio ovatus*, *Epilobium angustifolium*, *Sambucus racemosa*, and other nitrophytes are found on cleared or limed spruce stands in the vicinity.

The tree canopy at the Coulissenhieb site is still relatively dense (ca. 360 trees ha^{-1}, LAI of spruce trees 6.5–7.0). Thus the light-dependent grass *Calamagrostis villosa* could not yet establish itself as the main dominant species of the understorey vegetation, which is today commonly found in upper montane spruce forests (800–1,051 m a.s.l.) of the Fichtelgebirge (Koppisch 1994; Heindl and Betz 1995; Betz 1998). The current dramatic invasion of *Calamagrostis* in all mountainous regions in central Europe (Fiala 1996) is supposed to be promoted (1) by increased nitrogen pollution inputs to the stands and (2) increased light penetration through the canopy due to forest decline and defoliation (Betz 1998). The grass species, rather intolerant to other plant species of the understorey, needs a high water supply or high rainfall. At lower elevations with less precipitation, spreading is favoured by wet soil conditions; thus the occurrences are restricted to wells, margins of bogs, and along brooks. *Calamagrostis villosa* is a summer green rhizomatous grass species which – at Coulissenhieb – produces an annual litter of dead leaves and culms up to 1,075 kg ha^{-1}, with a C/N ratio of ca. 24.8 (corresponding to 474.0 kg C and 19.1 kg N ha^{-1}; Gerstberger, unpubl. data). The total underground biomass (mainly rhizomes) at Coulissenhieb varies from 773–1,880 kg ha^{-1}. In open habitats and on sites with damaged tree cover due to forest decline *C. villosa* reaches aboveground biomass production of up to 3,492 kg ha^{-1} (Koppisch 1994; Betz 1998). Thus *Calamagrostis* litter decomposition, nitrogen release, and humus accumulation play an important role in the cycling of nitrogen in the ecosystem.

2.1.4 Climate

The location of the Lehstenbach catchment on the upwind side of the Waldstein ridge is influenced by topographically caused precipitation. Easterly winds enter the region through the upper Eger/Ohre River valley from the northwest part of the Czech Republic. Particularly in winter, high-pressure cold air from the easterly lying Bohemian basin can reach the Waldstein region directly through the Eger/Ohre valley. According to the effective climate classification by Köppen, in the modification by Trewartha (Hupfer 1996), the region is a continental temperate climate (Dc). Because of the high precipitation sums, the climate has a maritime character. According to Henning and Henning (1977), the climate is also defined as 'moist-continental' (Eiden et al. 1989).

Extensive climatological measurements are available for the region. A list of all relevant climate and precipitation stations of the German Meteorological Service and the University of Bayreuth are given in Foken (2003a). Unfortunately, all hilltop stations (Ochsenkopf 1,023 m a.s.l. and Waldstein 879 m a.s.l.) were closed in approx. 1980 and the station on the Schneeberg (1,053 m a.s.l.) operated only during World War II (Holzapfel 1949).

The different climate elements were investigated primarily for the normal climate period 1961–1990 and 1971–2000 (Foken 2003a). The climate data for the Waldstein/Weidenbrunnen site and the upper Eger River valley are given in Tables 2.3 and 2.4.

The Waldstein region is rather cool in comparison to other regions in Bavaria or to Upper Franconia (Vollrath 1977, 1978, 1979; Reichel 1979). For instance, the average monthly temperature in all months is approx. 3 K lower compared with Bayreuth (the next large town ca. 25 km southwest of the Fichtelgebirge), in winter even more so, caused by the cold easterly wind in the upper Eger River valley. The yearly temperature amplitude in the Waldstein region was 18.3 K for 1971–1990. The typical vertical temperature gradient of the Fichtelgebirge mountains was 0.61 K/100 m (1971–1990).

The extreme temperatures are in winter approx. –20 °C and in summer approx. 30 °C. The precipitation has a maximum in December (11 % of the yearly sum for Waldstein/Weidenbrunnen) with a second maximum in July (9.5 %; Table 2.4). The maximum precipitation event in the period 1961–2000 was 87.1 mm on 21 July 1992. The precipitation gradient in the Fichtelgebirge is approx. 60–70 mm/100 m and thus in agreement with other mountain regions.

The accuracy of precipitation measurements is one of the serious problems in the study of mountain regions. Because of the many possible errors, only measurements with the standard rain gauge by Hellmann (200 cm^2 opening, for Waldstein/Weidenbrunnen 500 cm^2 opening) were used. Note that precipitation data are always published in their uncorrected form. The correction must be done by the user. The main errors are the evaporation error due to moisture and the wind error (Richter 1995). Evaporation error is approx. 2–4 % in the mountains and 4–8 % in flat lands with a maximum in summer (higher evaporation, more interrupted rain periods in flat lands). The wind error depends on the measuring station and possible shadow effects by buildings and trees. For moderately shaded stations (7–12° horizontal shadow effect), the error is approx. 3–15 %. This approximation is only valid for rain. Snow as well as mixed rain and snow precipitation cause larger errors of up to 70 % for heavy snowfall, mainly due to the influence of wind. In the region under investigation approx. 20–30 % of the annual precipitation is snow or mixed precipitation (Richter 1995). Taking all the errors into account for the Lehstenbach region, the mean annual precipitation is approx. 11–12 % higher than the measured precipitation. The distribution over the year is given in Table 2.5. The high error in the precipitation measurement does not allow a more detailed analysis of the precipitation distribution.

Wind data in the mountain region are heterogeneous and not very representative. The prevailing wind direction at the site Weidenbrunnen (tower) in the

Table 2.3. Mean monthly and annual temperatures in the Waldstein region (Fichtelgebirge; Foken 2003a) and Ebrach (Steigerwald; data from German Meteorological Service)

	Jan	Feb	March	April	May	June	July	Aug	Sept	Oct	Nov	Dec	Annual
1961–1990													
Voitsumra	–3.6	–2.0	0.8	4.8	9.1	12.5	14.1	13.1	10.5	6.1	1.4	–1.7	5.4
Waldstein/ Weidenbrunnen	–4.2	–3.1	0.2	4.3	9.0	12.3	14.1	13.7	10.5	5.8	0.2	–2.9	5.0
1971–2000													
Voitsumra	–2.9	–1.7	1.5	4.9	9.6	12.6	14.6	13.8	10.4	6.0	1.2	–1.3	5.8
Waldstein/ Weidenbrunnen	–3.6	–3.0	0.8	4.3	9.6	12.3	14.4	14.5	10.5	5.7	0.2	–2.2	5.3
1971–2000													
Ebrach/Steigerwald	–0.6	0.0	3.7	7.2	12.3	15.0	17.0	16.5	12.6	7.9	3.1	0.7	7.9

Table 2.4. Monthly and annual precipitation sums in the Waldstein region (Fichtelgebirge; Foken 2003a) and Ebrach (Steigerwald; data from German Meteorological Service)

	Jan	Feb	March	April	May	June	July	Aug	Sept	Oct	Nov	Dec	Annual
1961–1990													
Weißenstadt	93.4	73.2	79.4	79.1	80.2	98.6	93.5	94.5	76.9	76.2	95.8	118.9	1,059.8
Waldstein/ Weidenbrunnen	102.0	79.9	86.6	86.3	94.2	105.1	104.5	99.1	80.8	83.2	104.5	129.8	1,156.2
1971–2000													
Weißenstadt	99.9	76.2	82.5	67.0	71.5	98.5	107.4	85.4	79.1	85.7	91.7	121.0	1,065.8
Waldstein/ Weidenbrunnen	101.8	79.8	86.5	86.2	83.4	106.4	127.0	91.3	83.1	83.1	104.4	129.6	1,162.5
1971–2000													
Ebrach/Steigerwald	58.8	48.2	59.5	57.3	67.1	87.5	84.1	63.8	61.5	64.3	64.4	70.4	786.9

Table 2.5. Mean annual precipitation correction (%) for moderate shadowed stations of the regional class V (Fichtelgebirge area) according to Richter (1995)

Area	Jan	Feb	March	April	May	June	July	Aug	Sept	Oct	Nov	Dec	Annual
V, <700 m	17	18	16	13	10	9	9	8	10	11	13	15	12

Table 2.6. Relative humidity (Fichtelberg-Hüttstadel), fog days, and number of days with inversion at Waldstein/Weidenbrunnen (Foken 2003a)

	Jan	Feb	March	April	May	June	July	Aug	Sept	Oct	Nov	Dec	Annual
Relative humidity (%)													
1961–1990	89	86	82	76	74	74	74	76	81	84	89	90	81
1971–2000	90	86	82	76	73	75	74	76	82	85	90	91	82
Number of fog days													
1998–2000	21	15	18	10	12	7	17	11	16	22	24	24	195
Number of days with inversion													
1997–1999	12	11	6	9	13	9	6	13	17	8	6	6	114

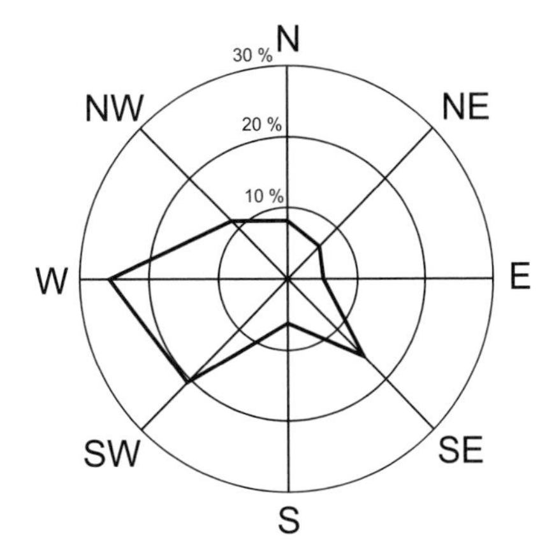

Fig. 2.3. Distribution of wind directions at Waldstein's Weidenbrunnen site (773 m a.s.l. plus 32 m above ground), 1 April 2001 to 1 April 2002. (Courtesy of Wrzesinsky 2003)

Lehstenbach catchment is from the west and southwest (Fig. 2.3). The meteorological station at the Waldstein/Weidenbrunnen site might have a reduced wind speed due to its location in a clearing. It is likely that the data of the station Hof-Hohensaas north of the Fichtelgebirge provide more realistic wind speeds for open sites with an annual average of about 3.5 m s^{-1}, a maximum in winter with 4.5 m s^{-1} in December, and a minimum of 2.8 m s^{-1} in August.

The relative air humidity of the Lehstenbach area is strongly influenced by the frequency of fog. There is a strong seasonality, with high values in autumn and winter and low values in spring and summer, with an annual average of about 80 % relative humidity in the mountains (Table 2.6).

Since September 1997, measurements of fog (i.e. air visibility) are available for the Weidenbrunnen site, measured at 20 m height above the forest. Fog was defined as the condition when the visibility was 10 min below 1,000 m (Wrzesinsky 2001). On the basis of these data, fog day statistics were calculated for the years 1998–2000 (Table 2.6). In the period from April 2001 to March 2002, the measured fog precipitation of 125 mm contributed about 8 % to the total precipitation of the Weidenbrunnen site (Wrzesinsky 2003).

Temperature inversions are particularly important in mountain valleys because of the strong influence on air pollution (Foken 2003b). In the years 1997–1999, the temperature difference between the stations Waldstein/Pflanzgarten and Voitsumra (5.64 km south of Waldstein/Pflanzgarten; lying 150 m lower in altitude) was investigated. The largest temperature difference was measured in January 1998 as 12 K. Because of often incomplete data at the Voitsumra station, the statistics given in Table 2.6 are based on the comparison of the gradients of the station Waldstein/Pflanzgarten and the stations Voitsumra and Bayreuth/Botanical Garden, respectively (Neuner 2000). Most of the inversions have a duration of only 1–2 h. The longest inversion situation was 15 h. This means that the inversion height in high-pressure areas in wintertime is often higher than 700 m a.s.l.

2.1.5 Soil Conditions of the Lehstenbach Catchment

2.1.5.1 Soils

The podzolic soils of the upper parts of the Fichtelgebirge developed from deeply weathered (up to 30 m) granite or gneiss bedrock. The weathering took place in the Tertiary age under tropical climatic conditions. A mosaic of Haplic Podzols, Cambic Podzols, and Cambisols (FAO classification) can be found in the Lehstenbach catchment area. Most of the soils are overlaid with a relatively thick humus layer. In topographic depressions and on saddles, fens and bogs have developed since the Pleistocene. About 30 % of the soils in the catchment area are wetland soils (Fig. 2.2).

2.1.5.2 Soil Properties of the Main Experimental Site Coulissenhieb

The forest floor is mor type, well stratified, approximately 8.5 cm thick, and consists of a litter (Oi), a fermented (Oe), and a humified (Oa) horizon. The soil texture (Table 2.7) is sandy loam to loam (US Soil Taxonomy) with a relatively high clay content in the Bh horizon. The content of rock fragments is about 10–25 vol% in the A and B horizons and increases in the C horizon to 50–75 vol%. High contents of organic matter in the A and B horizons are responsible for the relatively large water-holding capacity and porosity in these horizons. In turn, the bulk density is relatively small. In the C horizon, the bulk density increases whereas the water-holding capacity and the porosity of the soil decrease. Furthermore, the hydrologic conductivity is much smaller in the C horizon in comparison to the overlaying ones. Therefore, the C horizon is a potential stagnic horizon. The soil at the site Coulissenhieb is

Table 2.7. Texture, water content at different tensions, bulk density, and hydrologic conductivity (kf value) of the soil at Coulissenhieb site. *WC* Water content in vol% at a definite tension; *P* porosity volume in vol%; *dB* bulk density in g cm^{-3}; *kf* hydrologic conductivity (for saturated soil) in cm s^{-1}; *n.d.* not determined

Horizon	Depth (cm)	Sand (%)	Silt (%)	Clay (%)	WC at 60 hPa	300 hPa	15 bar	P	dB	kf
EA	0–10	51.6	38.0	10.4	47.2	38.9	7.8	61.9	0.97	0.008
Bh	10–12	34.0	49.6	16.4	n.d.	n.d.	n.d.	n.d.	n.d.	n.d.
Bs	12–30	44.7	44.8	10.4	50.5	40.8	15.2	70.9	0.73	0.018
Bw	30–55	45.8	43.4	10.8	31.0	25.8	9.6	48.2	1.36	0.011
C1	55–70	56.4	34.0	9.6	28.0	21.6	7.7	38.0	1.64	0.002
C2	>70	50.8	38.0	11.2	n.d.	n.d.	n.d.	n.d.	n.d.	n.d.

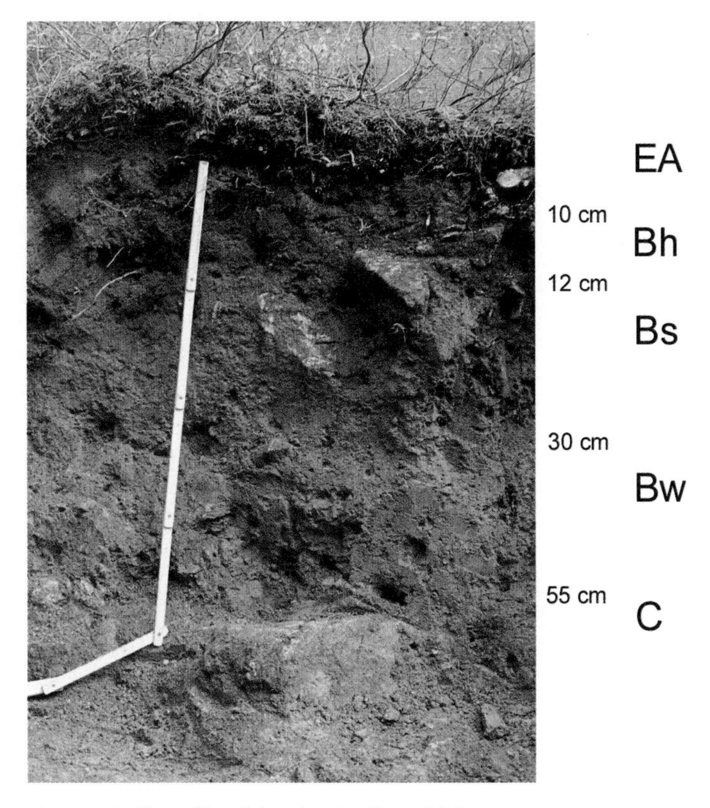

Fig. 2.4. Soil profile of the site Coulissenhieb

classified as a Haplic Podzol (FAO), Orthic Spodosol (US Soil Taxonomy), or Braunerde-Podsol (German classification; Fig. 2.4).

The pH values in the soil profile (sampled in 2000) are extremely acidic (see Table 2.8). They decrease from the litter (3.6) to the Oa horizon (2.6) and re-increase from the A horizon (2.9) to the Bw horizon (4.3). These extremely acidic conditions are representative for Haplic Podzols in this region and developed on bedrock with low contents of base cations (Guggenberger 1992). Only in the forest floor, with a high cation exchange capacity (CEC), are considerable amounts of exchangeable Ca and Mg stored. The B and C horizons are more or less completely depleted of these exchangeable cations. In contrast to Ca and Mg, 200 kg K ha^{-1} is stored in exchangeable form in the mineral soil. The low amounts of exchangeable base cations in the mineral soil are also reflected by a low base saturation of the CEC. In turn, contents and stored amounts of Al are high. In the mineral soil, Al is by far the dominant exchangeable cation (up to 97 % of the CEC).

The contents and amounts of pedogenic oxides reflect the dynamics of ongoing podzolization at the site Coulissenhieb, with highest contents and

Table 2.8. pH value, element content in 1 M NH_4Cl extract (exchangeable cations), calculated CEC_{eff}, and base saturation at the Coulissenhieb site (Kalbitz 2001). CEC_{eff} Effective cation exchange capacity; *BS* base saturation of the CEC

Horizon	pH H_2O	pH $CaCl_2$	Ca	Mg	K	Na (mmol$_c$ kg^{-1} soil)	Al	H^+	CEC_{eff}	BS (%)
Oi	4.50	3.60	38.9	6.94	6.64	2.30	163.3	9.59	245.8	22.7
Oe	3.80	2.90	84.6	11.2	10.6	2.07	56.9	59.6	237.4	46.2
Oa	3.50	2.60	111.4	18.4	21.8	2.30	5.16	104.3	274.2	56.8
EA	3.70	2.90	1.31	0.00	1.03	0.00	75.6	16.5	97.6	2.54
Bh	3.80	3.30	1.82	0.99	1.97	0.00	208.2	9.35	246.3	2.01
Bs	4.40	3.90	0.00	0.00	0.97	0.00	130.8	1.42	137.5	0.80
Bw	4.50	4.30	0.00	0.00	0.92	0.00	41.4	0.12	42.5	2.39
C1	4.50	4.20	0.00	0.00	0.77	0.00	31.0	0.08	31.8	2.42
C2	4.50	4.10	0.00	0.00	1.12	0.00	31.3	0.37	32.9	3.60

stocks in the B horizons. Furthermore, the portion of amorphous Fe is highest in the Bs horizon, reflecting both the formation of Fe oxides and the potential of this horizon to adsorb large amounts of organic matter (see Table 2.9; Kaiser and Zech 2000).

The contents and stocks of S in the soil are high, probably caused by the high S deposition up to the 1980s (Matzner et al., this Vol.). Although the stocks of inorganic sulfate are high in the deep mineral soil, the organic S pool is much higher than the inorganic one. In the B horizons, organic S exceeds the inorganic S stocks by a factor between 2 and 22.

The C and N contents decrease with increasing depth (Table 2.9). The C/N ratio of the forest floor is between 21 and 25 and similar to many forest ecosystem in central Europe which have been exposed to high atmospheric deposition. Leaching of dissolved organic matter and its retention in the mineral soil are reflected in the contents and stocks of organic matter. The highest C and N concentrations in the mineral soil were found in the thin Bh horizon. The Bs horizon has even higher C and N concentrations than the A horizon. The Bs horizon stores 51.1 Mg C ha^{-1}. The C and N stocks in the forest floor of about 60 Mg C and 2.9 Mg N ha^{-1} are similar to other sites in the Fichtelgebirge and can be considered as high (Guggenberger 1992). However, the C and N stocks in the mineral soil up to 80 cm depth (106 Mg C and 6.6 Mg N ha^{-1}) are much higher than those of the forest floor, emphasizing the importance of the mineral soil as a potential C and N sink.

In summary, the soil properties of the site Coulissenhieb reflect the low content of base cations of the bedrock, the high atmospheric deposition in the past, and the ongoing podzolization. Extremely acidic pH values, low stocks of base cations, Al as the dominant exchangeable cation, and high stocks of S indicate the strong acidification of this site. The soil stores large amounts of organic matter.

Table 2.9. Stocks of C, N, S, pedogenic Fe oxides, and exchangeable cations in the soil horizons at the Coulissenhieb site (Kalbitz 2001). *ox-Fe* Oxalate-soluble iron; *dith-Fe* dithionite-soluble iron; *n.d.* not determined; *n.dt.* not detectable

Horizon	Depth (cm)	C-stock (mg ha^{-1})	N-stock (kg ha^{-1})	S-tot	S-inorg	S-org	ox-Fe	dith-Fe	Ca^{2+}	Mg^{2+}	K^{+}	Na^{+}	Al^{3+}
Oi	8.5–8	3.6	0.16	20.3	1.1	190.8	n.d.	n.d.	5.9	0.6	2.0	0.4	11.2
Oe	8–3	25.4	1.26	192.0			n.d.	n.d.	103.6	8.3	25.4	2.9	31.3
Oa	3–0	31.1	1.49	253.6	2.3	251.3	n.d.	n.d.	224.4	22.5	85.7	5.3	4.7
EA	0–10	27.4	1.20	245.2	4.4	240.8	387	1,174	18.4	n.dt.	28.4	n.dt.	478.2
Bh	10–12	8.5	0.38	43.3	1.9	41.4	311	2,308	3.4	1.1	7.3	n.dt.	176.4
Bs	12–30	51.1	3.62	442.1	84.3	357.8	9,590	15,579	n.dt.	n.dt.	36.3	n.dt.	1,120.4
BvCv	30–55	20.7	1.23	393.6	141.6	252.1	4,040	15,765	n.dt.	n.dt.	89.1	n.dt.	917.0
C1	55–70	1.5	0.14	n.d.	379	2,506	n.dt.	n.dt.	20.3	n.dt.	188.6		
C2	70–80	0.9	0.09	n.d.	268	2,018	n.dt.	n.dt.	19.7	n.dt.	127.0		
Total		170.2	9.57	1,590.1	235.6	1,334.2	14,975	39,350	355.7	32.5	314.2	8.6	3,054.8

2.2 The Steinkreuz Catchment in the Steigerwald Hillsides

2.2.1 Brief Overview of the Region

The Steigerwald is a gently rolling hillside region of the district of Unterfranken between 250 and 490 m a.s.l., which has its highest elevations at the steep western edge and a shallow slope towards the east (Fig. 2.5). The forested Steigerwald is dissected by three west- to east-directed valleys dividing the Steigerwald into four low mountain ridges. The geological formations of the Steigerwald belong mainly to upper Triassic Keuper sandstones ('Blasen' sandstone, 'Coburg' sandstone, and 'Burg' sandstone). The sandstones themselves are intermingled with clayey-silty and water-impermeable layers as is the case with the underlying Lehrberg layer (Emmert 1985).

Forestry is the predominant land use form in the western part of the Steigerwald region. Towards the east, the role of agriculture increases. Partly due to soil properties, partly for historical reasons, most of the forest area is built up by beech (*Fagus sylvatica*) and oak (*Quercus petraea*) stands in the

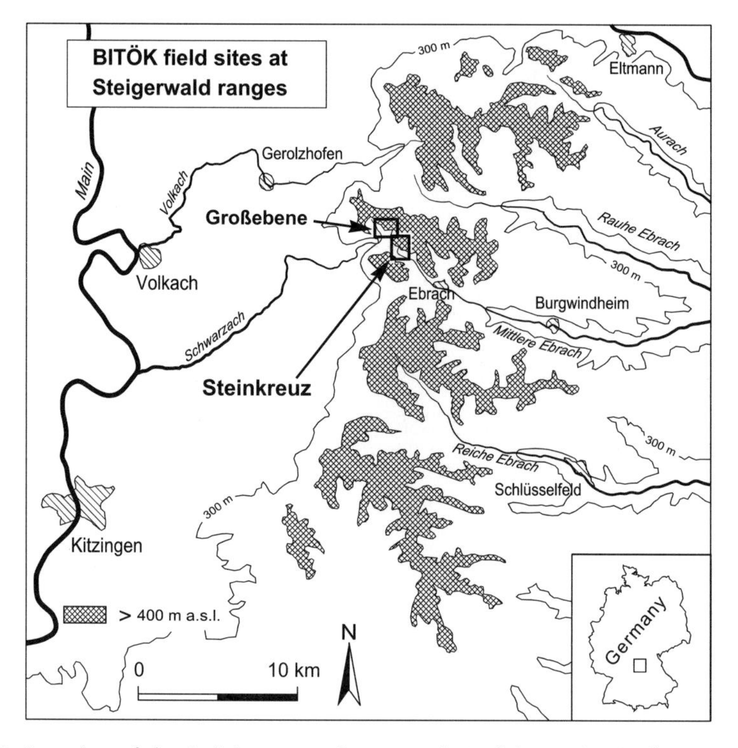

Fig. 2.5. Location of the Steinkreuz catchment and Großebene site in the Steigerwald hillsides (northern Bavaria)

Table 2.10. Characteristic parameters of the two investigation sites in the Steinkreuz catchment (after Fleck 2002). *LAI* Leaf area index; *n.d.* not determined

Site	Investigated since	Area (ha)	Altitude (m a.s.l.)	Exposition	Slope (°)	Stand age (years)	Number of trees (ha^{-1})	Max. stand height (m)	Understorey cover (%)	LAI trees ($m^2 m^{-2}$)	Soil depth (cm)	Humus layer (cm)	Soil pH ($H_2$0) in 0–5 cm	C/N of humus layer
Steinkreuz	1994	1.2	440	SSE	5.5	150	358	39	5–10	6.2	50–80	2.4±0.9	3.65±0.3	15.1±2.3
Großebene	1997–2000	0.3	460	SE	2	120	526	30	<1	6.1	n.d.	2.6±0.9	3.72±0.2	15.2±2.7

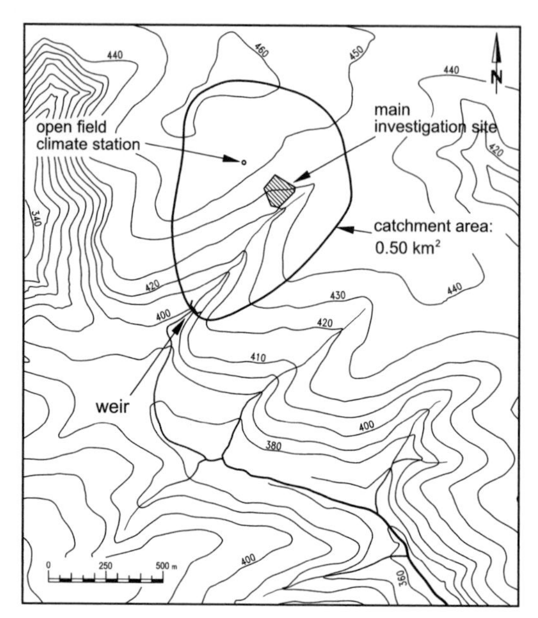

Fig. 2.6. The Steinkreuz catchment and main investigation site

western part, giving place to Scots pine (*Pinus sylvestris*) in the east (Lingmann 1952; Welss 1985). In order to investigate a broad-leaved forest, a small catchment was selected in the western part of the Steigerwald (433–445 m a.s.l.; area: 0.50 km^2), within the forest management unit 'Steinkreuz' (Fig. 2.6), a region poor in calcareous or dolomitic sediments. Basic parameters of the Steinkreuz catchment with the two experimental sites Steinkreuz and Großebene are given in Tables 2.1 and 2.10 (Lischeid and Gerstberger 1997; Fleck 2002).

The catchment is free from geological disturbance zones and leakage through the aquifer. A series of aquitards were identified in the underground by soil radar, drilling six 3-m holes at the site, and by sequential water analysis of the holes, the springs, and the brook. The aquitards dip at a rate of 2.5 % towards the southeast. The water of the small brook is caught by a V-notch weir and the outflow is measured and chemically analysed biweekly to estimate output fluxes of elements with the outflow (Lischeid and Gerstberger 1997).

2.2.2 Climate

Meteorological measurements (air temperature and moisture, open field precipitation and deposition, wind direction and velocity, net radiation) in the Steinkreuz catchment have been performed since the end of 1994 at an open field site, which was established after a severe wind throw in 1992 in the vicinity of the main investigation site (Lischeid 2001). Within a circle of about 25 m in the neighborhood of the small meteorological tower (powered with photovoltaic cells), all upgrowing trees are periodically removed. Additional meteorological data from three other sites within 7 km distance are available (Ebrach: Tables 2.3 and 2.4; Schmerb, and Kleingressingen).

The Steigerwald hillsides belong in the moderately warm and relatively moist climate belt in the transition zone between a maritime and continental climate (BayFORKLIM 1996), conditions that favor hardwood forests, with dominating beech (*Fagus sylvatica*) in central Europe. The difference in elevation (200 m) between the western lowlands and the Steigerwald ranges results in lower mean temperatures (depression approx. 1–1.5 °C) and higher annual precipitation (increase of ca. 100 mm) for the hillsides.

The mean air temperature between 1995 and 2000 at Steinkreuz was about 7.9 °C. This value lies at the upper range of long-term measurements of the Ebrach climate station (1951–1980; distance to Steinkreuz: 4 km; BayFORKLIM 1996), which may be an effect of the relatively warm 1990s in comparison to former decades. The minimum air temperature at Steinkreuz of –16.4 °C was recorded on 29 December 1996, the maximum of 36.3 °C on 12 August 1998.

The long-term mean of the precipitation at Ebrach between 1971 and 2000 is 787 mm year^{-1}. The same amount (mean) was observed at Steinkreuz in the years 1995–2000. The interannual variation of rain is relatively high and ranges between 665 mm (1997) up to 920 mm (1995) (Lischeid 2001). The precipitation tends to peak in summer with highest amounts in July (mean: 105 mm) and is lowest in winter (January 39 mm; Fig. 2.7). At Ebrach a complete snow cover was reported for 48 days year^{-1} (long-term observations); at Steinkreuz no continual measurement of snow cover was performed, but complete snow cover is only observed during short periods.

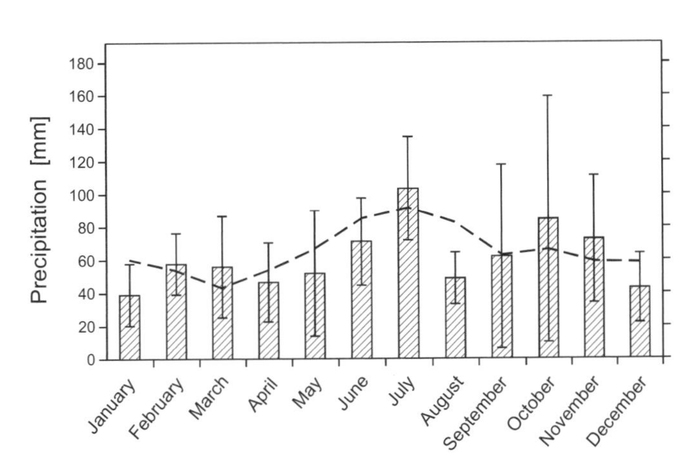

Fig. 2.7. Monthly sums of open field precipitation in the Steinkreuz catchment 1995–2000 (mean value and standard deviation) and mean value in Ebrach for the 1971–2000 period (*dashed line*). (Lischeid 2001)

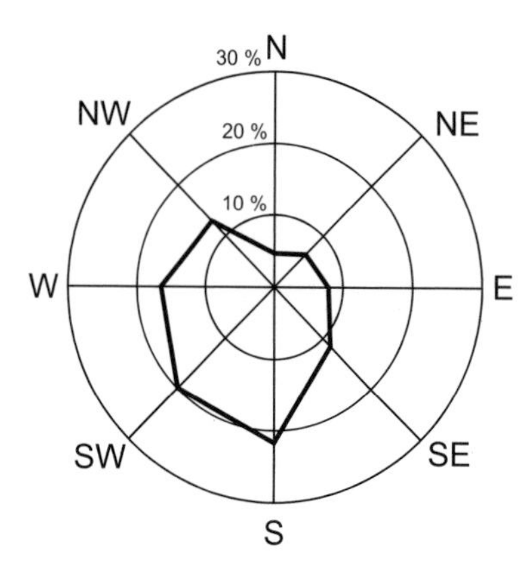

Fig. 2.8. Wind directions for wind velocities >1 m s^{-1} from 1 January 1999 to 31 December 2000 at Steinkreuz site. (Lischeid 2001)

The prevailing wind direction in the Steinkreuz catchment is from the southwest (Fig 2.8): 58 % of all wind directions (with a velocity of more than 1 m s^{-1}) were in the western to southern sector.

2.2.3 Vegetation of the Site Steinkreuz

The main investigation site Steinkreuz with a fenced area of 1.29 ha (at about 358 m a.s.l.) was established by natural regeneration about 150 years ago and is composed of different age classes of deciduous trees of about 74 % of beech (*Fagus sylvatica*), 25 % sessile oak (*Quercus petraea*), and 1 % hornbeam (*Carpinus betulus*); the oldest of them are about 100–140 years old and they have a breast height diameter of about 0.77 m (beech) and 0.28 m (oak). The trees reach heights of 28–32 m and their basal area extends to 43.5 m^2 ha^{-1}, with an annual increase of 1.31 m^2 ha^{-1}, which is contributed mainly by the large trees. All 462 trees (358 trees ha^{-1}) are numbered and mapped in a GIS (AutoCAD). In addition, along the brook in the valley bottom, alder (*Alnus glutinosa*), common ash (*Fraxinus excelsior*), sycamore (*Acer pseudo-platanus*), Scots pine (*Pinus sylvestris*), and planted, alien common larch (*Larix decidua*) can be found.

The composition of tree species in the smaller site Großebene (at about 526 m a.s.l.), approx. 300 m northwest of the site Steinkreuz, is about 66 % *Fagus sylvatica* and 34 % *Quercus petraea* with an age of about 120 years (Fleck 2002).

Due to the dense canopy of the broad-leaved forest, the two sites lack a shrubby layer – except for a few *Sambucus racemosa* shrubs – and have only a sparse understorey vegetation. Only in small openings caused by fallen trees does dense beech regrowth occur. The understorey vegetation cover is about 5–10 % of the area and is best developed shortly before leaf unfolding of the trees in May. It consists mainly of a few moderately acidophytic species such as *Luzula albida, Deschampsia flexuosa, Calamagrostis arundinacea,* and *Oxalis acetosella,* thus characterizing the stand a *Luzulo albida-Fagetum* (Gerstberger 2001). At the site Großebene the understorey stratum is composed mostly of the geophyte *Anemone nemorosa.*

2.2.4 Soil Conditions

2.2.4.1 Soils

The main parent materials for soil genesis are sandy and loamy materials from the upper Keuper, resulting in Dystric Cambisols as the dominant soil type in the catchment area (Fig. 2.9). Hydromorphic soils (Gleyic and Stagnic Cambisols) occur over about 10 % of the catchment area, mainly in the upper parts, with two-layered soils and clay layers below 40 cm depth and along the brook

In the Steinkreuz catchment the forest floor is mainly moder type. Only a few areas are covered with mor-type humus. Humus layers were found to be 1.5–3.5 cm thick.

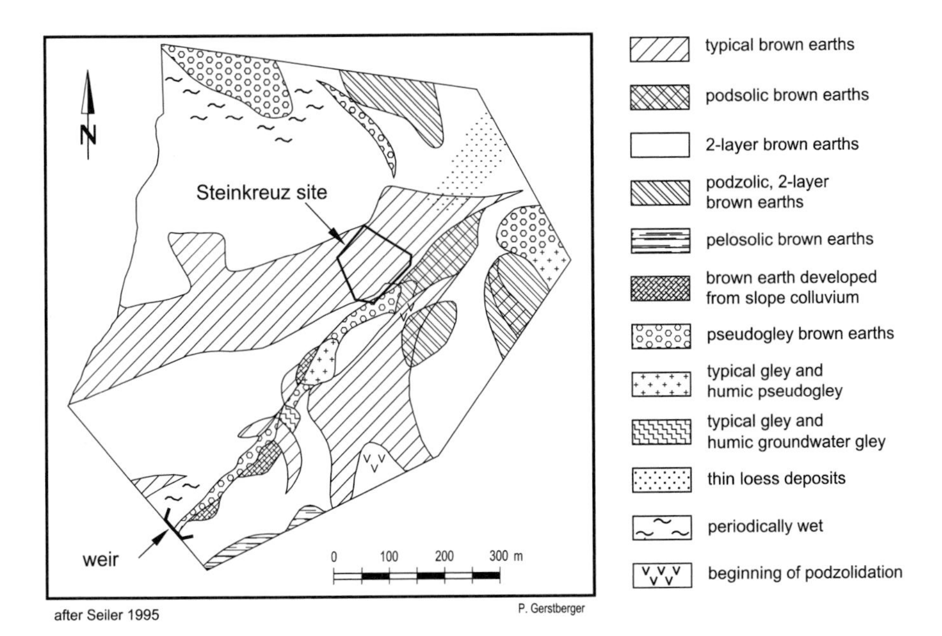

Fig. 2.9. Soils of the Steinkreuz catchment

2.2.4.2 Soil Properties of the Main Experimental Site Steinkreuz

The forest floor is mor type, approximately 3 cm thick, and consists of a litter (Oi), a fermented (Oe), and a humified (Oa) horizon. The thickness of the Oa horizon is highly variable at the study site. In the upper mineral soil, the soil texture is loam to sandy loam, whereas below 85 cm loamy sand occurs. The content of rock fragments increases continuously with depth from less than 2 vol% in the A horizon to more than 75 vol% at 125 cm depth. Below 125 cm the content of rock fragments decreases to 25–50 vol%. The water-holding capacity decreases with depth, whereas the bulk density increases. The Bw2 and Bw3 horizons have some stagnic properties, illustrated by the low porosity and the high bulk density (Table 2.11). The soil at the site Steinkreuz can be classified as a Dystric Cambisol (FAO), Umbric Inceptisol (US Soil Taxonomy), or pseudovergleyte Braunerde (German classification) (Fig. 2.10).

The pH values in the soil profile are acidic: they decrease from the Oi (4.7) to the Ah horizon (3.2) and re-increase from the Bw1 horizon (3.8) to the C horizon (4.1–4.2). The forest floor is characterized by a high CEC. It decreases considerably with increasing soil depth up to the Bw2 horizon as a result of the decreasing content of soil organic matter and increasing acidification. Also, the base saturation of the CEC decreases, reaching about 2 % in the Bw1 and Bw2 horizons. There are no measurable amounts of exchangeable Ca and Mg in the Bw1 and Bw2 horizons. In turn, Al saturation reaches 95 % in these horizons. The re-increase in the CEC, base saturation of the CEC, and of contents and stocks of exchangeable base cations from the Bw3 horizon (50 cm depth) to the C horizon indicates the depth of the current acidification line,

Table 2.11. Texture, water content at different tensions, bulk density, and hydrologic conductivity (kf value) of the soil at the Steinkreuz site. *WC* Water content in vol% at a definite tension; *P* porosity volume in vol%; *dB* bulk density in g cm^{-3}; *kf* hydrologic conductivity (for saturated soil) in cm s^{-1}

Horizon	Depth (cm)	Sand (%)	Silt (%)	Clay (%)	WC at 60 hPa	300 hPa	15 bar	P	dB	kf
A	0–5	47.2	38.8	14.0	39.2	27.7	18.6	57.1	1.06	0.004
Bw	5–24	54.7	35.2	10.0	29.9	21.3	6.8	45.0	1.44	0.002
Bw2	24–50	51.6	38.4	10.0	27.4	19.5	6.8	40.9	1.56	0.001
Bw3	50–80	57.4	32.5	10.0	27.8	20.2	10.2	35.7	1.70	0.012
2C	80–85	51.9	38.1	10.0						
3C	85–115	57.1	36.9	6.0	28.4	23.1	11.1	37.6	1.65	0.001
4C1	115–125	84.2	13.4	2.4						
4C2	>125	85.9	12.1	2.0						

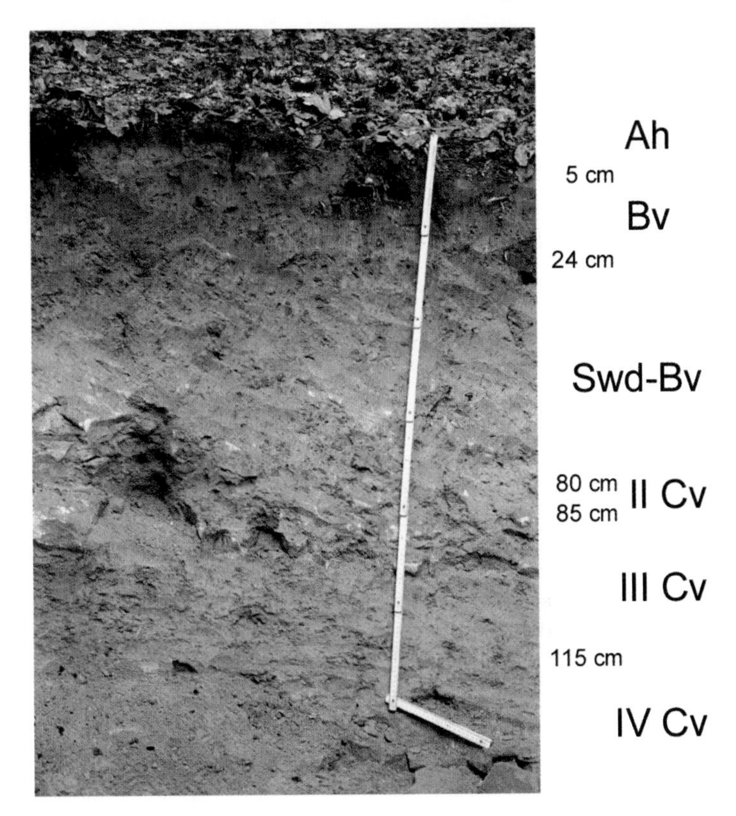

Fig. 2.10. Soil profile of the site Steinkreuz

which is located at about 50 cm depth. The stored nutrients in the Bw3 and C horizons can be used by the trees because roots could be detected at up to 115 cm depth (Table 2.12).

The highest contents and amounts of pedogenic oxides are located in the Bw horizon (Table 2.12), providing favourable conditions for the adsorption of organic matter even in the upper mineral soil (Kaiser and Zech 2000). These contents and stocks are relatively low in the deeper mineral soil, reflecting the sandy texture of these horizons.

The C and N contents decrease with increasing depth (Table 2.13). The C/N ratio of the forest floor is between 18 and 21 and should be representative for many deciduous forest ecosystems in central Europe with a moder-type forest floor. It decreases with depth of the mineral soil. The stocks of organic matter in the forest floor of about 13 Mg C ha^{-1} are in the same range as in many European forested sites (Gärdenäs 1988; mean: 19 Mg C ha^{-1}, standard deviation: 22 Mg C ha^{-1}). The stocks of organic matter are much larger in the mineral soil (79 Mg C ha^{-1}) than in the forest floor, highlighting the importance of the mineral soil as a potential sink for organic matter (Table 2.14).

Summarizing the soil properties of the site Steinkreuz, strong soil acidification has already reached a depth of 50 cm. Base cations are only abundant in the forest floor and the A horizon. However, in deeper horizons (>50 cm), the soil contains sufficient stocks of exchangeable cations to ensure proper nutrition of the trees.

Comparing both experimental sites, the soil of the site 'Steinkreuz' is less acidified than that of 'Coulissenhieb' as a result of different bedrock and dif-

Table 2.12. Element content in 1 M NH_4Cl-extract (exchangeable cations), calculated CEC_{eff}, and base saturation at the Steinkreuz site (Kalbitz 2001). CEC_{eff} Effective cation exchange capacity; *BS* base saturation of the CEC

	Ca	Mg	K	Na	Al	H^+	CEC_{eff}	BS
	(mmol$_c$ kg^{-1} soil)							(%)
Oi	367.3	85.2	37.5	1.86	2.91	0.00	567.9	86.7
Oe	373.2	63.5	25.3	3.24	4.58	0.00	564.7	82.5
Oa	90.4	13.3	7.51	1.38	16.1	21.1	172.3	65.7
A	21.8	5.00	3.25	0.00	37.0	12.6	87.8	34.5
Bw	0.00	0.00	0.86	0.00	42.6	0.88	45.3	2.06
Bw2	0.00	0.00	0.92	0.00	30.5	0.40	32.2	2.88
Bw3	4.41	6.12	1.58	0.00	27.9	0.73	41.1	29.5
2C	42.5	57.6	6.11	0.00	42.3	1.16	149.8	70.9
3C	34.7	37.8	4.94	0.89	37.6	0.68	116.7	67.2
4C1	29.3	22.2	3.51	0.00	13.5	0.49	69.0	79.8
4C2	30.5	22.1	3.57	0.00	12.6	0.56	69.3	81.0

Table 2.13. pH values, C and N content, and contents of pedogene oxides at the Steinkreuz site (Kalbitz 2001). *TOC* Total organic carbon; *TON* total organic nitrogen; Fe_{ox} and Al_{ox} oxalate-soluble iron and aluminum respectively; Fe_d dithionite-soluble iron

	pH H_2O	pH $CaCl_2$	TOC (g kg^{-1})	TON	TOC/TON (mg kg^{-1})	Fe_{ox}	Fe_d	Al_{ox}
L	5.3	4.7	445	20.4	21.8			
Oe	5.1	4.5	413	21.8	18.9			
Oa	4.1	3.4	205	11.5	17.8			
A	3.9	3.2	66.9	4.3	15.6	1,338	1,441	650
Bw	4.3	3.8	10.9	0.6	18.2	1,599	2,709	1,013
Bw2	4.6	3.9	3.5	0.2	17.5	466	1,491	877
Bw3	4.9	4.0	1.6	0.2	8.0	580	1,084	743
2C	5.2	4.1	1.4	0.3	4.7			1,233
3C	5.2	4.1	1.4	0.2	7.0	335	737	
4C1	5.5	4.2	0.7	0.1	7.0	143	384	904
4C2	5.5	4.2	0.8	0.1	8.0	121	458	

Table 2.14. Stocks of C, N, pedogenic Fe oxides, and exchangeable cations in soil horizons at the Steinkreuz site. *ox-Fe* Oxalate-soluble iron; *dith-Fe* dithionite-soluble iron; *n.d.* not determined; *n.dt.* not detectable

Horizon	Depth (cm)	C stock (mg ha^{-1})	N stock (kg ha^{-1})	ox-Fe	dith-Fe	Ca^{2+}	Mg^{2+}	K^+	Na^+	Al^{3+}
Oi	3–2	1.97	0.07	n.d.	n.d.	33	5	7	0.2	0.1
Oe	2–0.5	12.82	0.73		n.d.	332	33	46	3.8	6.6
Oa	0.5–0				n.d.					
A	0–5	34.9	2.2	698	752	228	32	66	n.dt.	174
Bw	5–24	27.1	1.5	3,980	6,746	0	0	84	n.dt.	955
Bw2	24–50	7.1	0.4	945	3,023	0	0	73	n.dt.	557
Bw3	50–80	4.1	0.5	1,479	2,764	225	190	157	n.dt.	641
2C	80–85	0.6	0.1	189	376	351	289	99	n.dt.	157
3C	85–115	1.9	0.3	456	1,003	947	626	263	27.9	460
4C1	115–140	1.4	0.2	295	792	1,213	555	283	n.dt.	250
4C2	140–160	1.3	0.2	200	756	1,010	443	230	n.dt.	187
Total		93.19	6.2	8,242	16,212	4,339	2,173	1,308	31.9	3,387.7

ferent atmospheric deposition during the last decades. Organic matter stocks are about twice as high at the site Coulissenhieb then at Steinkreuz, highlighting the great importance of Podzols for C storage (Batjes 1996).

References

Batjes NH (1996) Total carbon and nitrogen in the soils of the world. Eur J Soil Sci 47:151–163

BayFORKLIM (1996) Klimaatlas von Bayern. Bayerischer Klimaforschungsverbund, Munich

Betz H (1998) Untersuchungen zur Ausbreitungsökologie des Wolligen Reitgrases (*Calamagrostis villosa* (Chaix.) J. F. Gmel.). Bayreuther Forum Ökol 59:1–207

Eiden R, Förster J, Peters K, Trautner F, Herterich R, Gietl G (1989) Air pollution and deposition. In: Schulze E-D, Lange OL, Oren R (eds) Forest decline and air pollution. Ecological studies 77. Springer, Berlin Heidelberg New York, pp 57–103

Emmert U (1985) Geologische Karte von Bayern, 1:25.000. Blatt 6128 Ebrach. Bayerisches Geologisches Landesamt, Munich

Fiala K (1996) Estimation of annual production and turnover rates of underground plant biomass in *Calamagrostis villosa* and *Deschampsia flexuosa* stands. In: Fiala K (ed) Grass ecosystems of deforested areas in the Beskydy Mts. Institute of Landscape Ecology, Brno, pp 91–96

Firbas F, von Rochow M (1956) Zur Geschichte der Moore und Wälder im Fichtelgebirge. Forstwiss Centralbl 75:367–380

Fleck S (2002) Integrated analysis of relationships between 3D-structure, leaf photosynthesis and branch transpiration of mature *Fagus sylvatica* and *Quercus petraea* trees in a mixed forest stand. Bayreuther Forum Ökol 97:1–182

Foken T (2003a) Lufthygienisch-Bioklimatische Kennzeichnung des oberen Egertales. Bayreuther Forum Ökol 100:1–118

Foken T (2003b) Angewandte Meteorologie, Mikrometeorologische Methoden. Springer, Berlin Heidelberg New York

Gärdenäs AI (1988) Soil organic matter in European forest floors in relation to stand characteristics and environmental factors. Scand J For Res 13:274–283

Geiger R (1994) Vorkommen und Vergesellschaftung der Moorspirke (*Pinus mugo* ssp. *rotundata*) in Nordostbayern. Diploma Thesis, University of Bayreuth

Gerstberger P (2001) Vegetationskundliche und forstliche Charakterisierung des Wassereinzugsgebietes "Steinkreuz". Bayreuther Forum Ökol 90:132–136

Guggenberger G (1992) Eigenschaften und Dynamik gelöster organischer Substanzen (DOM) auf unterschiedlich immissionsbelasteten Fichtenstandorten. Bayreuth Bodenkundl Ber 26:1–164

Heindl B, Betz H (1995) Charakterisierung der Bodenvegetation im Einzugsgebiet Lehstenbach. In: Manderscheid B, Göttlein A (eds) Wassereinzugsgebiet Lehstenbach – das BITÖK-Untersuchungsgebiet am Waldstein (Fichtelgebirge, NO-Bayern). Bayreuther Forum Ökol 18:49–63

Henning I, Henning D (1977) Klimatologische Wasserbilanz von Deutschland. Ann Meteorol 12:119–123

Holzapfel R (1949) Über die Temperatur im deutschen Mittelgebirge. Meteorol Rundsch 2:33–34

Hupfer P (1996) Unsere Umwelt: Das Klima. Teubner, Stuttgart

Kaiser K, Zech W (2000) Dissolved organic matter sorption by mineral constituents of subsoil clay fractions. J Plant Nutr Soil Sci 163:531–535

Kalbitz K (2001) Bodenkundliche Charakterisierung der Intensiv-Messfläche, Coulissenhieb. In: Gerstberger P (ed) Waldökosystemforschung in Nordbayern: Die BITÖK-Untersuchungsflächen im Fichtelgebirge und Steigerwald. Bayreuther Forum Ökol 90:27–36

Koppisch D (1994) Nährstoffhaushalt und Populationsdynamik von *Calamagrostis villosa* (Chaix.) J. F. Gmel., einer Rhizompflanze des Unterwuchses von Fichtenwäldern. Bayreuther Forum Ökol 12:1–187

Lingmann (1952) Waldvegetationsgeschichtliche Betrachtungen über den Steigerwald. Der Frankenbund - Bundesbriefe 4(4):8–10; 4(5):11–14

Lischeid G (2001) Das Klima am Westrand des Steigerwaldes. Bayreuther Forum Ökol 90:169–173

Lischeid G, Gerstberger P (1997) The Steinkreuz catchment as a BITÖK main investigation site in the Steigerwald region: experimental setup and first results. Bayreuther Forum Ökol 41:73–81

Neuner C (2000) Dokumentation zur Erstellung der meteorologischen Eingabedaten für das Modell BEKLIMA. Universität Bayreuth, Abteilung Mikrometeorologie, Arbeitsergebnisse 11

Reichel D (1979) Wuchsklima-Gliederung von Oberfranken auf pflanzen-phänologischer Grundlage. Berichte der Akademie für Naturschutz und Landespflege, Laufen, pp 73–75

Reif A (1989) The vegetation of the Fichtelgebirge: origin, site conditions, and present status. In: Schulze ED, Lange OL, Oren R (eds) Forest decline and air pollution - a study of spruce (*Picea abies*) on acid soils. Ecological studies 77. Springer, Berlin Heidelberg New York, pp 8–22

Reif A, Leonhardt A (1991) Die Wald- und Forstgesellschaften im Fichtelgebirge. Hoppea Denkschr Regensb Bot Ges 50:409–452

Richter D (1995) Ergebnisse methodischer Untersuchungen zur Korrektur des systematischen Meßfehlers des Hellmann-Niederschlagsmessers. Ber Dtsch Wetterdienstes 194:1–93

Schulze E-D, Lange OL, Oren R (eds) (1989) Forest decline and air pollution. Ecological studies 77. Springer, Berlin Heidelberg New York

Seiler J (1995) Bodenkundliche Charakterisierung der Einzugsgebiete "Steinkreuz/ Erlensumpf" bei Ebrach und "Runderbusch" bei Schrappach im nördlichen Steigerwald. BITÖK, Bayreuth, 41 pp

Vollrath H (1977) Temperaturvergleich Fichtelgebirge, Vogtland, Obermainisches Hügelland. Der Siebenstern 46:82–89

Vollrath H (1978) Die Anomalie der Lufttemperatur im Umkreis von Fichtelgebirge und Vogtland. Ber Naturwiss Ges Bayreuth 16:289–308

Vollrath H (1979) Gibt es ein bayerisches Sibirien? Eine volkstümliche Hyperbel im Lichte einer klimatologischen Untersuchung. Siebenstern 48:93–95

Wedler M, Scharfenberg H-J, Zimmermann R (1997) Evapotranspiration from understory in forest ecosystems. BITÖK-Forschungsbericht 1996. Bayreuther Forum Ökol 41:191–196

Welss W (1985) Waldgesellschaften im nördlichen Steigerwald. Diss Bot 83

Wrzesinsky T (2001) Chemie und Meteorologie des Nebels im Fichtelgebirge. Bayreuther Forum Ökol 90:75–83

Wrzesinsky T (2003) Direkte Messung und Bewertung des nebelgebundenen Eintrgs von Wasser und Spurenstoffen in ein montanes Waldökosystem. PhD Thesis, University of Bayreuth, 109 pp

II The Changing Evironment

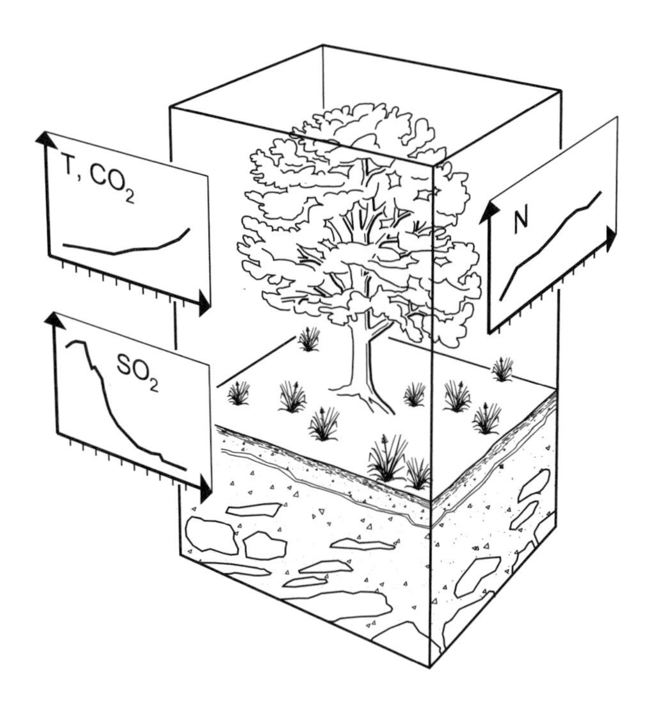

3 Trace Gases and Particles in the Atmospheric Boundary Layer at the Waldstein Site: Present State and Historic Trends

O. KLEMM

3.1 Introduction

The atmosphere plays a key role in ecosystem functioning. Its general status and development is mainly driven by large-scale factors and processes. Climate change as a global and hemispherical phenomenon (macro-scale) is an example. On smaller scales, the atmosphere interacts intensively with terrestrial ecosystems, and thus plays a key role in ecosystem processes. An important example is the input of liquid water through precipitation, which is mainly driven by regional scale (meso-scale) processes. On even smaller scales (micro-scales), the exchange of nutrients and pollutants between the biosphere and the atmosphere strongly affects either one. Therefore, the interaction between the atmospheric boundary layer and the vegetation is a main focus of ecosystem research. Several chapters in this book deal with related topics, such as turbulent exchange of carbon dioxide and water vapor (Chap. 9), ozone (Chap. 12), biogenic volatile organic compounds (Chap. 13), deposition of nutrients such as nitrogen, as quantified through ecosystem balance methods (Chap. 14), and more. This chapter reviews recent and historic data of gas and particulate concentrations from the Bayreuth Institute for Terrestrial Ecosystem Research (BITÖK) experimental research site Waldstein in the Fichtelgebirge mountain range. For further details concerning applied techniques, data structure and quality control strategies, the interested reader is referred to Klemm and Lange (1999)and Held et al. (2002a, b).

The data record exhibits a large heterogeneity of data completeness, ranging from 18 years time series for sulfur dioxide and ozone, through short-term collection of a few samples for analysis, for example, of heavy metal concentrations in fog. It is not in the scope of this chapter to evaluate the toxicological risks of air quality nor to quantify pathways of the forest ecosystem nutrient cycles, because these issues are related to fluxes rather than concentrations.

Ecological Studies, Vol. 172
E. Matzner (Ed.), Biogeochemistry of Forested Catchments in a Changing Environment

3.2 Sites

BITÖK operates an ecosystem research site in the Fichtelgebirge mountain range, NE Bavaria. Several experimental plots lie within the Lehstenbach catchment (Gerstberger et al., this Vol.). Most of the atmospheric chemistry routine data have been collected since 1993 at the Pflanzgarten site, at 50°08′40″N, 11°51′55″E, 765 m a.s.l., in a 100 × 200-m forest clearing surrounded mainly by Norway spruce (*Picea abies*). Other data, particularly those of fog chemistry, were collected at a meteorological walk-up tower at the Weidenbrunnen site, at 50°08′32″N, 11°52′04″E, 775 m a.s.l., in a Norway spruce plantation, about 18 m high, planted in about 1945. Both sites are located close to the water divide of the Lehstenbach catchment, and are often called the Waldstein site, according to the nearby mountain peak.

Earlier data originate from a forest decline research project between 1985 and 1993, at a Norway spruce forested site close to the village of Warmensteinach (Wagental catchment; Schulze et al. 1989), located 18 km to the south of the Waldstein site, at an altitude of 760 m a.s.l.

3.3 Trends in Sulfur Dioxide

Sulfur dioxide (SO_2) is one of the most prominent air pollutants. Its relevance originates from its direct toxicity to the vegetation, and from its contribution to the acidity of clouds and precipitation. SO_2 has been emitted into the atmosphere in large amounts worldwide, mainly through combustion of sulfur-rich fossil fuel. In North America and Europe, emissions have been significantly reduced over the past 20 years. Figure 3.1 shows the emissions of SO_2 in Germany and the neighboring Czech Republic between 1985 and 2000, revealing a reduction of emissions by almost 90 % between 1985 and 2000. The Waldstein site lies about 30 km west of the German/Czech border. Therefore, emissions from the Bohemian Basin will also be measurable at the Waldstein site if the winds are from the east.

Figure 3.2 shows percentiles of the SO_2 mixing ratio as measured in the Fichtelgebirge between 1985 and 2002. Two data sets from Warmensteinach (1985–1993) and Waldstein (1994–2002) were combined.

A statistical analysis of the two data subsets showed that the structure is similar and there are no objections against merging them into one combined data set (Klemm and Lange 1999). Apparently, the proximity of the locations and the similar altitude above sea level lead to a good comparability of the two sites with respect to trace gases that are advected through meso-scale processes. The medians (50 % percentiles) are not shown in Fig. 3.2 because they were often below the detection limit (2 ppb) of the analyzers. The data, as

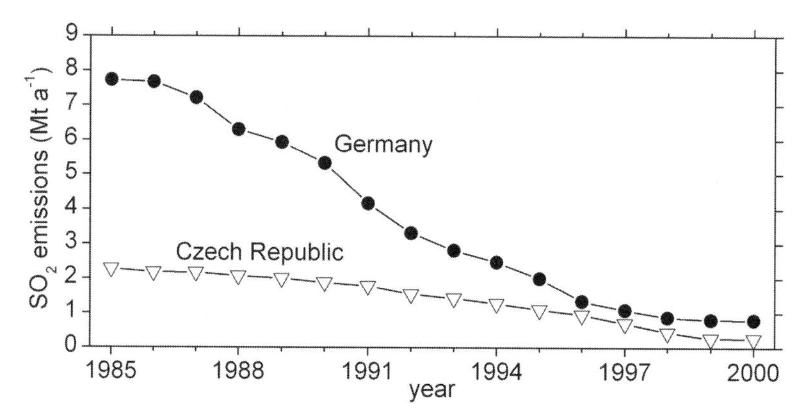

Fig. 3.1. Emissions of sulfur dioxide (SO_2) from Germany (*solid dots*; before 1990: west plus east Germany; source: Umweltbundesamt 2002) and the Czech Republic (*open triangles*; with today's borders; source: Vestreng and Klein 2002)

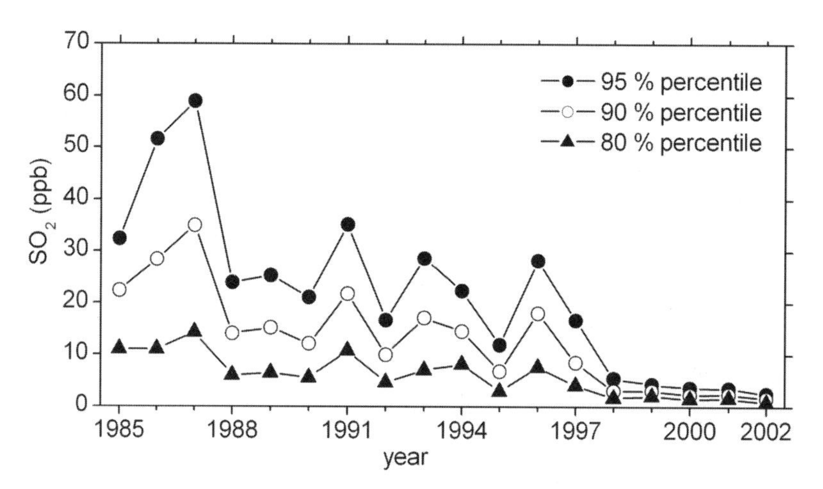

Fig. 3.2. Percentiles of SO_2 mixing ratios from the Fichtelgebirge research sites. SO_2 was measured with standard pulsed gas phase fluorescence analyzers which zero and span calibration every 23 h. Hourly means are displayed. A 90 % percentile equal to 22.3 ppb means that all measured data are below 22.3 ppb

presented in Fig. 3.2, exhibit a clear and statistically significant decrease of the SO_2 mixing ratio in the Fichtelgebirge. In light of the decrease of the SO_2 emissions in Germany, the Czech Republic (Fig. 3.1) and other countries of central Europe (not shown), the decrease is to be expected. However, the steep drop of observed mixing ratios from 1997–1998 is striking. Peak concentrations used to appear in episodic events, dominated by easterly winds. Such episodes decreased dramatically in magnitude and frequency. Figure 3.3 shows the last

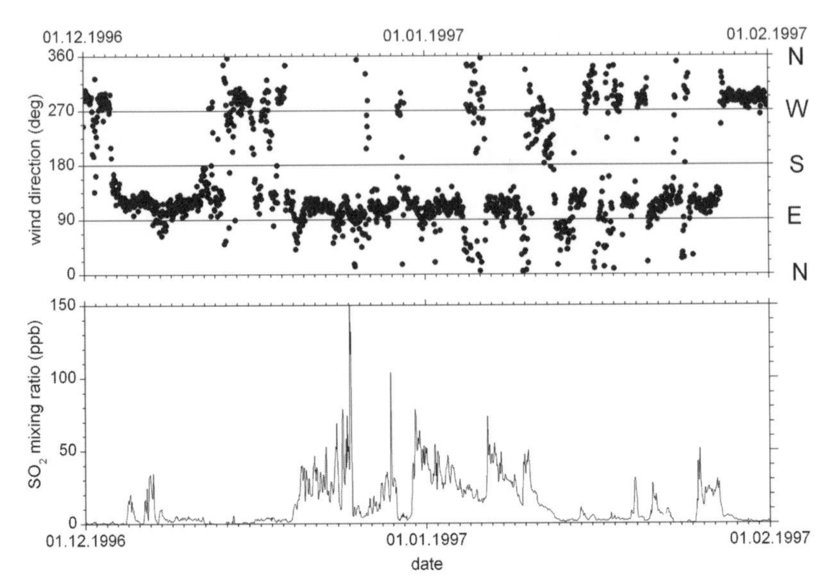

Fig. 3.3. The last period of high SO_2 mixing ratios as measured at the Waldstein site in December 1996 and January 1997. *Above* Wind direction; *below* hourly means of SO_2 mixing ratio

episode with mixing ratios up to over 100 ppb, which occurred around Christmas 1996. Episodes with SO_2 mixing ratios peaking around 50 ppb have hardly occurred since 1998. Even with easterly winds, the ambient SO_2 concentrations are often below the detection limits of the employed analyzer.

3.4 Trends in Nitrogen Gases

For nitrogen oxides (NO_x, which is the sum of NO and NO_2, if calculated on a mixing ratio basis) and ammonia (NH_3) the emissions have undergone a less dynamic development since 1985. The NO_x emissions decreased by 50% between 1986 and 1999 (Umweltbundesamt 2002). However, the emissions, as displayed in Fig. 3.4, show only those from anthropogenic sources, while biogenic sources such as soil emissions are not included. These are probably in the same order as the anthropogenic emissions; strong variations over the years should not be expected for the biogenic emissions either. The annual means of the ammonia emissions only show a minor decrease around the year 1990.

The database for both NO_x and NH_3 (Fig. 3.4, below) is much less complete than that for SO_2. The median annual NO_x mixing ratio has been between 2.7 and 5.4 ppb since 1994. The moderate trend of the emissions (top panel) is not reflected in the measured mixing ratio data. Relatively low median and 90%

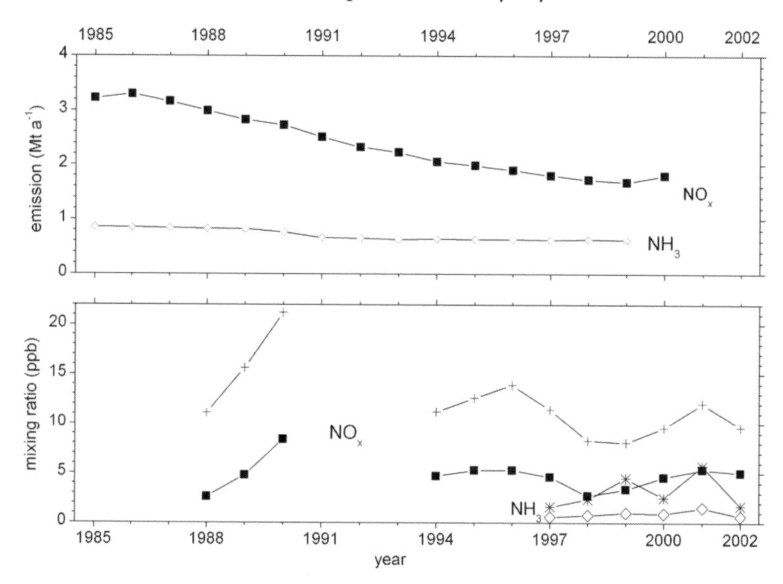

Fig. 3.4. *Above* Emissions of NO_x and NH_3 from anthropogenic sources in Germany (before 1990: west plus east Germany; source: Umweltbundesamt 2002). *Below* Mixing ratio medians (*solid squares* for NO_x, *open diamonds* for NH_3) and 90% percentiles (*pluses* for NO_x, *asterisks* for NH_3), calculated from hourly means

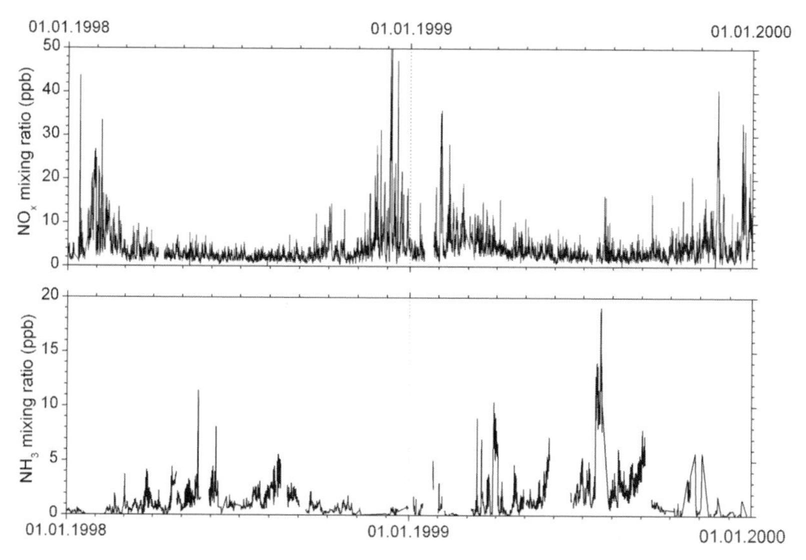

Fig. 3.5. Two-year record of hourly averages of mixing ratio of NO_x (*above*) and NH_3 (*below*) at the Waldstein site in the Fichtelgebirge mountains, Germany. Nitrogen oxides were measured with standard two-channel gas phase chemiluminescence analyzers, detecting NO after reaction with O_3. In the second channel, NO_2 is reduced to NO with a molybdenum converter before detection. Gas phase ammonia was measured with a horizontal continuous-flow wet annular denuder as described by Wyers et al. (1993), which was calibrated at weekly intervals

percentile mixing ratios in 1998 and 1999 appear to be caused by meteorological phenomena rather than the emission pattern. We hypothesize that larger depths of the boundary layer during 1998 and 1999 led to higher dilution of NO_x and thus to lower mixing ratios. This could also explain why the SO_2 mixing ratios experienced a strong step downward in 1998.

For NH_3, a time series of over 5 years has been assembled at the Waldstein site (Fig. 3.4, below). Yearly medians of hourly mixing ratios were between 0.6 ppb (1997) and 1.5 ppb (2001), with no significant trend established. A 2-year section of the time series of NO_x and NH_3 is shown in Fig. 3.5. For NO_x, the mixing ratios are higher during winter than during summer, due to lower depth of the atmospheric boundary layer. For NH_3, the pattern is different: higher mixing ratios occur during the summer months when the emissions from farming are higher.

3.5 Trends in Ozone

In the troposphere (the lowest 10 km of the atmosphere), ozone is a key indicator of the presence of photochemically active air masses. Ozone has two major sources in the troposphere. One is down-mixing from the stratosphere, where O_3 is formed via photolysis of O_2; the other is photochemical production through a series of reactions from its precursors, nitrogen oxides and volatile organic compounds (VOCs). The latter mechanism may lead to photochemical episodes. Tropospheric ozone formation results from the photolysis of NO_2 by solar light with wavelengths λ <400 nm:

$$NO_2 + h\nu \rightarrow NO + O\left({}^3P\right) \tag{1}$$

$$O\left({}^3P\right) + O_2 \rightarrow O_3 \tag{2}$$

With no NO_2 and its photolysis, there is virtually no tropospheric ozone formation. NO and O_3 react to form NO_2,

$$NO + O_3 \rightarrow NO_2 + O_2 \tag{3}$$

and a photochemical equilibrium among NO, NO_2, O_3 and the solar radiation is established rapidly, which accounts for higher NO_2/NO ratios at night and lower NO_2/NO ratios during the day. The NO_2 in the lower atmosphere has its origin mainly in reaction (3), and the NO that feeds reaction (3) originates from emissions from combustion sources (such as road traffic) and biogenic emissions. Virtually no NO_2 is directly emitted from these sources. NO plays a double role because (1) NO is an essential precursor of regional ozone formation as it produces NO_2 which is photolyzed to produce O_3, and because (2)

NO is an important sink of O_3 because it reacts rapidly with atmospheric O_3 (reaction 3). This leads to the situation where the highest O_3 concentrations are normally not observed at the times and locations where the emissions of its precursor, NO, are highest.

As most of the oxides of nitrogen are emitted into the atmosphere as NO, reactions (1)–(3) will not lead to a net formation of O_3. Another mechanism that oxidizes NO to NO_2 without consumption of O_3 is needed. In polluted air masses, the most important pathway is the attack on a VOC by the hydroxyl radical (OH) and abstraction of an H atom. The resulting alkyl radicals (R•) react with oxygen to form alkyl peroxy radicals (RO_2•) which oxidize NO:

$$RH + OH \rightarrow R\bullet + H_2O \tag{4}$$

$$R\bullet + O_2 \rightarrow RO_2\bullet \tag{5}$$

$$RO_2\bullet + NO \rightarrow RO\bullet + NO_2 \tag{6}$$

Therefore, reactive hydrocarbons are also important precursors for tropospheric ozone. However, the reaction expressed through Eq. (4) is slower than reactions (1)–(3), and NO_2 is formed within minutes and hours through the oxidation by RO_2•. In the intervening time, atmospheric transport may lift air masses by hundreds of meters and/or advect them horizontally by up to many kilometers. This contributes to the 'paradox' phenomenon that the ozone concentrations are often not highest where the emissions of the precursors are highest.

For these reasons, the relatively high altitudes of the research sites in the Fichtelgebirge mountain range (Warmensteinach and Waldstein, 760–775 m a.s.l.) and their distance from major emission sources (cities over 10,000 inhabitants: over 30 km; major autobahn: about 10 km) are prone to exhibit relatively high concentrations of O_3, at least during summer.

Figure 3.6 shows that the O_3 time series between 1985 and 2002 exhibits a peculiar pattern: From 1985–1994, there was a clear and statistically significant increase in the O_3 mixing ratio (cf. Klemm and Lange 1999). The increase is evident not only for the high percentiles, representing the peak mixing ratios during summer, but also for the background and winter values (median and 10% percentile). This indicates that the observed pattern is evident on the meso- or even macro-scale. The causes for this increase could lie in an increase in the global emissions of O_3 precursors (NO and VOC).

For the peak O_3 mixing ratios (95% percentile, Fig. 3.6), and in particular for the high mixing ratios (95% percentile–median), a decline can be interpreted beginning in 1994. This may have resulted from the reduction of emissions of O_3 precursors (see Fig. 3.4 for NO_x, for VOC: reduction by 15% between 1992 and 1994, further reduction by 12% between 1994 and 1996; Umweltbundesamt 2002), but causes of the increase before 1994 still remain

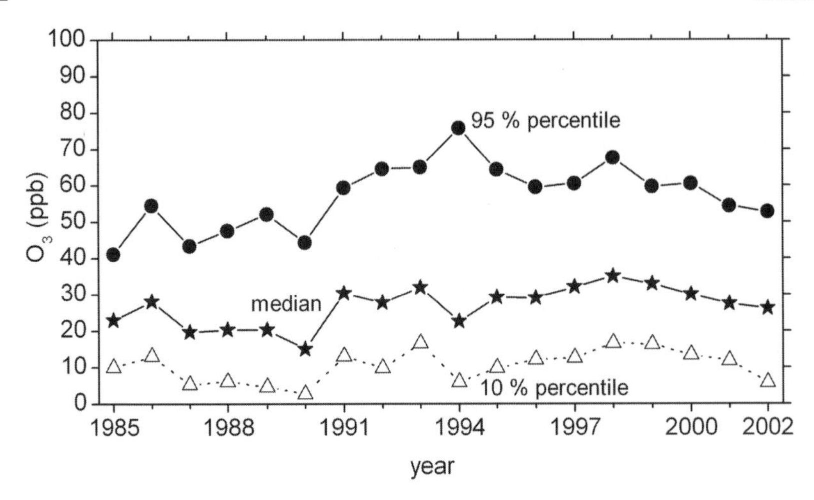

Fig. 3.6. Percentiles of O_3 mixing ratios (computed from hourly means) from the Fichtelgebirge research sites. The O_3 mixing ratio in air is measured by means of standard UV-absorption ozone monitors, with automatic zero and span checks performed every 23 h. Major calibration and standardization into the regional and federal ozone network were performed at least every 12 months

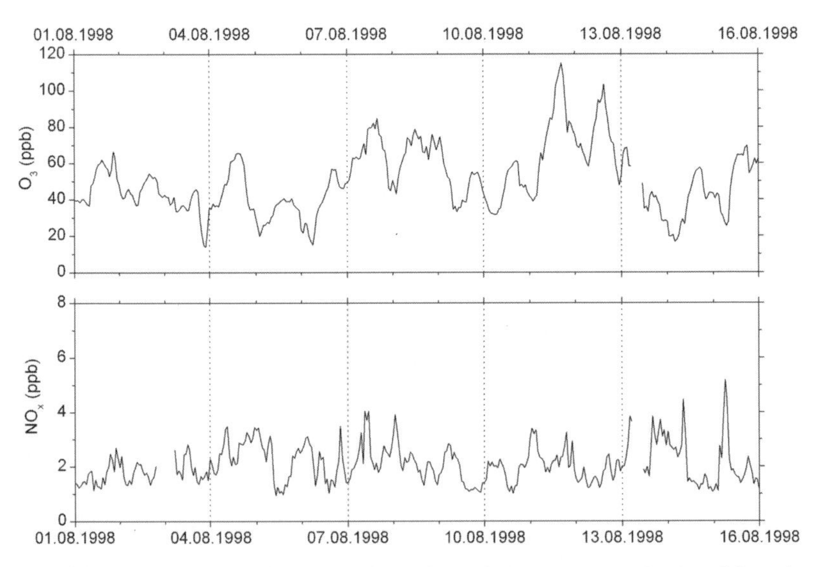

Fig. 3.7. The last photochemical episode with peak O_3 mixing ratios reaching almost 120 ppb, measured at the Waldstein site in August 1998. *Above* O_3 mixing ratio; *below* NO_x mixing ratio

unclear. During reduction of emissions, a transition of the chemical regime of ozone production, from one of NO_x limitation to VOC limitation (for details see Klemm et al. 2000), may have occurred. However, this is unlikely given the low NO_x mixing ratio at the Waldstein site (Fig. 3.4). Figure 3.7 also shows that the highest ozone mixing ratios of 115 ppb were reached at low NO_x mixing ratios of 2.5 ppb, far below the regime of NO_x limitation of tropospheric O_3 production.

For the low percentiles (10 % percentile) of ozone, a decrease since 1998 is evident. This decrease in O_3 since 1998 can probably not be explained through emission patterns of NO_x or VOC in Germany. It may be hypothesized that a change of the statistics of the meso-scale circulation patterns in central Europe, driven by global change processes, might lead to the advection of air masses with lower O_3 with higher frequency, but this is highly speculative at this point.

3.6 Trends in Fog Chemistry

Nitrogen oxides and sulfur dioxide are oxidized in the troposphere's gas and liquid phases to form HNO_3 and H_2SO_4. These are strong acids with high water solubilities and therefore lead to severe acidification of cloud and precipitation water, and eventually to the 'acid rain' phenomenon (see, for example, Schulze et al. 1989). The reduction of emissions of nitrogen oxides (Fig. 3.4) and, in particular, of SO_2 in central Europe (Fig. 3.1), leads to a reduction of the potential of acidity production. Although a number of studies including rain analysis have been performed in the Fichtelgebirge region since the early 1980s, no database now exists that allows an analysis of the trends of rain composition. For fog, the database is better developed. Figure 3.8 shows the pH over a period of more than 15 years. First, a large data gap between 1988 and 1996 is evident. No fog has been collected during this period for chemical analysis. The data have been carefully filtered (Klemm 2001) and only results from active fog water collectors, and from similar altitudes in the Fichtelgebirge region, are presented in Fig. 3.8. The scatter of the data is large. Pooling the data subsets for the earlier and later periods, respectively, yields data subsets that are not normally distributed or describable with other parametric distributions.

Visual data inspection in Fig. 3.8 suggests that there is a strong trend towards higher pHs in fog over the time period of about 15 years, with the median shifting from pH=3.54 to pH=4.24. The concentrations of sulfate decreased by 32 % (median; 61 % for the 5 % percentile, 72 % for the 95 % percentile, data not shown), over the same time period, while the nitrate and ammonium ion concentrations did not decrease. The time series data of fog chemistry are problematic in various aspects: First, the high variability of the

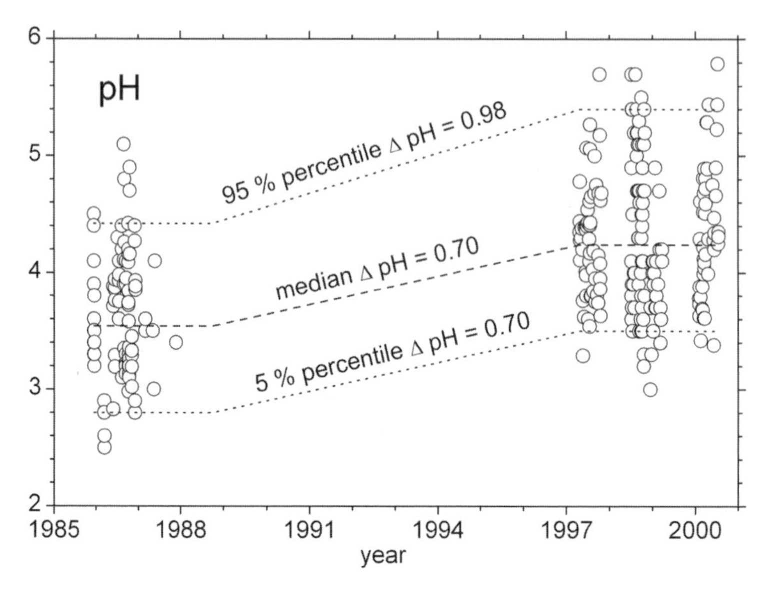

Fig. 3.8. pH in fog water samples in the Fichtelgebirge between 1985 and 2000

results and the application of non-parametric tests yielded trend results that were statistically not significant. Second, the liquid water content of the clouds was not measured or even estimated before 1990, so that only liquid water concentrations could be analyzed and no correction into atmospheric concentrations (e.g., ion content per cubic meter of air) could be performed. Third, a joint analysis of fog water chemistry with rainwater chemistry was not possible for historic data. Nevertheless, the data indicate that the relative importance of nitrate and ammonium, compared to that of sulfate in fog water, increased strongly (Klemm 2001). The present fog water chemistry at the Waldstein site is discussed in more detail by Wrzesinsky et al. (this Vol.).

3.7 Aerosol Particles

The emissions of aerosol particles from anthropogenic emissions, in particular from power and industrial plants, have decreased dramatically over the past 30 years in central Europe, so we would expect a decrease in the measured concentrations in the Fichtelgebirge mountain range. However, the historic database of field experiments is not well documented and allows no detailed analysis.

Physics and chemistry of atmospheric aerosol particles are complex issues. The sizes of particles range from about 3 nm (aerodynamic diameter) for

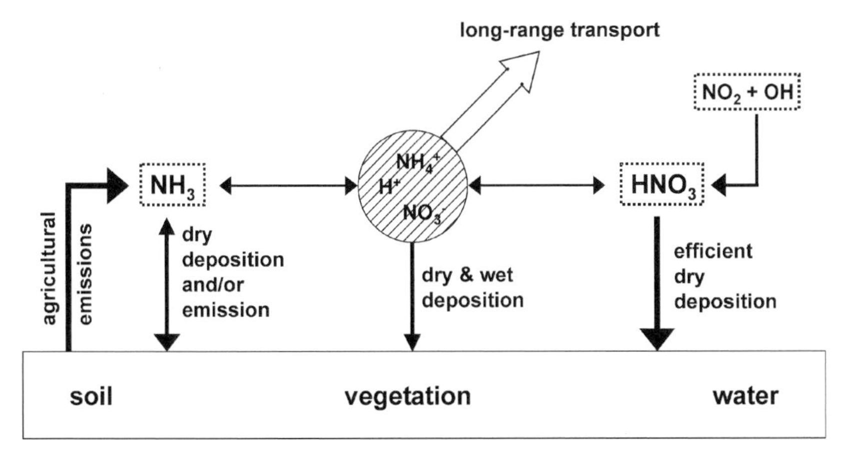

Fig. 3.9. Schematics of the processes and pathways that lead to reduced (NH_3 and NH_4^+) and oxidized (HNO_3 and NO_3^-) nitrogen in the atmospheric boundary layer. *Circle in centre* represents an aerosol particle (solid or liquid), gases are drawn as *dotted boxes*, *single arrows* represent fluxes and *double arrows* bidirectional fluxes that tend to establish thermodynamic equilibrium. (Held et al. 2002b, with permission)

new-borne secondary particles, through several tens of micrometers diameter for primary, large particles from erosion. Not only does the composition range with size but also individual particles of one size may be of different composition (*external mixture*).

Figure 3.9 shows a schematic of the interaction of oxidized and reduced inorganic nitrogen between the gas and particulate phases. A particle is displayed in the centre, containing H^+, NH_4^+ and NO_3^- ions, and interacts with gaseous NH_3 and HNO_3. Air temperature and humidity are important factors controlling the thermodynamics of the exchange. All gases originate from different sources and exhibit a high atmospheric reactivity. Deposition processes compete with the aerosol interaction. The particle itself may be deposited or undergo long-range atmospheric transport. From an ecological point of view, the deposition process is the most interesting to be quantified, because it leads to input of nitrogen which is a crucial nutrient for the vegetation, and because it contributes to the acidification of soils. However, it is not only the complexity of atmospheric chemistry, but also the deposition process itself that is hard to quantify as of now. First steps using a combination of eddy covariance technique with time-of-flight mass spectroscopy were successful (Held et al., submitted), but quantitative deposition fluxes have not yet been computed.

During intensive field campaigns from 1998–2002, the aerosol particle dynamics and chemistry at the Waldstein site were studied in detail (Held et al. 2002a, b). Some of the results are presented by Steinbrecher et al. (this Vol.), while others still require further analysis. Figure 3.10 shows an example of

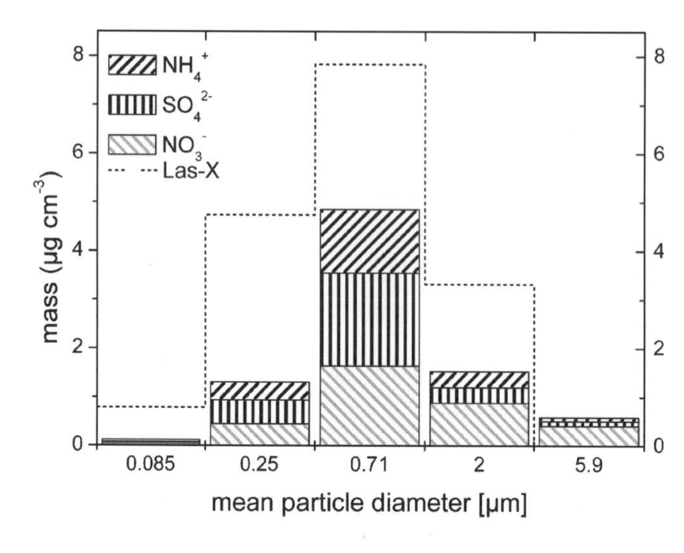

Fig. 3.10. Particle mass distribution of NH_4^+, NO_3^- and SO_4^{2-} and total mass distribution calculated from particle number distributions measured with a Las-X particle spectrometer, both for 4 February 2000. For the largest particles, no total particle mass (Las-X) was calculated due to experimental limitations. (Held et al. 2002b, with permission)

size-resolved sampling (with an impactor) and analysis of inorganic ions. The ions NH_4^+, NO_3^- and SO_4^{2-} make up about 90% of the ion charge balance. Nitrate clearly dominates in the larger aerosol particles of a few micrometers diameter. This is probably due to nucleation of HNO_3 onto existing particles (e.g. sea salt particles) during their atmospheric transport. On the other hand, NH_4^+ dominates in the smaller size range. Figure 3.10 also shows that there is one maximum of the mass size distribution around 1 µm aerodynamic diameter. This maximum corresponds to the maximum of the 'accumulation range' of atmospheric aerosol particles. These particles exhibit the longest atmospheric life time and the shape of the size distribution (Fig. 3.10) indicates that the particle population represents an 'aged' air mass, i.e. an air mass that has traveled tens of hours (or a few days) with these particles incorporated, before arriving at the Waldstein research site.

3.8 Conclusions

Environmental change is to a large extent defined and driven by atmospheric change. In this chapter, the concentrations of trace substances in the atmospheric boundary layer at the Waldstein site in the Fichtelgebirge mountain range, NE Bavaria, Lehstenbach catchment, were reviewed. Parts of the data basis reach back to the mid-1980s, so that long time series could be presented for sulfur dioxide (SO_2) and (O_3). Other data, such as those for aerosol particles, are less complete. The results showed that the Waldstein, located within the mountain range at about 775 m a.s.l., can be characterized as a central European, remote background site:

- The median O_3 mixing ratio is 26 ppb, with yearly 95% percentiles of hourly averages hardly exceeding 70 ppb. Photochemical episodes with high O_3 are associated with low NO_x mixing ratios in the order of a few parts per billion.
- The median NO_x mixing ratios were between 4.5 and 5.4 ppb within the past 3 years (2000–2002). Highest NO_x mixing ratios occurred during the winter months when the atmospheric boundary layer is less deeply developed.
- Median ammonia mixing ratios were between 0.6 and 1.5 ppb, with highest values (over 5 ppb) occurring during the spring and summer months, when the emissions from farming activities are highest.
- Although new formation of aerosol particles through nucleation processes seems to occur at the Waldstein site during summer when the emission of VOCs from the vegetation is high (Steinbrecher et al., this Vol.), the aerosol particle size distributions indicate that most of the time 'aged' air masses are advected to the site.
- Trend analysis exhibited a strong decrease of atmospheric SO_2. Periods with high mixing ratios hardly occur any more since 1998. For NO_x, no clear trend is established. The causes for the O_3 pattern with increasing mixing ratios until the mid-1990s and a decrease since then remain unclear so far. The acidity and sulfate content of fog decreased, while the relative contribution of NO_3^- and NH_4^+ to the fog water ion balance increased.

References

Held A, Hinz K-P, Trimborn A, Spengler B, Klemm O (2002a) Chemical classes of atmospheric aerosol particles at a rural site in central Europe during winter. J Aerosol Sci 33:581–594

Held A, Wrzesinsky T, Mangold A, Gerchau J, Klemm O (2002b) Atmospheric phase distribution of oxidized and reduced nitrogen at a forest ecosystem research site. Chemosphere 48:697–706

Held A, Hinz K-P, Trimborn A, Spengler B, Klemm O (2003) Towards direct measurement of turbulent vertical fluxes of compounds in atmospheric aerosol particles. Geophys Res Lett 30:216–219

Klemm O (2001) Trends in fog composition at a site in NE Bavaria. In: Proc 2nd Int Conf on Fog and Fog Collection, 15–20 July, St John's, Newfoundland, Canada

Klemm O, Lange H (1999) Trends of air pollution in the Fichtelgebirge mountains, Bavaria. Environ Sci Pollut Res 6:193–199

Klemm O, Stockwell WR, Schlager H, Krautstrunk M (2000) NO_x or VOC limitation in east German ozone plumes? J Atmos Chem 35:1–18

Schulze E-D, Lange OL, Oren R (1989) Forest decline and air pollution: a study of spruce (*Picea abies*) on acids soils. Ecological studies 77. Springer, Berlin Heidelberg New York

Umweltbundesamt (2002) Umweltdaten Deutschland 2002. http://www.umweltbundesamt.de

Vestreng V, Klein H (2002) Emission data reported to UNECE/EMEP: quality assurance and trend analysis and presentation of WebDab, MSC-W status report 2002. http://webdab.emep.int

Wyers GP, Otjes RP, Slanina J (1993) A continuous flow denuder for the measurement of ambient concentrations and surface-exchange fluxes of ammonia. Atmos Environ 27A:2085–2090

4 Climate Change in the Lehstenbach Region

T. Foken

4.1 Introduction

The Waldstein region with the Lehstenbach catchment is located in the northwestern part of the horseshoe-shaped Fichtelgebirge mountains in northeastern Bavaria (Germany). The present climate is described by Foken (2003), Foken and Lüers (2003) and Gerstberger et al. (this Vol.). Due to the dramatic changes in the global climate over the past 20–30 years (IPCC 2001), investigations of temperature and precipitation trends are also of special interest for ecosystem research. While temperature trends in Germany are more or less uniform, large regional differences in precipitation trends were observed. An increase is assumed for southern Bavaria (BayFORKLIM 1999), while in Saxonia (only 50–100 km NE of the Fichtelgebirge) a significant decrease was found (Küchler 2003). Because the Waldstein region is located between these two areas and the region, moving from west to east, experiences a change from maritime to continental climate, investigations of climate change are of special interest. Some of these changes, like the reduced number of days with snow cover, are even visible to the non-specialist. In the following text, the investigations by Foken (2003) of climate change in the Fichtelgebirge and upper Eger River valley focus on the Lehstenbach catchment and are supplemented by recent investigations of trends in spring precipitation and winter snow cover.

4.2 Change of Climatological Elements

The different climate elements were investigated primarily for the climate normal period 1961–1990 and compared with the new normal period 1971–2000. In addition, trends of the last 40 years were investigated for several climate elements.

Ecological Studies, Vol. 172
E. Matzner (Ed.), Biogeochemistry of Forested Catchments in a Changing Environment

4.2.1 Temperature

Figure 4.1 shows the temperature trend for Bayreuth of +1.1 K for the 150 years from 1851–2000, with the last 30–40 years showing a significant increase in temperature. This is in agreement with long-term temperature trends in Germany and very similar to global trends (IPCC 2001). The temperature trends found in the recent climate of Eastern Upper Franconia were

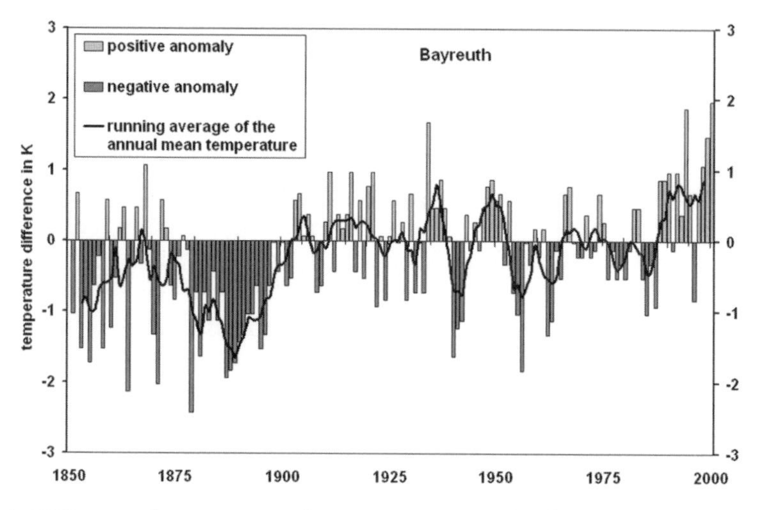

Fig. 4.1. Differences between annual mean temperature in the period 1851–2000 and normal mean temperature in the period 1961–1990 in Bayreuth. (Foken 2003)

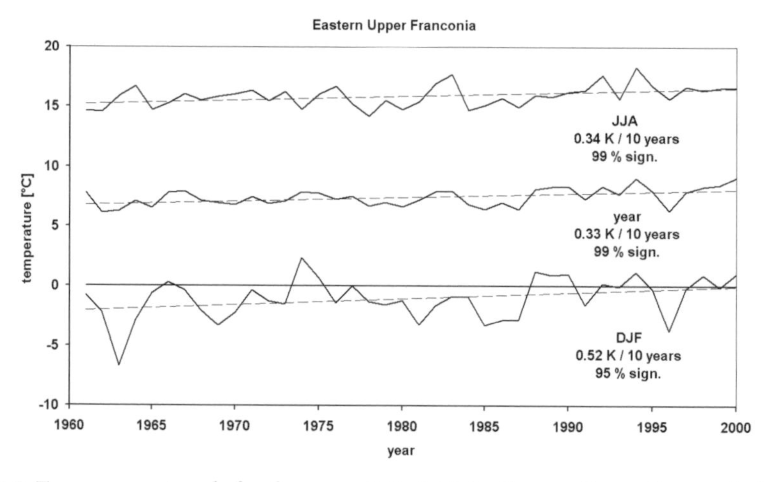

Fig. 4.2. Temperature trends for the years 1961–2000 for Eastern Upper Franconia for the summer (*above*; *JJA* June–July–August), full year (*middle*) and winter (*below*; *DJF* December–January–February). (After Foken 2003)

analysed for a synthetic-temperature time series from the stations at Hof-Hohensaas, Bayreuth, and Weiden for the years 1961–2000 (Foken 2003; Fig. 4.2). Temperature trends of 0.33 and 0.52 K/10 years in the winter months were found. For Fichtelberg-Hüttstadel, the mean winter minima increased by 0.51 K/10 years. The findings are similar across Germany. During the last century, the temperature increased by only 0.9 K, which is less than rates recorded for the last 40 years which showed a more dramatic increase. Of particular interest is the comparison of the climate periods 1961–1990 and 1971–2000 (Fig. 4.3).

The warming, especially in the winter, is related to a reduction of the day-to-day temperature variability due to higher minimum temperatures. A significant increase of approximately 0.5 K/10 years was found in the minimum temperature, which is in good agreement with the findings in Switzerland (Rebetez 2001). For Fichtelberg-Hüttstadel, while there was no significant change found in the number of frost days (minimum <0.0 °C), for ice days (maximum <0.0 °C) the value in winter declined from 48 (1961–1990) to 40 (1971–2000) and the value for the entire year declined from 60 (1961–1990) to 49 (1971–2000). For the summer period in this location, August showed an increase of approximately 20 % at all altitudes for summer days (maximum ≥25 °C), which corresponds to an increase of the mean number of summer days from 3.6 to 4.5 for Fichtelberg-Hüttstadel.

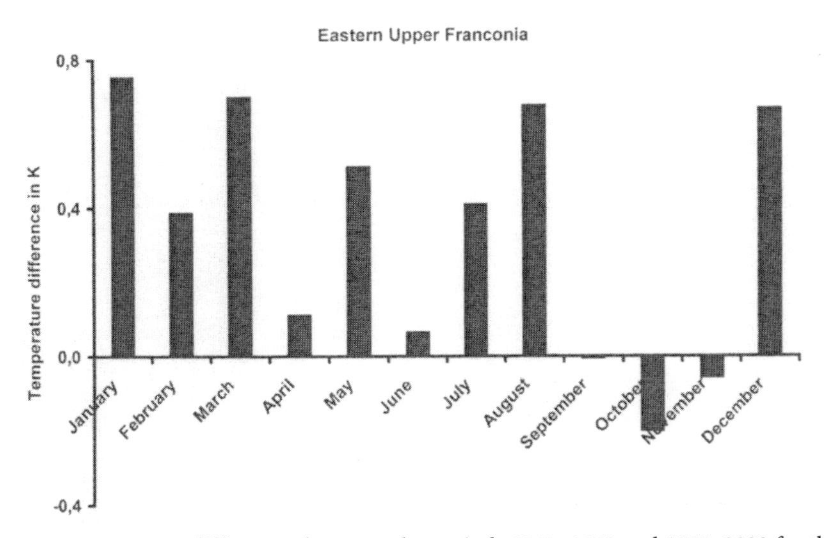

Fig. 4.3. Temperature differences between the periods 1961–1990 and 1971–2000 for the synthetic temperature series of the stations Hof-Hohensaas, Bayreuth, and Weiden. (Foken 2003)

4.2.2 Precipitation

Contrary to the findings for temperature, no significant changes were found in annual precipitation. In the Fichtelgebirge, there is an increase of only 19 mm/10 years to the annual sum and 16 mm/10 years to the winter sum (Foken 2003). The latter may be caused by the higher percentage of liquid precipitation in winter due to higher temperatures. For liquid precipitation, a

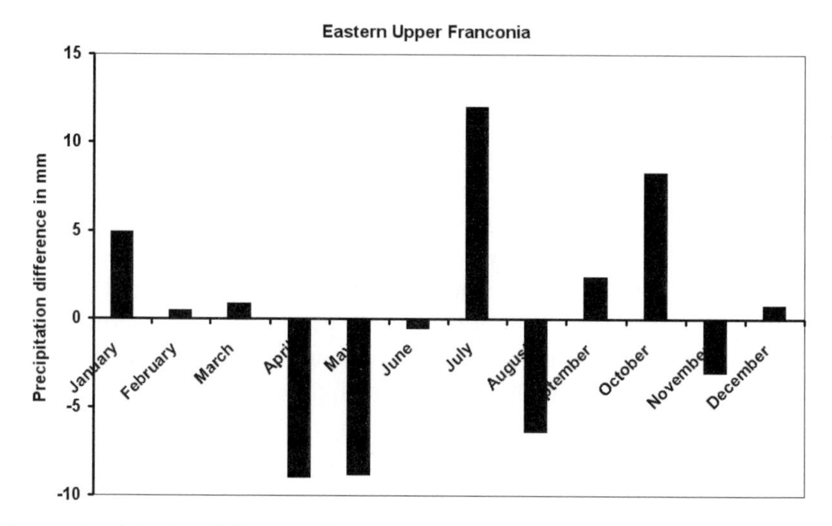

Fig. 4.4. Precipitation differences between the periods 1961–1990 and 1971–2000 for the synthetic precipitation series of five stations of the Fichtelgebirge. (Foken 2003)

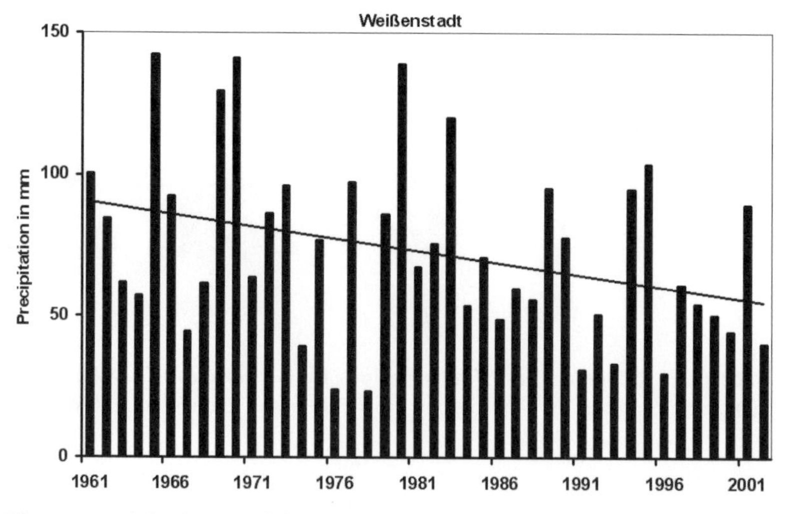

Fig. 4.5. Precipitation trend for Weißenstadt in April for the period 1961–2002

lower measuring error is typical (Richter 1995), which may explain this small increase, especially in wintertime. This is an interesting fact in light of an extreme reduction in precipitation in Saxony (Küchler 2003). Nevertheless, in the Fichtelgebirge, precipitation decreases significantly in spring and August, while it increases noticeably in July and October (Fig. 4.4).

The decline in spring precipitation, especially in April and May, is significant. Figure 4.5 illustrates the trend for the station at Weißenstadt in April. Besides the high interannual variability, a significant trend of 4 mm/10 years was found for the period 1961–2002. This reduced precipitation occurs during the growing season of most plants, therefore influencing several agricultural crops.

4.2.3 Snow Cover

According to Köppen's definition of climate classification, a winter snow cover may be assumed if the coldest month has a temperature lower than –3.0 °C (Essenwanger 2001). Accordingly, the climate period 1961–1990 has such January temperatures for Waldstein/Pflanzgarten (–4.2 °C), Fichtelberg-Hüttstadel (–3.4 °C), and at some places east of the Fichtelgebirge. In the period 1971–2000 the temperature in January averages only –3.6 °C at Waldstein/Pflanzgarten (Fichtelberg-Hüttstadel –2.6 °C). If in the future the climate continues to change, the reduction of the winter snow cover may increase. For the period 1951/1952–1995/1996, such results were found for Baden-Württemberg, where a 40 % decrease was found in the snow cover at low-elevation sites (Günther and Rachner 2000), and sites like the Waldstein region which experienced a 20–30 % decrease.

One analysis (Seifert, pers. comm.) has revealed a significant decrease, approximately 10 days/10 years, in the number of days with snow cover. Much more impressive is the decrease in days with snow cover of ≥30 cm (alpine skiing) and ≥15 cm (Nordic skiing) also by approximately 10 days/10 years, reaching a very low level in recent years (Fig. 4.6).

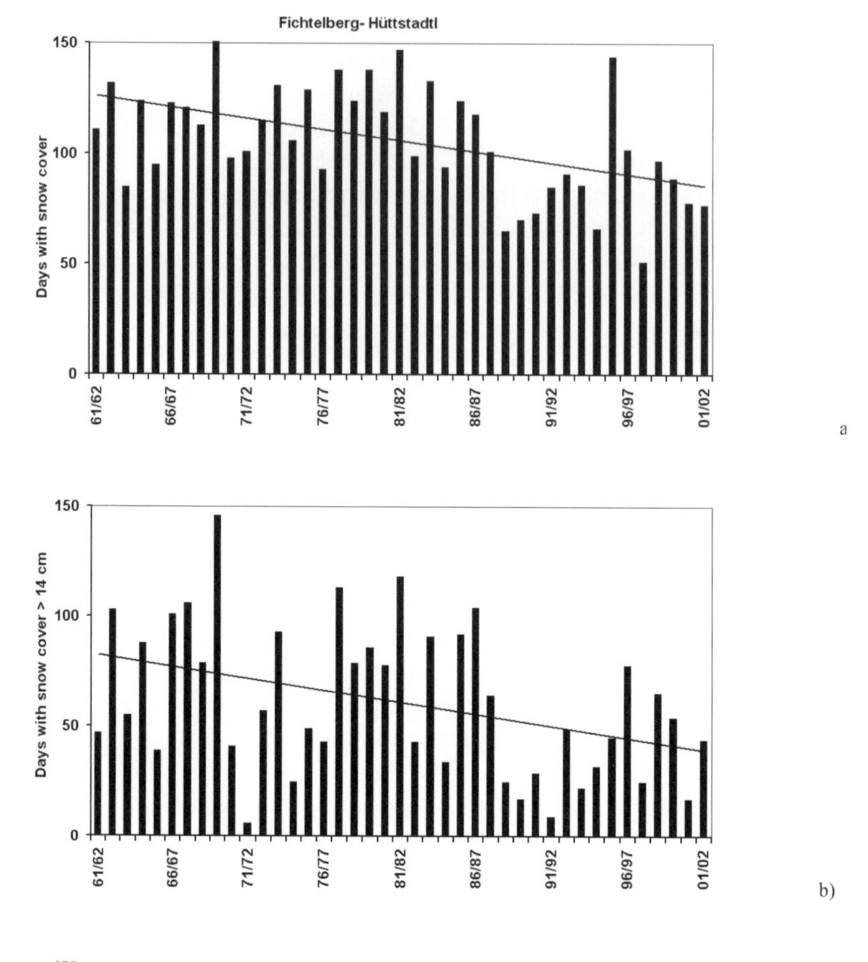

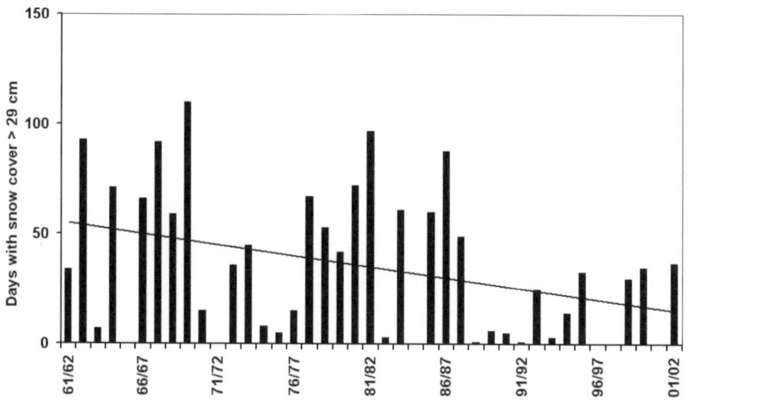

Fig. 4.6a–c. Number of days with snow cover at Fichtelberg-Hüttstadtl (1961–2002; Seifert, pers. comm.). **a** Number of days with snow cover; **b** number of days with snow cover >14 cm; **c** number of days with snow cover >29 cm

4.3 Consequences of Climate Change

The climate changes found in the Lehstenbach region have significant consequences for both ecology and tourism. The impact up to now of climate changes on winter sport and spa activities has been previously discussed (Foken 2003). For the climate of spa towns, a decrease in cold stress days and an increase in heat stress days (only in summer) were found. For winter tourism, a critical number of 50 days with working ski lifts are assumed for economic reasons. For 6 years during the period 1961/1962–1971/1972, this was the case, but for the period 1991/1992–2001/2002, it was not. Similar results were found for Switzerland (Elsasser and Bürki 2002), with a big impact on snow tourism (Elsasser and Messerli 2001).

It is obvious that rising spring temperatures influence the start of the growing season (Chmielewski and Rötzer 2002). Because different plants have different levels of air temperatures that launchs their growing seasons, more detailed analysis is necessary. For most plants, the growing season closes before the end of September where a decrease of the autumn temperature was found. However, because the ecosystems are presently not nitrogen stressed but are instead influenced by high nitrogen deposition rates, the impact may be different, with climate playing a stronger role in the growing process as demonstrated by high production rates in warm years. These findings must be taken into account if the ecosystem is to be changed by forest conversion.

4.4 Conclusions

The Fichtelgebirge and Lehstenbach catchment are located in a part of Europe where a significant change from oceanic to continental climate is occurring. However, precipitation rates are higher and temperatures are lower than in other mountain regions in Germany with altitudes between 400 and 1,000 m a.s.l. The region is characterized by the same effects of climate change as other mean European landscapes, but, until now, without changes in annual precipitation sums. The winter is significantly warmer, with a strong influence on snow cover. All these effects greatly influence life in the mountain region as well as the ecosystem. The changes are significant and cannot be discussed as normal variations and noise. In any future discussion of trends and changes in ecosystem parameters, ecosystem research of the past 20 years and that of the future must take these changes into account. Knowledge of the extent of climate change is of immense significance for all projects in the region.

Acknowledgements. The author gratefully acknowledges the work of more than 20 students, who over the course of 1998–2003 prepared the climatological basis of this chapter. This research was supported by the Bavarian State Ministry of Regional Development and Environmental Problems (contract no. 111450) and by the Federal Ministry of Education and Research (contract no. PT BEO51-0339476 D).

References

BayFORKLIM (1999) Klimaänderungen in Bayern und ihre Auswirkungen. Bayerischer Klimaforschungsverbund, Munich

Chmielewski FM, Rötzer T (2002) Annual and special variation of the beginning of the growing season in Europe in relation to air temperature changes. Clim Res 19:257–264

Elsasser H, Bürki R (2002) Climate change as a threat to tourism in the Alps. Clim Res 20:253–257

Elsasser H, Messerli (2001) The vulnerability of the snow industry in the Swiss Alps. J Mountain Res Dev 21:335–339

Essenwanger OM (2001) Classification of climates. Elsevier, Amsterdam

Foken T (2003) Lufthygienisch-Bioklimatische Kennzeichnung des oberen Egertales. Bayreuther Forum Ökol 100:68 + XLVIII

Foken T, Lüers J (2003) Klimawandel in Oberfranken. Terra Nostra 6:129–135

Günther T, Rachner M (2000) Langzeitverhalten von Schneedeckenparametern, Ergebnisse aus KLIWA. KLIWA-Ber 1:68–80

IPCC (2001) Climate change 2001. The scientific basis. Cambridge University Press, Cambridge

Küchler W (2003) Zur regionalen Klimaentwicklung in Sachsen. DMG Mitt 1:16–17

Rebetez M (2001) Changes in daily and nightly day-to-day temperature variability during the twentieth century for two stations in Switzerland. Theor Appl Clim 69:13–21

Richter D (1995) Ergebnisse methodischer Untersuchungen zur Korrektur des systematischen Meßfehlers des Hellmann-Niederschlagsmessers. Ber Dtsch Wetterdienstes 194:93

III Vegetation Response

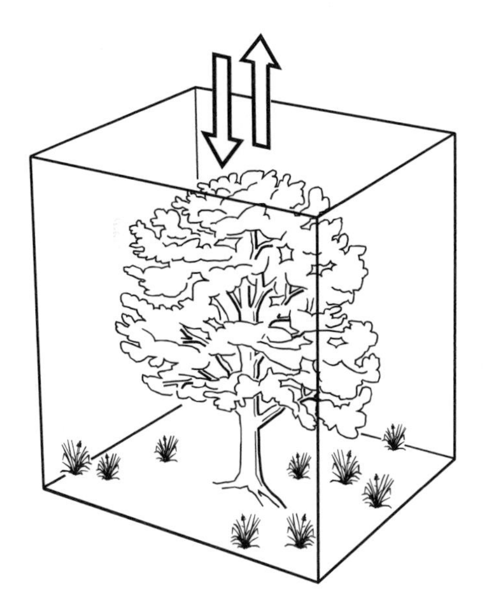

5 Atmospheric and Structural Controls on Carbon and Water Relations in Mixed-Forest Stands of Beech and Oak

B. KÖSTNER, M. SCHMIDT, E. FALGE, S. FLECK, and J.D. TENHUNEN

5.1 Introduction

The natural vegetation of central Europe is dominated by European beech (*Fagus sylvatica*) increasingly mixed with pedunculate oak (*Quercus robur*) in dry lowlands and sessile oak (*Quercus petraea*) in lower montane regions (Walter and Breckle 1994). Companion species are hornbeam (*Carpinus betulus*) and lime (*Tilia platyphyllos*, *T. cordata*). Since many tree species disappeared during the glacial periods, even natural deciduous forests in central Europe are relatively species-poor (Ellenberg 1982; Mayer 1984). Within broad-leaved species, forest management has concentrated on oak and beech for many years, even though the variety of species used for wood production is now slightly increasing. Today, additional benefits of forest functions like air and water quality, flux control (nitrogen, carbon), biodiversity and recreation are considered. More mixed-deciduous forests are being re-established, now reaching an area of 44% in Germany (Smaltschinski 1990; Krüger et al. 1994). Despite their increasing importance, comparably little information on the physiology and ecology of mixed stands is available. Ecological benefits expected from mixed-forest stands include higher structural diversity, higher physical stability, higher diversification in the use of resources, higher resistance to herbivory and pests, and more balanced response to environmental change (e.g., Cannel et al. 1992; Kelty et al. 1992; Larson 1992; Thomasius 1992; Pretzsch 2003). However, the properties and greater potentials of mixed-species stands do not necessarily include higher production, especially timber yield, because benefits of mixed stands depend very much on species-specific properties and environmental conditions (Roy 2001). Under specific constraints, monospecific stands may grow better than mixed-species forests (Pretzsch 2003). The production in mixed stands was only increased when specific resource limitations could be compensated by the presence of an adapted species (Kelty 1992) or when species of different ecological amplitude were combined (Pretzsch 2003). In the long term, however, the higher genetic

Ecological Studies, Vol. 172
E. Matzner (Ed.), Biogeochemistry of Forested Catchments in a Changing Environment

potential of mixed-species stands should provide greater adaptability to changing environment and reproductive success.

Exchange processes between the forest canopy and atmosphere belong to the most intensively studied forest ecosystem functions (e.g., Geiger 1961; Monteith 1965; Waring and Running 1998). Knowledge of atmospheric exchange has increased significantly during the last 15 years, when eddy-covariance techniques were improved and adopted in relation to forest canopies (Baldocchi et al. 1988; Valentini 2003). While micrometeorological techniques are used to determine spatially integrated net ecosystem fluxes over extended forest canopies, it is not possible to discern physically from biologically controlled fluxes (interception/transpiration) or fluxes from different ecosystem compartments (soil/plant respiration, canopy transpiration/soil evaporation). Thus, additional measurements and modelling are required to analyse and explain total fluxes at the ecosystem level (Ehman et al. 2002). Gas-exchange measurements at the leaf level need to be scaled-up to the tree crown or forest canopy level (Beyschlag et al. 1995; Tenhunen et al. 2001; Baldocchi and Amthor 2001). Between gas-exchange measurements at small scales and micrometeorological measurements at larger scales, sap flow measurements at the branch, tree and plot level are used to validate gas-exchange models (Falge et al. 2000; Fleck 2002) and to interpret total fluxes of eddy covariance measurements (Köstner et al. 1992; Granier et al. 1996a, 2003). Sap flow measurements are the only technique providing fluxes at the whole-tree level under natural atmospheric conditions, discerning tree water flux from other sources, and directly enabling its analysis with respect to tree size, age and species.

In forest ecology, structure refers to the number and mass of plant parts, individuals and stands as well as to age and species (Oliver 1992). It is further related to spatial, temporal and functional change. Compared to studies on functional changes of plants with environmental variables, relatively less is known about changes with structural stand parameters (Peck and Mayer 1996). Structure-related parameters in vegetation–atmosphere studies and models are often confined to leaf area index (LAI), vegetation height and/or roughness length (Monteith 1975; Oke 1987; Waring and Running 1998). However, the importance of these parameters may change significantly with stand dynamics, aging and species composition (e.g., Magnani et al. 2000; Köstner et al. 2001; Wilson et al. 2001). For instance, increasing tree height reduced transpiration and photosynthesis independently of leaf area (Ryan and Yoder 1997; McDowell et al. 2002). Despite increasing leaf area, canopy transpiration, carbon uptake and water-use efficiency (WUE, carbon uptake/transpiration) decreased in old Norway spruce stands with lower stand density (Köstner et al. 2002). Structural dynamics are related to forest growth and management practices. Understanding structural controls and limitations on carbon flows of forest is therefore essential for managing and accounting for carbon sinks (Schulte et al. 2001; Valentini 2003). Forest inventories are used for spatial scaling of carbon sinks (Myneni et al. 2001). Therefore, structures essential in physiology (e.g., LAI) and management (e.g., basal area) need to be linked for interpretation and management of forest carbon sinks. Another important issue is the interaction of water and carbon with increased nitrogen input (Schulze 2000). Nitrogen availability and spatial distribution of N within tree crowns provide an important basis of detailed structure-related gas-exchange models (Fleck et al., this Vol.). It is known that species-, size- and age-specific behavior is relevant for maximum transport capacities and exchange rates of plants and stands (e.g., Schulze et al. 1994; Ryan et al. 1997; Binkley et al. 2002). The fact that the range of maximum leaf conductances decreases strongly at the community level indicates that structure has an important balancing role (Körner 1996). However, it still remains difficult to predict how physiological capacities interacting with environmental conditions and structure integrate at the stand level.

The scope of the study was to quantify and analyse contributions of the deciduous tree species *Fagus sylvatica* and *Quercus petraea* to total water use of the forest canopy. It was questioned how atmospheric variables control canopy transpiration (E_c) and canopy conductance (g_c) in the short term, and how structural control related to species, tree and stand structure affects exchange processes between canopy and atmosphere in the long term. A three-dimensional stand gas-exchange model was applied as an analytical tool to study the potential range of variation in forest gas-exchange related to species composition and structural change without variation of other site conditions. The combination of both approaches will allow new insights into the behavior and quantitative valuing of two important tree species with respect to functional characteristics relevant to physiology and gas-exchange modeling, but also with respect to structures important for forestry. Further, improved understanding at the tree and stand level should provide the basis for better process understanding at the catchment and landscape level as well as for management decisions.

5.2 The Study Sites Steinkreuz and Großebene

The presented studies were mainly conducted in the Steigerwald, a forested area in northern Bavaria, Germany, between the cities of Bamberg and Würzburg, comprising around 1,000 km^2 with elevations of 300 m a.s.l. in the east to approx. 500 m a.s.l. in the west (Gerstberger et al., this Vol.). In contrast to most forest regions in Germany, where natural deciduous forest has been replaced by coniferous monocultures, many forest stands in the Steigerwald are autochthon. Owned by the archbishops of Bamberg and Würzburg during the medieval age, only coppicing systems around villages had been allowed (Sperber and Regher 1983). The originally dominating species were probably

Quercus petraea, *Carpinus betulus*, *Fagus sylvatica*, *Sorbus torminalis* and *Betula pendula* (Welß 1985). Later on, forest management tried to maintain relatively natural mixed broad-leaved stands, favoring especially regrowth of beech and oak (Klöck 1980; Sperber and Regher 1983). Today, the Steigerwald represents an important region of deciduous old-growth forests with valuable timber resources (Franz et al. 1993; Pretzsch 1993; Krüger et al. 1994). According to the climate index of Ellenberg (Q=mean temperature of the warmest month/annual sum of precipitation $\times$ 1,000), the considered study sites belong to the area where oak, beech and hornbeam should be most abundant, with higher vitality of beech (Q=21–25; Q at Steinkreuz = 23 in 1999), while regions more dominated by oak are related to Q values between 26 and 30 (Hofmann 1968; Ellenberg 1982).

5.2.1 Structural Characteristics of Study Sites

Measurements were conducted at the site Steinkreuz and the site Großebene, 1.3 km from Steinkreuz. The site Großebene was selected because of its higher portion of oak. Although of similar age, maximum tree height at Großebene was 10 m less than at Steinkreuz. This may be because trees at Großebene were growing on a plateau, whereas trees at Steinkreuz were growing on a slight slope. Further, soil depth was lower at Großebene (up to 40 cm) than at Steinkreuz (up to 80 cm), while soil pH (3.7) and C/N of the humus layer (15.16) were similar (Fleck 2002; Gerstberger et al., this Vol.). Thus, owing to edaphic reasons, forestry may have favored oak at this site. Additionally, an area of pure beech (beech plot) was selected at Steinkreuz for comparison with the mixed stands. Biometric characteristics were determined in a fenced area of 1.29 ha at Steinkreuz and in 0.31 ha at Großebene (Fleck and Schmidt 2001; Gerstberger et al., this Vol.). LAI was estimated seasonally by optical methods (LAI-2000, Li-Cor, Corvallis, USA), litter fall and allometric relationships adopted to the site (Fleck et al., this Vol.). Overall, Steinkreuz was dominated by beech reaching 68 % of stand basal area (SBA), whereas Großebene was more dominated by oak forming the upper canopy and reaching 64 % of SBA (Table 5.1). As is quite common for the region, oak at the sites is periodically defoliated by the feeding of caterpillars of *Tortrix viridana*. Insect outbreaks may last 4–5 years separated by 7-year periods of latency (Horstmann 1984). Increasing damage by insect defoliation in oak stands is generally observed in Germany and is connected with high N content of leaves and oak decline symptoms (Thomas and Blank 1996; Thomas 1998).

Sample trees for sap flow measurements were selected according to species composition and size distribution of trees (Fig. 5.1A, B). At Steinkreuz, the largest beech trees were found contributing most to SBA. Additionally, a significant number of smaller suppressed trees of the same age were present. Oak trees at this site occupied medium diameter at breast height (DBH) and

Table 5.1. Structural characteristics of tree density, stand basal area (*SBA*), stand sapwood area (*SSA*) and projected leaf area index (*LAI*) at the study sites pure beech plot, mixed beech–oak Steinkreuz and mixed oak–beech Großebene in the Steigerwald. Additionally, at Steinkreuz, four individuals of hornbeam (*Carpinus betulus*) were present which are not included in the biometric analysis

	Beech plot Beech	Steinkreuz Beech	Steinkreuz Oak	Steinkreuz Total	Großebene Beech	Großebene Oak	Großebene Total
Tree density (n ha^{-1})	279	266	87	353	347	179	526
Percentage of total	100	75.4	24.6	100	66.0	34.0	100
SBA (m^2 ha^{-1})	39.8	22.5	9.9	32.4	10.7	17.1	27.8
Percentage of total	100	69.4	30.6	100	38.5	61.5	100
SSA (m^2 ha^{-1})	32.8	18.6	2.2	20.8	8.9	3.7	12.6
Percentage of total	100	89.4	10.6	100	70.6	29.4	100
LAI (m^2 m^{-2})	7.2	4.4	2.2	6.6	2.6	3.8	6.4
Percentage of total	100	66.7	33.3	100	40.6	59.4	100

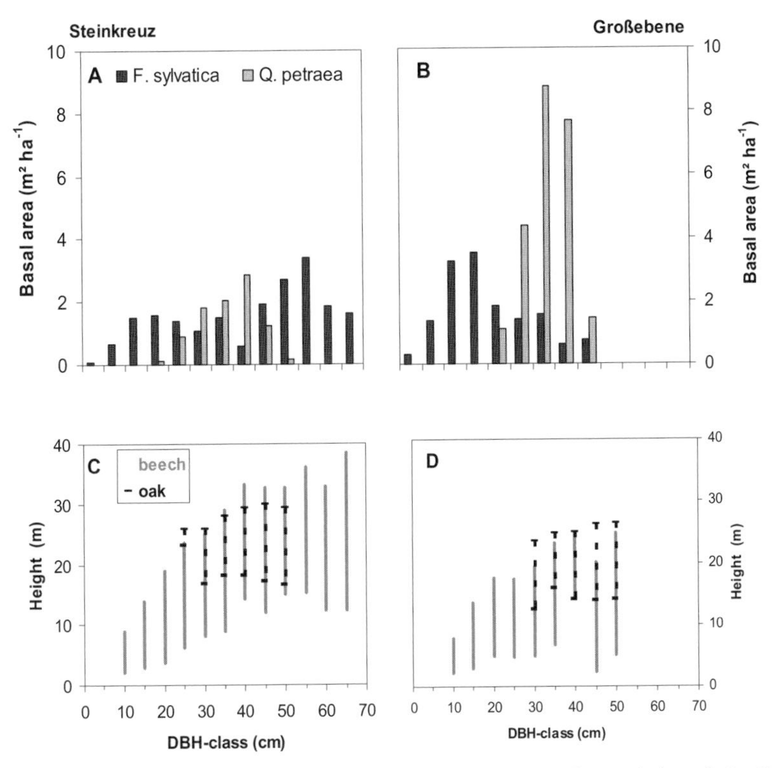

Fig. 5.1. Frequency distribution of tree basal area at site Steinkreuz (**A**) and Großebene (**B**) related to DBH classes as well as tree height and crown length at Steinkreuz (**C**) and Großebene (**D**), respectively

Table 5.2. Structural characteristics of sample trees (mean and standard deviation, SD) from sap flow plots at beech plot, Steinkreuz and Großebene. Projected leaf area of trees (A_l, m^2) was derived from stem basal area (A_b, cm^2) as follows: beech: $A_l=0.674 \times A_b^{0.81}$; $r^2=0.89$; oak: $A_l=0.616 \times A_b^{0.85}$; $r^2=0.854$. Sapwood area (A_s) was calculated according to equations in Fig. 5.3

Plot/species		n	DBH (cm)	Height (m)	A_b (cm^2)	A_s (cm^2)	A_l (m^2)	A_l/A_s ($m^2\ cm^{-2}$)
Beech plot								
Dominant	Mean	3	56.77	33.0	2,582	2,145	444.2	0.21
	SD		9.89	2.8	871	721	123.0	0.01
Suppressed	Mean	3	18.48	17.1	270	226	71.6	0.32
	SD		1.89	4.5	54	45	11.7	0.01
Steinkreuz beech								
Dominant	Mean	3	54.90	33.8	2,415	2,007	420.8	0.21
	SD		9.51	2.1	782	648	113.4	0.01
Suppressed	Mean	3	20.17	18.3	355	297	86.7	0.32
	SD		8.25	4.1	290	242	57.8	0.04
Steinkreuz oak								
Dominant	Mean	3	39.67	29.8	1,246	282	270.6	0.97
	SD		4.51	0.9	279	75	52.0	0.08
Großebene beech								
Intermediate	Mean	5	32.00	22.5	830	692	176.5	0.26
	SD		6.36	2.2	328	273	56.8	0.02
Suppressed	Mean	4	17.90	18.8	260	218	68.9	0.32
	SD		3.80	1.6	108	91	23.4	0.03
Großebene oak								
Dominant	Mean	5	41.48	26.4	1,360	312	291.7	0.94
	SD		3.64	0.9	234	64	43.1	0.06
Intermediate	Mean	3	30.50	24.7	731	149	171.9	1.16
	SD		0.75	0.6	36	9	7.3	0.02

height (Fig. 5.1C), but dominated basal area (Fig. 5.1B) and the upper canopy at Großebene (Fig. 5.1D). Overall, 32 trees were measured including 21 and 11 trees of beech and oak, respectively. Range and mean size of sample trees are described in Table 5.2.

5.2.2 Structural Scaling Parameters

The most important functional parameters for scaling-up fluxes from leaf and tree to stand level are leaf area (A_l) and cross-sectional sapwood area (A_s). For instance, LAI is used as a scaling factor in BIGLEAF models (e.g., Monteith 1965), and spatial leaf area densities are used in complex 3D models (Falge et al. 2000). In this study, A_l of individual trees was estimated by allometric relationships based on the literature and adopted by results from tree harvests at Großebene (Fleck et al., this Vol.). Tree- and stand-level sap flow is most appropriately scaled-up by A_s (Köstner et al. 1998b). In addition to stem cores, computer tomography (CT; see Habermehl and Ridder 1993) was applied to beech in order to visualize the complete cross section of sapwood in standing trees (Fig. 5.2). Determination of A_s in diffuse-porous beech is especially difficult, for the species usually does not form heartwood (Hillis 1987). In oak, spatial distribution of wood density was more irregular. Therefore, only stem cores were used for sapwood determination. Comparable inhomogeneity of CT pictures had been observed in oak under air pollution (Raschi et al. 1995). At Steinkreuz, the irregular change of wood density may be related to higher variation in annual growth due to periodic defoliation by caterpillars of *Tortrix viridana* (see above).

Strong relationships between tree DBH and A_s were found for species at both sites. As observed earlier in Norway spruce (Alsheimer et al. 1998), different methods of sapwood determination agreed well in beech (Fig. 5.3). In contrast to ring-porous oak, where sapwood was restricted to the outer ring of 1–2 cm, sapwood depth of large beech reached up to 20 cm. However, significant sap flow activity was confined to approx. 6 cm behind the cambium (Schmidt et al. 2000; Granier et al. 2003).

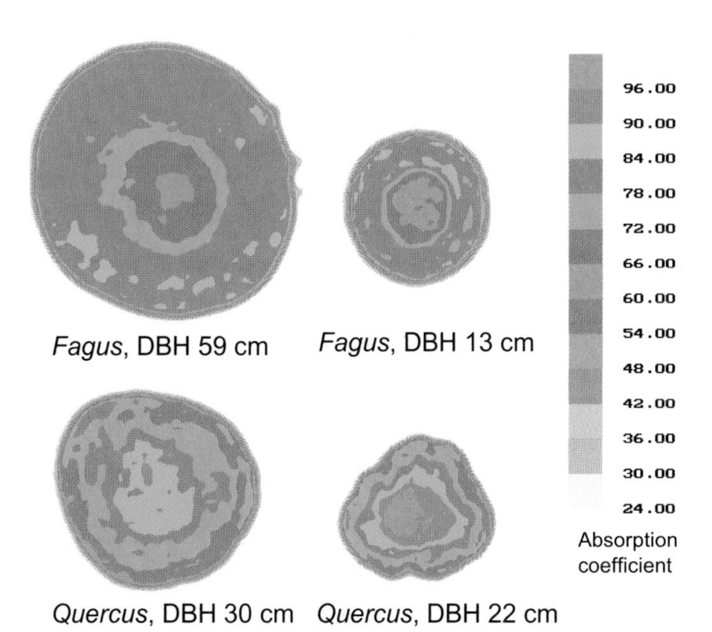

Fig. 5.2. Spatial changes in wood density within the cross section of tree stems at breast height derived from computer tomography (CT). *Colors* refer to relative absorption of X-rays. In diffuse-porous beech, highest wood densities (absorption classes from 72–90) are influenced by water content and typically related to sapwood, whereas in oak, the outer ring of sapwood represents less dense material than the heartwood

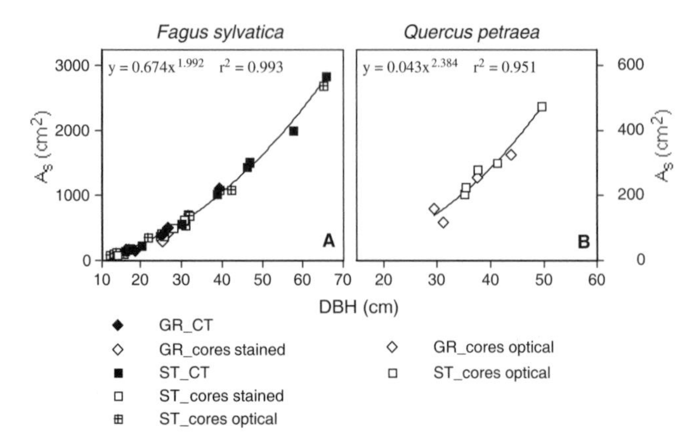

Fig. 5.3. Relationships between tree DBH and cross-sectional sapwood area (A_s) at breast height determined from stem cores and computer tomography (*CT*) for beech (**A**) and oak (**B**), respectively. Stem cores represent means of two samples per tree. Sapwood depth of stem cores was determined by staining with bromcresol-green (Burrows 1980). In fresh cores, optical differentiation of the wetter and therefore darker sapwood was also possible. For beech (**A**) the equation of the curve is $y=0.674x^{1.992}$; $r^2=0.993$, for oak (**B**) the equation is $y=0.043x^{2.384}$; $r^2=0.951$

In the present study, canopy transpiration E_c was derived from scaled-up tree sap flow applying the method according to Granier (1985), using one to three sensors per tree in oak and up to nine sensors per tree in beech (different expositions and sapwood depths: 0–2, 2–4, 4–6 cm behind cambium). A function describing the decline in sap flux density with increasing sapwood depth was applied to scale measured flow rates in beech to the tree level (Schmidt et al. 2000; Granier et al. 2003). For scaling-up to the stand level, sap flow rates from trees of different size were weighted according to tree size distribution (see Fig. 5.1; Schmidt, in prep.).

5.3 Analysis of Canopy Transpiration and Conductance by Means of Sap Flow Measurements

5.3.1 Atmospheric Controls

Knowledge of atmospheric controls of water vapor exchange between vegetation and atmosphere is essential to predict evapotranspiration (transpiration plus soil evaporation), a significant component of the catchment water balance. The relevant controlling components are available energy, e.g., net radiation (R_n), saturation vapor pressure deficit of the air (D), the aerodynamic (g_a) and canopy conductance (g_c) of the related vegetation (Monteith 1965,

1979; McNaughton and Jarvis 1983). Within these parameters, the estimation of g_c is most difficult, because it represents the vertical and horizontal mean of stomatal conductance for the entire stand depending on atmospheric variables, plant internal genetic and physiological properties as well as on canopy structure (Shuttleworth 1989). Because the underlying and interacting mechanisms determining g_c are not sufficiently understood, it is usually empirically determined from gas-exchange measurements in cuvettes scaled-up to the canopy by models (e.g., Beyschlag et al. 1995; Falge et al. 2000) or from evaporation measurements applying the inverse Penman–Monteith equation (e.g., Jarvis and Stewart 1979; Kelliher et al. 1995). Tree sap flow represents the only approach deriving g_c directly from entire tree crowns under natural atmospheric conditions (Köstner et al. 1992; Granier et al. 1996a, b). However, assumptions on tree water storage are sometimes needed when g_c is derived from short-term measurements (Köstner et al. 1998a, b; Ewers and Oren 2000).

Seasonal changes in tree and stand water use at the sites Großebene and Steinkreuz were continuously monitored by sap flow measurements during the years 1998–2000 (Schmidt et al. 2000; Schmidt, in prep.). Here we present

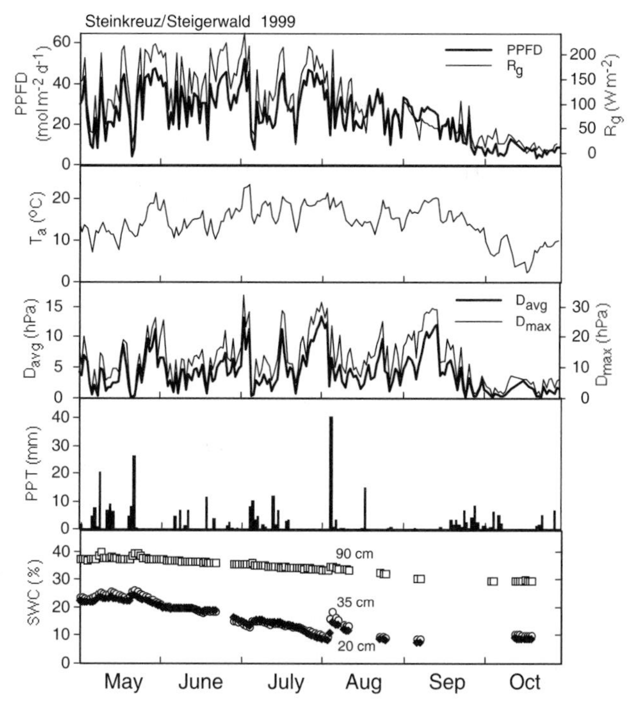

Fig. 5.4. Seasonal changes in climatic variables (*PPFD* photosynthetic photon flux density; R_g global radiation; T_a air temperature; D_{avg} 24-h average of vapor pressure deficit; D_{max} daily maximum D; *PPT* precipitation) and soil water content (*SWC*) at the site Steinkreuz during 1999. Climatic variables were measured at 5-m height above regrowth in a forest gap. Precipitation was supplemented by data from the German Weather Service (station Ebrach). SWC was measured by TDR probes in 20, 30 and 90 cm soil depth below the canopy at the site Steinkreuz. (Courtesy of Department of Geohydrology, BITÖK)

daily canopy transpiration estimates (E_c) based on sap flow measurements of the year 1999. At the same time, meteorological data were recorded at a climate station within a forest gap adjacent to Steinkreuz (Gerstberger et al., this Vol.). Atmospheric conditions throughout the growing season were typical when compared to long-term climate data, with highest radiation during June and July (Fig. 5.4). Radiation was reduced on cloudy and rainy days and was significantly decreasing at the end of the season in September and October. Air temperature (T_a) as well as vapor pressure deficit of the air (D), the driving force of transpiration, were highest during July and August, reaching maximum values of 33 °C and 32 hPa for T_a and D, respectively. Precipitation (PPT) occurred less frequently during the second part of the vegetation period. Hence, soil water content (SWC) decreased continuously during the season and only strong rain events (5 Aug) showed significant effect on SWC in upper soil layers.

The upper limit of potential evaporation (E_{potmax}) was described by net radiation ($\lambda E=R_n$), assuming similar reflectance of the forest regrowth in the gap and the old-growth canopy, and neglecting energy losses by soil heat flux and storage accounting for ca. 5–10 % of total available energy (Fig. 5.5). Formulations of potential evaporation usually result in somewhat lower values taking into account diabatic (Priestley-Taylor equation for equilibrium evap-

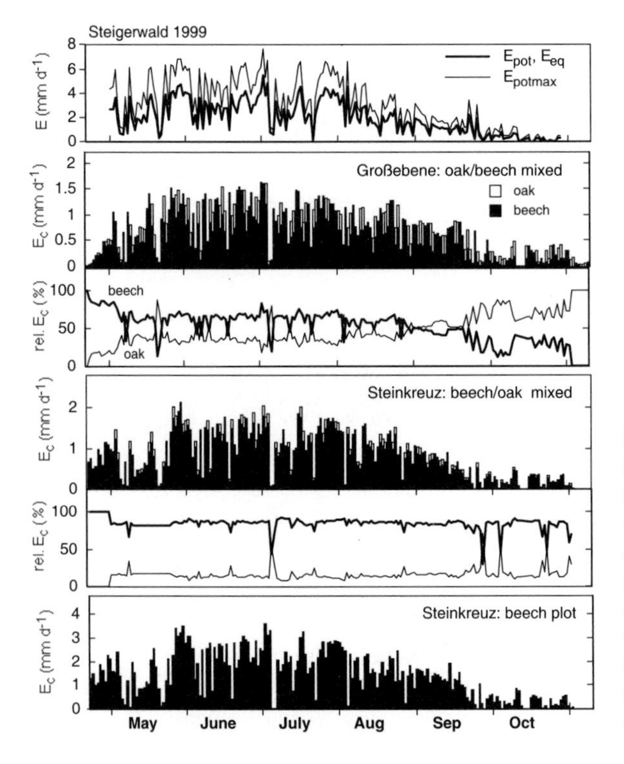

Fig. 5.5. Seasonal changes in daily potential evaporation (E_{pot}, E_{eq}, E_{potmax}) and canopy transpiration (E_c) derived from scaled-up sap flow of oak and beech at the three study sites as well as relative E_c (*rel.* E_c) indicating contribution of oak and beech to total E_c of the stands

oration, E_{eq}) or both diabatic and adiabatic components (Penman equation for potential evaporation, E_{pot}) of latent heat (Monteith and Unsworth 1990). For E_{eq}, the following formula was applied (Monteith and Unsworth 1990):

$$E_{eq} = \left(\Delta R_n / (\Delta + \gamma)\right) * \lambda^{-1}$$

where Δ (Pa K^{-1}) is the rate of change of water vapor pressure with temperature, γ (Pa K^{-1}) is psychrometric constant and λ (J kg^{-1}) is latent heat of vaporization.

Penman evaporation was computed according to the equation described in Waring and Running (1998):

$$E_{pot} = E_{eq} + D * g_a / \zeta$$

where D (kPa) is the saturation vapor pressure deficit of the air, g_a (mm s^{-1}) aerodynamic conductance and the factor ζ summarizing temperature-dependent physical constants: $\zeta = \varrho_a \times (\Delta/\gamma + 1) \times G_v \times T_K$, where ϱ_a is density of air (kg m^{-3}), G_v is gas constant for water vapor (0.462 m^3 kPa kg^{-1} K^{-1}), T_K is air temperature in kelvins, for Δ and γ see above.

Aerodynamic conductance was estimated according to Businger (1956) and Monteith et al. (1965) using the following equation: $g_a = k^2 \times u(z)/[\ln((z-d)/z_0)]^2$, where k is Karman constant (0.41), u is wind speed, z is measurement height, d is displacement height and z_0 is zero plane displacement.

Wind speed (u) measured at 5 m height in the forest gap was transferred to estimated u at height z above the old-growth forest by multiplying by a factor of 2.5. The empirical factor was found for a forest gap of comparable size when compared with adjacent tower measurements well above the canopy at 42 m height (Goldberg et al. 2003). For displacement height d and roughness length z_0 estimates of 26 and 3.9 m, respectively were applied.

Computed g_a ranged from 0.04–0.93 m s^{-1}. Overall, it turned out that the aerodynamic term of the Penman equation had no significant effect on E_{pot} and thus E_{pot} and E_{eq} differed by less than 0.1 mm day^{-1}. This was also independent of the empirical correction of u for gap and above-canopy measurements. Relatively high values of g_a and thus minor limiting effects on evaporation are typical for rough forest canopies (Jarvis and Stewart 1979). Using 24-h averages of R_n, daily rates of E_{pot} increased from May and June up to maximum rates in July (5.5 mm day^{-1}), while daily and mean monthly rates decreased significantly from August to October (Fig. 5.5A). Seasonal fluxes (May to October) reached 356.5 and 561.1 mm $season^{-1}$ for E_{pot} and E_{potmax}, respectively.

Seasonal E_c started around 22 April in beech and about 4 days later in oak. Accordingly, sap flow ceased earlier in beech (4 November) than in oak (12 November) at the end of the season. In general, the seasonal course of actual E_c followed E_{pot} with lower ratios of E_c/E_{pot} in May (leaf unfolding) and higher ratios in September/October. On average, E_c/E_{pot} increased from the oak-dominated site Großebene (0.42) to mixed Steinkreuz (0.49) and the pure beech plot (0.81). At low absolute rates (rainy days), E_c often exceeded E_{pot} (potentially stem refilling). At the beech plot, E_c could also reach similar or higher values than E_{pot} on sunny days. High daily E_c was observed before on a small beech plot (360 m^2) within a spruce stand. There, E_c of beech reached up to 7 mm day^{-1} under non-limiting soil water conditions, while maximum E_c of the surrounding spruce forest was confined to 3.5 mm day^{-1} (Köstner 2001).

Highest daily E_c was found during June and July, reaching maximum rates of 1.6, 2.1 and 3.7 mm day^{-1} for Großebene, Steinkreuz and the beech plot, respectively (Fig. 5.5B, D, F). Generally, oak contributed less than beech at the stand level (0.54 mm day^{-1} at Großebene, 0.33 mm day^{-1} at Steinkreuz), although oak dominated the canopy at Großebene. On seasonal average, oak contributed 39 % to total E_c at Großebene and 14 % at Steinkreuz. Total seasonal sums of E_c increased with increasing portion of beech in the stands, reaching 153, 178 and 292 mm $year^{-1}$ at Großebene, Steinkreuz, and the beech plot, respectively (Table 5.3). Thus, relative E_c of the species (related to total E_c of the stands) remained lowest in oak during most of the season except after rainfall, when the canopy was wet (Fig. 5.5C, E). This could be explained by faster drying of the oak canopy. After rain, sap flow recovered first in oak, followed by dominant beech and was most delayed in subcanopy beech. At Großebene, relative E_c of oak exceeded that of beech at the end of the season when soil water content was decreasing and leaf shedding proceeded faster in beech (Fig. 5.5C).

Most important atmospheric drivers of E_c are photosynthetic photon flux density (PPFD) and D. Being still limited by leaf development, daily E_c increased linearly with the daily sum of PPFD during spring (April/May;

Table 5.3. Overview of cumulative and maximum flux rates of canopy transpiration (E_c) throughout the season (end of April to beginning of November) of species and stand totals at the study sites. Fluxes are related to leaf area index (*LAI*), stand basal area (*SBA*) and stand sapwood area (*SSA*)

Flux	Beech plot Beech	Steinkreuz Beech	Steinkreuz Oak	Steinkreuz Total	Großebene Beech	Großebene Oak	Großebene Total
E_c (mm $season^{-1}$)	291.8	153.4	24.3	177.7	92.7	60.1	152.8
Percentage of total	100	86.3	13.7	100	60.7	39.3	100
E_{cmax} (mm day^{-1})	3.65	1.82	0.33	2.14	1.35	0.54	1.63
E_c/LAI (kg m^{-2} $season^{-1}$)	40.5	34.9	11.0	26.9	35.7	15.8	23.9
E_{cmax}/LAI (kg m^{-2} day^{-1})	0.51	0.41	0.15	0.32	0.52	0.14	0.25
E_c/SBA (kg m^{-2} $season^{-1}$)	7.3	6.8	2.5	5.5	8.7	3.5	5.5
E_c/SSA (kg m^{-2} $season^{-1}$)	8.9	8.2	11.0	8.5	10.4	16.2	12.1

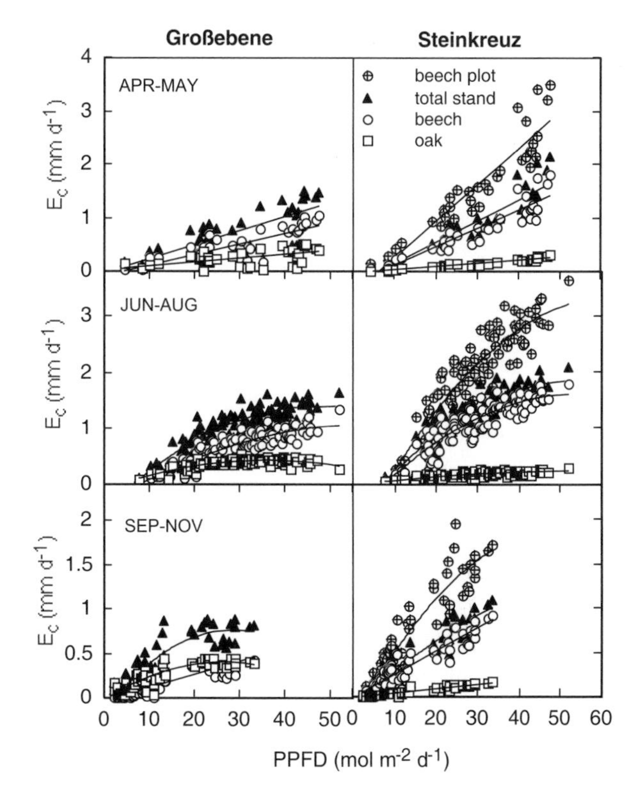

Fig. 5.6. Relationships between daily cumulative photosynthetic photon flux density (*PPFD*) and daily sums of canopy transpiration (E_c) of the study sites for different periods of the season. Values for oak, beech and total mixed stands are shown

Fig. 5.6). Leaf development of oak lagged behind beech. During summer, saturation of E_c at highest daily PPFD (53 mmol m^{-2} day^{-1}) was less pronounced in beech. In the autumn, when maximum daily sums of PPFD were restricted to 60 % of summer values, E_c did not reach saturating rates, which was comparable to spring time.

Similarly to PPFD, linear dependencies of E_c on daily (24-h) average D (D_{avg}) were observed at the beginning of the season (Fig. 5.7), whereas E_c followed a saturation curve during the summer and autumn. In some cases, during August E_c was reduced compared to E_c in June and July, although PPFD and D_{avg} were similar. Obviously, additional limiting effects (reduced photosynthetic or hydraulic capacities) became apparent. Therefore, the regression curves in summer are only based on values from June and July. Coefficients of determination were highest during summer and autumn, with on average somewhat lower values for oak (r^2=0.77) compared to beech (r^2=0.82) but similar values for regressions with PPFD (r^2=0.78) and D_{avg} (r^2=0.80). The non-linear functions were applied to compute maximum E_c (E_{cmax}) of the stands (see below). Environmental variables at which E_{cmax} has been reached were lower in oak (PPFD 34/45 mmol m^{-2} day^{-1}; D_{avg} 7/7.5 hPa) than in beech (PPFD >=53 mmol m^{-2} day^{-1}; D_{avg} 9/10 hPa).

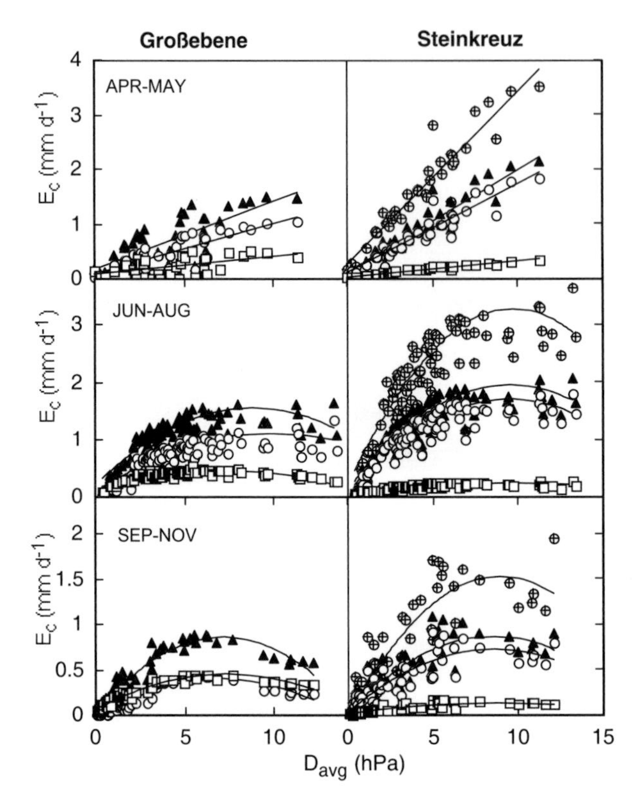

Fig. 5.7. Relationships between daily average vapor pressure deficit of the air (D_{avg}) and daily sums of canopy transpiration (E_c) of the study sites for different periods of the season (24- h D_{avg} is related to half-hourly D_{max} of the day as follows: $D_{max} = 3.904 \times D_{avg}^{0.85}$)

Canopy conductance was derived from E_c and D according to Monteith et al. (1965):

$$g_c = \lambda * \gamma / (\varrho_a * c_p) * E_c / D$$

where c_p (J kg^{-1} K^{-1}) is specific heat of air; for other parameters see equations above.

For parameterization of g_c in the Penman-Monteith equation (Monteith 1965) based on daily values, it is helpful to derive g_c directly from the complete Penman-Monteith equation:

$$g_c = \left[\left(\Delta / \gamma \left((R_n - \lambda E_c) / \lambda E_c \right) - 1 \right) * 1 / g_a + \varrho_a * c_p / (\lambda E_c * \gamma) * D \right]^{-1}$$

If g_a is large as is typical for rough forests, the first part of the term (radiation term) becomes very small. It was therefore suggested to neglect the radiation term of the equation for forests (e.g., Jarvis and Stewart 1979). At our sites, g_c derived from the complete Penman-Monteith equation amounted on average to 94% of g_c only derived from the second part (ventilation term) of the equation. Thus, g_c is sufficiently described by the second

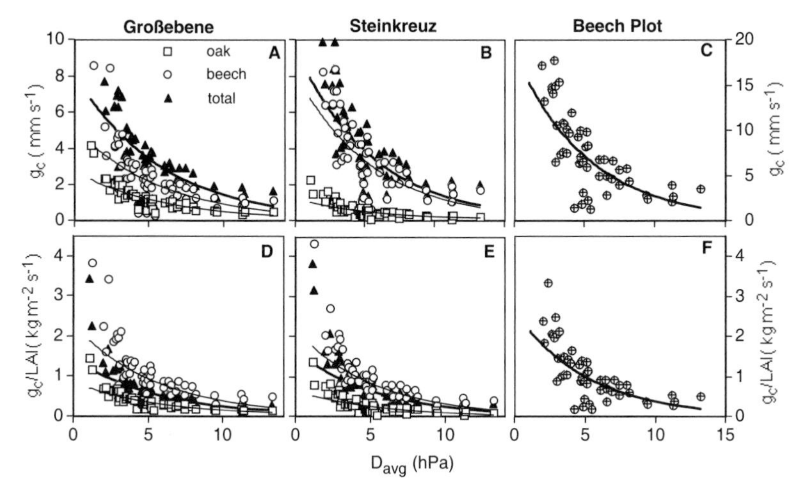

Fig. 5.8. Relationships of daily average vapor pressure deficit of the air (D_{avg}) with canopy conductance (g_c) derived from daily E_c using the inverse Penman-Monteith equation and with g_c per LAI for the sites Großebene (**A, D**), Steinkreuz (**B, E**) and beech plot (**C, F**), respectively. Values for the period June and July for oak, beech and total mixed stands are shown. Note different scale of g_c for the beech plot (**C**)

term analogous to leaf conductance in well-ventilated cuvettes (cf. Köstner et al. 1992).

At all sites in the Steigerwald, a similar decline in g_c with increasing D was observed (Fig. 5.8A–C). The strong non-linear decline is typical for forest sites and has been described both with micrometeorological (e.g., Shuttleworth 1989) and sap flow measurements (Granier et al. 19996a, 2003). Maximum canopy conductance was highest at the beech plot followed by Steinkreuz and Großebene. When g_c was related to LAI (Fig. 5.8D–F), the ratio oak/beech at Großebene decreased from 0.53 to 0.37, increasing the difference by 16%, while at Steinkreuz the ratio oak/beech changed from 0.15 to 0.31, decreasing the difference by 16%. Thus, mean leaf-level conductance of beech was higher at Großebene. Lower g_c/LAI of beech and oak at Steinkreuz may be related to higher amounts of shade-adapted leaves (e.g., in subcanopy trees).

5.3.2 Structural Controls

Fluxes from plants are usually related to leaf area (gas-exchange techniques at leaf and twig level) or LAI (micrometeorological techniques, stand-level models). At regional scales, NDVI (normalized difference vegetation index) derived from remote sensing is most common. However, allometric relationships between tree size (DBH, SBA, height) and leaf area vary with species and site conditions and, hence, the same LAI or NDVI may represent very different vegetation structures. Therefore, additional information on land use, species composition, vegetation height, density, etc. is needed for adequate functional interpretation of LAI or NDVI (Waring and Running 1998). The sap flow approach primarily addresses structural parameters such as A_s, A_l, DBH and SBA, the latter being also very common in forest management and inventory data. Thus, sap flow measurements functionally link different structural characteristics related to physiology (LAI) and forest management (SBA). In the following, both types of structural parameters will be addressed.

Daily water use of individual trees (E_t) reached maximum values of 345 and 60 kg tree^{-1}day^{-1} for beech and oak, respectively. For beech, relationships between maximum tree water use (E_{tmax}) and tree size were similar at all sites. For all beech trees combined (n=21), high coefficients of determination were observed using tree DBH (r^2=0.975) or tree basal area, A_b (r^2=0.982) as the independent variable. Only E_{tmax} of canopy trees (height ≥20 m) was related to tree height (r^2=0.877, n=15; see Fig. 5.9A). While eight oak trees were measured at Großebene, only three oak trees could be evaluated at Steinkreuz due to technical problems. Therefore, additional data on oak from 1996 (Schäfer 1997) were included. All oak trees combined, the relationship of tree water use with height was less pronounced (r^2=0.649, n=15). Further, the whole range of tree height was much smaller in oak.

Although differences in flux rates per leaf area are expected to decrease between trees of different size, an increase in E_{tmax} per leaf area (E_{tmax}/A_l) with tree height was observed in beech (Fig. 5.9B). This indicates that high physiological capacity at the leaf level is maintained even in large tree crowns keeping a high number of leaves (cf. Fleck et al., this Vol.). However, it has to be considered that small trees were growing below the canopy of large trees under limiting light conditions. This is confirmed by clear differences in stable isotope signals measured along the canopy profile of a dominant and suppressed beech at Steinkreuz. The $\delta^{13}C$ values ranged from –27‰ (crown top) to –30‰ (crown base) in the dominant tree while it was more or less stable at around –32‰ in the suppressed tree (Dawson et al., unpubl.). The high negative $\delta^{13}C$ in leaves of the subcanopy tree indicates high internal CO_2 concentration (c_i) related to low carboxylation capacity. The more shade-adapted leaf area of subcanopy trees required less sapwood area for water transport, while the more transpiring dominant trees kept larger sapwood area and water storage capacities (cf. Köstner et al. 1998a; Phillips et al. 2003). Consequently, the leaf area/sapwood area relationship (A_l/A_s) was significantly declining with tree height.

Values of E_{tmax}/A_l observed at the Steigerwald sites agreed with E_{tmax}/A_l (0.5–0.6 kg m^{-2} day^{-1}) described for two beech trees of comparable size (DBH 34 cm) and age (100 years) measured with a modified sap flow system according to Cermák (Köstner et al. 1998b) in a mixed oak–beech stand in northern

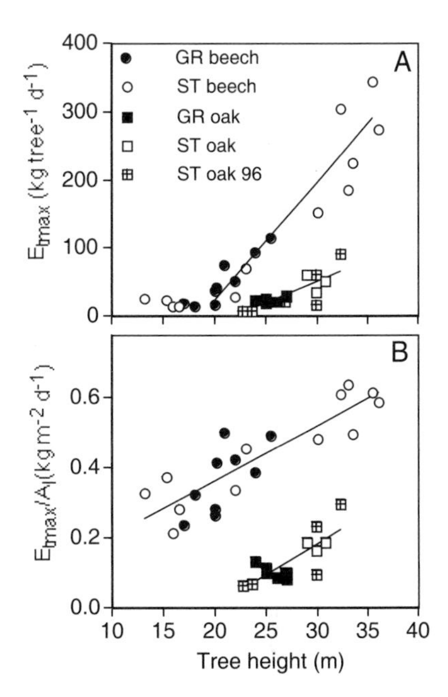

Fig. 5.9. Relationships between tree height and maximum canopy transpiration (E_{tmax}, **A**) of beech (*circles*) and oak (*squares*) as well as between tree height and E_{tmax} per leaf area of trees (E_{tmax}/A_l, **B**). Values of oak at Steinkreuz are supplemented by data from 1996 (Schäfer 1997). Linear regressions are highly significant for both species (beech: $E_{tmax}=17.22 \times \text{height}-321.25$; $r^2=0.88$; $E_{tmax}/A_l=0.016 \times \text{height}+0.042$; $r^2=0.745$; oak: $E_{tmax}=6.41 \times \text{height}-142.47$; $r^2=0.65$; $E_{tmax}/A_l=0.018 \times \text{height}-0.351$; $r^2=0.590$)

Germany (Backes 1996). At this site, oak trees were older (ca. 200 years) and larger (DBH ca. 54 cm) than oak at the Steigerwald sites with similar or higher values of E_{tmax}/A_l (0.5–1.0 kg m^{-2} day^{-1}) compared to beech at the same site. The relatively low E_{tmax}/A_l of oak in the Steigerwald may be explained by competition with beech and periodic defoliation by caterpillars of *Tortrix viridana* (Horstmann 1984). Thus, growth of oak proceeds rather irregularly, which was also indicated by inhomogenous cross sections of CT pictures. If sap flow was confined to the youngest annual tree rings of oak, standard sap flow sensors of 2 cm length may have been too long to give accurate results (Clearwater et al. 1999). Nevertheless, in an earlier study, agreement was found between sap flow rates in oak determined with the 2-cm sensor according to Granier and the modified method according to Cermák (see Granier et al. 1996a).

At the stand level, relationships between E_{cmax} and LAI or SBA were investigated. In Fig. 5.10, results from the Steigerwald sites are compared with results from other deciduous and coniferous forests in central Europe. Within the range of LAI for beech from 5–9 m^2 m^{-2}, E_{cmax} of pure beech stands increased with LAI while no increase was observed for Norway spruce (*Picea abies*) stands (Fig. 5.10A). Highest LAI of 9 m^2 m^{-2} was found on a small area (360 m^2) of densely planted beech within a spruce stand. In extended beech forests LAI is typically lower. E_{cmax} of the mixed deciduous sites from the Steigerwald ranged within those of coniferous stands, whereas E_{cmax} of a 32-year-old oak stand (Bréda et al. 1993) was comparable to the beech stands. A

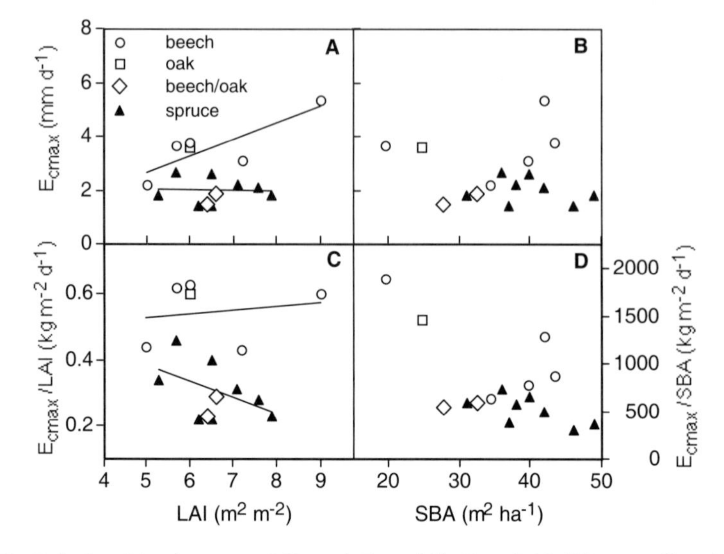

Fig. 5.10. Relationships between LAI and E_{cmax} (**A**), E_{cmax}/LAI (**C**), as well as between stand basal area (*SBA*) and E_{cmax} (**B**) and E_{cmax}/SBA (**D**) for monospecific stands of beech, oak and spruce as well as for the mixed stands at the Steigerwald site (for data from other sites see Bréda et al. 1993; Alsheimer et al. 1998; Köstner 2001; Granier et al. 2003). Each data point represents the seasonal maximum value of one stand

similar pattern is found when E_{cmax} is related to SBA, with the exception of the younger beech (Granier et al. 2000) and oak stand showing high flux rates at relatively low SBA of 20/25 m^2 ha^{-1} (Fig. 5.10B). E_{cmax} related to LAI (E_{cmax}/LAI) remained stable over the whole range of LAI in beech stands, but decreased with increasing LAI in spruce stands (Fig. 5.10C). As demonstrated for six spruce stands in the Lehstenbach catchment (Alsheimer et al. 1998), the decrease in E_{cmax}/LAI is also associated with an increase in stand age, decrease in stand density, higher aggregation of needle mass and less light interception per needle area (Falge et al. 2000; Köstner et al. 2001). Maintenance of high E_{cmax}/LAI at high LAI in beech supports the hypothesis that beech is characterized by very effective space capture and adaptation of leaves to light environment (Niinemets 1995; Leuschner 2000; Fleck 2002). This contradicts with general BIGLEAF model assumptions that maximum transpiration rates stabilize at LAI greater than 4 (McNaughton and Jarvis 1983; Kelliher et al. 1995). Because E_{cmax} of different sites refers to comparable D, the results are also valid for g_{cmax}.

Maximum flux rates related to stand basal area (E_{cmax}/SBA) are obviously limited in spruce stands of high SBA (old stands) but not in beech (Fig. 5.10D). This means that consequences of management activities (maintaining/removing biomass) on atmospheric exchange depend very much on species and age. At the same SBA, water flux rates in old stands of beech are twice as high as in old spruce stands. Despite lower SBA of younger stands, they usually exhibit higher flux rates than old stands. Up to now, information on fluxes of mixed-species stand is not sufficient to draw general conclusions. In the species-poor deciduous stands of the Steigerwald, beech dominated flux rates at the stand level.

5.4 Analysis of Carbon and Water Relations by Structural Variation in Model Systems

Structural and compositional effects on forest gas-exchange are difficult to analyse in the field. As mentioned earlier, direct measurements are only possible for tree water use, while CO_2-exchange measurements have to be performed at the leaf/twig (cuvettes) or whole ecosystem (eddy-covariance technique) level including soil respiration. Further, it is not possible to find and to study a sufficient number of stands in the field changing gradually in structure but not in environmental variables. Therefore, detailed gas-exchange models depending on 3D stand structure and related microclimate are instruments to study potential effects of structure on stand gas exchange.

5.4.1 Model Description and Parameterization

The STANDFLUX model allows simulation of net photosynthesis and transpiration of twigs, individual trees and stands in relation to their 3D structure and composition (Falge et al. 2000). It integrates 3D information on stand structure and vertical information on stand microclimate to compute spatial light interception of leaves, canopy net photosynthesis (A_c) and stand respiration. The model consists of (1) a twig gas-exchange module (PSN6), (2) a 3D single-tree light interception and gas-exchange module and (3) a 3D forest canopy module. Gas-exchange of twigs is described according to Harley and Tenhunen (1991) combining enzyme kinetics of Rubisco (Farquhar et al. 1980) and the empirical model for stomatal conductance of Ball et al. (1987). Light interception of foliage elements within the canopy is based on the given spatial distribution of foliage, branches and tree stems, modeled according to Ryel et al. (1990). Canopy conductance for water vapor is derived from A_c and D taking into account leaf boundary layer conductance and the different diffusivity of CO_2 and H_2O.

In order to simulate forest gas-exchange depending on structure and species composition, structural gradients with changing abundance of oak and beech were established. To avoid unrealistic biometric compositions (relationships between DBH, height, crown size, stand basal area, etc.) structural information was derived from an existing data set of 30 measurement

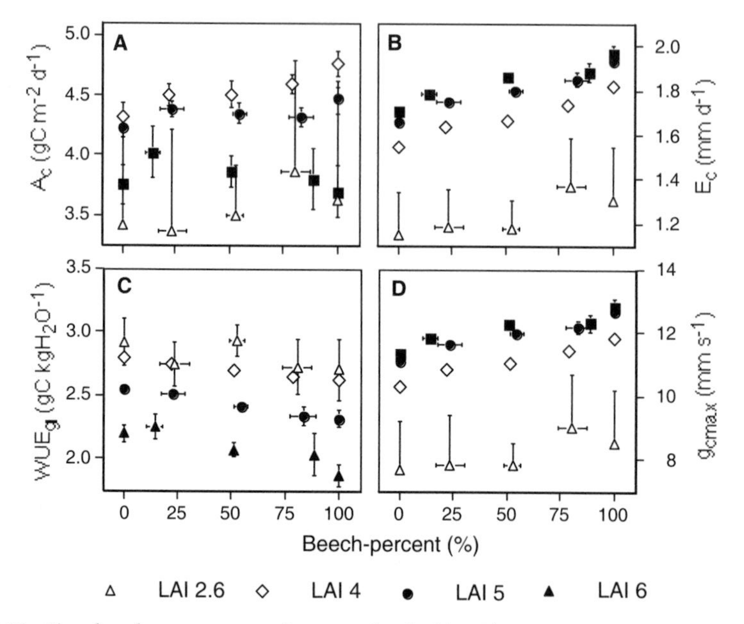

Fig. 5.11. Simulated net canopy photosynthesis (A_c, **A**), canopy transpiration (E_c, **B**), water-use efficiency of gas exchange (WUE_g, **C**) and maximum canopy conductance of water vapor (g_{cmax}, **D**) for structural gradients from pure oak (beech percent=0) to pure beech (beech percent=100) grouped by mean LAI (2.6, 4, 5 and 6 $m^2\,m^{-2}$) of 150 model stands. Data points represent mean values of individual model runs for the same percentage of beech (n=3–14). *Standard error bars* are indicated; non-visible error bars are smaller than symbols

plots (Falge, unpubl.) in a mixed broad-leaved stand (Walker Branch, Tennessee). A set of 150 model stands was created, gradually changing in the percentage (related to SBA) of oak and beech in combination with changing LAI from 2.2–6.5 or SBA from 12–59 $m^2\,ha^{-1}$. Model runs were performed with given data sets of meteorological variables (FLUXNET database, cf. Baldocchi et al. 2001) measured above the forest canopy of the Walker Branch site (Fig. 5.11). The climatic drivers used in the model represented slightly more humid but in general comparable atmospheric conditions to the site Steinkreuz. Model parameters for leaf gas-exchange (PSN6) were based on porometer measurements of beech and oak at the site Großebene (Fleck 2002). At similar nitrogen concentrations, leaf photosynthesis of oak reached only ca. 70 % of photosynthetic rates in beech leaves.

5.4.2 Results from Structural Gradients of Model Systems

Model results of different stand compositions are presented with respect to concurrent variation in LAI (Fig. 5.11) or SBA (Fig. 5.12). Individual model stands were grouped to mean LAI of 2.6 (range 2.2–3.1), 4 (3.5–4.3), 5 (4.5–5.3) and 6 (5.6–6.5). Highest rates of canopy net photosynthesis were found at intermediate LAI (4–5), while A_c was decreased at very low but also at high LAI (Fig. 5.11A). The decrease in A_c at high LAI is mainly caused by increased leaf respiration rates. The model physiology did not consider different respiration response of sun and shade leaves which could have led to overestimation of respiration at the canopy level. High variation at mean LAI of 2.6 was related to great changes in predicted A_c with LAI from 2–3.

Different from A_c, canopy transpiration E_c increased with LAI and beech-percent (Fig. 5.11B). These findings are supported by measurements based on sap flow (Fig. 5.10). Water-use efficiency (i.e., gas-exchange efficiency, $WUE_g=A_c/E_c$) decreased with increasing LAI and increasing percentage of beech (Fig. 5.11C). Lowest WUE_g was therefore found in stands of pure beech at relatively high LAI. Maximum canopy conductance was comparable to E_c with highest values reached in stands of pure beech at high LAI. Decreasing water- and light-use efficiency with increasing LAI is intensified with increasing proportion of beech in the mixed beech–oak stands.

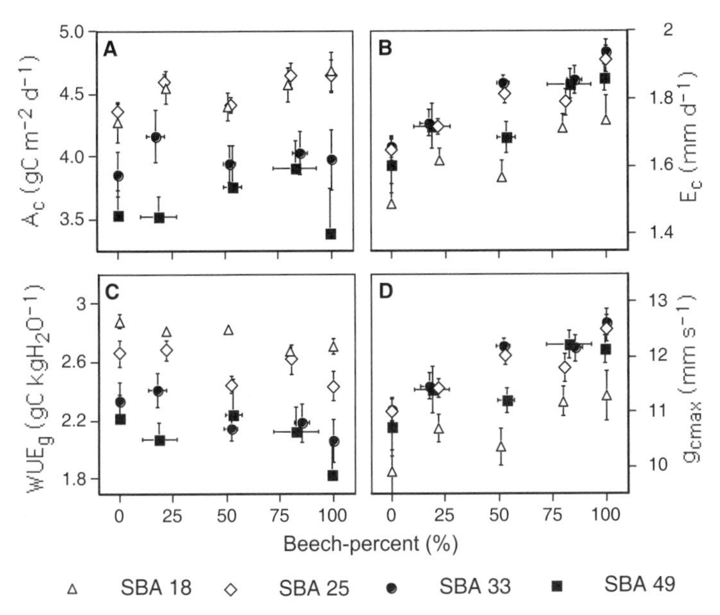

Fig. 5.12. Simulated net canopy photosynthesis (A_c, **A**), canopy transpiration (E_c, **B**), water-use efficiency of gas exchange (WUE_g, **C**) and maximum canopy conductance of water vapor (g_{cmax}, **D**) for structural gradients from pure oak (beech percent=0) to pure beech (beech percent=100) grouped by mean stand basal area (SBA) (18, 25, 33 and 49 m^2 ha^{-1}) of 150 model stands. Data points represent mean values of individual model runs for the same percentage of beech (n=4–15). *Standard error bars* are indicated; non-visible error bars are smaller than symbols

In general, changes in simulated fluxes with LAI were comparable to changes with SBA using groups of 18 (range 12–20), 25 (22–28), 33 (31–37) and 49 (44–59) m^2 ha^{-1} (Fig. 5.12). Highest A_c was related to low SBA (18–25 m^2 ha^{-1}) and increasing beech percent (Fig. 5.12A). It should be noticed that reduction in A_c was found within a range of SBA (33–49 m^2 ha^{-1}) typical for managed sites. This could mean that some stand structures within this range are operating with reduced efficiency of carbon gain. Again comparable with results based on sap flow (Fig. 5.10B), a pronounced increase of E_c and g_{cmax} with beech percent was observed with highest values at intermediate and high SBA (Fig. 5.12B, D). Results in Fig. 5.12C indicate that stands with lower SBA and a high proportion of oak should grow most efficiently under restricted water availability. However, for comprehensive understanding of these findings, the importance of further structural characteristics related to SBA needs to be known as well as the relationship between net carbon uptake and forest growth (timber yield). Additional information on stem growth and on the relationship between tree growth and sap flow (WUE_t, transpiration efficiency) could help to value model results with respect to management decisions.

5.5 Synopsis

For many years, vegetation–atmosphere studies have focused on appropriate physically based formulations of atmospheric exchange processes, while differences in absolute numbers of parameters or flux rates often were signified as 'only site-specific'. With increasing importance of regionalization and management of ecosystem functions beyond biomass production, the connection of atmospheric exchange with structural characteristics of land-use types and traits of species was questioned (e.g., Waring and Running 1998). The presented study combined measurement and modeling of canopy water flux and stand gas-exchange in order to describe fluxes between atmosphere and forests in relation to species composition and structural characteristics of individuals and stands.

It could be demonstrated that independent of beech–oak compositions, beech played the dominant role at all sites. Beech occupied 70 % of SBA and 67 % of LAI at site Steinkreuz, but contributed 86 % to seasonal E_c and seasonal average g_c. Even at site Großebene, where beech only kept 39 % of SBA and 41 % of LAI, contribution to seasonal E_c and g_c reached 61 and 63 %, respectively. For E_{cmax} and g_{cmax}, differences to oak were even larger. Seasonal E_c per LAI or per SBA of oak reached only 38 % of beech. Thus, almost threefold more water transpired through the leaf area of beech than of oak.

Within beech in mixed sites, similar E_c/LAI and g_c/LAI was found and somewhat higher values of E_c/SBA at Großebene. Within oak, approx. 30 %

higher E_c/LAI, g_c/LAI and E_c/SBA occurred at Großebene compared to Steinkreuz. Hence, oak at the beech-dominated site Steinkreuz may be outcompeted in the long run. This is supported by the visual impression that oak at Großebene was better developed than oak at Steinkreuz. At the whole-stand level, exactly the same seasonal E_c/SBA was found at Großebene and Steinkreuz while E_c/LAI and g_c/LAI were somewhat higher at the beech-dominated site Steinkreuz.

At the pure beech plot with highest LAI and SBA, not only absolute values of E_c and g_c were increased compared to the mixed stands, but also values related to LAI or SBA were higher compared to beech in mixed stands (provided that allometric relationships between DBH and A_l did not differ between sites). Potentially, dense beech canopies create their own microenvironment, promoting higher flux rates. A significant profile of $\delta^{18}O$ in leaf tissue along the crown of beech at Steinkreuz supports the assumption that crown internal air humidity was not evenly distributed (Dawson et al., unpubl.). It would follow that general assumptions on forest microclimate being well mixed are not always valid (cf. Aussenac and Ducrey 1977). Potential consequences on g_a and g_c require further investigations.

Overall, observations on the mixed-species stands did not contradict earlier qualitative assumptions on the natural abundance and vitality of beech and oak in this region (Hofmann 1968; Ellenberg 1982). However, oak seemed to be even less competitive than previously estimated. Periodic defoliation by caterpillars may increasingly depress growth and recovery of that species. Also, mixing of single oak trees between beech did not shield them from severe insect attack. Leaf N concentrations (2.1–3.5 %) were comparable with elevated values of other sites in Germany (Thomas and Kiehne 1995). However, as observed at Großebene, N use of leaf photosynthesis of oak reached only 70 % of beech. As light response curves at leaf and stand level demonstrated, limitations of photosynthesis and transpiration were reached earlier in oak than in beech. Beech and oak trees of comparable size kept a similar amount of N in their crowns (ca. 500 g N/crown for a tree of 40 cm DBH); however, E_{cmax}/N of oak crowns was reduced to 30 % of beech.

As pointed out in several investigations, sessile oak is the more drought-tolerant species (Bréda et al. 1993; Aranda et al. 2000). Thus, under significant soil-water restrictions connected with reduced g_c and growth rates, higher soil water extraction rates (Leuschner 1993) and higher specific hydraulic conductivity of the ring-porous oak should be advantageous. This is confirmed at the southern distribution limit of mixed oak–beech stands, where leaf gas-exchange rates and conductance as well as recovery from drought were higher in sessile oak than in beech (Aranda et al. 2000). However, at high or moderate soil-water availability, beech is obviously able to sufficiently supply its large amount of foliage with water and nutrients. Based on growth measurements, it is assumed that the ecological amplitude of beech towards dry conditions is underestimated (Felbermeier 1994). As previously explained, the vitality of beech should be decreased at a climate index of 30 or higher (Ellenberg 1982). Even under these site conditions, higher growth rates of beech compared to sessile oak have been observed (Bonn 2000).

Reasons for the successful development of beech at moderate or dry sites may be not only a previous underestimation of the ecological niche, but also changed boundary conditions. At Steinkreuz, N sequestration in biomass reached 14 kg ha^{-1} $year^{-1}$. The soil organic N pool was a similar important sink (Matzner et al., this Vol.). Investments for space capture and maintenance of leaves are lower in beech than in oak (Hagemeier 2002). Thus, under increased atmospheric N input, N is more effectively used for growth in beech than in oak. Another indication of the long-term response of beech to N input may be the generally increasing leaf mass per area (which is connected with increasing N content of leaves) when values from older literature are compared with recent publications (Fleck 2002). Furthermore, beech leaves developed several days earlier in spring. Under changing climate with earlier onset of the growing season this may be another advantage for beech.

Conclusions on the atmospheric and structural controls of mixed beech–oak stands in the Steigerwald have to consider the specific constraints of oak under the given physical and chemical climate and biological interactions at the study sites. Model systems only included N-related differences in leaf photosynthesis of beech and oak at the sites, but excluded a priori differences in structure and LAI. Under these conditions, flux rates and conductance of oak-dominated stands were only slightly reduced compared to beech. At high LAI, light interception per leaf area was reduced and leaf respiration increased in beech, resulting in decreasing A_c and WUE. Thus, advantages or disadvantages of various species compositions depended on absolute LAI. Nevertheless, the strong decrease in WUE at high LAI predicted by the model still needs to be confirmed by measurements.

Generally, the following can be concluded:

- Water use of the mixed stands and simulated carbon gain depended on species-specific capacities at given structure (relative abundance) and environmental conditions (e.g., N input, biotic stress).
- Higher N-use efficiency of beech occurred at the leaf level and especially at the crown level. This seems to be more related to crown morphology and spatial distribution of N within crowns of beech.
- Oak is more susceptible to shading, and repeated thinning around oak trees in mixed stands is required (Sperber, pers. comm.; cf. Spiecker 1983; Pretzsch 2003). The N-use efficiency of this specific oak management should be questioned, especially under high N input. Oak growing in groups may be less out-competed than single-tree mixtures with beech.
- Concerning N and C pools, beech and oak provide similar above-ground storage capacities at the tree level, but beech may have more advantage at high SBA.

Future investigations of atmospheric exchange should compare extended monospecific beech canopies with stands consisting of species patches. Extended beech canopies could be limited by available energy, reduced horizontal exchange and less diverse use of resources.

Acknowledgements. This work was supported by the German Ministry for Education and Research and the European Union through the CarboEurope projects CarboEuroflux and CarboData.

References

Alsheimer M, Köstner B, Falge E, Tenhunen JD (1998) Temporal and spatial variation in transpiration of Norway spruce stands within a forested catchment of the Fichtelgebirge, Germany. Ann Sci For 55:103–124

Aranda I, Gil L, Pardos JA (2000) Water relations and gas exchange in *Fagus sylvatica* L. and *Quercus petraea* (Mattuschka) Liebl. in a mixed stand at their southern limit of distribution in Europe. Trees 14:344–352

Aussenac G, Ducrey M (1977) Etude bioclimatique d'une futaie feuillue (*Fagus sylvatica* L. et *Quercus sessiliflora* Salisb.) de l'Est de la France. I. Analyse des profils microclimatiques et des caractéristiques anatomiques et morphologiques de l'appareil foliaire. Ann Sci For 34(4):265–284

Backes K (1996) Der Wasserhaushalt vier verschiedener Baumarten der Heide-Wald-Sukzession. Dissertation, Göttingen, Germany

Baldocchi DD, Amthor JS (2001) Canopy photosynthesis: history, measurements, and models. In: Roy J, Saugier B, Mooney HA (eds) Terrestrial global productivity. Academic Press, San Diego, pp 9–31

Baldocchi DD, Hicks BB, Meyers TP (1988) Measuring biosphere–atmosphere exchanges of biologically related gases with micrometeorological methods. Ecology 69(5):1331–1340

Baldocchi DD, Falge E, Gu L, Olson R, Hollinger D, Running S, Anthoni P, Bernhofer C, Davies K, Fuentes J, Goldstein A, Katul G, Law B, Lee X, Mahli Y, Meyers T, Munger W, Oechel W, Paw U K, Pilegaard K, Schmid H, Valentini R, Verma S, Vesala T, Wilson K, Wofsy S (2001) FLUXNET: a new tool to study the temporal and spatial variability of ecosystem-scale carbon dioxide, water vapor and energy flux densities. Bull Am Meteor Soc 82:2415–2435

Ball JT, Woodrow IE, Berry JA (1987) A model predicting stomatal conductance and its contribution to the control of photosynthesis under different environmental conditions. In: Biggins J (ed) Proc 7th Int Congr on Photosynthesis, Providence, Rhode Island, 10–15 Aug 1986. Nijhoff, Dordrecht, pp 221–224

Beyschlag W, Ryel RJ, Caldwell MM (1995) Photosynthesis of vascular plants: assessing canopy photosynthesis by means of simulation models. In: Schulze E-D, Caldwell MM (eds) Ecophysiology of photosynthesis. Springer study edition. Springer, Berlin Heidelberg New York, pp 409–430

Binkley D, Stape JL, Ryan MR, Barnard HR, Fownes J (2002) Age-related decline in forest ecosystem growth: an individual-tree, stand-structure hypothesis. Ecosystems 5:58–67

Bonn S (2000) Konkurrenzdynamik in Buchen/Eichen-Mischbeständen und zu erwartende Modifikationen durch Klimaänderungen. Allg Forst J Ztg 171(5–6):81–88

Bréda N, Cochard H, Dreyer E, Granier A (1993) Water transfer in a mature oak stand (*Quercus petraea*): seasonal evolution and effects of a severe drought. Can J For Res 23:1136–1142

Burrows LE (1980) Differentiating sapwood, heartwood, and pathological wood in live mountain beech. Protection forestry report 172. New Zealand Forest Service, Forest Research Institute, Christchurch, New Zealand

Businger JA (1956) Some remarks on Penman's equation for the evaporation. Neth J Agric Sci 4:77–80

Cannell NGR, Malcolm DC, Robertson PA (1992) The ecology of mixed species stands of trees. Blackwell, Oxford

Clearwater MJ, Meinzer FC, Andrade JL, Goldstein G, Holbrook NM (1999) Potential errors in measurement of nonuniform sap flow using heat dissipation probes. Tree Physiol 19:681–687

Ehman JL, Schmid HP, Grimmond CSB, Randolph JC, Hanson PJ, Wayson CA, Cropley FD (2002) An initial intercomparison of micrometeorological and ecological inventory estimates of carbon exchange in a mid-latitude deciduous forest. Global Change Biol 8:575–589

Ellenberg H (1982) Vegetation Mitteleuropas mit den Alpen. Ulmer, Stuttgart

Ewers BE, Oren R (2000) Analyses of assumptions and errors in the calculation of stomatal conductance from sap flux measurements. Tree Physiol 20(9):579–589

Falge E, Tenhunen JD, Ryel R, Alsheimer, Köstner B (2000) Modelling age- and density-related gas exchange of *Picea abies* canopies in the Fichtelgebirge, Germany. Ann Sci For 57:229–243

Farquhar GD, von Caemmerer S, Berry JA (1980) A biochemical model of photosynthetic CO_2 assimilation in leaves of C_3 species. Planta 149:78–90

Felbermeier B (1994) Arealveränderungen der Buche infolge von Klimaänderungen. Allg Forstz 49(5):222–224

Fleck S (2002) Integrated analysis of relationships between 3D-structure, leaf photosynthesis, and branch transpiration of mature *Fagus sylvatica* and *Quercus petraea* trees in a mixed forest stand. Bayreuther Forum Ökol 97:183

Fleck S, Schmidt M (2001) Biometrie, Kronenarchitektur, Blattphotosynthese und Kronendachtranspiraiton von Buche und Eiche im Einzugsgebiet 'Steinkreuz'. In: Gerstberger P (ed) Waldökosystemforschung in Nordbayern. Die BITÖK-Untersuchungsflächen im Fichtelgebirge und Steigerwald. Bayreuther Forum Ökol 90:137–146

Franz F, Röhle H, Meyer F (1993) Wachstumsgang und Ertragsleistung der Buche. Allg Forstz 6:262–267

Geiger R (1961) Das Klima der bodennahen Luftschicht. Vieweg, Braunschweig

Goldberg V, Baums A, Häntzschel J (2003) Klima, Boden und Landnutzung. In: Bernhofer Ch (ed) Exkursions- und Praktikumsführer Tharandter Wald. Tharandter Klimaprotokolle 6:15–26

Granier A (1985) Une nouvelle méthode pour la mesure du flux de sève brute dans le tronc des arbres. Ann Sci For 42:81–88

Granier A, Biron P, Bréda N, Pontailler JY, Saugier B (1996a) Transpiration of trees and forest stands: short and long-term monitoring using sapflow methods. Global Change Biol 2:265–274

Granier A, Biron P, Köstner B, Gay LW, Najjar G (1996b) Comparisons of xylem sap flow and water vapour flux at the stand level and derivation of canopy conductance for Scots pine. Theor Appl Clim 53:115–122

Granier A, Biron P, Lemoine D (2000) Water balance, transpiration and canopy conductance in two beech stands. Agric For Meteorol 100:291–308

Granier A, Aubinet M, Epron D, Falge E, Gudmundsson J, Jensen NO, Köstner B, Matteucci G, Pilegaard K, Schmidt M, Tenhunen J (2003) Deciduous forests: carbon and water fluxes, balances and ecophysiological determinants. In: Valentini R (ed) Fluxes of carbon, water and energy of European forests. Ecological studies, vol 163. Springer, Berlin Heidelberg New York, pp 55–70

Habermehl A, Ridder H-W (1993) Anwendungen der mobilen Computer-Tomographie zur zerstörungsfreien Untersuchung des Holzkörpers von stehenden Bäumen. Holz Roh Werkstoff 51:1–6

Hagemeier M (2002) Funktionale Kronenarchitektur mitteleuropäischer Baumarten am Beispiel von Hängebirke, Waldkiefer, Traubeneiche, Hainbuche, Winterlinde und Rotbuche. Diss Bot 361:154

Harley PC, Tenhunen JD (1991) Modeling the photosynthetic response of C_3 leaves to environmental factors. In: Boote KJ, Loomis RS (eds) Modeling crop photosynthesis – from biochemistry to canopy. CSSA Spec Publ 19, Sect 2. American Society of Agronomy and Crop Science Society of America, Madison, pp 17–39

Hillis WE (1987) Heartwood and tree exudates. Springer, Berlin Heidelberg New York

Hofmann W (1968) Vitalität der Rotbuche und Klima in Mainfranken. Feddes Repertorium 78(1–3):135–137

Horstmann K (1984) Untersuchungen zum Massenwechsel des Eichenwicklers, *Tortrix viridana* L. (Lepidoptera, Tortricidae), in Unterfranken. Z Angew Entomol 98:73–95

Jarvis PG, Stewart J (1979) Evaporation of water from plantation forests. In: Ford ED, Malcolm DC, Atterson J (eds) The ecology of even-aged forest plantations. In: Proc Meeting Division I. IUFRO, Edinburgh, pp 327–349

Kelliher FM, Leuning R, Raupach MR, Schulze E-D (1995) Maximum conductances for evaporation from global vegetation types. Agric For Meteorol 73:1–16

Kelty MJ (1992) Comparative productivity of monocultures and mixed-species stands. In: Kelty MJ, Larson BC, Oliver CD (eds) The ecology and silviculture of mixed-species forests. Kluwer, Dordrecht, pp 125–141

Kelty MJ, Larson BC, Oliver CD (1992) The ecology and silviculture of mixed-species forests. Kluwer, Dordrecht

Klöck W (1980) 30 Jahre Standortserkundung in Unterfranken. Allg Forstz 16:431

Körner C (1996) The response of complex multispecies systems to elevated CO_2. In: Walker B, Steffen W (eds) Global change and terrestrial ecosystems. IGBP Book Ser 2. Cambridge University Press, Cambridge, pp 20–42

Köstner B (2001) Evaporation and transpiration from coniferous and broad-leaved forests in central Europe – relevance of patch-level studies for spatial scaling. Meteorol Atmos Phys 76:69–82

Köstner BMM, Schulze E-D, Kelliher, FM, Hollinger DY, Byers JN, Hunt JE, McSeveny TM, Meserth R, Weir PL (1992) Transpiration and canopy conductance in a pristine broad-leaved forest of *Nothofagus*: an analysis of xylem sap flow and eddy correlation measurements. Oecologia 91:350–359

Köstner B, Falge EM, Alsheimer M, Geyer R, Tenhunen JD (1998a) Estimating tree canopy water use via xylem sapflow in an old Norway spruce forest and a comparison with simulation-based canopy transpiration estimates. Ann Sci For 55:125–139

Köstner B, Granier A, Cermák J (1998b) Sap flow measurements in forest stands: methods and uncertainties. Ann Sci For 55:13–27

Köstner B, Tenhunen JD, Alsheimer M, Wedler M, Scharfenberg HJ, Zimmermann R, Falge E, Joss U (2001) Controls on evapotranspiration in a spruce forest catchment of the Fichtelgebirge. In: Tenhunen JD, Lenz R, Hantschel R (eds) Ecosystem approaches to landscape management in central Europe. Ecological studies147. Springer, Berlin Heidelberg New York, pp 379–415

Köstner B, Falge E, Tenhunen JD (2002) Age-related effects on leaf area/sapwood area relationships, canopy transpiration, and carbon gain of *Picea abies* stands in the Fichtelgebirge/Germany. Tree Physiol 22:567–574

Krüger S, Mößmer R, Bäumler A (1994) Der Wald in Bayern. Ergebnisse der Bundeswaldinventur 1986–1990. Landesanstalt für Wald und Forstwirtschaft, Freising

Larson BC (1992) Pathways of development in mixed-species stands. In: Kelty MJ (ed) The ecology and silviculture of mixed-species forests. Kluwer, Dordrecht, pp 3–10

Leuschner C (1993) Patterns of soil water depletion under coexisting oak and beech trees in a mixed stand. Phytocoenologia 23:19–33

Leuschner C (2000) Changes in forest ecosystem function with succession in the Lüneburger Heide. In: Tenhunen JD, Lenz R, Hantschel R (eds) Ecosystem approaches to landscape management in central Europe. Ecological studies 147. Springer, Berlin Heidelberg New York, pp 517–568

Magnani F, Mencuccini M, Grace J (2000) Age-related decline in stand productivity: the role of structural acclimation under hydraulic constraints. Plant Cell Environ 23:251–263

Mayer (1984) Europäische Wälder. Uni-Taschenbücher, Stuttgart

McDowell N, Barnard H, Bond BJ, Hinckley T, Hubbard R, Ishii H, Köstner B, Meinzer, FC, Marshall JD, Magnani F, Phillips N, Ryan MG, Whitehead D (2002) The relationship between tree height and leaf area:sapwood area ratio. Oecologia 132:12–20

McNaughton KG, Jarvis PG (1983) Predicting effects of vegetation changes on transpiration and evaporation. In: Kozlowski TT (ed) Water deficits and plant growth, vol 7. Academic Press, New York, pp 1–47

Monteith JL (1965) Evaporation and environment. In: Fogg GE (ed) The state and movement of water in living organisms. Symp Soc Exp Biol 19. Academic Press, New York, pp 205–234

Monteith JL (1975) Vegetation and the atmosphere, vols 1, 2. Academic Press, London

Monteith JL, Szeicz G, Waggoner PE (1965) The measurement and control of stomatal resistance in the field. J Appl Ecol 2:345–355

Monteith JL, Unsworth MH (1990) Principles of environmental physics, 2nd edn. Arnold, London

Myneni RB, Dong J, Tucker CJ, Kaufmann RK, Kauppi PE, Liski J, Zhou L, Alexeyev V, Hughes MK (2001) A large carbon sink in the woody biomass of northern forests. Proc Natl Acad Sci USA 98(26):14784–14789

Niinemets Ü (1995) Distribution of foliar carbon and nitrogen across the canopy of *Fagus sylvatica*: adaptation to a vertical light gradient. Acta Oecol 16:525–541

Oke TR (1987) Boundary layer climates, 2nd edn. Routledge, London

Oliver CD (1992) Similarities of stand structures and stand development processes throughout the world – some evidence and applications to silviculture through adaptive management. In: Kelty MJ (ed) The ecology and silviculture of mixed-species forests. Kluwer, Dordrecht, pp 11–26

Peck A, Mayer H (1996) Einfluß von Bestandesparametern auf die Verdunstung von Wäldern. Forstw Centralbl 115:1–9

Phillips NG, Ryan MG, Bond BJ, McDowell NG, Hinckley TM, Cermák J (2003) Reliance on stored water increases with tree size in three species in the Pacific Northwest. Tree Physiol 23:237–245

Pretzsch H (1993) Struktur und Leistung naturgemäß bewirtschafteter Eichen-Buchen-Mischbestände in Unterfranken. Allg Forstz 6:281–284

Pretzsch H (2003) Diversität und Produktivität von Wäldern. Allg Forst J Ztg 174(5–6): 88–98

Raschi A, Tognetti R, Ridder HW, Béres C, Fenyvesi A (1995) The use of computer tomography in the study of pollution effects on oak trees. Agric Med Spec Vol 298–306

Roy J (2001) How does biodiversity control primary productivity? In: Roy J, Saugier B, Mooney HA (eds) Terrestrial global productivity. Academic Press, San Diego, pp 169–186

Ryan MG, Yoder BJ (1997) Hydraulic limits to tree height and tree growth. BioScience 47:235–242

Ryan MG, Binkley D, Fowness JH (1997) Age-related decline in forest productivity: pattern and process. Adv Ecol Res 27:213–262

Ryel RJ, Barnes PW, Beyschlag W, Caldwell MM, Flint SD (1990) Plant competition for light analyzed with a multispecies canopy model. I. Model development and influence of enhanced UV-B conditions on photosynthesis in mixed wheat and wild oat canopies. Oecologia 82:304–310

Schäfer K (1997) Wassernutzung von *Fagus syvatica* und *Quercus petraea* im Wassereinzugsgebiet Steinkreuz im Steigerwald, Bayern. Thesis, University of Bayreuth, Germany

Schmidt M, Köstner B, Tenhunen JD (2000) Bedeutung von Lichtklima und Blattflächenentwicklung für die Wasser- und CO_2-Flüsse des Kronendaches entlang eines Baumarten-Struktur-Gradienten im Steigerwald. BITÖK Forschungsbericht 1998–2000. Bayreuther Forum Ökol 78:35–46

Schulte A, Böswald K, Joosten R (2001) Weltforstwirtschaft nach Kyoto. Wald und Holz als Kohlenstoffspeicher und regenerativer Energieträger. Shaker, Aachen

Schulze E-D (2000) Carbon and nitrogen cycling in European forest ecosystems. Ecological studies 142. Springer, Berlin Heidelberg New York

Schulze ED, Kelliher FM, Körner C, Lloyd J, Leuning R (1994) Relationships between plant nitrogen nutrition, carbon assimilation rate, and maximum stomatal and ecosystem surface conductances for evaporation: a global ecology scaling exercise. Annu Rev Ecol Sys 25:629–660

Shuttleworth WJ (1989) Micrometeorology of temperate and tropical forests. Philos Trans R Soc Lond Ser B 324:299–334

Smaltschinski T (1990) Mischbestände in der Bundesrepublik Deutschland. Forstarchiv 61(4):137–140

Sperber G, Regher A (1983) Vorratspflege in Unterfranken am Beispiel des Steigerwaldes. Allg Forstz 39:1020–1024

Spiecker H (1983) Durchforstungsansätze bei Eiche unter besonderer Berücksichtigung des Dickenwachstums. Allg Forst J Ztg 154(2):21–36

Tenhunen JD, Falge E, Ryel R, Manderscheid B, Peters K, Ostendorf B, Joss U, Lischeid G (2001) A flux model hierarchy for spruce forest ecosystems. In: Tenhunen JD, Lenz R, Hantschel R (eds) Ecosystem approaches to landscape management in central Europe. Ecological studies 147. Springer, Berlin Heidelberg New York, pp 417–462

Thomas FM, Blank R (1996) The effect of excess nitrogen and of insect defoliation on the frost hardiness of bark tissue of adult oaks. Ann Sci For 53:395–406

Thomas FM, Kiehne U (1995) The nitrogen status of oak stands in northern Germany and its role in oak decline. In: Nilsson LO, Hüttl RF, Johansson UT (eds) Nutrient uptake and cycling in forest ecosystems. Kluwer, Dordrecht, pp 671–676

Thomasius H (1992) Prinzipien eines ökologisch orientierten Waldbaus. Forstw Centalbl 111:141–155

Valentini R (2003) Fluxes of carbon, energy and water of European forests. Ecological studies 163. Springer, Berlin Heidelberg New York

Walter H, Breckle S-W (1994) Ökologie der Erde, vol 3. Fischer, Stuttgart

Waring RH, Running SW (1998) Forest ecosystems. Analysis at multiple scales. Academic Press, San Diego

Welß W (1985) Waldgesellschaften im nördlichen Steigerwald. Diss Bot 83:174

Wilson KB, Baldocchi DD, Hanson PJ (2001) Leaf age affects the seasonal pattern of photosynthetic capacity and net ecosystem exchange of carbon in a deciduous forest. Plant Cell Environ 24:571–583

6 Impacts of Canopy Internal Gradients on Carbon and Water Exchange of Beech and Oak Trees

S. FLECK, M. SCHMIDT, B. KÖSTNER, W. FALTIN, and J.D. TENHUNEN

6.1 Introduction

6.1.1 The Scaling Problem in Gas-Exchange Models

Identifying the relevance of forest structure for stand photosynthesis and transpiration is one of the remaining challenges in plant physiological ecology. While leaves and their stomata are the causal agents of stand transpiration and canopy conductance, and their position and orientation are known to be decisive for their gas-exchange contribution, the spatial distribution pattern of leaves inside a forest cannot yet be considered in stand gas-exchange models. Canopy conductance of a mature tree is created by the conductances of about 10^{11} stomata (Larcher 2001; Fleck 2002, p. 49) which act depending on their local micrometeorological conditions (light, humidity). A process-oriented representation of forest canopy structures in stand gas-exchange models is, thus, still hindered by the enormous complexity of canopies, which makes it nearly impossible to assess and model them in detail.

Current stand gas-exchange models therefore usually incorporate rougher descriptions of the forest canopy, mostly based on regular, predefined grids like layers (in 2D models) or (in 3D models) based on adaptable geometric forms like ellipsoids and regularly segmented cylinders which describe the positions and dimensions of single tree crowns (Cescatti 1997; Falge 1997). A main assumption of such models is spatial homogeneity throughout these volumes, which allows us to represent each physiological property of the leaf area in the segment by one average value per segment and to build flux and conductance calculations on that. This assumption is an approximation to reality which includes some unknown degree of spatial variability. Very high spatial variability is at least reported for the structure-dependent light climate in forest canopies (Baldocchi and Collineau 1994) and leaf properties may be assumed to be adapted to this high variability. However, the light climate variability cannot be reproduced with the employed light models unless they are based on a fine-scaled structure representation.

Ecological Studies, Vol. 172
E. Matzner (Ed.), Biogeochemistry of Forested Catchments in a Changing Environment

The effect of structural complexity on the accuracy of stand gas-exchange models may become large and difficult to handle wherever non-linear relationships are involved. The non-linear dependence of photosynthesis and transpiration on climate variables does not allow for averaging of the driving climate variable over space and time before calculating the flux, since this may result in a deviation of 40 % in the flux calculation (Jarvis 1995). This is due to the varying effect that a constant difference in the driving variable has on the resulting flux. While a 100-µmol/(m^2*s) increase in photosynthetic photon flux density (PPFD) at lower light levels may double the photosynthesis rate, it does not have any effect under high light conditions. Averaging, thus, is only allowed after the separate flux calculation for each individual value of the driving variable(s), i.e., for each individual leaf.

The necessity to keep models simple has led to two solutions for this upscaling dilemma. The more conservative approach simply employs effective parameters instead of averages for the representation of spatial compartments which are selected to produce the correct average flux when given to the non-linear equation. More sophisticated models perform separate flux calculations for classes of leaves in the compartments with the same physiological properties and the same light situations and average afterwards the flux for the compartment (de Pury and Farquhar 1997; Falge 1997). Both approaches have specific parameterization problems that could be overcome with a higher resolution in the structure description. While effective parameters may be supposed to lie close to the average parameter value, when the compartments become smaller and therefore more homogenous in physiological and microclimatological properties, the relative amount of leaves in the different physiological and microclimatological classes cannot accurately be estimated without a structure description that allows us to derive the spatial variability in light climate.

Actually the rapid developments in information technology lead to a new situation, where the necessity to keep models simple becomes less and less important. Modern computers provide the capacity to store and evaluate larger and larger amounts of data, and it may be assumed that this development will soon principally allow us to model a forest with all single leaves represented. Once the possibility exists, the realization of such a leaf-true stand model seems to be just a question of time.

The bottleneck for the development of a leaf-true stand model will probably be the necessity to perform the adequate structure-related physiological measurements. While the structure representation can be built on emerging tools for automatic structure assessment (Tanaka et al. 1998) and architecture modeling (Kurth and Sloboda 1999; Godin 2000), the establishment of structure–function relationships cannot be automated to that extent.

The scope of this study is, therefore, to investigate an intermediate level of spatial integration that may be realized with the existing computer capacities, in order to pre-investigate the relevant measurement and modeling strategies for a leaf-true stand model. Instead of representing every leaf, the study approaches structural complexity on the level of branches and their leaf clouds, assuming that their distribution in the crown is – in a first approximation – decisive for the spatial distribution of light interception. While a mature beech crown may have around 200,000 leaves, the investigated branches count between 500 and 4,000 leaves in their leaf clouds. Unfortunately, the necessary high resolution of these measurements limited the number of trees investigated to three mature beeches and one oak and prevented statistical evaluation among trees, so that the study must be considered as a detailed exemplary analysis of relationships between structure and function that could similarly occur with other oaks and beeches.

The study shall:

- Assess the spatial variability of leaf properties and their link to light climate and canopy structure.
- Investigate whether a branch-based model is sufficiently fine-scaled to allow flux modeling based on averaging after separate flux calculations for branches.
- Investigate whether the links between the fine-scaled structure, meteorology and gas-exchange may principally be reduced to a mechanistic model at the leaf level.
- Provide structure-related photosynthesis and transpiration measurements for analysis with a fine-scaled model.

6.1.2 Representation of the Investigated Trees

Most of the measurements were conducted on one *Fagus sylvatica* tree (beech Gr12) and one *Quercus petraea* tree (oak Gr13) in the mixed stand Großebene in the Steigerwald (49°52′N, 10°28′E) at an elevation of 460 m a.s.l. (Fig. 6.1). Further measurements originated from two beech trees (Bu38 and Bu45) in the pure stand Buchenallee (Fichtelgebirge, 50°03′N, 11°52′E). While the general attributes of the mountain ranges Steigerwald and Fichtelgebirge are to be found in Gerstberger et al. (this Vol.), the mixed stand Großebene is described in Köstner et al. (this Vol.).

The investigation area 'Buchenallee' is located on the south slope of the Schneeberg mountain with an inclination of 13.5°, ranging from 900–915 m a.s.l. The partly podzolized brown earth in this area has $pH_{(H2O)}$ values of around 4.75 in the upper 5 cm of the mineral soil. Low temperatures and high precipitation (950–1,300 mm) are typical for the climate in the upper Fichtelgebirge. About 20 % of precipitation is deposited during the prevalent foggy weather (Wrzesinsky and Klemm 2000). Slow growth and a high occurrence of damage by pathogens suggest that beech is at its altitudinal limit at the Schneeberg. The oldest beeches in the stand are 120 years old, but are not higher than 26 m. However, the optically determined leaf area index (LAI) was

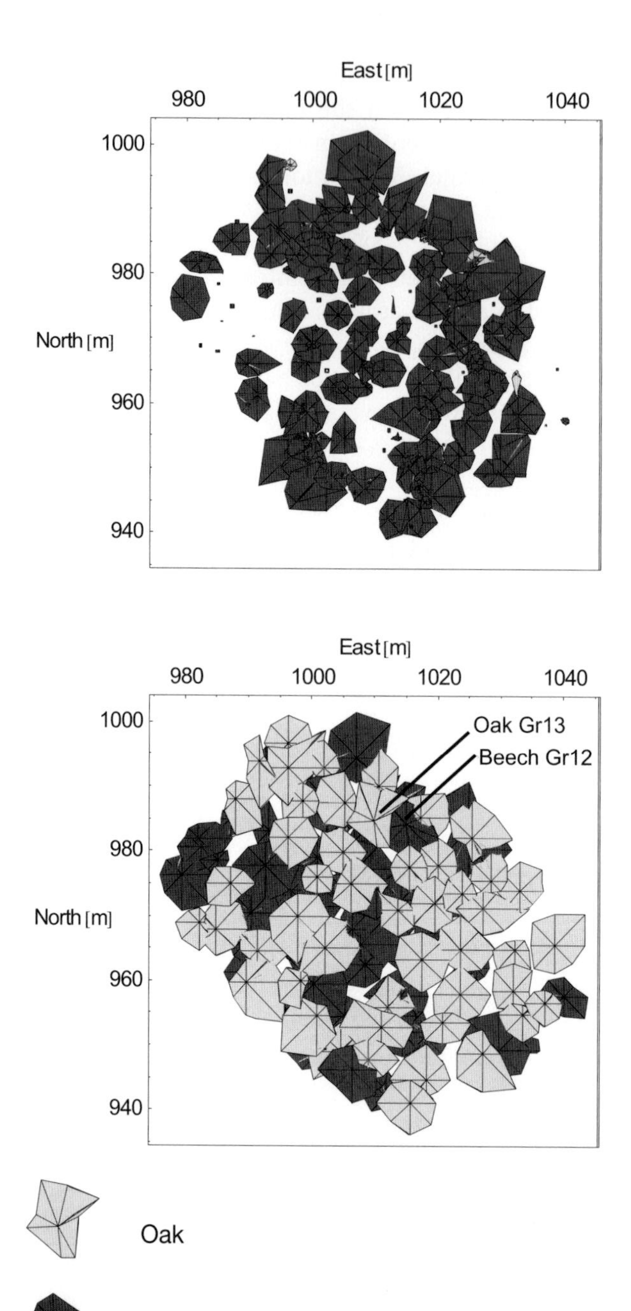

Fig. 6.1. Simulated view vertically to the floor from 10 m height (*above*) and from above the stand (*below*). While oak trees occupy the upper canopy layer, they are nearly not participating at all in the crown layer below 10 m above the floor. Position of the two investigated trees in the Großebene stand is indicated

rather high (8.1), which may be due to lateral irradiation on the south slope. More detailed descriptions are given by Fleck (2002).

While stand density and age of the Großebene stand (526 trees ha^{-1}, 120 years) are similar to the Buchenallee (524 trees ha^{-1}, 120 years), its maximum height is 30 m and the LAI is 6.3. Both stands originate from natural regeneration, but actually natural regeneration is not possible for oaks in the Großebene due to extremely low light levels (compare Leuschner 1998), while it was possible at least 10 years ago for beech trees. The stand has, therefore, a layered structure, with the two species occupying different height ranges (Fig. 6.1).

The representation of the single trees is based on the fact that the distribution of leaves in tree crowns is usually clumped due to their attachment to branches. Since the overwhelming majority of leaves of the investigated tree species are – in a mature developmental stage – located near the distal parts of the branches, the leaves of each branch form a characteristic leaf cloud with high LAI, while the supporting branch and the space around the leaf cloud is basically free of leaves. Nearly all leaves of both species were concentrated in leaf clouds. The distribution of gaps between leaf clouds may be assumed to be decisive for the spatial light distribution in the tree crown, when the influence of the branch system and the much smaller gaps in the leaf clouds are neglected for a first approximation. The branches were mostly hidden by leaves when viewed from above and LAI-2000 measurements from below yielded an area index of 1.2 in the winter but more than 6 in the summer.

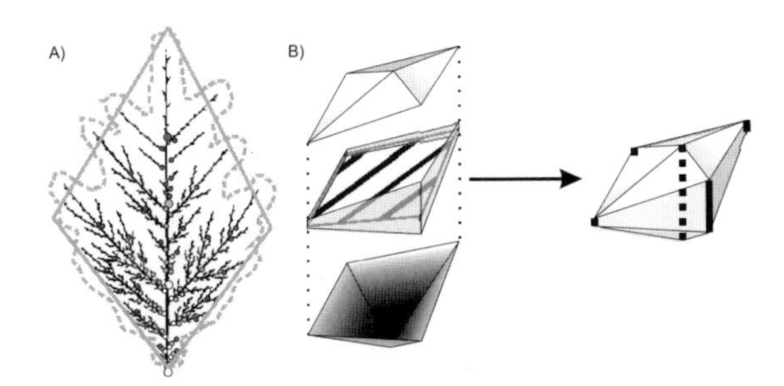

Fig. 6.2. Definition of corner points of leaf cloud enveloping polyhedrons. **A** Determination of maximum plane area in the leaf cloud of a beech branch (*dotted line*) and approximation of this area by a quadrangle. **B** Construction of a leaf cloud enveloping polyhedron on the basis of such a quadrangle (*hatched area*) and vertical distance measurements at its four corner points (*thick lines*) and one central point of the quadrangle (*thick dotted line*). The central point is chosen where leaf cloud extension vertical to the plane area is maximum. While the corner point locations were measured with a theodolite, distance measurements orthogonal to the leaf cloud plane were performed standing on a high-lift with a meter stick (Fleck 2002)

The spatial extension of leaf clouds was described by leaf cloud enveloping polyhedrons (Fig. 6.2). This approximate leaf cloud description is part of the model CRISTO for the representation of tree structures in 3D maps using the software Mathematica (Fleck 2002). The representations are based on theodolite measurements from the ground, with a second person on some device (ladder, high-lift, canopy crane) in the crown, holding a reflector at the positions in question. The problem of tree sway influencing accuracy of the resulting description of leaf clouds was small, since it became apparent that the trees typically experience wind speeds that are too low to move leaf clouds.

6.2 Crown Structures of Mature Beech and Oak Trees

6.2.1 Crown Forms

A full description of the 3D arrangement of leaf clouds in the crown was achieved for beech Bu38 (139 leaf clouds), oak Gr13 (88 leaf clouds) and beech Gr12 (66 leaf clouds). The graphic representation with CRISTO (Fig. 6.3) shows the different heights, crown forms and crown lengths of the trees, as well as the relative orientation of leaf clouds to each other and their leaf area density (leaf area per leaf cloud volume), which is necessary for a branch-based light model. Leaf area density was calculated based on leaf area measurements and allometric relationships (see below). All trees were part of the uppermost canopy layer of their stand.

Though the outside view of these tree representations suggests a more or less even distribution of leaves in the crowns, the tree crowns turned out to be naturally layered and structured around coherent cavities. Typical for all three crowns was a central leafless space below a more or less dense cupola of leaf clouds in the uppermost crown layer which forms a roof above it. Beech leaf clouds were often grouped side by side forming horizontal leaf layers. A vertical cross-section through the digital tree representation in depth of the stem shows that such layers were also in oak Gr13 oriented around coherent, layer-like cavities that were parallel to the ground surface (Fig. 6.4).

6.2.2 Allometric Relationships on Branch and Stem Level

Allometric relationships are suited to provide the basis for leaf area density calculations of leaf clouds, when, as in the case of beech Gr12, only a part of the whole tree shall be harvested and all branch diameters can be measured. They turned out to be relatively constant on the branch scale among three

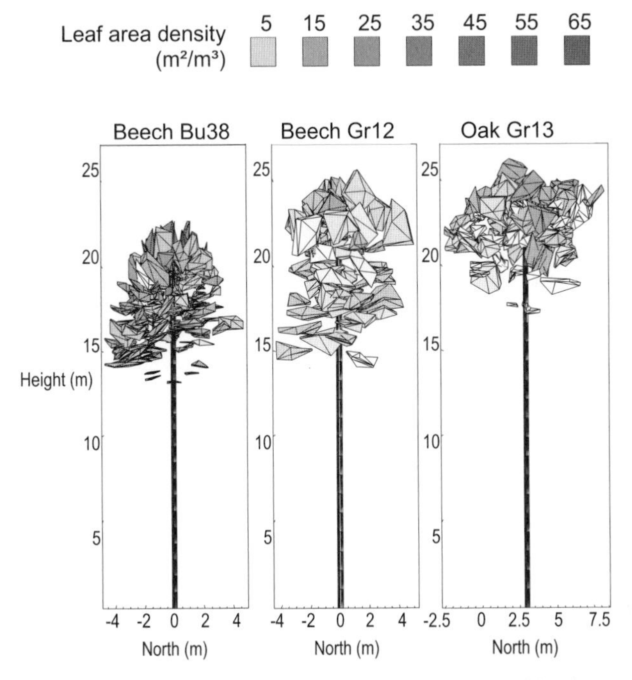

Fig. 6.3. Three-dimensional maps of tree crown structures and leaf area densities of leaf cloud enveloping polyhedrons in the canopies of beech Bu38, beech Gr12 and oak Gr13. Though beech Bu38 is smaller than the "Großebene" trees, it is a dominant tree in its stand "Buchenallee". The *y axis* of the graphs represents height above floor (m). Origin of the coordinate system for oak Gr13 is the same as for beech Gr12, so that the stem base point of oak Gr13 is (–4.79|2.96|0.14) (east/north/height), while other stem base points are situated in the origin of their coordinate system

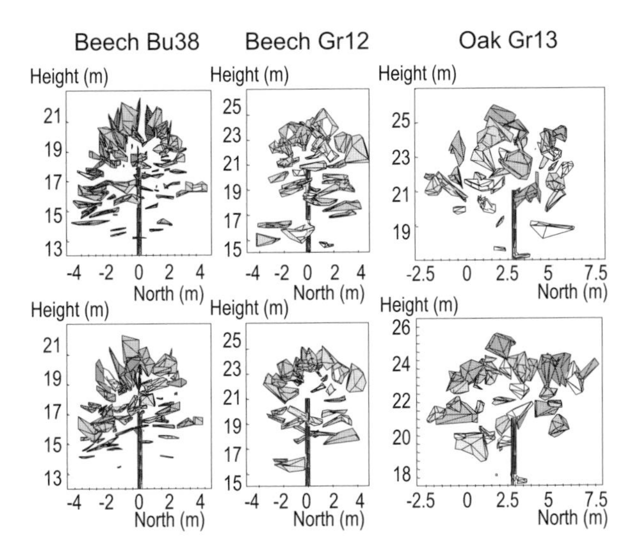

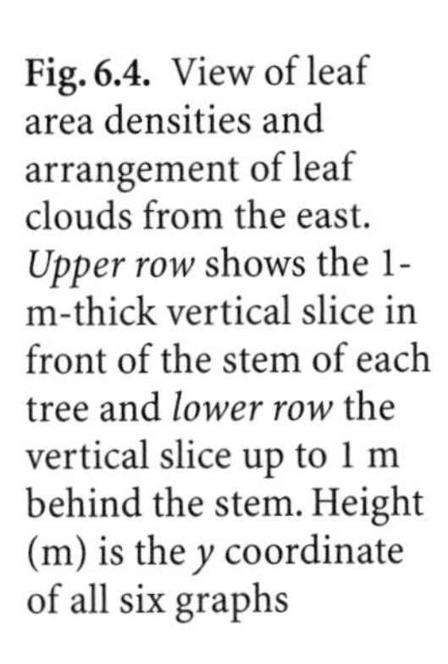

Fig. 6.4. View of leaf area densities and arrangement of leaf clouds from the east. *Upper row* shows the 1-m-thick vertical slice in front of the stem of each tree and *lower row* the vertical slice up to 1 m behind the stem. Height (m) is the *y* coordinate of all six graphs

investigated beeches and consistent (r^2 >0.9 for leaf area and r^2 >0.8 for leaf dry mass) throughout the branches of each beech tree separately. The ratio between leaf area and branch basal area varied between 2900 and 4400, and that between leaf mass (in grams) and branch basal area (in square centimeters) ranged from 19 to 31 for the beech trees. The same ratios were 4100 and 32.7 for the investigated oak, but with the less high r^2 values of 0.74 and 0.81, respectively.

The cross-sectional area of branches was reduced by 19 % (beech, r^2=0.93) and 16 % (oak, r^2=0.94) between the base of a leaf-cloud-supporting branch and the summed cross sections 1 m further. The use of allometric relationships on the branch scale requires due to this reduction a consistent definition for the position to measure branch basal area, which was in this case 5 cm below the first ramification point of a leaf cloud, because the value directly at the ramification point was often influenced by thicker bark or other kinds of irregularity.

The available data on the scale of whole crowns and stems for *Quercus petraea*/*Quercus robur* (Burger 1947, and own measurements) and for *Fagus sylvatica* (Burger 1945; Pellinen 1986; Krauss and Heinsdorf 1996; Bartelink 1997, and own measurements) show some variation among the older trees in the area-based and mass-based relationships, which appeared to be

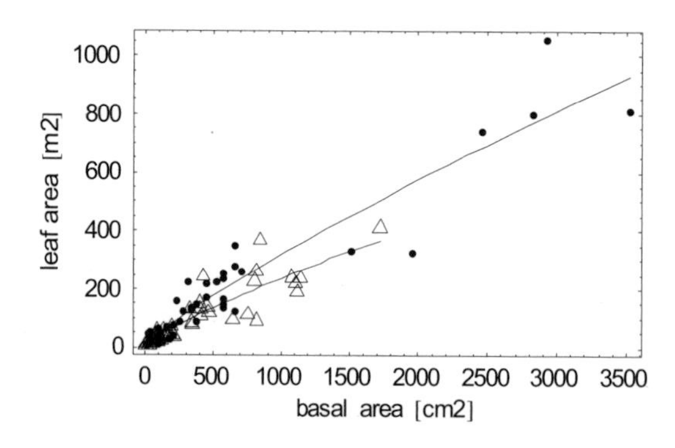

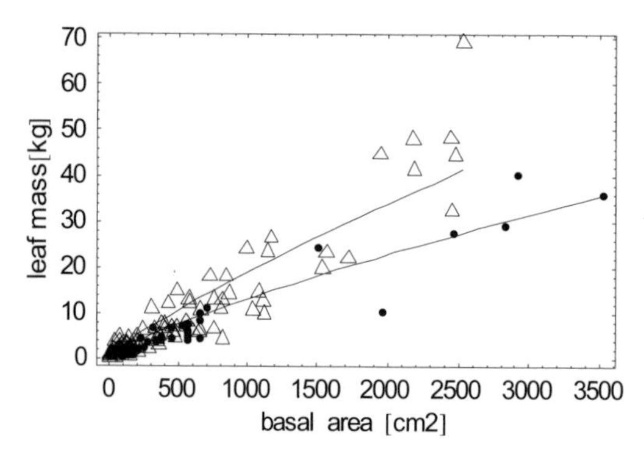

Fig. 6.5. Allometric relationships at the crown level that are based on literature data of 52 oaks and 45 beeches (107 beeches for the mass-based relationship). The equations are leaf area (m^2) = 0.876 × basal area (cm^2) ^ 0.81, leaf mass (kg) = 0.06 × basal area (cm^2) ^ 0.835 for beech trees (*open triangles*) and leaf area (m^2) = 0.871 × basal area (cm^2) ^ 0.854, leaf mass (kg) = 0.052 × basal area (cm^2) ^ 0.8 for oaks (*filled circles*)

stand-specific. Comparison with data from ten other species (Rogers and Hinckley 1979; Martin et al. 1998) shows empirically that the ratio between basal area and appending leaf mass or leaf area decreases generally with increasing stem diameter, so that the allometric relationship slightly deviates from linear (Fleck 2002, p. 37). This decrease reflects the decreasing functionality for water transport of the wood of older stems, which is associated with cavitation of vessels and the formation of heartwood. Since a linear relationship between basal area and appertaining leaf area and mass is not expected under these conditions, the relationships were approximated by power functions with exponent below 1 instead of linear functions (Fig. 6.5).

Species-specific variations in the allometric relationship between leaf mass and leaf cloud volume may be supposed to have an influence on interspecific competition in mixed stands (Schulze et al. 1986). But, the visual impression that oak leaf clouds appear to have less leaf area per volume was not confirmed: the range of leaf area/leaf cloud volume and leaf mass/leaf cloud volume was for the neighboring trees of both species approximately the same (compare Fleck 2002, pp. 39–45).

6.2.3 Vertical and Azimuthal Distribution of Leaf Area in the Crowns

The leaf distribution in the crown is decisive for the degree of clumping and the ability to describe parts of the crown by homogeneous leaf area densities in a light model. The calculated leaf area distribution in horizontal layers of 10 cm height confirms the pattern visible from the cross sections through virtual trees. Assuming that the leaf area of leaf clouds is evenly distributed throughout each leaf cloud volume, the leaf cloud enveloping polyhedrons were in CRISTO sectioned along horizontal planes in 10 cm distance and the leaf area was distributed according to volume proportions between the 10-cm layers. Figure 6.6 shows that the resulting vertical distributions of leaf area had characteristic gaps and peaks for the three investigated crowns. Instead of having a more or less constant amount of leaves in each layer, the trees had a naturally layered structure with two or three leaf-rich layers that were separated by one or two layers with low amount of leaves. Light extinction by the upper leaf clouds could be the reason for these gap layers in the crown, while the next leaf layer below may survive because it can better profit from penetrating light due to the larger distance to shadow-casting leaf clouds from the upper layer. The findings support to some extent the forester's purely structural concept of sun-crown and shade-crown, which is sometimes obvious at other latitudes, but has not yet been proven with 3D measurements for mid-European forest trees.

The amount of leaf area per layer is partly a consequence of the horizontal extension of tree crowns at different heights, so that the effect on volumetric

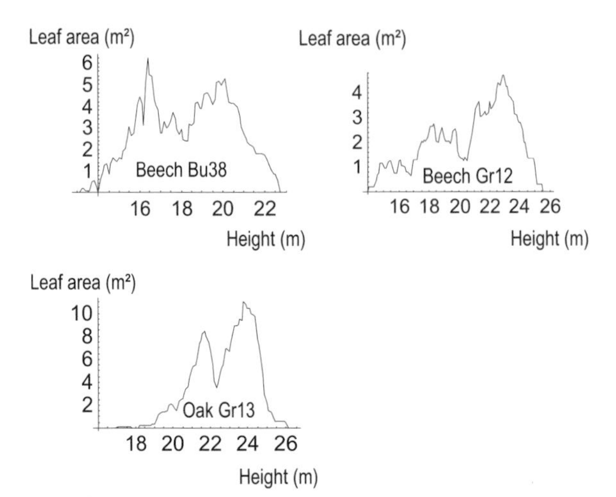

Fig. 6.6. Leaf area in height layers of 10 cm thickness of the investigated canopies. Original bar chart was approximated with a line connecting the midpoints of each 10-cm layer, because 20 bars per height meter are difficult to discern from a graph of this size. The value on the *y*-axis integrates therefore over a height range of 10 cm

leaf area density, which is the quantity with impact on light penetration through that layer, cannot directly be deduced from the vertical distribution of leaf area. A layer-wise investigation of leaf area densities (referring to layer volumes) shows that the beeches achieved their highest leaf area densities in the relatively small layer in the uppermost meter of the crown, while the investigated oak had its highest leaf area density 2–3 m below the crown apex and very high layer volumes in the uppermost meters (Fleck 2002, p. 48).

An azimuthal preference in leaf area distribution was found for the neighbouring oak and beech in the Steigerwald (Großebene). While oak tended to build high leaf area densities in a direction towards competing neighbor trees (probably due to wind-induced twig breaks when crowns touch each other), the beech crown grew into gaps between tree crowns and had the highest leaf area density there (Fleck 2002, p. 70).

6.2.4 Relative Amount of Gap Volumes and Wood

The volume of leaf cloud enveloping polyhedrons was only a relatively small part of the whole volume of canopy layers. Layer volume was defined as the area of the convex hull of all vertically projected polyhedron corner points of a layer times height of that layer (usually 1 m). It emerged that on average between 82 and 90 % of a canopy layer consists of leafless gaps between leaf clouds. More than 90 % were reached in all mature trees in the lower layers of the shade crown, while the lowest gap proportions (58–72 %) were found in the uppermost layers of oak and beech from the Großebene stand. This indicates a highly clumped distribution of leaves in the layers and has probably an influence on the light climate which differs from that of a layer with uniform distribution.

The projected area of woody organs in leaf clouds amounted to about 6% of the leaf area of leaf clouds and has therefore only marginal influence on the light climate in the crown.

6.3 Leaf Properties

6.3.1 Relative Irradiance and Leaf Mass per Area (LMA)

Relative irradiance above a certain leaf in the crown – as measured with hemispherical photography – expresses the amount of light that reaches this leaf due to the interaction of the surrounding canopy structure and the sun. The relevance of light for photosynthesis suggests a physiological adaptation to the light environment of single leaves which may be used for a light-model-driven physiological parameterization of single leaves, given that a highly detailed structure description is provided.

Leaf mass per area (LMA) of two beeches could be shown to be strongly light dependent. Since the highest influence on the light climate of a leaf often comes from its neighboring leaves, hemispherical photography had to consider this by exactly placing the fish-eye lens at the former location of a leaf blade without destruction of the surrounding architecture. Figure 6.7 shows the strong relationship ($r^2 \geq 0.89$) between LMA and relative irradiance for two mature beeches from the stand Buchenallee and the combined relationship based on data from both trees ($r^2 = 0.88$).

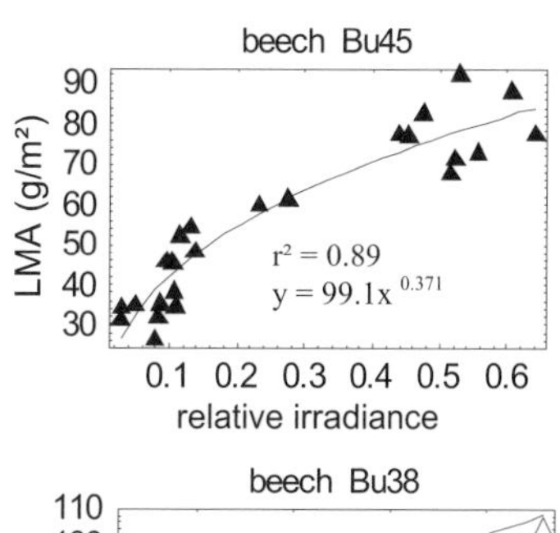

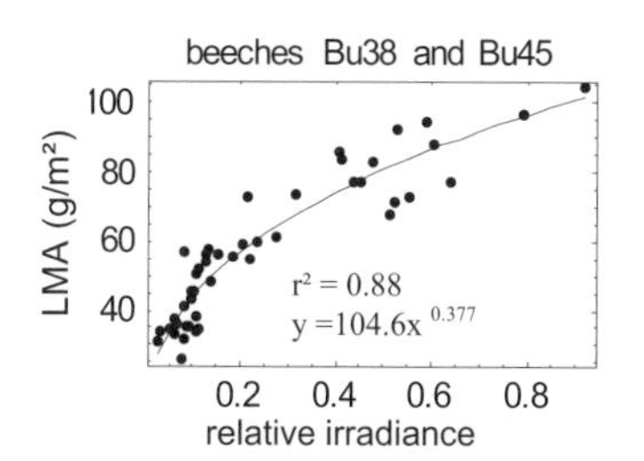

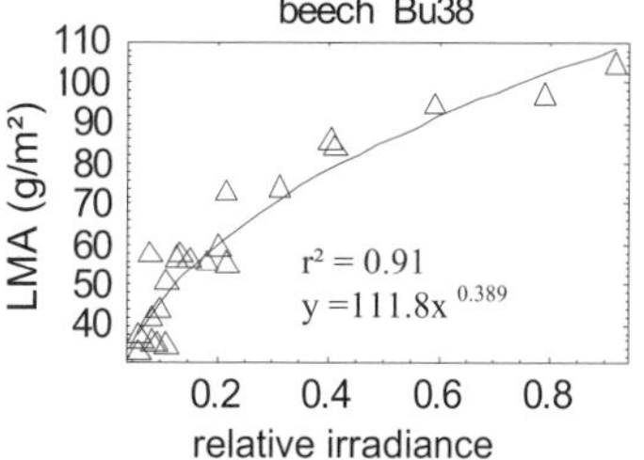

Fig. 6.7. Non-linear regressions for the LMA vs. relative irradiance relationship for two beech trees from the Buchenallee stand (Bu38 and BU45). Spot checks from the Großebene beech trees agreed well with these regressions. (Fleck 2002)

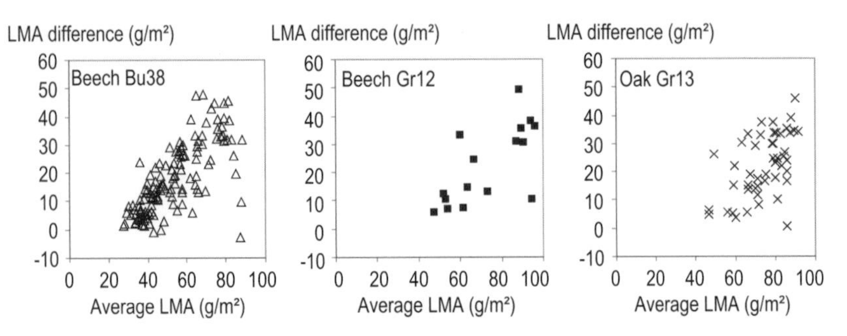

Fig. 6.8. LMA ranges (differences between proximal and distal part of a leaf cloud) of leaf clouds from two beech trees (Bu38, Gr12) and one oak (Gr13) versus average LMA of the leaf clouds

Unfortunately it is not possible to parameterize all leaves in a leaf cloud with the same LMA value. Figure 6.8 shows the increasing difference between leaves of the proximal part and the distal part of leaf clouds with increasing average LMA of the leaf cloud.

6.3.2 Leaf Angles

Leaf angles to the horizon are another decisive parameter for the light interception of leaves. While the single angles of each leaf need to be described by a structural model, leaf angle distributions of height layers have been shown to be ellipsoidal distributions in the case of the investigated beech trees (Fleck 2002, pp. 86–88). This confirms earlier findings by Campbell and Norman (1989), who described leaf angle distributions with one-parametric ellipsoidal distribution functions. The single parameter k of these functions modifies them from rather even distributions with a broad maximum at relatively high inclinations (15–30°; k=2) to very skew distributions with a peak-like maximum at relatively low inclinations (14–16°; k=5). The amount of steep inclinations above 60° was always very small, but a bit more frequent in the uppermost meter, which is supposed to be due to higher light intensities, but influenced the distributions only marginally. Instead, the variation of leaf angle distributions with height in the canopy appeared to be mainly driven by architectural constraints rather than micro-environmental conditions. As visible in Fig. 6.9, the k values seem to have a clearer dependence on vertical distance to the crown apex than on the relative irradiance measured at the location of the leaves.

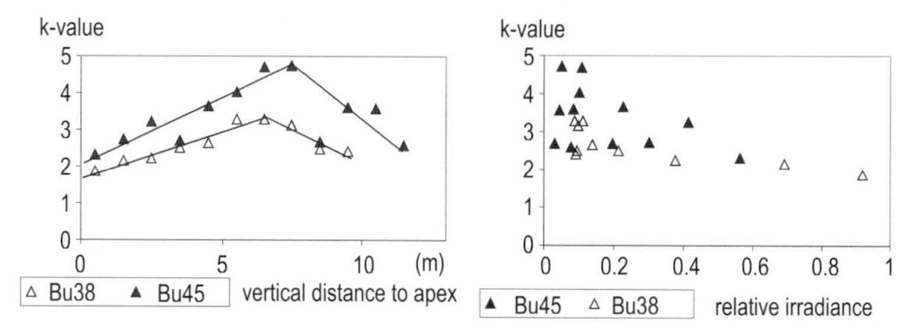

Fig. 6.9. Dependence of k value of the approximated ellipsoidal distribution of vertical distance to the tree's apex (*left*) and irradiance (*right*). *Trend lines* were drawn by hand and equal $y=0.24x+1.67$ and $y=-0.38x+5.85$ in Bu38

6.3.3 Leaf Nitrogen

A higher dry mass per leaf area on its own does not directly imply a physiological difference, since the gas-exchange physiology of leaves depends mainly on concentration and activity of the involved nitrogen-rich metabolites, with Rubisco and chlorophyll as the most relevant and abundant compounds. If LMA increases without additional production of these compounds (nitrogen dilution) leaf physiology is expected to change only marginally. Generally, total plant averages of leaf nitrogen concentrations (per gram of dry weight) have been shown to be largely linearly correlated to averages of LMA^{-1} (Schulze et al. 1994).

A strong linear relationship between LMA and leaf nitrogen per area was found for the leaves from six beeches and the investigated oak (Fig. 6.10). The correlations were nearly identical for both species and indicate that no nitrogen dilution occurred with the investigated trees. The relationship between relative irradiance and LMA thus implies a direct relationship between relative irradiance and nitrogen per area, as indicated in Fig. 6.10. The range of measured values contains among 'normal' values also very high nitrogen contents that are actually the highest published values for leaves of mature trees of the investigated species [van den Burg (1990) summarizes older measurements, while Bauer et al. (1997) published values of mature beech trees along a European transect.] This might partly have to do with the fact that leaves from the true uppermost part of a mature tree canopy are difficult to get and rarely analyzed. It is also a hint to the very good nitrogen supply of the trees due to the higher nitrogen depositions. At Steinkreuz the flux of mineral nitrogen with throughfall was about 15 kg N ha^{-1} $year^{-1}$ (Matzner et al., this Vol.).

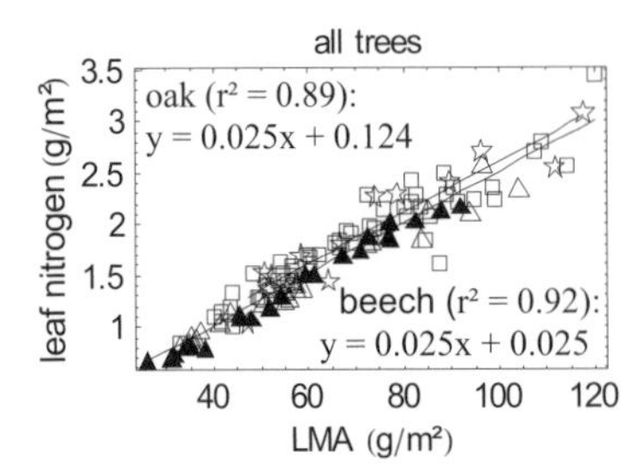

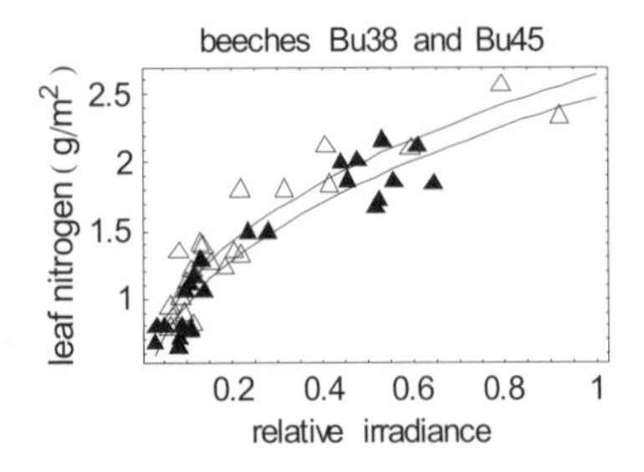

Fig. 6.10. Area-related nitrogen content of leaves versus leaf mass per area (*left*) and relative irradiance (*right*). *Open squares* indicate leaves from two beeches in the Steinkreuz stand, beech Gr12 and three additional beeches in the Großebene stand; *open stars* indicate oak Gr13. Data points of beeches Bu38 (*open triangles*) and Bu45 (*solid triangles*) are averages of at least three neighboring leaves; r^2 for these data only against LMA was 0.97 and 0.99, respectively. Relationships to relative irradiance were $y=2.64x^{0.383}$ ($r^2=0.88$) for beech Bu38, $y=2.48x^{0.405}$ ($r^2=0.9$) for beech Bu45 and $y=2.54x^{0.389}$ ($r^2=0.87$) for both beech trees

6.3.4 Day Respiration

The aim of a mechanistic leaf-oriented model is to derive not only the material basis of leaf functioning (LMA, N), but also photosynthesis parameters on the leaf scale that determine the fluxes. The relevance of photosynthesis parameters depends always on the used model, which was in this study the Harley/Tenhunen model of leaf photosynthesis (Harley and Tenhunen 1991) which integrates the biochemical leaf model of Farquhar and von Caemmerer (1982) with the stomatal model of Ball et al. (1987). The parameters R_d (day respiration), J_{max} (electron transport capacity) and Vc_{max} (carboxylation capacity) were derived from photosynthesis measurements (A/C_i curves) on leaves at known locations in the two neighboring beech and oak crowns at the stand Großebene with the Li-Cor 6400 system (Fleck 2002, p. 95). The three parameters and their temperature dependence are decisive for the biochemical part of the model, while the stomatal reaction to environment and gas-exchange is mainly dependent on the parameter g_{fac} (coefficient of stomatal sensitivity).

The temperature dependence of day respiration of the investigated beech and oak leaves was separately investigated for nitrogen classes of leaves and showed a stronger exponential increase of respiration rates when nitrogen content was higher. The comparison of rates at the standard temperature of 298 K also shows higher rates for the high nitrogen classes, but the variation between the classes was still too high and does not permit to derive a functional relationship (see Fig. 6.11).

The concept of exponential Arrhenius equations to model the temperature dependence of respiration has been questioned by Atkin et al. (2000), who

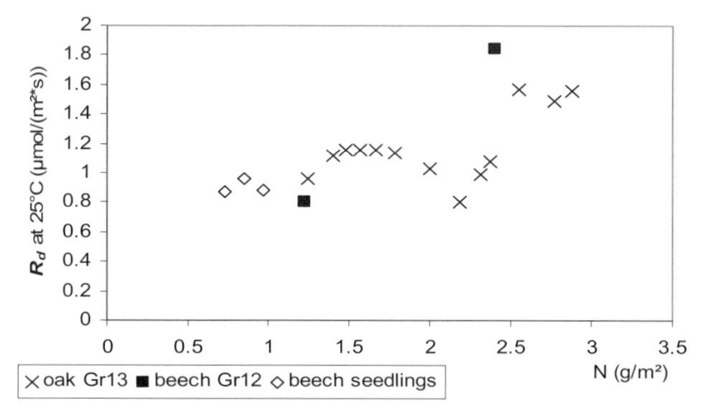

Fig. 6.11. Variation of day respiration at 298 K with nitrogen per leaf area. Each point represents an interpolation between R_d measurements on at least six leaves of the same nitrogen class at different temperatures. Interpolation between the temperatures was done with an exponential Arrhenius equation. The beech seedlings represent data from Forstreuter and Strassenmeyer as published in Medlyn et al. (1999)

observed a sudden decrease in respiration rates of leaves of *Eucalyptus pauciflora* above a threshold value of around 300 K. Further investigations on larger numbers of leaves may clarify whether additional factors (for example, daytime) might have an influence on day respiration rates.

6.3.5 Electron Transport Capacity (J_{max}) and Carboxylation Capacity (Vc_{max})

While day respiration represents the loss of assimilates in the form of CO_2, electron transport capacity J_{max} and carboxylation capacity Vc_{max} are leaf-specific physiological constants that characterize the biochemical equipment for CO_2 assimilation of a leaf.

The dependence of J_{max} and Vc_{max} values at a standard temperature of 298 K on the nitrogen content of leaves of both species emerged to be strictly nonlinear for both species with $r^2 \geq 0.96$ (Fig. 6.12). The high r^2 values may be interpreted as indicators for the mechanistic dependence of carboxylation and electron transport on nitrogen and light adaptation of the leaf.

In contrast to the expectations, both quantities did not increase continually with increasing nitrogen content of the leaves, but reached a nitrogen saturation above threshold values of 2.2 g N m^{-2} (beech) and 2.3 g N m^{-2} (oak, Vc_{max}). This and the high leaf nitrogen contents indicate an oversupply of nitrogen in the investigated leaves, since they contain more nitrogen than useful to enhance photosynthesis.

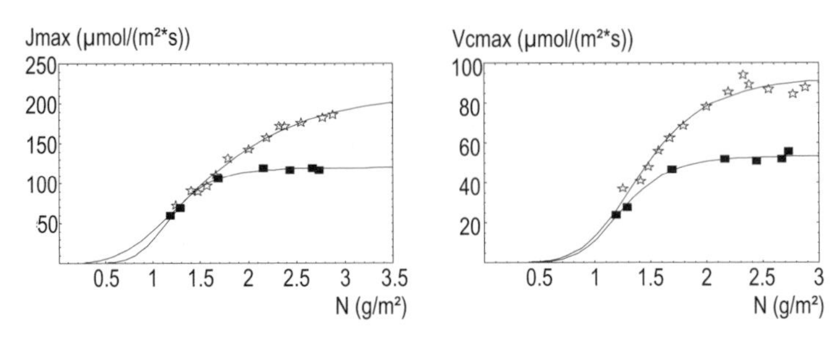

Fig. 6.12. Variation of J_{max} and Vc_{max} at 25 °C of leaves of oak Gr13 (*open stars*) and beech Gr12 (*solid squares*) with nitrogen content per area. Each data point represents the temperature interpolated value from nitrogen classes of leaves. The observed saturation at higher nitrogen contents was described with an approximation function of the type $y=ax^b/(x^b+c)$. Coefficients a, b and c were 120.3, 5.94 and 3.02 (J_{max}, $r^2=0.99$) and 53.9, 5.85 and 3.68 (Vc_{max}, $r^2=0.99$) for beech Gr12. Coefficients for oak Gr13 were 221.1, 3.07 and 4.44 (J_{max}, $r^2=0.99$) and 93.7, 4.93 and 5.9 (Vc_{max}, $r^2=0.96$), respectively

Comparable investigations on the relationship between leaf nitrogen and photosynthesis capacity (Harley et al. 1992; Niinemets and Tenhunen 1997; Porte and Lousteau 1998; Le Roux et al. 1999; Medlyn et al. 1999; Saito and Kakubari 1999; Kazda et al. 2000) usually approximate the found relationships with a linear function, assuming that additional nitrogen always increases leaf physiological activity. Our database differs from these studies in that it contains leaves with unprecedented high nitrogen content. A re-evaluation of the mentioned data sources reveals that evidence for a nitrogen saturation of photosynthesis may be found there as well (Fleck 2002, p. 120).

6.3.6 Stomatal Sensitivity (g_{fac}) vs. Nitrogen

The relationship between stomatal conductance to H_2O and the product of net assimilation (A), relative humidity at the leaf surface (rh) and the inverse of the CO_2 concentration at the leaf surface (C_a) is expected to be linear (Ball et al. 1987) and the slope of this relationship (stomatal sensitivity, g_{fac}) is an important parameter in the Harley/Tenhunen model, linking photosynthesis and transpiration. Our investigations shall clarify to what extent it depends on structural light adaptation of the investigated leaves.

A variation of the stomatal sensitivity of oak Gr13 leaves and beech Gr12 leaves with nitrogen per area was observed. G_{fac} of beech leaves was generally lower and varied between 6.1 and 13, while it was generally higher in oak leaves (9.9–16.3 mol H_2O/mol air). The nitrogen relationships of g_{fac} showed fundamentally difference trends for oak Gr13 and beech Gr12 (Fig. 6.13). Shade leaves of beech were relatively sensitive to changes in the $A \times rh/C_a$

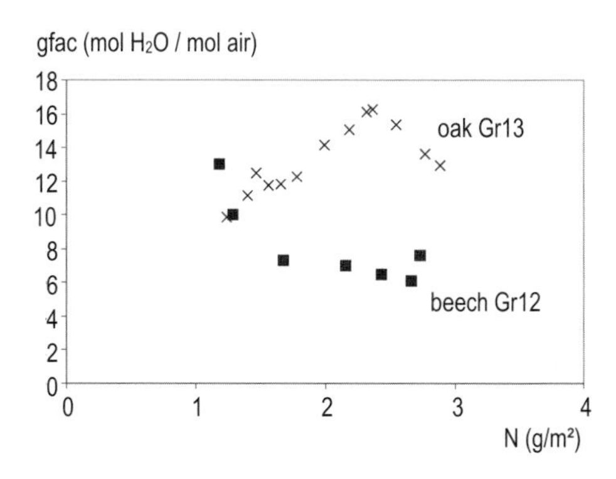

Fig. 6.13. Relationships between the Ball/Woodrow/Berry coefficient of stomatal sensitivity (g_{fac}) and leaf nitrogen for nitrogen classes of leaves of both mature trees in the Großebene stand

ratio, but the sensitivity coefficient g_{fac} steeply decreased with increasing nitrogen content to a constant value around 7, which was maintained for leaves between 1.7 and 2.7 g N m^{-2}. Nearly the opposite trend was found in oak leaves: g_{fac} of extreme shade leaves was lowest (9.8 g m^{-2}) and increased linearly with increasing nitrogen content up to a maximum value of 16.3 at 2.4 g m^{-2}. Above this value, g_{fac} of oak Gr13 leaves tended to decrease. These findings may indicate a species-specific difference in the dependence of stomatal sensitivity on structural light adaptation which meets with other differences between oak and beech (Fleck 2002, pp. 143–145).

6.4 A Modeling Study on the Gas-Exchange of Leaf Clouds

6.4.1 Modeling Leaf Mass Per Area of Leaf Clouds with the Light Model STANDFLUX-SECTORS

Is a model on the branch scale sufficiently fine-scaled to model the light distribution in the canopy with an accuracy that allows us to derive leaf properties of the branches? The light model STANDFLUX-SECTORS (Faltin 2004) is a version of the light model STANDFLUX (Falge 1997; see Köstner et al., this Vol.) which enables a fine-scaled structure representation on the level of leaf clouds by approaching their form with radial sections of cylinders and rings (see Fleck 2002). The model was applied in this study to derive average and maximum LMA values of six leaf clouds of beech Gr12 using a stand-specific modification of the relationship between relative irradiance and LMA (see equations in Fig. 6.7). The modification corrects with a constant factor for the higher maximum value of LMA in the uppermost part of the canopy of the Großebene beech, thereby conserving the overall shape of the relationship. The six leaf clouds were chosen for branch sap-flow measurements (Schmidt,

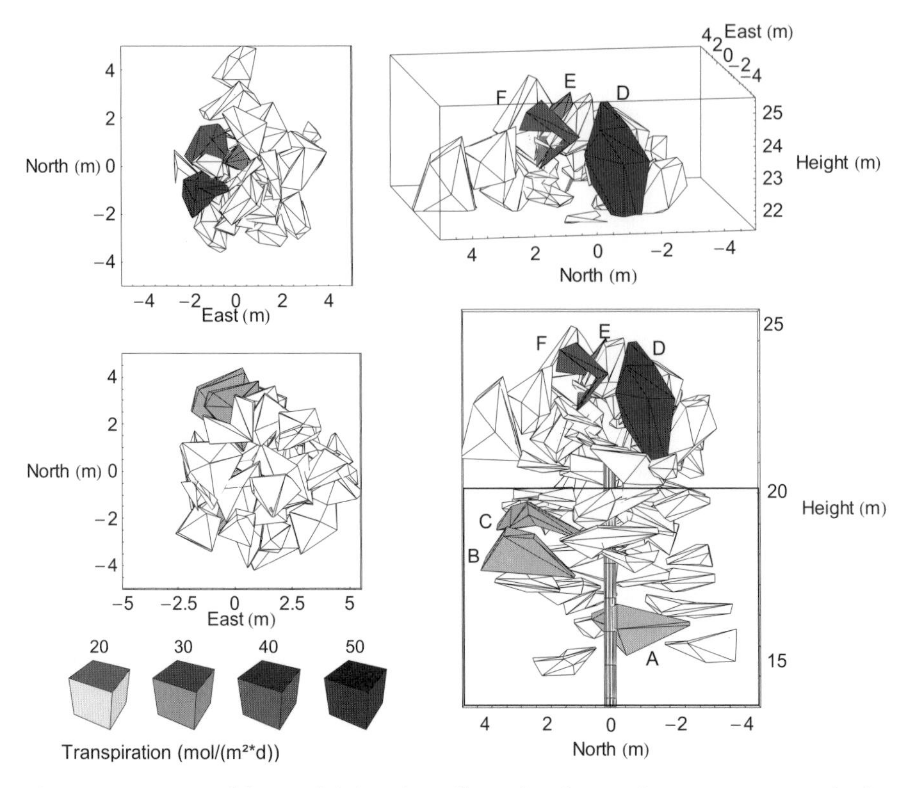

Fig. 6.14. 3D maps of the spatial situation of branches for sap-flow measurements in the canopy of beech Gr12 (M. Schmidt, thermal dissipation method). *Left* Vertical view from above the tree on three measured sun leaf clouds (*D*, *E*, *F* – *above*) and on three measured shade leaf clouds (*A*, *B*, *C*) from above the shade crown (*below*). *Right* Side views of upper canopy from west above (*above*) and of the whole canopy from west direction (*below*). Leaf cloud *F* was inserted as one of the two last branches directly below the apex leaf cloud that is visible between the *E* and *F*. Measured average transpiration rates of the six leaf clouds during the period 19 June to 2 July 1998 are indicated by intensity of their *color*

Department of Plant Ecology, University of Bayreuth) to represent three leaf clouds of the sun crown and three leaf clouds of the shade crown. Their position in the crown is visible on the virtual tree map in Fig. 6.14.

Total leaf mass per total leaf area of these leaf clouds (=average LMA) was modeled based on the *average* relative irradiance in a matrix point layer directly above each leaf cloud, while the maximum LMA of leaves occurring in each leaf cloud was modeled based on the *maximum* relative irradiance from that matrix point layer. Both were in good agreement with the measured values of harvested leaves of these leaf clouds with root mean square errors of 10.28 and 10.3 g m^{-2}, respectively, which equals mean absolute errors of +1.5

and –6.1 g m^{-2} (Fleck 2002, pp. 116–117). This indicates that the branch-oriented light model is able to predict the light climate as accurately as necessary to model relevant leaf cloud properties.

6.4.2 Measured Daily Courses of Leaf Cloud Sap Flow vs. Calculated Photosynthetic Photon Flux Density (PPFD)

Given that the resolution of the combined structure and light model is sufficient to adequately model the light climate in the whole crown, what can we say about the role of PPFD above leaf clouds in their transpiration rates? Can a simpler PPFD-transpiration relationship at the branch scale in combination with a complex light distribution and the structure of the vessel system serve as an explanation for the high variability that we find on the scale of tree canopies?

Daily courses of sap flow were recorded during the period from 19 June to 2 July 1998 and are compared in Figs. 6.15 (shade crown) and 6.16 (sun crown) with the modeled PPFD values above each leaf cloud. Some observations may be summarized concerning the course of transpiration rates of sun and shade leaf clouds during the investigation period:

- The daily course of transpiration rates of sun leaf clouds was generally smoother than that of shade leaf clouds: while sun leaf clouds D, E and F had wide and round daily peaks of transpiration (Fig. 6.16), the peaks in the course of transpiration rates from shade leaf clouds A, B and C were more pointed (Fig. 6.15).
- When a continuous increase in irradiance was given, transpiration rates of leaf cloud D and the other sun leaf clouds steeply increased early in the morning and then approached a maximum value at noon. Discontinuities in the irradiance increase with parallel discontinuities of the usual increase of water vapor pressure deficit (VPD) between leaf mesophyll and air in the morning – probably due to dew fall or smaller rain events – caused a decrease in transpiration rates even if irradiances just stayed constant, which may be observed on 19., 22 and 26 June on all sun and shade leaf clouds.
- The shade leaf clouds often did not start to transpire on these days, when such a discontinuity occurred, which may indicate that a critical light or VPD level was not yet reached before that time. Their later start is one reason for the more pronounced peaks in transpiration.
- Another reason was that the course of PPFD above these leaf clouds consists also of pointed peaks, thereby inducing high irradiances for a short time.
- The decrease of irradiance after this maximum peak is often (leaf clouds A and C during 19 to 26 June) but not in all cases (leaf cloud B, 20, 21 and 25 June) accompanied by a sharp decrease of transpiration rates. Opposite to the other shade leaf clouds, irradiance above leaf cloud B again increases in

the afternoon, and opposite to A and C, the transpiration rates above leaf cloud B stay high or even increase (21 and 25 June) during the time of decreasing irradiance, thereby ignoring the sharp but short decrease in irradiance.

- In sun leaf clouds, a series of high transpiration rates was achieved on most days between 10:00 and 14:00, which was partly independent of light. The transpiration rates of leaf cloud D, for example, do not completely follow

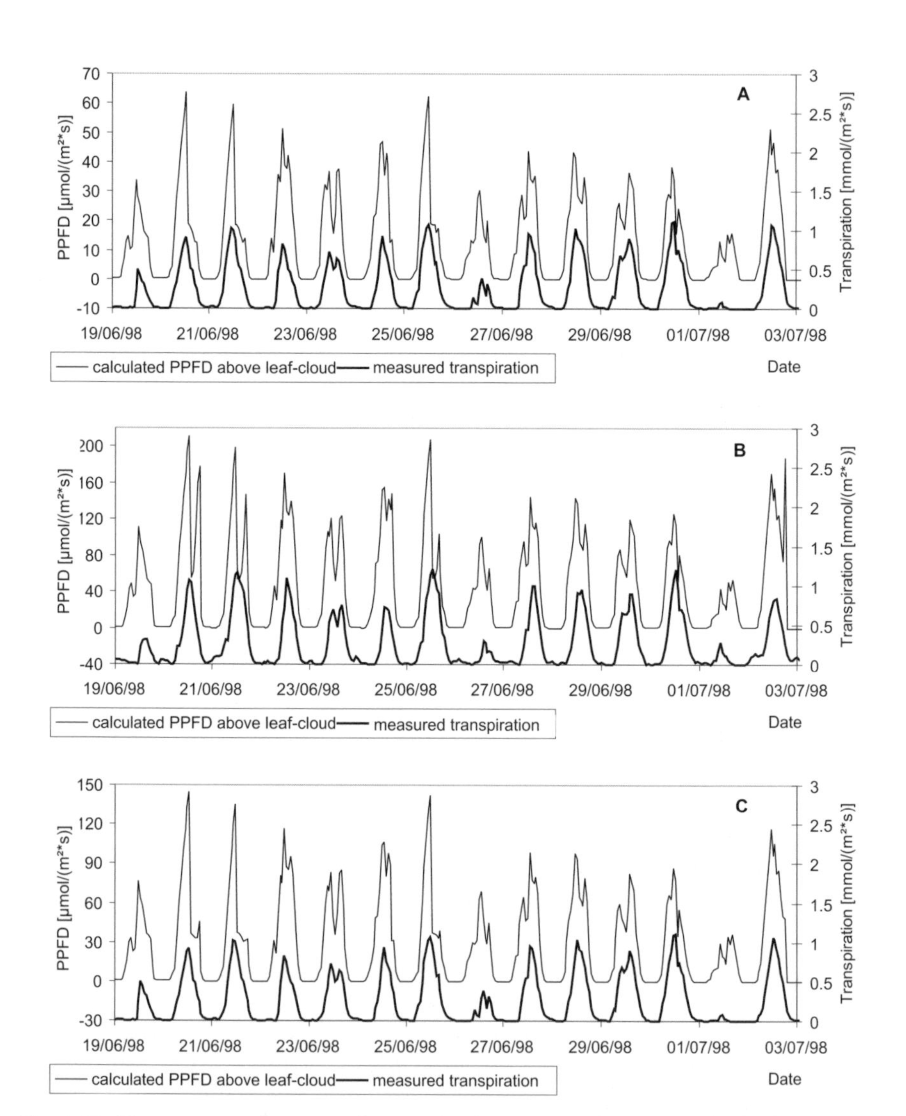

Fig. 6.15. Time course of measured transpiration rates and average of calculated PPFD of matrix point layer directly above the investigated shade leaf clouds (clouds *A*, *B* and *C*) during the investigation period

the course of irradiation on 22, 24 and 28 June. More or less strong decreases in irradiance and VPD did not lead to reductions in transpiration on these days.

- If irradiance increases in the afternoon parallel to VPD, a reaction of transpiration rate is often visible in sun and shade clouds, which is best seen on 24 June (all leaf clouds).

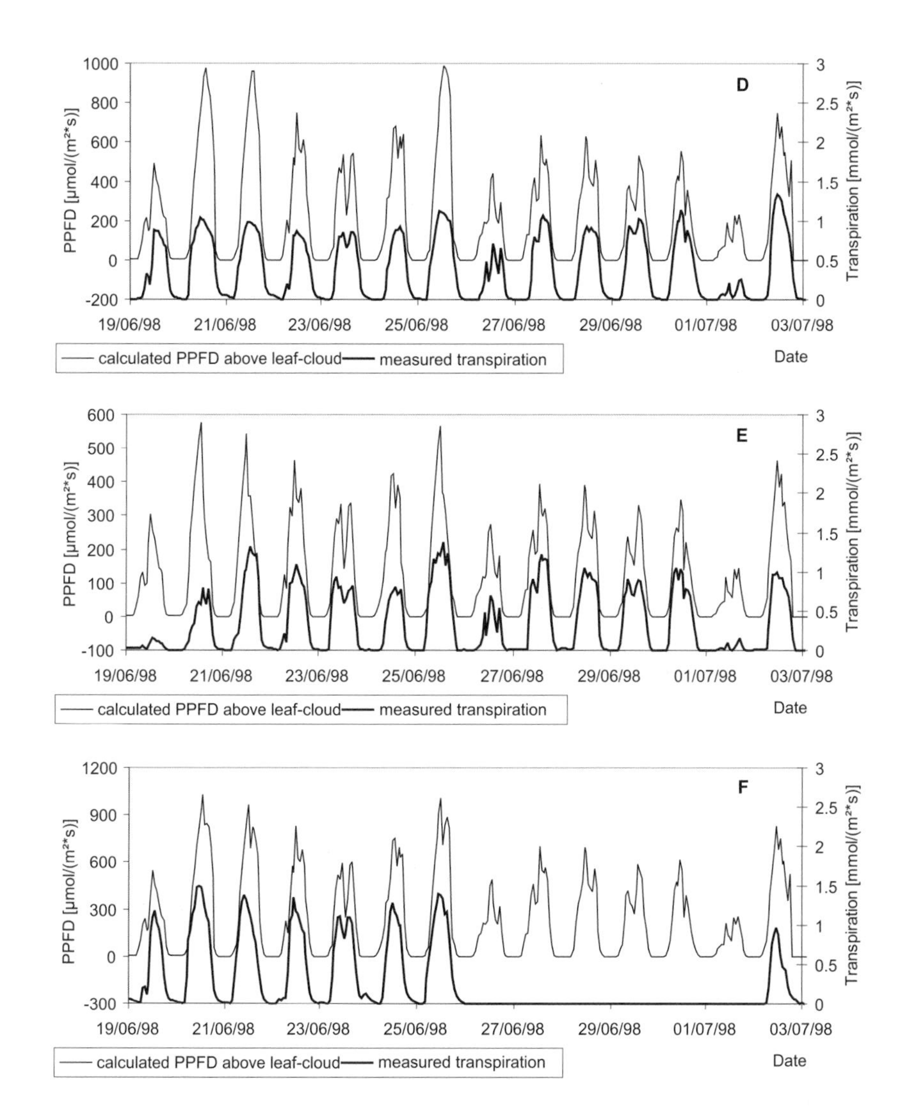

Fig. 6.16. Time course of measured transpiration rates and average of calculated PPFD directly above the investigated sun leaf clouds (clouds *D*, *E* and *F*) during the investigation period. Transpiration peaks are generally wider than those of the shade leaf clouds. A gap exists in the sap-flow measurements of leaf cloud *F* due to technical problems

6.4.3 Links Between Branch Transpiration and PPFD or VPD

Differences between sun leaf clouds and shade leaf clouds may also be observed in the direct comparison between branch transpiration and the meteorological driver variables PPFD and VPD. The easiest way to relate these to transpiration is a linear regression to both quantities. This led for the relationship to VPD to r^2 values between 0.54 and 0.8, based on the assumption that VPD is a spatially constant value throughout the crown. The relationship to modeled PPFD values in a point-layer above each leaf cloud yielded r^2 values between 0.58 and 0.79 (Fleck 2002, pp. 138–140). If the slope of both regression lines is interpreted as the sensitivity of branch transpiration to VPD and PPFD, the sensitivity to VPD was generally higher in sun leaf clouds, while that to PPFD was higher in shade leaf clouds. The latter shows a continuous trend from shade leaf clouds to sun leaf clouds (Fig. 6.17) which may similarly be observed in the stomatal sensitivity coefficient g_{fac} of shade and sun leaves (cf. Fig. 6.13).

Considering that branch transpiration is finally nothing else than the sum of transpiration rates of all appending leaves, it might be that more integrated variables that summarize over time show clearer relationships due to compensation effects. This was true for two variables that integrate over the measurement period of 2 weeks. The average branch transpiration rates during this time were well correlated with the cumulated quantum sums of the matrix point layers above leaf clouds (Fig. 6.18). Furthermore, the total amounts of transpired water of the branches during 14 days were very well correlated with their total leaf masses (Fig. 6.18). These encouraging results may partly be due to the low number of repetitions and need further investigation over longer time periods and more branches.

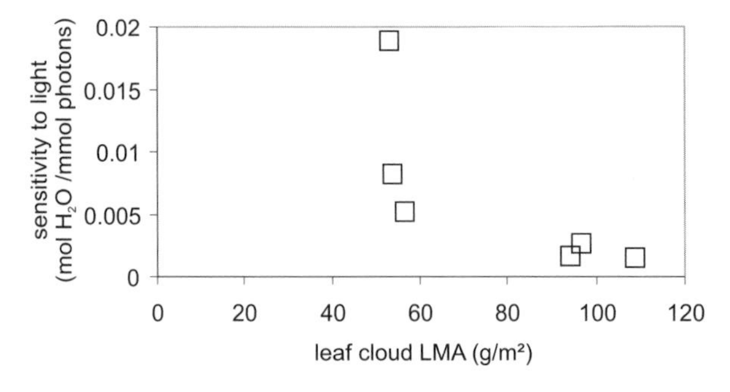

Fig. 6.17. Light sensitivity of transpiration in relation to average leaf cloud LMA of the six measured branches. Light sensitivity represents the slope of a linear regression of measured branch transpiration rates on modeled PPFD above each leaf cloud (hourly time step)

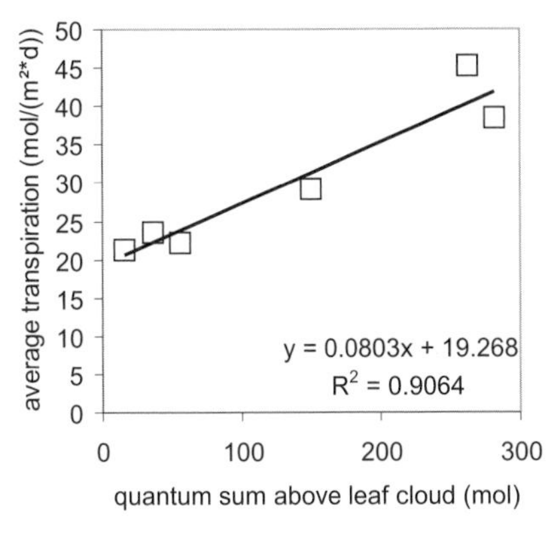

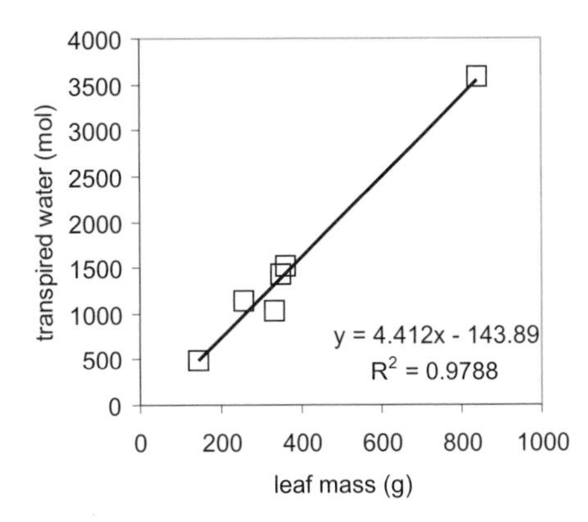

Fig. 6.18. Correlation between measured average transpiration rate over the whole investigation period and modeled quantum sum in the matrix point layer above each of the six leaf clouds (*left*) and correlation between the sum of transpired water in the period 19 June to 2 July 1998 and total leaf dry mass of the leaf clouds (*right*)

6.5 Discussion

The most relevant parameters for the calculation of fluxes and conductances on the leaf scale appear to be mechanistically determined by canopy fine-structure and the resulting light distribution. Tight relationships were found between relative irradiance and LMA (Fig. 6.7) and between LMA and nitrogen per area (Fig. 6.10). This is in accordance with earlier findings of several other authors on other species (e. g. Kull and Niinemets 1993) and may be seen as a general principle. Artefacts that seem to be exceptions to this rule can easily be created by the measurement method. It must be considered that the light climate of even two neighboring leaves can strongly vary due to the large impact of shading by objects in close proximity (other leaves, twigs), so that a hemispherical photo has to be taken exactly at the position of a specific leaf blade without destruction of the directly surrounding structure elements. This is not always possible and due to the size of the lens requires often tedious and exact positioning work that needs a safe stand of the investigator in the canopy.

LMA and nitrogen content appear on top of this to be the key to leaf physiological capacities like carboxylation capacity (Vc_{max}) and electron transport capacity (J_{max}) for beech as well as for oak (Fig. 6.12). Other important physiological properties like day respiration (Fig. 6.11) and stomatal sensitivity (Fig. 6.13) may as well be linked to these quantities, but the relationship was less obvious and needs further investigation.

The relationship between measured photosynthesis capacities and the partly very high nitrogen content of the investigated leaves saturated at high nitrogen contents, thereby showing that these leaves contain more nitrogen than useful for their photosynthesis. This is at least an indication of the high nitrogen supply, but could as well be seen as a hint to the fate of the nitrogen surplus. Nitrogen content of litter samples in the Steinkreuz was nevertheless not higher than normal (Berg and Gerstberger, this Vol.).

The apparent mechanistic dependence of leaf gas-exchange capacities on the structure-induced light distribution in the crown via the chain leaf surrounding structure–relative irradiance–LMA–nitrogen–photosynthesis parameters [$J_{max}(T)$, $Vc_{max}(T)$, $R_d(T)$] provides the theoretical basis for a leaf-oriented gas-exchange and conductance model of canopies. The input of meteorological drivers (mainly light, VPD and temperature T) to such a fine-scaled model would principally allow us to calculate photosynthesis rates, conductances and transpiration rates of all single leaves, and, thus, also canopy conductance, but it is extremely dependent on the accuracy of the appertaining structure description. Though the idea of a model that depends on something that appears to be not assessable is not very convincing, it must be considered that the latter statement is not new to gas-exchange modeling (see Introduction). All kinds of stand gas-exchange models are extremely dependent on the accuracy of their structure description – the difference to a leaf-oriented model is just that the structure dependence is not always obvious and explicitly modeled, but implicit in general assumptions over average structural properties. An accurate fine-structure description of tree canopies is not yet available. However, an artificial leaf-oriented model, based on a leaf-oriented structure scenario, even now provides the chance to analyze the effect of structure on gas-exchange and conductance of canopies, since it can build on mechanistic relationships and the enormously enhanced computer power.

Spatial variability of LMA and, thus, leaf physiological properties within single leaf clouds was nearly as high as at the level of whole tree crowns (Fig. 6.7). The stronger the light gradient in a leaf cloud was, the higher the variability in LMA values. The resulting variability in physiological reactions of leaves is probably the major cause of non-parallel developments in the time-course of climate variables and branch transpiration. Figures 6.15 and 6.16 show that this non-parallelism is at the branch scale not less than at the level of canopies (compare Köstner et al., this Vol.).

Explanatory variables for branch transpiration properties are anyway to be found at the leaf scale. So was stomatal sensitivity of beech leaves (different from oak leaves) for shade leaves higher than for sun leaves (Fig. 6.13). The higher light sensitivity of sun leaf cloud transpiration in comparison to shade leaf cloud transpiration (Fig. 6.17) is probably a direct consequence of sun and shade leaf proportions in the investigated leaf clouds.

The modeling results show that a branch-based structure description may be sufficiently fine-scaled to model the light climate that is relevant for average leaf cloud structural properties. The good relationship between modeled light variables and measured leaf mass per area of leaf clouds was probably due to the major impact of branch-oriented leaf clumping on light distribution in the canopy. Clumping of leaves to leaf clouds concentrated the leaves in less than 20 % of the canopy volume.

While the dependence of branch conductance on spatially integrating structural quantities remained unclear, temporally integrating quantities (over the 14 days of measurement) showed strong linear relationships between average branch transpiration rate and integrated irradiance, as well as between total amount of transpired water and branch leaf mass. However, the short investigation period without major weather changes does not allow us to draw conclusions about the general validity of these concepts.

A mechanistic explanation of branch transpiration and conductance was not possible based on the branch-oriented structure and light modeling. Non-parallel developments in the time-courses of light and transpiration (Figs. 6.15 and 6.16) and high variability in the direct regression at the level of branches (Fig. 6.17) indicate a similar complexity to that at the level of canopies (Köstner et al., this Vol.). There was no obvious reduction in complexity of the light–transpiration relationship, though structural complexity was expected to be lower at the branch scale than at the canopy scale.

Anyway we interpret the complexity of this relationship as induced by the heterogeneity of micrometeorological and physiological situations in a leaf cloud and not as a complexity of the mechanism itself, since the well-described mechanism at the leaf level does not show complexity to this extent.

The scaling problem of gas-exchange models for the calculation of canopy conductance may, thus, not be solved by branch-oriented models without extra assumptions about the distribution of light in leaf clouds, the frequency of micrometeorological and physiological situations of leaves (leaf classes) and their change over time. However, these assumptions appear to be as difficult to prove and to assess as their relatives at the level of canopies, as may be expected due to the large range of LMA values in each leaf cloud.

Due to the low number of repetitions that such a detailed investigation entails, it may not prove the possibility or impossibility of branch-based transpiration modeling. However, the results allow us to conclude that the insecurity in gas-exchange models that is induced by structural complexity cannot be removed by a medium-fine (branch-based) structure description. The full gain of a finer structure description can only be achieved with a very fine leaf-oriented stand model. The existence of tight mechanistic relationships at the leaf level, the enormously increased capacity of computers and the development of methods for automatic structure description and modeling are encouraging hints that this challenge will be faced in the near future.

6.6 Conclusions

- Beech and oak leaves contained slightly more nitrogen than useful to enhance photosynthesis capacities, indicating the whereabouts of nitrogen surpluses.
- Leaves in mature beech and oak canopies are concentrated in leaf clouds comprising less than 20 % of the canopy volume.
- The clumped structure of mature beech canopies (leaf clouds and gaps) is decisive for the light distribution in the canopy, as far as light is responsible for average leaf properties (LMA) of the leaf clouds.
- The mechanistic dependence of water conductance on structure and micrometeorological variables is at the level of branches as complex as at the level of canopies. The scaling problem of gas-exchange models may, thus, not be solved at the level of branches.
- From the decisive properties for water conductance at the level of leaves. Vc_{max} and J_{max} are strongly dependent on the surrounding fine structure, expressed as relative irradiance. Strong correlations between relative irradiance, LMA, leaf nitrogen and photosynthesis capacities provide the basis for an alternative method to upscale conductance. Stomata sensitivity was, however, not as clearly derivable from structural properties and shows an opposite trend for both species throughout the canopy.

Acknowledgements. This work was mainly funded by the German Bundesministerium für Forschung und Technologie (PT BEO-0339476 C) and additionally supported by the EU projects CARBODATA and CARBOEUROFLUX.

References

Atkin OK, Evans JR, Ball MC, Lambers H, Pons TL (2000) Leaf respiration of snow gum in the light and dark. Interactions between temperature and irradiance. Plant Physiol 122:915–923

Baldocchi D, Collineau S (1994) The physical nature of solar radiation in heterogeneous canopies: spatial and temporal attributes. In: Caldwell MM, Pearcy RW (eds) Physiological ecology. A series of monographs, texts, and treatises: exploitation of environmental heterogeneity by plants. Ecophysiological processes above- and belowground. Academic Press, San Diego

Ball JT, Woodrow IE, Berry JA (1987) A model predicting stomatal conductance and its contribution to the control of photosynthesis under different environmental conditions. In: Biggens J (ed) Progress in photosynthesis research. Proc 7th Int Photosynthesis Congress, vol 4. Martinus Nijhoff, Dordrecht

Bartelink HH (1997) Allometric relationships for biomass and leaf area of beech (*Fagus sylvatica* L). Ann Sci For 54:39–50

Bauer G, Schulze E-D, Mund M (1997) Nutrient contents and concentrations in relation to growth of *Picea abies* and *Fagus sylvatica* along a European transect. Tree Physiol 17:777–786

Burger H (1945) Holz, Blattmenge und Zuwachs – die Buche. Mitt Schweiz Anst Forstl Versuchswesen 26:419–468

Burger H (1947) Die Eiche. Mitt Schweiz Anst Forstl Versuchswesen 25:211–279

Campbell GS, Norman JM (1989) The description and measurement of plant canopy structure. In: Russell G, Marshall B, Jarvis PG (eds) Society for Experimental Biology seminar series, vol 31. Plant canopies: their growth, form and function. Cambridge University Press, Cambridge

Cescatti A (1997) Modelling the radiative transfer in discontinuous canopies of asymmetric crowns. I. Model structure and algorithms. Ecol Model 101:263–274

De Pury DGG, Farquhar GD (1997) Simple scaling of photosynthesis from leaves to canopies without the errors of big-leaf models. Plant Cell Environ 20:537–557

Falge E (1997) Die Bedeutung der Kronendachtranspiration von Fichtenbeständen (*Picea abies* (L.) Karst.) mit unterschiedlichen Modellierungsansätzen. Bayreuther Forum Ökol 48:549–550

Faltin W (2004) Scaling up water and CO_2 fluxes of forested areas from stand to footprint level by means of an individual-based 3D model. Dissertation, University of Bayreuth, Germany

Farquhar GD, von Caemmerer S (1982) Modeling of photosynthetic response to environmental conditions. In: Lange OL, Nobel PS, Osmond CB, Ziegler H (eds) Encyclopedia of plant physiology, 12B. Physiological plant ecology, vol 2. Springer, Berlin Heidelberg New York

Fleck S (2002) Integrated analysis of relationships between 3D-structure, leaf photosynthesis, and branch transpiration of mature *Fagus sylvatica* and *Quercus petraea* trees in a mixed forest stand. Bayreuther Forum Ökol 97

Godin C (2000) Representing and encoding plant architecture: a review. Ann For Sci 57(5/6):413–438

Harley PC, Tenhunen JD (1991) Modeling the photosynthetic response of C_3 leaves to environmental factors. In: Boote KJ (ed) CSSA special publication, no 19. Modeling crop photosynthesis – from biochemistry to canopy. American Society of Agronomy and Crop Science Society of America, Madison

Harley PC, Thomas RB, Reynolds JF, Strain BR (1992) Modelling photosynthesis of cotton in elevated CO_2. Plant Cell Environ 15:271–282

Jarvis PG (1995) Scaling processes and problems. Plant Cell Environ 8:1079–1089

Kazda M, Salzer J, Reiter I (2000) Photosynthetic capacity in relation to nitrogen in the canopy of a *Quercus robur*, *Fraxinus angustifolia* and *Tilia cordata* flood plain forest. Tree Physiol 20:1029–1037

Krauss HH, Heinsdorf D (1996) Herleitung von Trockenmassen und Nährstoffspeicherungen in Buchenbeständen. Forschungsbericht im Auftrag der Landesforstverwaltung Brandenburg, Forstliche Forschungsanstalt Eberswalde. Forstliche Forschungsanstalt Eberswalde, Eberswalde, Germany

Kull O, Niinemets Ü (1993) Variation in leaf morphometry and nitrogen concentration in *Betula pendula* Roth., *Corylus avellana* L. and *Lonicera xylosteum* L. Tree Physiol 12:311–318

Kurth W, Sloboda B (1999) Tree and stand architecture and growth described by formal grammars. I. Non-sensitive trees. II. Sensitive trees and competition. J For Sci 45:16–30/53–63

Larcher W (2001) Ökophysiologie der Pflanzen. Ulmer, Stuttgart

Le Roux X, Grand S, Dreyer E, Daudet F (1999) Parameterization and testing of a biochemically based photosynthesis model for walnut (*Juglans regia*) trees and seedlings. Tree Physiol 19:481–492

Leuschner C (1998) Mechanismen der Konkurrenzüberlegenheit der Rotbuche. Ber Reinhard-Tüxen-Gesellsch 10:5–18

Martin JG, Kloeppel BD, Schaefer TL, Kimbler DL, McNulty SG (1998) Aboveground biomass and nitrogen allocation of ten deciduous southern Appalachian tree species. Can J For Res 28:1648–1659

Medlyn BE, Badeck FW, de Pury DGG, Barton CVM, Broadmeadow M, Ceulemans R, de Angelis P, Forstreuter M, Jach ME, Kellomäki ME, Laitat E, Marek M, Philippot S, Rey A, Strassemeyer J, Laitinen K, Liozon R, Portier B, Wang K, Jarvis PG (1999) Effects of elevated CO_2 on photosynthesis in European forest species: a meta-analysis of model parameters. Plant Cell Environ 22:1475–1495

Niinemets Ü, Tenhunen JD (1997) A model separating leaf structural and physiological effects on carbon gain along light gradients for the shade-tolerant species *Acer saccharum*. Plant Cell Environ 20:845–866

Pellinen P (1986) Biomasseuntersuchungen im Kalkbuchenwald. Dissertation, Universität Göttingen, Germany

Porte A, Loustau D (1998) Variability of the photosynthetics of mature needles within the crown of a 25-year-old *Pinus pinaster*. Tree Physiol 18:223–232

Rogers R, Hinckley TM (1979) Foliar weight and area related to current sapwood area in oak. For Sci 25:298–303

Saito H, Kakubari Y (1999) Spatial and seasonal variations in photosynthetic properties within a beech (*Fagus crenata* Blume) crown. J For Res 4:27–34

Schulze E-D, Küppers M, Matyssek R (1986) The roles of carbon balance and branching pattern in the growth of woody species. In: Givnish TJ (ed) On the economy of plant form and function. Cambridge University Press, Cambridge, pp 585–602

Schulze E-D, Kelliher FM, Körner C, Lloyd J, Leuning R (1994) Relationships among maximum stomatal conductance, carbon assimilation rate, and plant nitrogen nutrition: a global ecology scaling exercise. Annu Rev Ecol Syst 25:629–660

Tanaka T, Yamaguchi J, Takeda Y (1998) Measurement of forest canopy structure with a laser plane range-finding method – development of a measurement system and applications to real forests. Agric For Meteorol 91(3–4): 149–160

Van den Burg J (1990) Foliar analysis for determination of tree nutrient status – a complication of literature data. "De Dorschkamp", Institute for Forestry and Urban Ecology, Wageningen, the Netherlands

Wrzesinsky T, Klemm O (2000) Summertime fog chemistry at a mountainous site in central Europe. Atmos Environ 34:1487–1496

7 Soil CO_2 Fluxes in Spruce Forests – Temporal and Spatial Variation, and Environmental Controls

J.-A. Subke, N. Buchmann, and J.D. Tenhunen

7.1 Introduction

The predicted changes to the climate in temperate zones (IPCC 2001) are prone to alter physiological processes of both carbon sequestration and carbon release of terrestrial ecosystems. Whether temperate ecosystems will act as sources or sinks of carbon under altered environmental conditions depends on the way in which the balance of these processes is shifted in the short term, and in the long term. Soils of forest ecosystems are generally regarded as crucial to this balance since real long-term storage of carbon occurs only in the soil (Goodale et al. 2002). In temperate forest ecosystems, over 60 % of carbon stocks are in the soil (IPCC 2000), and changes in climate, as well as in land use, are likely to alter the input to and release from soils, thus changing their potential for C storage.

Soil CO_2 efflux includes the CO_2 originating from microbial decomposition of fresh litter, older soil organic matter (SOM) and root exudates (heterotrophic respiration), as well as CO_2 originating from autotrophic growth and maintenance respiration from roots. The rate of decomposition of SOM depends on environmental conditions within the soil (mainly soil temperature and moisture) and on the composition and availability of substrates for decomposition (i.e., above- and below-ground litter input). Rhizosphere respiration (i.e., autotrophic root respiration and heterotrophic respiration resulting from the decomposition of root exudations), on the other hand, shows a strong dependence on above-ground processes (Högberg et al. 2001), so that independent dynamics of the CO_2 evolution rate from these two soil compartments are likely.

For abiotic factors controlling soil CO_2 evolution, the dependence on soil temperature is well documented (e.g., Raich and Schlesinger 1992; Lloyd and Taylor 1994), while soil moisture effects, which are commonly reported from more arid ecosystems, have only rarely been found in temperate forest soils (Davidson et al. 1998; Epron et al. 1999). Even if soil moisture limitation has gone undetected in past experiments (possibly owing to the rare occurrence

Ecological Studies, Vol. 172
E. Matzner (Ed.), Biogeochemistry of Forested Catchments in a Changing Environment

of extreme water stress in temperate ecosystems), it is critical to include the soil moisture effect on soil CO_2 evolution if predictive landscape models are to accommodate possible changes in precipitation patterns as well as increases in temperature. Laboratory experiments can help to understand the qualitative influence of abiotic factors, but they cannot form the basis of C flux modelling under field conditions. Field sampling is indispensable for obtaining data on which to base environmental dependencies and future predictions, but the question of how these data are obtained is crucial. In order to capture rare events of low soil moisture, and also to distinguish long- and short-term effects on the soil CO_2 efflux, continuous measurements are necessary. At the same time, a representative spatial measuring scheme is crucial to capture the spatial variability of soil CO_2 efflux rates (Davidson et al. 2002). This includes variability between stands, caused by different land-use or management regimes, but also at a smaller scale, owing to the heterogeneous distribution of roots and heterotrophic respiration substrates in the soil. While permanently installed soil CO_2 measuring facilities that record the efflux rate are more complex than portable chamber systems, it is typically not feasible to measure the efflux rate at many different locations. Portable chamber systems, on the other hand, require the presence of an experimenter, thus generally limiting the sampling frequency to feasible intervals of several weeks to months, but allowing a representative sampling within a given area. The sampling strategy therefore represents a trade-off between spatial or temporal resolution of CO_2 efflux averages (Savage and Davidson 2003), except when a dual-method approach is used (this study).

This chapter reports on the efflux dynamics of CO_2 from the forest floor of a mature spruce stand at diurnal to seasonal time scales. In order to capture all relevant conditions (of both the diurnal and the seasonal cycle), a continuous sampling strategy with open dynamic soil chambers was adapted. For comparison of different forest stands located within the Lehstenbach watershed that differ in age and therefore forest structure, a sampling strategy with a state-of-the-art portable chamber system was employed. This dual approach allowed the assessment of temporal variation and environmental dependency of soil CO_2 efflux found by the continuous chamber system at a larger spatial scale.

7.2 Measuring Sites

Measurements were carried out in four Norway spruce stands [*Picea abies* (L.) Karst] located in close proximity to each other within the Lehstenbach watershed. The soil type of all stands is a cambic podzol over granite bedrock, characterized by low pH values of around 3.5–4.2. The organic soil horizons (O_i, O_e and O_a) are well stratified and of the moder type (Gerstberger et al.,

Table 7.1. Characteristics of the four forest stands included in this study. *N.D.* No data

Site	Age (years)	LAI ($m^2 m^{-2}$)	Thickness of soil horizons (cm)		
			O_i	O_e	O_a
Weidenbrunnen	47	10.4	1.3±0.3	5.1±0.4	3.6±0.8
Weidenbrunnen 3	87	N.D.	1.5±0.1	4.8±0.7	4.2±0.7
Weidenbrunnen 2	111	6.0	1.0±0.1	5.2±0.6	7.2±0.6
Coulissenhieb	146	6.6	0.9±0.1	5.9±0.2	4.4±0.2

this Vol.). Some characteristics of the different stands are given in Table 7.1. All forest stands are managed and had virtually no coarse woody debris on the forest floor. Ground cover consisted mainly of the grass species *Deschampsia flexuosa* and *Calamagrostis villosa* and the ericaceous dwarf shrub *Vaccinium myrtillus*, all of which occur in monospecific patches. The degree of ground cover ranged from very sparse (Weidenbrunnen) to a closed cover of *D. flexuosa* and *C. villosa* in Weidenbrunnen 2, with a ground LAI of up to 2.5.

7.3 Temporal Variation of Soil CO_2 Efflux

7.3.1 Continuous Measuring Approach

In order to assess the temporal variation of the soil CO_2 efflux, and its dependence on environmental variables, an open dynamic soil chamber was installed in the Weidenbrunnen 2 stand in April 1999 for continuous measurements of soil CO_2 efflux over the entire growing season. The open dynamic chambers (Fig. 7.1) allow continuous measurements of the soil CO_2 efflux rate without requiring the presence of an experimenter. Particular care was taken to eliminate pressure differences between the chamber space and the ambient atmosphere, as this has been found to be a considerable source of error for chamber measurements (Kanemasu et al. 1973). The basic design was adapted from Rayment and Jarvis (1997), and chamber tests showed that the chamber operation artificially reduced the internal chamber pressure by only about 10 mPa, which would result in only a negligible mass flow of air from soil pores. Soil temperature inside the chambers was only slightly increased at the soil surface during day time and an effect of soil moisture was not discernible (for a detailed description of the chamber design and tests of the impact of the soil environment, see Subke 2002).

The instantaneous soil CO_2 efflux was measured from five collars sequentially once every hour by automatically directing both measuring and refer-

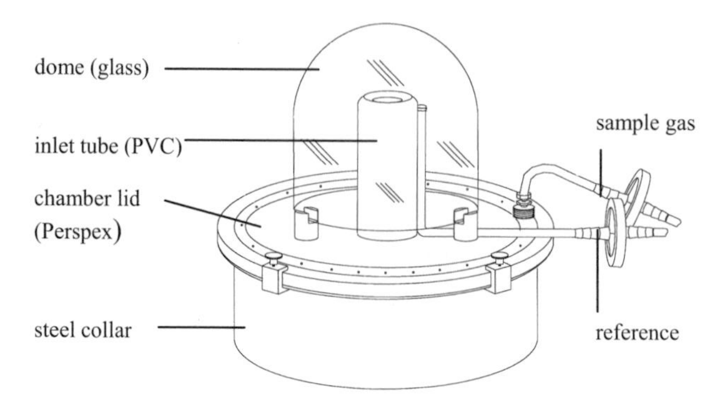

Fig. 7.1. Open dynamic soil chamber (diameter 20 cm, collar height 8 cm). Air is drawn from the lateral canal visible in the chamber lid, which is connected to the chamber volume through perforations. Ambient air follows the pressure gradient ($\Delta P \cong 10$ mPa) through the centrally mounted inlet aperture. The dome mounted over the inlet prevents Ventouri-type suction in the aperture from wind passing horizontally over its opening

ence gas through a vapour trap to a differential Infrared Gas Analyser (BINOS 100, Fisher-Rosemount, previously Leybold Heraeus, Hanau, Germany). Switching intervals of 10 min between chambers allowed ample time for the gas path common to all chambers to be flushed. The CO_2 concentration was recorded every minute, and the mean of the last three values for each 10-min interval was used for the soil CO_2 efflux calculation. Following specific stability requirements for the CO_2 concentration readings (relating to transient temperature conditions in the chamber, see Subke 2002), the soil CO_2 efflux was calculated according to:

$$F_{soil} = \frac{C_{dif} f k}{A} \quad (1)$$

where C_{dif} is the differential CO_2 signal (in µmol mol^{-1}), f the pump rate of measuring air from the chamber (in l min^{-1}), A the area of the soil chamber (0.315 m^2), and k a constant factor combining all conversions of C_{dif} from [µmol mol^{-1}] to [µmol m^{-3}] and for f from [l min^{-1}] to [m^3 s^{-1}].

Each chamber lid was moved between three rings regularly (generally every 2 or 3 days) to improve the spatial representation of the flux and reduce possible artifacts due to the prolonged presence of the chamber. The position of soil collars reflected the ground vegetation patches with two groups of collars (i.e., 2 × 3 collars), each located in patches of *Deschampsia* and *Calamagrostis* and one group in a 'nursery' of *Picea* saplings. All above-ground parts of the ground vegetation were removed within collars upon installation, and any regrowth was removed throughout the year. All 15 collars remained in the same location throughout the entire growing season. Measurements proceeded from mid-April to early December 1999 with larger gaps in the data set in July and in autumn owing to instrumental failure.

Temperature probes were installed close to the soil collars in five locations each at 5, 10 and 30 cm depth, while soil moisture probes were inserted into the organic horizon to between 2 and 8 cm depth (Theta Probes, Delta-T Devices Ltd., Cambridge, UK).

7.3.2 Seasonal and Diurnal Course of the Soil CO_2 Efflux and Dependencies on Environmental Variables

In 1999, hourly flux values ranged from 0.5 μmol m^{-2} s^{-1} in November to about 4.0 μmol m^{-2} s^{-1} in July, with diurnal variations as great as 50 % of peak values throughout the season. Most of the variation in soil CO_2 efflux rate could be explained by the variation in temperature at 5 cm depth alone, which is also documented by the general agreement between temperature and efflux curves in Fig. 7.2. Daily peaks in soil CO_2 flux generally preceded peaks in temperature at 5 cm depth, and occurred in the early afternoon. This is presumably due to the fact that the peak in soil surface temperatures, where a significant part of the CO_2 originates, also preceded the peak at 5 cm, owing to the lag in temperature propagation with increasing soil depth.

On the other hand, the influence of the soil water content (SWC) on the soil CO_2 efflux is less obvious, but can be seen for periods of low SWC values, as for

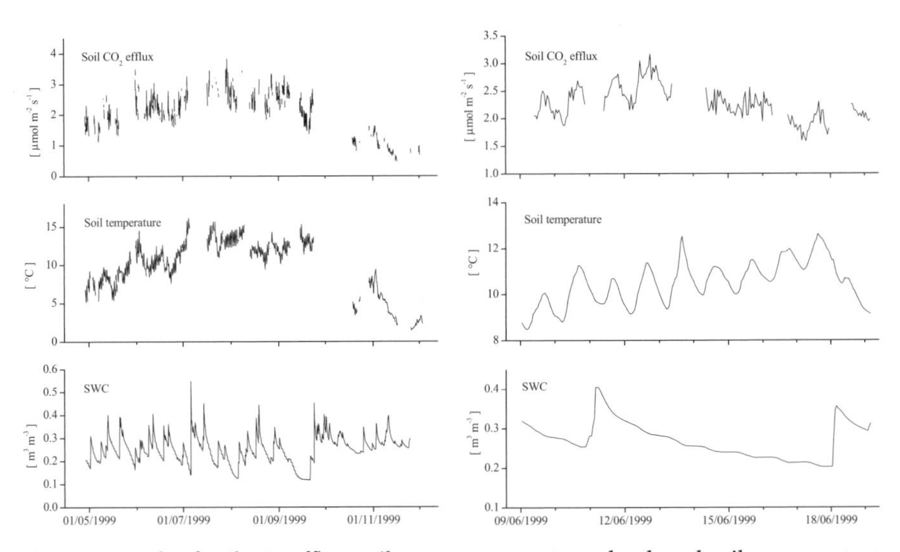

Fig. 7.2. Trends of soil CO_2 efflux, soil temperature at 5 cm depth and soil water content (SWC) in the organic layer for the growing season 1999 (*left*) and exemplified for 10 days in June 1999 (*right*) in Weidenbrunnen 2

example in mid-September 1999, or between 14 and 18 June 1999 (Fig. 7.2), when the CO_2 efflux values are suppressed despite high soil temperatures. However, the soil still responds to changes in temperature under low moisture conditions (e.g., on 17 June), which means that both temperature and soil moisture control soil CO_2 efflux independently. Rather than defining threshold values for SWC above which soil temperature can be said to control CO_2 efflux, and below which SWC becomes crucial, both variables should be considered simultaneously and independently. A thorough treatment of temperature and SWC dependence of the soil CO_2 efflux has been presented elsewhere (Subke et al. 2003) and only a short summary is given here. The fundamental temperature dependence can be best described using the model by Lloyd and Taylor (1994), while the inclusion of SWC limitation was best achieved by multiplication with a hyperbolic function after Bunnell et al. (1977), yielding:

$$F_{\text{soil}} = R_{\text{ref}} e^{E_0\left(\frac{1}{56.02} - \frac{1}{46.02+T}\right)} \frac{SWC}{SWC_{\frac{1}{2}} + SWC} \tag{2}$$

where T is the soil temperature at 5 cm depth, and SWC the volumetric soil water content in the organic layer, both of which have been measured simultaneous to the soil CO_2 efflux. Model parameters that were fitted by multivariate non-linear regression are: R_{ref}, the soil respiration at a reference temperature of 10 °C, E_0, an exponential factor related to the activation energy in the Arrhenius equation (Lloyd and Taylor 1994) and $SWC_{1/2}$, conceptually the volumetric water content at which 50 % of the optimal soil respiration occurs (i.e., soil respiration with no moisture limitations for a given temperature). For data fits, only readings obtained within 24 h of closing a chamber lid were considered, and readings were discarded following rain, if it occurred within this measuring interval, since rain could not enter the chamber. For the entire growing season, the best data fit was found for R_{ref}=3.57, E_0=403 and $SWC_{1/2}$=0.172 (Fig. 7.3). Running a simple temperature regression for the same data set (by fixing parameter $SWC_{1/2}$ to a value of 0) reduced the coefficient of determination considerably (r^2=0.83 for regression with SWC limitation compared to r^2=0.72 for temperature only).

For measurements using the portable system (see Sect. 7.4.1), the temperature dependency was best using an exponential model for the Weidenbrunnen 2 stand in 1998 ($y=1.03\ e^{0.105\ T5}$, r^2=0.80, p <0.0001). Using an Arrhenius type equation resulted in smaller coefficients of determination (r^2=0.63), and no soil moisture limitation was evident (Buchmann 2000). The better time resolution of the efflux by the continuous system is preferred here for the calculation of annual efflux sums (Sect. 7.3.3), but for the comparison among forest stands, where measurements took place on the same day, the results of the portable system provide a more adequate basis (Sect. 7.4.3).

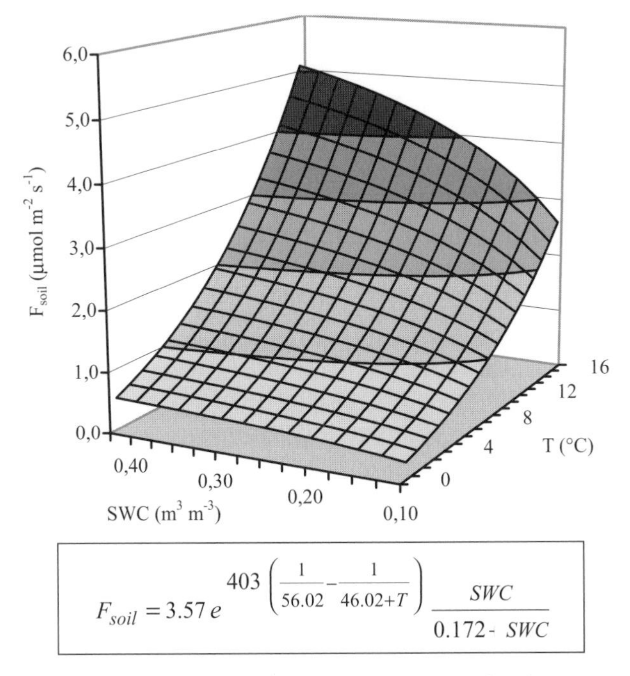

Fig. 7.3. Soil CO_2 efflux in Weidenbrunnen 2 as a function of soil temperature and SWC as obtained by data fits for the entire growing season in 1999

Expressed as Q_{10} (i.e., the factor by which a metabolic function increases following a change in temperature by 10 °C), the continuous data yielded a value of 2.4 for the change between 5 and 15 °C. This compares to the results of the portable chamber system, which measured a range of 2.4–3.2 for the four forest stands, and all Q_{10} values are typical for temperate forest ecosystems (Raich and Schlesinger 1992). However, as was pointed out in the Introduction, soil CO_2 efflux is the result of a variety of processes, and one single value for the temperature sensitivity may mask different responses to changes in temperature of the various soil compartments (i.e., roots, old SOM, fresh needle and root litter). Separating the soil CO_2 efflux from hetero- and autotrophic sources has considerable methodological problems (Hanson et al. 2000), and mutual interactions between the soil compartments may exist that would be ignored by an experimental separation (Subke et al. 2004). Some experiments have indicated different Q_{10} values for rhizosphere respiration compared to that of bulk soil (Boone et al. 1998), but more research is needed to address different responses of soil compartments to environmental change. Stable or radioactive isotope methods are most promising in this context, as this technique is non-intrusive and thus allows analysis of the individual and interactive responses of the soil compartments (Ehleringer et al. 2000; Hanson et al. 2000).

7.3.3 Interannual Variability of Soil CO_2 Efflux

Using Eq. (2) with the parameterization described for the seasonal data set, it is possible to calculate soil CO_2 efflux sums over periods for which the model drivers (T, SWC) are available. Long-term soil temperature records were obtained from the Weidenbrunnen research site, but the SWC of the organic layer has not been monitored continuously. However, using an ecosystem model developed in order to calculate short-term dynamics of heat, water and CO_2 exchange ($PROXEL_{NEE}$; Reichstein 2001), it was possible to simulate soil water conditions using meteorological model drivers, all of which were available from the same station at Weidenbrunnen. Parameters for $PROXEL_{NEE}$ were fitted using the SWC data recorded in 1999 as well as other stand-specific canopy parameters (see Appendix A in Subke 2002 for details of SWC simulation). A continuous data set of model drivers could thus be compiled for the period from 1 April 1997 to 31 March 2001. Since driver data were available for daily time steps for part of the data set, fluxes were calculated as daily averages. The data from 1999, which is resolved at hourly intervals (data gaps aside), showed that for estimates on a growing season time scale, the error of using daily averages of drivers is less than 0.2 %.

Aggregation of the modelled soil CO_2 efflux as well as temperature and rainfall data over 1-year periods reveals considerable interannual variability (Fig. 7.4, Table 7.2). If soil moisture effects in the flux data go undetected, as it

Table 7.2. Climate variables and total annual carbon loss through soil CO_2 efflux for the same period as in Fig. 7.4. Owing to the periods over which climate drivers were available, each year is calculated from 1 April to 31 March of the following year. The *right-hand column* states the overestimation that results if the soil CO_2 efflux is calculated with a simple temperature model with no representation of SWC limitation (see text for details)

	Annual mean T (°C)	Precipitation sum (mm $year^{-1}$)	Soil CO_2 efflux model (g C m^{-2} $year^{-1}$)	Overestimation by simple T model (%)
1997	5.9	572[a]	497	15.5
1998	6.0	1,300	566	2.6
1999	6.4	1,170	592	1.5
2000	6.8	945	586	5.3
Mean ± SD	6.3±0.4	996±319	560±43	6.2±6.3

[a] Precipitation data in 1997 is too low due to instrumental failure in winter, when soil moisture is not limiting, and between late June and late July; comparison with independent precipitation data showed that the modelled SWC underestimated actual rainfall by only 38 mm between 26 June and 22 July, but that the dry period in the second half of 1997 (Aug/Sept 1997; Fig. 7.4) is no artefact. The overestimate in the right-hand column for 1997 is therefore not likely to be substantially exaggerated

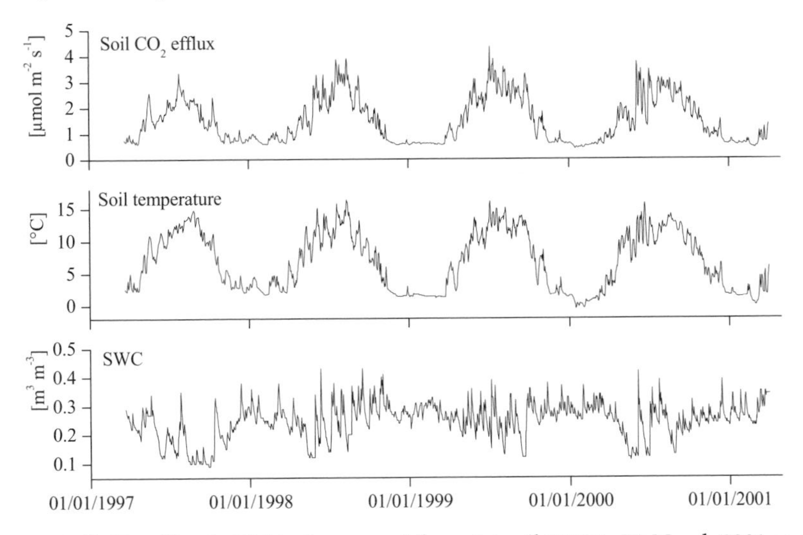

Fig. 7.4. Soil CO_2 efflux in Weidenbrunnen 2 from 1 April 1997 to 31 March 2001 as calculated using the soil CO_2 efflux model (see Fig. 7.3) with soil temperature at 5 cm depth and SWC in the organic layer as drivers

is likely if a discontinuous sampling strategy is adapted, simple temperature-driven models would overestimate the annual efflux sums. Using a model for temperature dependence only, and using only the soil CO_2 efflux data obtained for SWC >0.2 m^3 m^{-3} for model parameterization, a considerable overestimation of annual efflux sums was found for years with lower than average rainfall (about 15 %, Table 7.2). The results show that the soil CO_2 efflux is sensitive to soil moisture conditions also in temperate forest ecosystems. More specifically, the distribution of precipitation during the year and the interaction with seasonal temperature changes leads to considerable variation in the annual soil CO_2 efflux. Models simulating the C dynamics in forest ecosystems have to reflect this SWC limitation, especially if altered precipitation patterns are to be addressed in C budget projections.

7.4 Within- and Between-Stand Variation of Soil CO_2 Efflux

7.4.1 Portable Measuring Approach

For the soil CO_2 efflux comparison among forest stands, soil respiration rates were measured at monthly intervals during the growing season 1998 (April to October), using a portable soil respiration system (LI-6400 with connected soil chamber LI-6400–09; LiCor, Lincoln, Nebraska, USA). Four or five soil collars (10 cm high, 10 cm inner diameter) were installed in each of the stands to between 5 and 8 cm depth. To minimize disturbance due to collar installation, collars were installed at least 24 h before measurements were carried out. During measurements, the chamber was placed on a collar and the rate of increase over an interval of 10 ppm CO_2 around ambient concentrations was recorded in five consecutive observations. Soil temperatures were measured at the time of measurements next to a collar at 5, 10 and 15 cm depth, while the soil water content was recorded gravimetrically for each sampling date with three replicates per stand.

7.4.2 Within-Stand Variation

Variations within a stand of up to 40 % of peak rates were found during summer months from measurements with the portable respiration chamber (n=5 for each stand; Buchmann 2000), but could not be attributed to temperature variations within the stand. For the collars of the continuous system, variation of base respiration rates during the entire growing season (at standard temperature of 10 °C) was found to be 42 % of maximum values (maximum and minimum values of 3.00 and 1.75 μmol m^{-2} s^{-1} respectively; mean=2.18, SE=0.188, n=15). This variation could not be explained by a systematic influence of the ground vegetation type in which the collars were located. However, a significant positive correlation was found between the base respiration rate and the root density below collars of the continuous system (r^2=0.461, p=0.005, n=15; root biomass was sampled after flux measurements had finished; no such analysis was performed for the portable system). Root density is often dependent on the distance from tree stems in a stand, but also on small-scale heterogeneity in nutrient supply within the soil. The variation in soil CO_2 efflux of up to 40 % observed within a stand can be linked to differing autotrophic root contributions or to the heterogeneity of decomposition within the soil. Partitioning between both sources in all four stands suggested a major contribution of microbial respiration (>70 %; Buchmann 2000).

7.4.3 Between-Stand Variation

In contrast to the considerable within-stand variability, the difference in average soil CO_2 efflux was found to be not significantly different among the four forest stands (p=0.196; analysis of variance), even though the average rates of the Weidenbrunnen 2 stand (111 years old) tended to be slightly higher than those of the other stands, particularly in the second half of the growing season (Fig. 7.5). This similarity of soil CO_2 efflux over a range of stand ages (from 47–146 years) seemed surprising at first, since productivity both above- and below-ground is known to change with tree age. In addition, soil CO_2

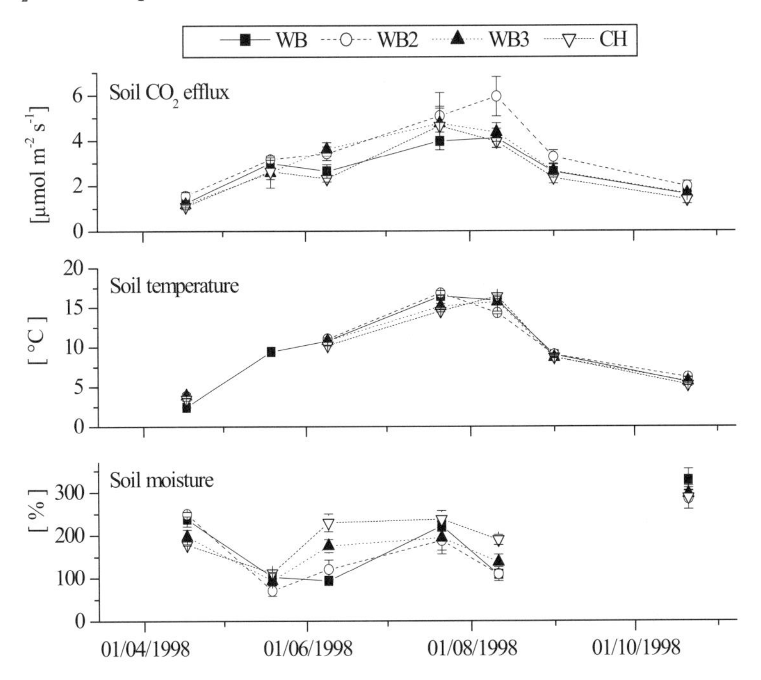

Fig. 7.5. Seasonal variation in soil CO_2 efflux rate, soil temperature at 5 cm depth and soil moisture in the O_f and O_h layers in the four forest stands in 1998. Soil CO_2 efflux measurements were made by the portable soil respiration system. Values are means ± SE, n=3–5. *WB* Weidenbrunnen; *CH* Coulissenhieb

efflux is closely linked to the gross productivity, since the main route of assimilated C returning to the atmosphere is via the soil (Janssens et al. 2001), either as auto- or heterotrophically respired CO_2. However, Mund et al. (2002) found similar rates of productivity for stands older than 70 years in a comparison of forest stands in the Fichtelgebirge (six stands, of which two – 'Weidenbrunnen' and 'Coulissenhieb' – are the same as in this study). The authors attribute this similarity in productivity to synergetic effects of forestry practice (early thinning and liming), atmospheric N deposition and increased mean annual temperatures over the past decades. The feedback among these environmental conditions, forest management practice, forest productivity and soil C dynamics is complex and a change in one of the factors may lead to a change in ecosystem C assimilation and release.

Although age-related changes in soil CO_2 production from autotrophic and heterotrophic sources may exist, it is probable that they partly compensate each other in managed forests and were therefore not detectable. For the age range covered in this study, no great quantitative error arises if the spatial average of fluxes is calculated across forest stands in the Lehstenbach watershed. The soil CO_2 efflux can therefore be directly compared to other ecosystem C fluxes (NEE, NPP) measured by eddy covariance above the canopy, for which the source area of the measured fluxes comprises more than one forest stand. Rebmann et al. (this Vol.) found that total ecosystem respiration was about two times that of the soil respiratory flux, with above-ground plant parts contributing the remaining fraction. This partitioning into respiratory components was supported by either measured or modelled respiration fluxes between April 1997 and December 1999 (Fig. 8.3 of Rebmann et al., this Vol.). Ecosystem models, which attempt to simulate processes within a forest ecosystem, cannot rely on the net flux measured above the canopy alone. Additional and independent flux measurements of individual components, as in this case soil respiration, are needed to constrain the total flux, and extract the individual contributions over which the eddy flux measurement integrates.

7.5 The Annual C Budget of the Forest Soil

A chronosequence study carried out in the Fichtelgebirge has shown that for spruce forests, C accumulation stabilizes after about 70 years both above and below ground (Mund et al. 2002). This means that for all stands in this study, apart from Weidenbrunnen 2, the mean C input into the soil is balanced by a C output of the same magnitude. The C input can be divided into above-ground litter fall (including needles, wood, bark and seeds), root net primary production and root exudates. C input through litter fall in the Coulissenhieb site averaged about 170 g C m^{-2} $year^{-1}$ (Berg and Gerstberger, this Vol.), i.e., about one third of the total amount of C lost from the soil calculated for the Weidenbrunnen 2 site. Below-ground NPP was not measured in the course of this study, but Mund et al. (2002) estimate annual fine-root NPP for stands older than 70 years to be between 90 and 176 g C m^{-2} $year^{-1}$. According to Schlesinger (1997), fast turnover rates of fine roots can be assumed, so that annual root litter production can be assumed to be of similar magnitude as above-ground litter input. Comparison of C input and C output therefore suggest that, on an annual basis, CO_2 efflux from the soil derives to two thirds from heterotrophic decomposition of litter (50 % each from above- and below-ground sources). The remaining part derives either directly from root growth and maintenance respiration or from microbial decomposition of root exudates. This fraction is supported by Buchmann (2000), who found that root exclusion in some collars reduced soil CO_2 efflux by about 30 %. Nevertheless, recent results indicate complex interactions between rhizosphere activity and litter decomposition, with root exudates leading to increased decomposition of SOM (Subke et al. 2004). This clearly indicates that when assessing the C budget of a forest soil, a better understanding of mechanisms underlying the observed C flux into and out of the soil processes are needed.

7.6 Conclusions

- Methods for soil CO_2 efflux measurements are critical if fluxes are estimated. Each technique has strengths as well as shortcomings; both chamber systems used in this study have since been validated in a comparative study and were shown to yield realistic efflux estimates (Pumpanen et al. 2004). Differences between methods are therefore likely to result from different time resolutions rather than systematic measuring errors.
- The combination of continuous stationary measurements on the one hand and mobile periodic measurements on the other hand combine the strengths of each system.
- The temporally highly resolved data set from the continuous measuring system forms a basis of a thorough analysis of environmental variables influencing soil CO_2 efflux.
- Dependence of the soil CO_2 efflux rate on soil temperature is well documented, but the existence of significant soil moisture limitations has only rarely been detected in temperate forest ecosystems. However, estimating the annual CO_2 flux based on continuous data shows that considerable error of up to 15 % may occur if soil moisture conditions are not considered in ecosystem models.
- Variations within a forest stand are considerable, but the mean efflux did not differ significantly between forest stands despite differences in age and structure. However, this similarity in soil CO_2 efflux for forest stands of different ages cannot be assumed automatically, since it is likely to be the result of site-specific management conditions.
- For soil C models, the individual processes of C input, storage and release have to be accounted for.
- The impact of forest management strategies on soil C dynamics could be inferred from these results, illustrating the importance of forestry practice for soil C sequestration.

Acknowledgements. This work was funded by the German Ministry for Education, Science, Research and Technology (BMBF, PT BEO-0339476 C) and the Deutsche Forschungsgemeinschaft (BU-1080/1). The authors are grateful for climatic data supplied by members of the Bayreuth Institute for Terrestrial Ecosystem Research (BITÖK, Bayreuth, Germany).

References

Boone RD, Nadelhoffer KJ, Canary JD, Kaye JP (1998) Roots exert a strong influence on the temperature sensitivity of soil respiration. Nature 396:570–572

Bunnell FL, Tait DEN, Flanagan PW, van Cleve K (1977) Microbial respiration and substrate weight loss. I. A general model of the influences of abiotic variables. Soil Biol Biochem 9:33–40

Buchmann N (2000) Biotic and abiotic factors controlling soil respiration rates in *Picea abies* stands. Soil Biol Biochem 32:1625–1635

Davidson EA, Belk E, Boone RD (1998) Soil water content and temperature as independent or confounded factors controlling soil respiration in a temperate mixed hardwood forest. Global Change Biol 4:217–227

Davidson EA, Savage K, Verchot LV, Navarro R (2002) Minimizing artifacts and biases in chamber-based measurements of soil respiration. Agric For Meteorol 113:21–37

Ehleringer JR, Buchmann N, Flanagan LB (2000) Carbon isotope ratios in belowground carbon cycle processes. Ecol Appl 10:412–422

Epron D, Farque L, Lucot É, Badot PM (1999) Soil CO_2 efflux in a beech forest: dependence on soil temperature and soil water content. Ann For Sci 56:221–226

Goodale CL, Apps MJ, Birdsey RA, Field CB, Heath LS, Houghton RA, Jenkins JC, Kohlmaier GH, Kurz W, Liu S, Nabuurs GJ, Nilsson S, Shvidenko AZ (2002) Forest carbon sinks in the northern hemisphere. Ecol Appl 12:891–899

Hanson PJ, Edwards NT, Garten CT, Andrews JA (2000) Separating root and soil microbial contributions to soil respiration: a review of methods and observations. Biogeochemistry 48:115–146

Högberg P, Nordgren A, Buchmann N, Taylor AFS, Ekblad A, Högberg MN, Nyberg G, Ottosson M, Read AJ (2001) Large-scale forest girdling shows that current photosynthesis drives soil respiration. Nature 411:789–792

IPCC (2000) Intergovernmental panel on climate change. In: Watson RT, Noble IR, Bolin B, Ravindranath NH, Verardo DJ, Dokken DJ (eds) Special report: land use, land-use change and forestry: summary for policymakers. Cambridge University Press, Cambridge

IPCC (2001) A report of working group I of the Intergovernmental Panel on Climate Change. Climate change 2001: the scientific basis. Cambridge University Press, Cambridge

Janssens IA, Lankreijer H, Matteucci G, Kowalski AS, Buchmann N, Epron D, Pilegaard K, Kutsch W, Longdoz B, Grünwald T, Montagnani L, Dore S, Rebmann C, Moors EJ, Grelle A, Rannik Ü, Morgenstern K, Oltchev S, Clement R, Gumundsson J, Minerbi S, Berbiger P, Ibrom A, Moncrieff J, Aubinet M, Bernhofer C, Jenson NO, Vesala T, Granier A, Schulze ED, Lindroth A, Dolman AJ, Jarvis PG, Ceulemans R, Valentini R (2001) Productivity overshadows temperature in determining soil and ecosystem respiration across European forests. Global Change Biol 7:269–278

Kanemasu ET, Powers WL, Sij JW (1973) Field chamber measurements of CO_2 flux from soil surface. Soil Sci 118:233–237

Lloyd J, Taylor JA (1994) On the temperature dependence of soil respiration. Funct Ecol 8:315–323

Mund M, Kummetz E, Hein M, Bauer GA, Schulze ED (2002) Growth and carbon stocks of a spruce forest chronosequence in central Europe. For Ecol Manage 171:275–296

Pumpanen J, Kolari P, Ilvesniemi H, Minkkinen K, Vesala T, Niiniströ S, Lohila A, Larmola T, Morero M, Pihlatie M, Janssens I, Curiel Yuste J, Grünzweig JM, Reth S, Subke JA, Savage K, Kutsch W, Østreng G, Ziegler W, Anthoni P, Lindroth A, Hari P (2004) Com-

parison of different chamber techniques for measuring soil CO_2 efflux. Agric For Meteorol (in press)
Raich JW, Schlesinger WH (1992) The global carbon dioxide flux in soil respiration and its relationship to vegetation and climate. Tellus 44B:81–99
Rayment MB, Jarvis PG (1997) An improved open chamber system for measuring soil CO_2 effluxes in the field. J Geophys Res 102(D24):28779–28784
Reichstein M (2001) Drought effects on carbon and water exchange in three Mediterranean ecosystems. Bayreuther Forum Ökol 89
Savage KE, Davidson EA (2003) A comparison of manual and automated systems for soil CO_2 flux measurements: trade-offs between spatial and temporal resolution. J Exp Bot 54:891–899
Schlesinger WH (1997) Biogeochemistry. Academic Press, London
Subke JA (2002) Forest floor CO_2 fluxes in temperate forest ecosystems. Bayreuther Forum Ökol 96
Subke JA, Reichstein M, Tenhunen JD (2003) Explaining temporal variation in soil CO_2 efflux in a mature spruce forest in southern Germany. Soil Biol Biochem 35:1467–1483
Subke JA, Hahn V, Battipaglia G, Linder S, Buchmann N, Cotrufo MF (2004) Feedback interactions between needle litter decomposition and rhizosphere activity. Oecologia 139:551–559

8 Carbon Budget of a Spruce Forest Ecosystem

C. REBMANN, P. ANTHONI, E. FALGE, M. GÖCKEDE, A. MANGOLD,
J.-A. SUBKE, C. THOMAS, B. WICHURA, E.-D. SCHULZE, J.D. TENHUNEN,
and T. FOKEN

8.1 Introduction

The investigation of carbon fluxes is of immense interest in ecosystem and climate research. Forest ecosystems may be a sink for anthropogenic carbon, if the assimilation is larger than the respiration. Alternatively, increasing temperatures due to climate change (IPCC 2001) may be a reason for increasing respiratory fluxes. While low-altitude spruce sites in Germany are significant carbon sinks (e.g. Bernhofer et al. 2003), sites above 600 m a.s.l. are only small sinks or may change their character by climate change. Therefore the Weidenbrunnen site in the Lehstenbach catchment was selected as a EUROFLUX site (Valentini et al. 2000) and was also used in the following CARBOEUROFLUX program for systematic investigations with respect to the data quality of turbulent fluxes. Overviews of the European carbon program and of the worldwide FLUXNET program are respectively given by Valentini (2003) and Baldocchi et al. (2001). All relevant references are also provided herein. Furthermore, the site was used for process studies to separate assimilation and respiration fluxes, and to study the exchange conditions between the forest and the atmosphere (Wichura et al., this Vol.). All of these studies were part of the ecosystem research of the Lehstenbach catchment, the main research area of the Bayreuth Institute of Terrestrial Ecosystem Research (BITÖK). The main results for the carbon dioxide flux measurements since 1997 are discussed in this chapter.

The carbon dioxide fluxes were measured using the well-established eddy covariance method which is based mainly on the EUROFLUX technology (Aubinet et al. 2000, 2003). For the direct measurement of the net ecosystem exchange (NEE), this method allows a high accuracy of about 10 % in contrast to the questionable results produced from observing the very large difference between respiration and assimilation rates. To achieve such results, the quality of both the data and the site were carefully checked. The most important problem was the best technology for filling data gaps (Falge et al. 2001a, b).

Ecological Studies, Vol. 172
E. Matzner (Ed.), Biogeochemistry of Forested
Catchments in a Changing Environment

Because of long periods during 1997–2002 without measurements, some monthly sums are missing and not gap filled. For comparison purposes, the different respiration rates of soil, wood and foliage were determined.

It should be noted that none of these program takes into account special effects of the energy exchange mechanisms, which were investigated by Wichura et al. (this Vol.). These effects are much more important for short-term investigations than for annual values. However, for a better understanding of the complicated interaction between assimilation and respiration, further investigations are necessary.

8.2 Measuring Site and Quality Control

The measurements were carried out at the 'Weidenbrunnen' site of the University of Bayreuth in the mountainous region of the Fichtelgebirge, 50°09′N, 11°52′E, 775 m a.s.l. at a height of 32 m a.g.l., above spruce forest (Picea abies, 50 years in 2003) with a canopy height of approx. 19 m. Additional eddy covariance measurements were performed at 22 m height on the same tower in the years 1997, 1998 and 1999 to investigate the influences of different footprints.

According to Foken (2003), the mean annual temperature is 5.0 °C (1961–1990) or 5.3 °C (1971–2000) and the mean yearly precipitation sum is 1,156 mm (1961–1990) or 1,162 mm (1971–2000). The climate has a continental character but with high precipitation in summer. The temperature increased by more than 0.25 K per 10 years in the last 40 years.

The quality of turbulent flux measurements at this site was systematically investigated in recent years and footprint information was taken into account (Foken et al. 2000). The interest in these measurements is twofold. Firstly, there is a desire to understand how the area of interest (e.g. spruce forest) affects the footprint of the measurements (different for day and night), and, secondly, there is a need to know for which footprint areas good data quality can be assumed. The developed method for the CARBOEUROFLUX program (Göckede et al. 2004), which was also used for the AMERIFLUX program (Foken et al. 2004), is based on land-use information of the surrounding terrain, given by input matrices. The integrated Schmid (1997) model determines characteristic dimensions defining the two-dimensional horizontal extension of each so-called effect-level ring. Using these dimensions, which sketch a discrete version of the source weight function, it is possible to assign a weighting factor to each of the cells of the matrix. In the next step, the land-use structure within the computed source area is analysed. The resulting weighting factors of the last source weight function are used to calculate the contribution of each type of land use to the total flux. Finally, images like that shown in Fig. 8.1a for the Weidenbrunnen site can be constructed which give

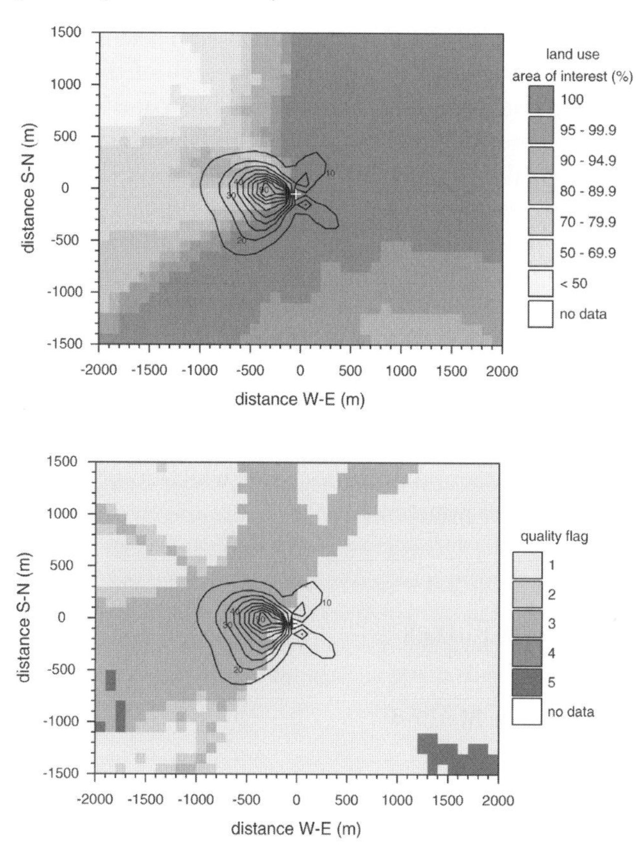

Fig. 8.1. a Quality analysis for the land use evaluation with the relative flux contribution for all stratification cases. Results were obtained with data from the Waldstein Weidenbrunnen site for the period May 1st to August 31st, 1998 (Göckede et al. 2004). The normalised three-dimensional source weight function is indicated by the 9 black isopleths ranging from 90 percent (innermost ring) to 10 percent. Greyscales indicate the percentage contribution of the land use type intended to be observed (spruce forest) to the total flux. **b** Same as Figure 8.1.a for the quality evaluation of the latent heat flux. Greyscales indicate the dominant quality flag for the latent heat flux during the observation period

as an example a flux distribution over a 4-month measuring period depending on the footprint. The colour of a grid element characterizes the contribution of the area of interest to the flux. Such pictures can help to determine the best wind directions or the best tower positions under study.

Results of all footprint calculations were combined with a data quality assessment to compute the overall performance of flux data quality for the specific site. Quality flags for each half-hourly flux were derived from stationarity tests and integral turbulence characteristics (Foken and Wichura 1996). The products of the procedure are two-dimensional matrices and graphs that form a combination of all the footprint analyses. These matrices show, for example, the dominating data quality class for each of the grid cells (mean value) of the matrix surrounding the tower, and its contribution to the total flux. As an example the data quality distribution for the latent heat flux λE of the Weidenbrunnen site is given in Fig. 8.1b. The lower data quality for the latent heat flux in western wind directions is most probably caused by a clearing, which is also indicated in the land-use distribution (Fig. 8.1a). The low data quality in the SWS direction (for stable stratification) is caused by the Waldstein mountain at a distance of 1.5 km. The equivalent figure for the carbon dioxide flux shows that all grid elements are of highest quality (flag 1), even though about 12 % of the investigated half-hourly flux data were assigned to a lower quality, attributable to instationarity and are rejected for the development of the parameterizations (see next chapter). However, there is obviously no wind direction or distinct area which results in flux sampling problems for CO_2 (figure not shown).

8.3 Calculation of Fluxes

8.3.1 Eddy Covariance Method

Since 1997, eddy covariance measurements have been carried out at the Weidenbrunnen site. The eddy covariance method has become a widely accepted tool for ecological studies in the last decade. This micrometeorological method provides a direct way to determine the fluxes of momentum τ, sensible and latent heat (H and λE) and carbon dioxide (F_{CO2}). Additional meteorological parameters were monitored at the tower site, such as air and soil temperatures, air humidity, net and global radiation, wind speed and direction and precipitation.

The eddy covariance measuring system consists of a sonic anemometer (Solent R2, Gill Instruments, Lymington, UK) for the high-frequency determination of the wind components u, v, w and the air (sonic) temperature, and a closed path analyser (LI6262, LI-COR, Lincoln, Nebraska, USA) for measuring CO_2 and H_2O concentrations with approx. 3 m tube length. The gas analyser was replaced by an open-path system (LI7500, LI-COR, Lincoln, Nebraska, USA) in May 2002. Data were recorded with a frequency of 20.8 Hz with a computer placed in a hut near the tower. The gas analyser was calibrated weekly using calibration gas for CO_2 and a dew point generator for H_2O near ambient and zero gases to account for overall sensor drift.

From the measurements, fluxes of momentum, sensible and latent heat and the carbon dioxide fluxes between the atmosphere and the forest were determined on a half-hourly basis in a post-processing procedure. Details concerning the system settings, data acquisition, determination of fluxes and corrections are in accordance with Moncrieff et al. (1997) and Aubinet et al. (2000). Corrections that were necessary because of an underestimation of the high frequencies in the turbulent spectrum were performed after Eugster and Senn (1995). The quality of the measured fluxes was checked with the procedures described in Foken and Wichura (1996), where stationarity is used as the main feature of quality.

A main goal of the measurements is the determination of annual sums of the net ecosystem exchange (NEE) of carbon dioxide between the forest and the atmosphere depending on climatic differences and for comparisons with other ecosystems. For the determination of the NEE of carbon dioxide, gap filling of the CO_2 fluxes was necessary. Gaps in the time series were mainly caused by power failures, ice coating in winter, instrument failures and during the calibration periods. Under weak wind conditions the turbulence may not be well developed at night and thus CO_2 fluxes were parameterized in case the friction velocity u_* was below a threshold of 0.3 m s^{-1}. For this, parameterizations were developed for each year separately, with the CO_2 fluxes rated as ‘high-quality’ data (stationary, no rain, no fog, no dewfall).

Parameterizations for the respiratory fluxes were developed depending on air temperature T in 2 m (3 m) with the modified Arrhenius equation according to Lloyd and Taylor (1994):

$$R_{eco} = R_{10}\, e^{E_0\left(\frac{1}{283.15-T_0} - \frac{1}{T-T_0}\right)} \tag{1}$$

With R_{10} representing the respiration rate at 10 °C (fitted parameter), E_0=308.56 K, a parameter derived from the activation energy and T_0=227.13 K, the temperature at which the activation takes place.

Daytime NEE values (F_{Cday}) were fitted to a Michaelis-Menten-type equation (Schopfer and Brennicke 1999) for temperature classes of 4 K each, depending on global radiation (R_G):

$$F_{Cday} = \frac{a \cdot R_G \cdot F_{Csat}}{a \cdot R_G + F_{Csat}} - R_{day} \tag{2}$$

Three parameters were determined from the regression analysis: the initial slope of the saturation curve, a; the saturation value, F_{Csat}; and the daytime respiration value, R_{day}. CO_2 fluxes showed a clear dependency on wind directions for daytime as well as night-time conditions. The parameterizations were thus developed depending on the wind directions. Annual sums were then calculated over all wind directions, representing the carbon exchange of the ecosystem in the footprint of the tower.

As flux measurements were performed at two different heights (32 and 22 m) from 1997–1999, NEE was additionally determined for this period as a combination of the two measuring heights to account for the different footprints under daytime and night-time conditions.

8.3.2 Determination of Total Ecosystem Respiration

A general problem of using long-term eddy covariance data to determine the integrated turbulent flux of CO_2 or the NEE of CO_2 is the accuracy of determining night-time fluxes (Baldocchi et al. 2000). In this regard, automated chambers provide independent estimates of soil, wood and foliage respiration, and are viewed as important tools to determine fluxes missed by the above-canopy eddy covariance system. In forest ecosystems continuous below-canopy flux measurements via eddy covariance can be applied as an additional estimate of soil and wood respiration (Law et al. 1999), but for stands with high tree and foliage density the turbulence below the canopy might be insufficient to apply this technique (Subke 2002). Chamber measurements have been performed in the surroundings of the eddy covariance tower for soil and wood respiration. Foliage respiration was determined with the canopy model GAS-FLUX (Falge et al. 2003) for the year 1997 and also calculated with an Arrhenius-type equation (e.g. Lloyd and Taylor 1994) for the years 1997, 1998 and 1999.

The ecosystem respiration was determined from eddy covariance measurements as follows: regression analysis was performed with the night-time fluxes (Eq. 1) using only data with high quality (stationarity and $u_* > 0.3$ m s^{-1}). The developed parameters were applied during daytime and gaps that had to be filled at night (R_{eco} from night-time extrapolation). As potential underestimation by the eddy covariance methods at night and extrapolation from night-time data to estimate daytime respiration may not be appropriate, an additional upscaling of ecosystem respiration was determined from the intercept (R_{day}) of the light-response curve (Michaelis-Menten equation, Eq. 2).

8.3.3 Soil Respiration

The soil CO_2 efflux was measured approx. 100 m westward of the Weidenbrunnen site, in a 111-year-old Norway spruce stand (1999) adjacent to the tower site over the entire growing season in 1999. Measurements were continuous, using an open dynamic chamber design which aimed to minimize artefacts during the measuring process. An overview of the chamber system is given in Subke et al. (this Vol.); a detailed description can be found in Subke (2002). Alongside the soil CO_2 efflux, soil temperature and soil water content (SWC) of the organic layer were recorded, to determine the relation between the CO_2 efflux on these environmental factors. A comparison of forest stands in the Lehstenbach catchment has shown that the soil CO_2 efflux is quite similar despite considerable differences in stand age and structure of the understorey (Buchmann 2000). It can therefore be assumed, for a first approximation at least, that the fluxes measured at this site, and the annual soil C efflux calculated from these fluxes, are representative for the forest plots in the catchment.

8.3.4 Wood Respiration

The estimation of woody respiration at the Weidenbrunnen site is based on continuous chamber measurements of stem carbon exchange, made between June and October 1995 near the eddy covariance tower (Mirschkorsch 1996). The CO_2 efflux of stems and non-green parts of branches was measured at nine spruce trees using a differential open flow-through IRGA (Minicuvette system, Walz, Effeltrich, Germany) system. The objectives of this research were to describe respiration rates as exponential functions of sapwood temperature, and to investigate seasonal patterns that are not correlated with temperature. July had the most significant respiration, while temperature response was less sensitive than in autumn months (Q10 of 1.7 vs. 3.4). Respiration on a surface-area basis was much smaller in branches than in stems. Considerable variability in the respiration rates was observed between the stems of individual trees. Stem and branch respiration on a wood-area basis was scaled up using Arrhenius equations (e.g. Lloyd and Taylor 1994) derived separately for the average stem of the stand (based on diameter at breast height) and branch:

$$R = R_{25}\, e^{\left(\frac{E_a}{R_{gas}}\left(\frac{1}{T_{ref}} - \frac{1}{T_K}\right)\right)} \tag{3}$$

R_{25}, the ecosystem respiration rate at T_{ref} (here 298 K), and E_a, the activation energy (in J mol^{-1}) are fitted parameters, R_{gas} is the gas constant (8.134 J K^{-1} mol^{-1}) and T_K the sapwood temperature (in K) in 1 cm depth. The coefficients for branch respiration were E_a=42,670 J mol^{-1} and R_{25}=1.07 µmol m^{-2} s^{-1}, those for stem respiration E_a=41,900 J mol^{-1} and R_{25}=1.13 µmol m^{-2} s^{-1}. From harvests of five trees and area determination at seven branches per tree, the total woody surface area of the stand was calculated as 3.5 m^2 m^{-2}, corresponding to an LAI of 1.1 m^2 m^{-2} (Falge et al. 1997). From these results an estimation of stand woody respiration yielded 320 g C $m^{-2}year^{-1}$ for 1997 (Falge et al. 2003).

8.3.5 Foliage Respiration

Respiration of foliage elements was determined with the traditional homogeneous-layered canopy model GAS-FLUX as described by Falge et al. (2003) for the year 1997. Additionally, the foliage respiration was determined for 1997–1999 with an Arrhenius equation with respect to the canopy temperature. Daily sums of the two different methods were compared for 1997 and a regression of these yielded an r^2 of 0.99. From this finding it became reasonable to use the Arrhenius equation for the years 1998 and 1999, multiplied by the slope of the regression line, as a more simple procedure.

8.4 Results and Discussion

8.4.1 Net Ecosystem Exchange (NEE)

The yearly sum of NEE was very similar for the years 1997, 1998, 1999 and 2001 (–55, –41, –35 and –28 g C m^{-2}, respectively). The latter numbers are determined from eddy covariance measurements if available or parameterized as described above, with on average 40 % of the data available from measurements (night-time fluxes with low friction velocities were also rejected). The cumulative curves of NEE for the years 1997–2001 are plotted in Fig. 8.2a, b. Lines only denote periods with mainly modelled values, markers denote values derived from measurements directly. The annual sum of the carbon exchange is very similar for all years, even though the shapes are different. In 1997, a very dry year, the carbon uptake was reduced in the summer months caused by water limitations. In August the ecosystem became already a source for CO_2 (see also Table 8.1). The spring and autumn months had lower temperatures compared to the other years and resulted in lower respiration rates. The years 1998, 1999 and 2001 were more similar, as also the climatic conditions were more comparable. These years were almost normal years concerning air temperatures and precipitation (Table 8.1). The start of the growing season (average daily temperature above 5 °C) was around 22 April in the years 1997, 1998, 1999 and 2001. The year 2000 was an extraordinary year concerning the climate: the growing season started about 1 week earlier and lasted until 3 November. This year was the warmest of the last 150 years in Germany and in the region. The annual mean temperature was 1.8 K higher than the normal value for the climate period 1961–1990. Except for July, all months were warmer than normal, particularly the spring (Foken 2003). The radiation input was higher in the months April, May and June 2000, compared with other years; that, of course, would enhance the photosynthetic uptake. Temperatures in July–September were of average magnitude and did not

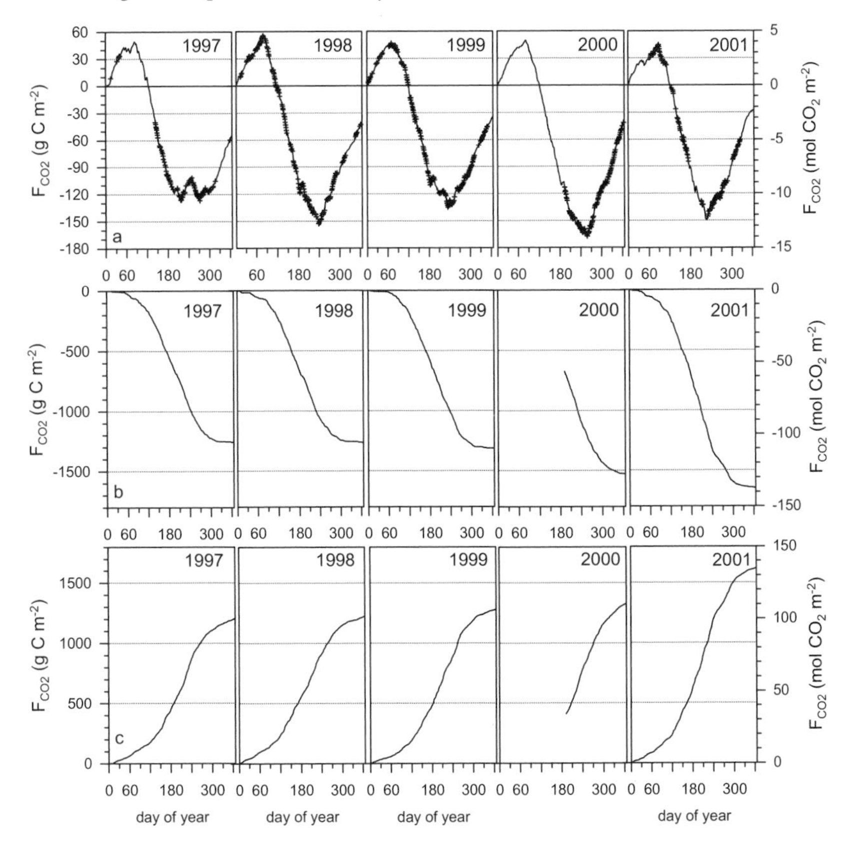

Fig. 8.2. a Net ecosystem exchange of CO_2 for 1997–2001; *lines with markers* denote predominantly measured data, *lines only* are predominantly modelled data. **b** Assimilation for 1997–2001 derived from NEE minus respiration. **c** Respiration for 1997–2001 derived from measured and modelled data

strengthen the respiration as, for example, in August 1997 or in autumn 1998, where high respiration rates compensated for higher uptake in spring and summer. As no data were available until July, an annual sum is difficult to determine. As a rough estimate to fill the missing data up to July, an average from the other years was used for the first part of the year and small gaps that occurred later in the year were parameterized. This would result in a very similar NEE of –40 g C m^{-2} for 2000. However, an earlier start of the growing season may have lead to an enhanced carbon uptake in spring compared to the average, which cannot be deduced due to the lack of available data. Using the fitted parameters from 2001 and the meteorological variables of 2000 for filling the data gaps in 2000 would result in a higher estimate of the carbon uptake (–115 g C m^{-2}). Similarly to our site in 2000, at the Norunda site in Sweden a higher NEE uptake is found at years with longer vegetation period and a high annual precipitation in the previous year (Lindroth et al. 2003).

Table 8.1. Monthly sums of net ecosystem exchange of carbon (*NEE*), soil CO_2 efflux (R_{soil}), latent heat flux (λE), global radiation (R_G), precipitation (*RR*) (for correction see Foken 2003) and mean air temperatures (T_{air}) in 30 m. *Values in parentheses* indicate more than 50 % of the data are modelled; NEE as described in text; λE with net radiation (related to Priestley-Taylor potential evaporation)

Month, year	NEE (g C m^{-2})	R_{soil} (g C m^{-2})	λE (MJ m^{-2})	R_G (MJ m^{-2})	RR (mm)	T_{air} (°C)
Jan 97	(26.1)	–	(17.5)	15.7	(32.7)	–4.0
Feb 97	(15.4)	–	(19.2)	88.9	(200.3)	0.4
March 97	(3.4)	–	(25.3)	188.8	(109.5)	2.9
April 97	(–34.8)	24.6	(80.9)	370.3	(66.6)	2.9
May 97	(–66.8)	53.8	(113.8)	566.1	(55.9)	10.0
June 97	–42.6	52.2	110.9	563.7	(70.7)	12.6
July 97	(–22.0)	75.4	113.2	461.3	(154.2)	14.0
Aug 97	14.2	74.0	132.7	550.9	(42.5)	17.4
Sept 97	–18.2	50.4	84.5	436.7	(29.4)	12.4
Oct 97	10.9	43.5	34.3	233.4	(70.6)	4.8
Nov 97	(28.5)	25.3	(7.6)	101.0	(31.8)	1.1
Dec 97	(31.2)	24.8	(6.5)	40.3	(101.9)	–1.3
Jan 98	(24.9)	26.5	(3.6)	80.9	(83.8)	–0.2
Feb 98	(15.0)	21.5	12.7	159.6	(23.4)	1.6
March 98	5.4	24.8	23.1	266.4	(119.2)	1.1
April 98	–47.1	35.1	67.1	366.5	(59.3)	6.3
May 98	–55.5	50.9	96.7	563.6	64.9	11.4
June 98	–55.4	77.4	147.4	540.0	117.1	14.2
July 98	–18.4	88.0	131.2	456.5	178.5	13.2
Aug 98	(–21.1)	93.3	165.0	532.9	91.3	15.1
Sept 98	29.3	69.8	57.6	250.7	197.6	10.1
Oct 98	(33.6)	48.6	20.1	89.2	(308.7)	5.4
Nov 98	(24.0)	24.9	(5.4)	49.4	(112.9)	–1.5
Dec 98	(24.2)	19.5	(9.5)	53.8	(67.1)	–2.2
Jan 99	(25.4)	19.8	(13.7)	59.8	(124.2)	–1.0
Feb 99	(17.8)	18.0	10.7	100.3	(136.8)	–3.6
March 99	–6.1	20.3	28.7	231.6	(93.6)	2.6
April 99	(–36.3)	33.1	70.3	366.1	(55.0)	5.7
May 99	–51.0	63.5	123.2	554.6	119.4	10.5
June 99	–55.5	72.3	131.9	519.5	108.6	12.1
July 99	(–14.0)	101.6	141.3	553.5	156.8	16.0
Aug 99	–9.5	88.9	136.5	481.2	114.8	14.5
Sept 99	16.7	80.4	98.6	385.8	105.5	14.7
Oct 99	23.6	48.3	24.6	186.7	(89.1)	5.7
Nov 99	(29.5)	27.0	(10.5)	76.5	(88.9)	0.0
Dec 99	(24.6)	21.9	2.9	42.3	(188.9)	–1.8

Month, year	NEE (g C m^{-2})	R_{soil} (g C m^{-2})	λE (MJ m^{-2})	R_G (MJ m^{-2})	RR (mm)	T_{air} (°C)
Jan 00	–	15.8	(4.0)	91.2	(110.6)	–2.9
Feb 00	–	15.8	(9.5)	121.6	(137.1)	0.4
March 00	–	22.7	(27.0)	208.9	(210.7)	1.7
April 00	–	39.0	(61.0)	453.3	(48.5)	7.8
May 00	–	49.4	(107.9)	622.7	46.5	12.7
June 00	–	72.6	(127.6)	681.5	105.7	15.0
July 00	(–37.7)	83.9	79.4	428.8	185.8	12.1
Aug 00	–15.9	87.4	121.2	522.3	53.6	16.3
Sept 00	10.5	73.7	52.6	321.9	65.8	11.2
Oct 00	38.6	58.9	21.7	162.5	(65.0)	8.0
Nov 00	29.5	32.9	9.3	99.5	(54.7)	3.0
Dec 00	42.4	27.2	6.0	72.9	(82.4)	–0.1
Jan 01	(23.0)	19.8	(2.6)	91.4	(91.3)	–2.8
Feb 01	(6.1)	16.9	(10.1)	164.1	(71.8)	–0.4
March 01	13.2	23.3	14.0	206.1	(196.1)	1.2
April 01	–39.7	–	37.7	343.2	(97.3)	3.9
May 01	(–51.1)	–	(98.8)	613.7	22.8	11.6
June 01	–43.4	–	81.7	486.3	107.1	11.4
July 01	(–36.8)	–	(108.9)	587.9	128.5	15.8
Aug 01	(–8.1)	–	(90.1)	511.4	27.8	16.6
Sept 01	21.9	–	29.6	194.5	171.4	8.4
Oct 01	(32.8)	–	34.6	213.5	53.3	10.6
Nov 01	32.4	–	11.0	92.6	(145.7)	0.6
Dec 01	(22.3)	–	(2.3)	59.0	(204.5)	–3.3
Jan 02	–	–	(2.0)	100.7	(109.5)	–1.8
Feb 02	–	–	(6.9)	112.2	(247.6)	1.6
March 02	–	–	(34.1)	279.2	(96.7)	2.6
April 02	–	–	54.0	381.5	(43.7)	4.8
May 02	–	–	111.8	471.3	(75.6)	11.2
June 02	–	–	162.9	619.7	(79.6)	15.0
July 02	–	–	162.8	506.9	(58.9)	15.4
Aug 02	–	–	(101.3)	413.9	(180.3)	16.2
Sept 02	–	–	104.2	327.8	(83.7)	10.1
Oct 02	–	–	70.7	153.3	(177.3)	5.2
Nov 02	–	–	(27.2)	80.7	(222.1)	2.5
Dec 02	–	–	(8.0)	60.5	(117.4)	–3.2

Figure 8.2 shows that NEE is the relatively small difference of two large flux terms (assimilation and respiration, Fig. 8.2b, c, respectively). The CO_2 assimilation is derived by subtracting the respiration from the NEE, whereas respiration is either measured or modelled values at night and modelled values by day. The cumulative respiration is similar for all years until the end of April, but with a steeper slope in 2001. High respiration rates in this year compensate the higher assimilation and result in the lowest net carbon uptake. The year 2002 seemed to be also a more productive year, but problems with a change in the employed gas analyser (open-path system in 2002 vs. closed path in 1997–2001) do not allow a reliable determination of the annual sum of the NEE (81 % of the half-hourly data had to be parameterized) as abundant fog conditions, dewfall and frequent precipitation caused problems in data quality with the open-path flux system.

Climatic variability between the years makes it difficult to distinguish clearly the effect of liming [with dolomite, $CaMg(CO_3)_2$] which took place in November 1999 in forest plots surrounding the tower site (but still within the footprint area), and in December 2001 at the tower site itself. Liming primarily increases respiratory activities, but also enhances photosynthetic CO_2 uptake by trees and understorey due to better nutrient availability within the soil.

As was pointed out before, both instrumental problems and unusual climatic conditions coincide with the exceptional years, so that a distinction between these factors, or in fact synergistic effects of both climate and soil chemistry, is impossible. Anyhow, the saturation values derived from the light dependencies of the CO_2 fluxes are highest in 2000, slightly lower in 2001 and generally lower in the years before the liming. An overview of all data is given in Table 8.1.

NEE values derived as a combination of the two measurement heights in the years 1997–1999 are slightly different from those derived only from measurements in 32 m (–116 g C m^{-2} in 1997, –111 g C m^{-2} in 1998 and –138 g C m^{-2} in 1999). The small differences in the annual NEE are potentially caused by respiration differences in the different night-time footprints.

8.4.2 Respiration

The soil CO_2 efflux was found to have a strong temperature dependence, best described by a well-established model based on the Arrhenius equation (Lloyd and Taylor 1994). Dependence of the CO_2 efflux on soil water content was found only for extremely low values. A combined formulation of the Lloyd and Taylor (1994) temperature dependence and the soil moisture limitation proposed by Bunnell et al. (1977) gave the best fit for the measured data (Subke et al., this Vol.). Using available data of soil temperature and soil water content, it was possible to calculate the soil CO_2 efflux for 1 April 1997 to 31 March 2001 on a daily basis. Annual sums vary between approximately 500 and 590 g C m^{-2} $year^{-1}$ (Subke et al., this Vol.), and monthly efflux sums show considerable variability with the seasonal trends of the years (Table 8.1, Fig. 8.3), owing mainly to variations in rainfall.

From the measurements in 1995 it became obvious that July had the most significant wood respiration, while temperature response was less sensitive than in autumn months (Q10 of 1.7 vs. 3.4). Wood respiration on a surface-area basis was much smaller in branches than in stems. Considerable variability in the respiration rates was observed between the stems of individual trees. From the developed Arrhenius equation (Eq. 3) and for the year 1997 an upscaling of stand woody respiration yielded 320 g C m^{-2} $year^{-1}$ for 1997 (Falge et al. 2003), and 322 and 325 g C m^{-2} $year^{-1}$ for 1998 and 1999, respectively. As no differentiation between wood growth and maintenance respiration was performed, and the bulk growth of the trees might not yet have been completed in June when the measurement period started, the above parameters represent a lumped respiration, and the annual sum might be overestimated. Separate analysis of the June and July–October data could yield parameters for a two-component functional model, in which the respiration is partitioned into growth and maintenance respiration (e.g. Amthor 2000; Vose and Ryan 2002).

Foliage respiration determined from model results (GAS-FLUX) for 1997 yielded an annual sum of 393 g C m^{-2} $year^{-1}$. Upscaling with the Michaelis-

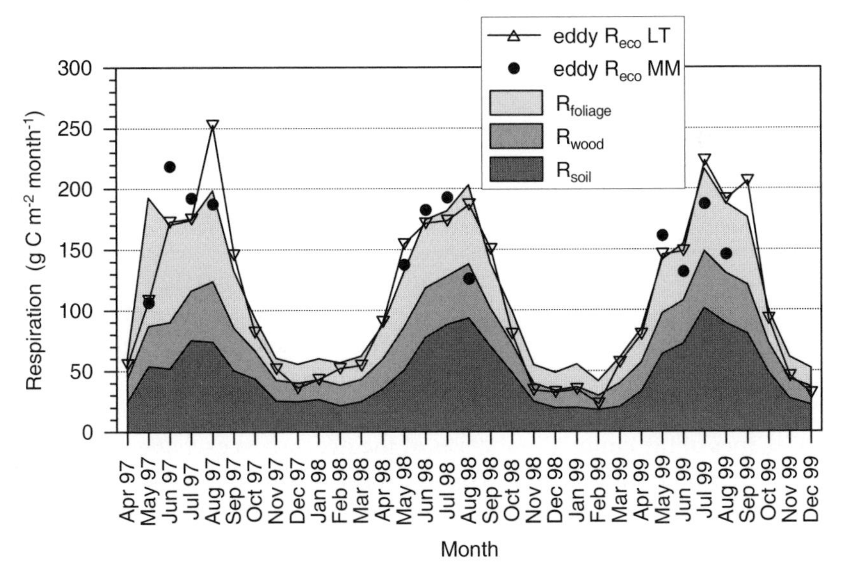

Fig. 8.3. Respiration rates estimated using soil chambers (Falge et al. 2003), wood chambers and foliage gas exchange modelling (Falge et al. 2003); eddy covariance measurements upscaled using night-time data (R_{eco} LT, *line with triangles*); eddy covariance measurements upscaled using intercept of light-response curve (R_{eco} MM, *dots*)

Menten equation for the years 1997 and 1998 yielded annual sums of foliage respiration of 398 and 412 g C m^{-2} $year^{-1}$. It becomes obvious that foliage respiration contributes with about 30 % to the overall ecosystem respiration.

In Fig. 8.3, all upscaled respiration components from soil, wood and foliage are plotted together with an estimate of whole ecosystem respiration from the eddy covariance data. If the extrapolation from night-time respiration data is used, a regression between the sum of the chamber measurements in combination with modelling results and the eddy covariance data yields an r^2 of 0.93 and a slope of the regression line of 0.99. Late spring, summer and early autumn months show a good agreement of the monthly sums. Only in August 1997 is the difference large. In this month a source of CO_2 was detected from the eddy covariance measurements. Monthly sums determined from the intercept (R_{day}) of the light response curve of assimilation show a stronger scatter, but also tend to be in the range of the different methods used for the determination of total ecosystem respiration.

8.5 Conclusions

The NEE determined with quality-controlled eddy covariance measurements yields useful data for ecosystem studies. The quality checks and gap filling procedures were important for the Weidenbrunnen site. Furthermore, a site-specific footprint analysis was helpful for data interpretation. Nevertheless, an acceptable number of high-quality data was shown to be necessary to calculate annual NEE sums, which was not possible in 2000 and 2002 due to missing data. The benefit of the eddy covariance technique is the high accuracy (approximately 10 % for periods with data gaps) in determining the very important NEE, as compared with other methods which allow us to only determine this parameter as a residual of the balance equation of other very large parameters.

In contrast to central European spruce sites located at low elevations (e.g. Bernhofer et al. 2003), the Weidenbrunnen site formed only a small sink of carbon in most years, as does the nearby Wetzstein site in Thuringia (Anthoni et al., 2004). Yearly sums of NEE were very similar for the years 1997, 1998, 1999 and 2001 (–55, –41, –35 and –28 g C m^{-2} $year^{-1}$, respectively). Despite different shapes of the cumulative curves of the net carbon exchange, most years had a very similar annual sum. The apparent high NEE uptake in 2000 was most likely caused by the high radiation input in spring, the high temperatures (Bayreuth: +1.8 K above the climatological normal 1961–1990) and less likely attributable to a liming procedure.

A comparison of respiration estimates from eddy covariance measurements and the upscaled chamber measurements (soil, wood and foliage) showed that for 1998 and 1999 the sum of the three components (1,300 and

1,333 g C m^{-2} $year^{-1}$) exceeded the estimates for ecosystem respiration by the night-time flux extrapolation method (1,224 and 1,279 g C m^{-2} $year^{-1}$) by only about 5 %. Assuming that soil CO_2 efflux sums for January to March 1997 (for which no estimates are available, see Table 8.1) are similar to those of the following years (about 60 g C m^{-2}), the annual CO_2 soil efflux in 1997 can be estimated at around 470 g C m^{-2}. The sum of soil, wood and foliage respiration for that year would thus be 1,183 g C m^{-2} $year^{-1}$, which is also in good agreement with the estimate based on the night-time flux extrapolation method (1,202 g C m^{-2} $year^{-1}$). This good agreement between chamber and eddy covariance estimates of whole ecosystem respiration provides confidence for employing the eddy covariance method at this site.

Acknowledgements. We are grateful to all who supported the field measurements. The project is funded by the European EUROFLUX, CARBOEUROFLUX and CARBODATA projects and by the Federal Ministry of Education, Science, Research and Technology (PT BEO-0339476 B, C and D).

References

Amthor JS (2000) The McCree–de Wit–Penning de Vries–Thornley respiration paradigms: 30 years later. Ann Bot 86:1–20

Anthoni PM, Knohl A, Rebmann C, Freibauer A, Mund M, Ziegler W, Kolle O, Schulze E-D (2004) Forest and agricultural land use dependent CO_2 exchange in Thuringia, Germany. Global Change Biol, accepted

Aubinet M, Grelle A, Ibrom A, Rannik Ü, Moncrieff J, Foken T, Kowalski AS, Martin PH, Berbigier P, Bernhofer Ch, Clement R, Elbers J, Granier A, Grünwald T, Morgenstern K, Pilegaard K, Rebmann C, Snijders W, Valentini R, Vesala T (2000) Estimates of the annual net carbon and water exchange of forests: the EUROFLUX methodology. Adv Ecol Res 30:113–175

Aubinet M, Clement R, Elbers J, Foken T, Grelle A, Ibrom A, Moncrieff H, Pilegaard K, Rannik U, Rebmann C (2003) Methodology for data acquisition, storage and treatment. In: Valentini R (ed) Fluxes of carbon, water and energy of European forests. Ecological studies 163. Springer, Berlin Heidelberg New York, pp 9–35

Baldocchi DD, Finnigan J, Wilson K, Paw UKT, Falge E (2000) On measuring net ecosystem carbon exchange over tall vegetation on complex terrain. Boundary-Layer Meteorol 96:257–291

Baldocchi D, Falge EH, Gu L, Olson R, Hollinger D, Running S, Anthoni P, Bernhofer C, Davis K, Evans R, Fuentes J, Goldstein A, Katul G, Law B, Lee XH, Malhi Y, Meyers T, Munger W, Oechel W, Paw UKT, Pilegaard K, Schmid HP, Valentini R, Verma S, Vesala T (2001) FLUXNET: a new tool to study the temporal and spatial variability of ecosystem-scale carbon dioxide, water vapor, and energy flux densities. Bull Am Meteorol Soc 82:2415–2434

Bernhofer C, Aubinet M, Clement R, Grelle A, Grünwald T, Ibrom A, Jarvis P, Rebmann C, Schulze E-D, Tenhunen JD (2003) Spruce forests (Norway and Sitka spruce, including Douglas fir): carbon and water fluxes and balances, ecological and ecophysiological determinants. In: Valentini R (ed) Fluxes of carbon, water and energy of European forests. Ecological studies 163. Springer, Berlin Heidelberg New York, pp 99–123

Buchmann N (2000) Biotic and abiotic factors controlling soil respiration rates in Picea abies stands. Soil Biol Biochem 32:1625–1635

Bunnell FL, Tait DEN, Flangan PW, Cleve KV (1977) Microbial respiration and substrate weight loss. I. A general model of the influences of abiotic variables. Soil Biol Biochem 9:33–40

Eugster W, Senn W (1995) A cospectral correction for measurement of turbulent NO_2 flux. Boundary-Layer Meteorol 74:321–340

Falge EM, Ryel RJ, Alsheimer M, Tenhunen JD (1997) Effects on stand structure and physiology on forest gas exchange: a simulation study for Norway spruce. Trees 11:436–448

Falge E, Baldocchi D, Olson R, Anthoni P, Aubinet M, Bernhofer C, Burba G, Ceulemans R, Clement R, Dolman H, Granier A, Gross P, Grunwald T, Hollinger D, Jensen NO, Katul G, Keronen P, Kowalski A, Lai CT, Law BE, Meyers T, Moncrieff H, Moors E, Munger JW, Pilegaard K, Rannik U, Rebmann C, Suyker A, Tenhunen J, Tu K, Verma S, Vesala T, Wilson K, Wofsy S (2001a) Gap filling strategies for defensible annual sums of net ecosystem exchange. Agric For Meteorol 107:43–69

Falge E, Baldocchi D, Olson R, Anthoni P, Aubinet M, Bernhofer C, Burba G, Ceulemans R, Clement R, Dolman H, Granier A, Gross P, Grunwald T, Hollinger D, Jensen NO, Katul G, Keronen P, Kowalski A, Lai CT, Law BE, Meyers T, Moncrieff H, Moors E, Munger JW, Pilegaard K, Rannik U, Rebmann C, Suyker A, Tenhunen J, Tu K, Verma S, Vesala T, Wilson K, Wofsy S (2001b) Gap filling strategies for long term energy flux data sets. Agric For Meteorol 107:71–77

Falge E, Tenhunen J, Aubinet M, Bernhofer C, Clement R, Granier A, Kowalski A, Moors E, Pilegaard K, Rannik Ü, Rebmann C (2003) A model-based study of carbon fluxes at ten European forest sites. In: Valentini R (ed) Fluxes of carbon, water and energy of European forests. Ecological studies 163. Springer, Berlin Heidelberg New York, pp 151–177

Foken T (2003) Lufthygienisch-Bioklimatische Kennzeichnung des oberen Egertales. Bayreuther Forum Ökol 100:69 + XLVIII S

Foken T, Wichura B (1996) Tools for quality assessment of surface-based flux measurements. Agric For Meteorol 78:83–105

Foken T, Mangold A, Rebmann C, Wichura B (2000) Characterization of a complex measuring site for flux measurements. In: Proc 14th Symp on Boundary Layer and Turbulence, Aspen, Colorado, 7–11 Aug, American Meteorological Society, pp 388–389

Foken T, Göckede M, Mauder M, Mahrt L, Amiro BD, Munger JW (2004) Post-field data quality control. In: Lee X, Massman WJ, Law B (eds) Handbook of micrometeorology: a guide for surface flux measurement and analysis. Kluwer, Dordrecht, pp 81–108

Göckede M, Rebmann C, Foken T (2004) A combination of quality assessment tools for eddy covariance measurements with footprint modelling for the characterisation of complex sites. Agric For Meteorol (in press)

IPCC (2001) Climate change 2001, the scientific basis. Cambridge University Press, Cambridge, 881 pp

Law BE, Ryan MG, Anthoni PM (1999) Seasonal and annual respiration of a ponderosa pine ecosystem. Global Change Biol 5:169–182

Lindroth A, Grelle A, Mölder M, Linderson M-L, Lankreijer H, Lagergren F (2003) Boreal forest evaporation: dependency on season, weather, soil moisture and stand age. Geophys Res Abstr 5:09215 (CD)

Lloyd J, Taylor JA (1994) On the temperature dependence of soil respiration. Funct Ecol 8:315–323

Mirschkorsch C (1996) Die Stamm- und Zweigrespiration eines jungen Fichtenbestandes (Picea abies (L.) Karst.) und die Bedeutung für den CO_2-Netto-Austausch. Diploma Thesis, University of Bayreuth

Moncrieff JB, Massheder JM, DeBruin H, Elbers J, Friborg T, Heusinkveld B, Kabat P, Scott S, Søgaard H, Verhoef A (1997) A system to measure surface fluxes of momentum, sensible heat, water vapor and carbon dioxide. J Hydrol 188/189:589–611

Schmid HP (1997) Experimental design for flux measurements: matching scales of observations and fluxes. Agric For Meteorol 87:179–200

Schopfer P, Brennicke A (1999) Pflanzenphysiologie. Springer, Berlin Heidelberg New York

Subke J-A (2002) Forest floor CO_2 fluxes in temperate forest ecosystems. Bayreuther Forum Ökol 96:119

Valentini R (2003) Fluxes of carbon, water and energy of European forests. Ecological studies 163. Springer, Berlin Heidelberg New York

Valentini R, Matteucci G, Dolman AJ, Schulze E-D, Rebmann C, Moors EJ, Granier A, Gross P, Jensen NO, Pilegaard K, Lindroth A, Grelle A, Bernhofer C, Grünwald T, Aubinet M, Ceulemans R, Kowalski AS, Vesala T, Rannik Ü, Bergigier P, Loustau D, Guomundsson J, Thorgeirsson H, Ibrom A, Morgenstern K, Clement R, Moncrieff J, Montagnani L, Minerbi S, Jarvis PG (2000) Respiration as the main determinant of carbon balance in European forests. Nature 404:861–865

Vose JM, Ryan MG (2002) Seasonal respiration of foliage, fine roots, and woody tissues in relation to growth, tissue N, and photosynthesis. Global Change Biol 8:182–193

9 Structure of Carbon Dioxide Exchange Processes Above a Spruce Forest

B. Wichura, J. Ruppert, A.C. Delany, N. Buchmann, and T. Foken

9.1 Introduction and ^{13}C Signatures

Several micrometeorological techniques, such as the flux-gradient method or the eddy covariance technique, offer the potential to measure net fluxes of water vapor, CO_2 and other trace gases exchanged between ecosystems and the atmosphere (e.g., Baldocchi and Meyers 1998). Subsequent data analyses allow the calculation of net ecosystem CO_2 exchange. These net fluxes, however, reflect the balance between different component fluxes. In the case of CO_2, two opposing fluxes contribute to this net flux: CO_2 uptake during photosynthesis and CO_2 release during respiration from above- and belowground organisms. Distinguishing among these components is critical to obtain insights into the processes underlying ecosystem responses to climate forcing (Buchmann 2002). This is because environmental parameters, such as temperature and soil moisture, differentially affect biological activities (e.g., Baldocchi et al. 2001).

Using stable carbon isotopes at the ecosystem scale can help to decipher the ecosystem signal in NEE (Keeling 1958; Buchmann et al. 1998). The CO_2 exchange of the biosphere with the atmosphere and its isotopic signature is

---

Parts of this paper have already been presented at Symposia of the American Meteorological Society:

Wichura B, Buchmann N, Foken T (2000) Fluxes of the stable carbon isotope ^{13}C above a spruce forest measured by hyperbolic relaxed eddy accumulation method. In: Proc 14th Symp on Boundary Layer and Turbulence, Aspen, Colorado., 7–11 Aug, American Meteorological Society, pp 559–562

Wichura B, Buchmann N, Foken T (2002) Carbon dioxide exchange characteristics above a spruce forest. In: Proc 25th Symp, Agricultural and Forest Meteorology, Norfolk, 20–24 May, American Meteorological Society, pp 63–64

Ruppert J, Wichura B, Delany AC, Foken T (2002) Eddy sampling methods, a comparison using simulation results. In: Proc 15th Symp on Boundary Layer and Turbulence, Wageningen, 15–19 July, American Meteorological Society, pp 27–30

Ecological Studies, Vol. 172
E. Matzner (Ed.), Biogeochemistry of Forested Catchments in a Changing Environment

influenced not only by the interactions of the turbulence regime but also by ecosystem physiology. While the turbulence regime will influence the mixing of CO_2 with different isotopic compositions between the biosphere and the atmosphere, ecosystem physiology will affect the carbon isotopic signatures ($\delta^{13}C$) of the biospheric flux and the magnitude of this flux. The main ecosystem processes that alter the isotopic signature of canopy CO_2 are assimilation as well as autotrophic and heterotrophic respiration, each carrying very specific ^{13}C signals integrated over different time spans.

Foliar $\delta^{13}C$ ratios are influenced by two key factors: (1) by the $\delta^{13}C$ of canopy CO_2, the source air for photosynthetic assimilation, and (2) by carbon discrimination (Δ_{leaf}) during photosynthesis. Large fractionation against ^{13}C is usually observed in C_3 plants during CO_2 fixation (O'Leary 1988), thus determining the isotopic imprint during biological CO_2 exchange. The plant foliage, stems and roots are subsequently input pools for the large pool of soil organic matter (SOM). The $\delta^{13}C$ values of SOM reflect these varying contributions, but are also altered during decomposition processes.

Since it is nearly impossible to determine the physiological contribution of each component of the ecosystem in the field, only eddy covariance measurements of carbon dioxide with simultaneous isotope analyses of those fluxes can solve the problem and offer the potential to obtain insights into processes within the ecosystem (Yakir and Wang 1996).

In order to measure ^{13}C fluxes, one needs to employ measurement methods sensitive enough to identify small fluxes. Because ^{13}C concentrations cannot be measured with a high sampling rate, which is necessary for eddy covariance measurements and can be used to measure carbon dioxide fluxes (Aubinet et al. 2000), only accumulation methods (Businger and Oncley 1990) can be applied. These methods allow samples to be over 30–60 min for ^{13}C concentration measurements. To increase the accuracy of the measurements, a hyperbolic relaxed accumulation method was proposed by Bowling et al. (1999). Based on these findings, one measurement technique was developed.

To determine the carbon dioxide and ^{13}C balance as well as the exact measurement of the fluxes, a well-mixed turbulent flow within and above the canopy is necessary. However, forests have a very complicated turbulence structure with counter gradients, decoupling, coherent structures and a mixing layer above the canopy. All of these factors can affect the fluxes and calculations of the balances. Therefore, a special focus of this investigation was the classification of the fluxes according to the mixing layer assumption by Raupach et al. (1996).

9.2 Balances of ^{13}C and CO_2

We assume a column of air extending from the soil surface to some height above the forest canopy. If there is no horizontal advection of CO_2, the column CO_2 mass balance is (cf. Lloyd et al. 1996)

$$M_i \frac{dC_i}{dt} = F_{oi} - F_{io} + R - A \tag{1}$$

where M_i is the number of moles of air in the column per unit ground area, C_i is the average mole fraction of CO_2 within the column, F_{oi} and F_{io} are the (one-way) fluxes of CO_2, per unit ground area, into and out of the column from and to the atmosphere above, respectively, R is the respiration rate and A the net rate of assimilation within the column per unit ground area. The $^{13}CO_2$ isotope mass balance for the column is

$$M_i \frac{d\mathfrak{R}_i C_i}{dt} = \mathfrak{R}_{oi} F_{oi} - \mathfrak{R}_{io} F_{io} + \mathfrak{R}_R R - \mathfrak{R}_A A \tag{2}$$

where $\mathfrak{R}_i$ is the average isotope ratio $^{13}CO_2/^{12}CO_2$ of CO_2 within the column, $\mathfrak{R}_{oi}$ and $\mathfrak{R}_{io}$ are the isotope ratios of CO_2, per unit ground area, entering and leaving the column from and to the atmosphere above, respectively, $\mathfrak{R}_R$ and $\mathfrak{R}_A$ are the isotope ratios $^{13}CO_2/^{12}CO_2$ of respired and assimilated CO_2, respectively, within the column per unit ground area. After replacement and rearrangement of Eqs. (1) and (2) (introducing δ notation for the $^{13}C/^{12}C$ isotope ratio: $\delta^{13}C=(\mathfrak{R}_{sample}/\mathfrak{R}_{standard}-1) \times 1000‰$, where $\mathfrak{R}_{sample}$ and $\mathfrak{R}_{standard}$ are the isotope ratios $^{13}C/^{12}C$ of the sample and of a standard (PDB for carbon), respectively), the isotope balance can be written as

$$M_i C_i \frac{d\delta_i}{dt} = F_{oi}\left(\delta_{oi} - \delta_i\right) - F_{io}\left(\delta_{io} - \delta_i\right) + R\left(\delta_R - \delta_i\right) - A\left(\delta_A - \delta_i\right), \tag{3}$$

where $\delta_{oi}=\delta_o$ and $\delta_{io}=\delta_i$ (i.e., the second term on the right-hand side of Eq. (3) vanishes) are the isotope ratios of CO_2, entering and leaving the column from and to the atmosphere above, respectively, assuming a well-mixed canopy layer. δ_R and δ_A are the isotope ratios of respired and assimilated CO_2 within the column per unit ground area. If the changes in the CO_2 mole fraction and in the $^{13}CO_2/^{12}CO_2$ isotope ratio are small compared to the fluxes of CO_2 and $^{13}CO_2$ (steady state), the left-hand sides of Eqs. (1) and (3) vanish and, after rearrangement, separate equations for R and A can be formulated:

$$R = \left(F_{oi} - F_{io}\right)\frac{\delta_A}{\delta_R - \delta_A} - \left(F_{oi}\delta_o - F_{io}\delta_i\right)\frac{1}{\delta_R - \delta_A}$$

$$A = \left(F_{oi} - F_{io}\right)\frac{\delta_R}{\delta_R - \delta_A} - \left(F_{oi}\delta_o - F_{io}\delta_i\right)\frac{1}{\delta_R - \delta_A}. \quad (4)$$

The first bracket terms of R and A in Eq. (4) represent the flux of CO_2 and can be measured by an eddy covariance system. The second bracket terms in Eq. (4) represent the flux of $^{13}CO_2$ and can be measured by a hyperbolic relaxed eddy accumulation system.

9.3 Hyperbolic Relaxed Eddy Accumulation (HREA) as a Relevant Measuring Method

The hyperbolic relaxed eddy accumulation method (HREA) is a modification of the relaxed eddy accumulation method (REA). In the REA application, sample air is accumulated at a constant flow rate to separate reservoirs, depending on the direction of the vertical wind velocity w. The up- and down-draft reservoirs of accumulated air are analyzed after the sampling interval for the mean concentrations $\overline{c^+}$ and $\overline{c^-}$ of the scalar c, respectively. The turbulent flux F_c is proportional to the concentration difference Δc_w (Businger and Oncley 1990):

$$F_c = \overline{w'c'} = b(\zeta)\sigma_w\left(\overline{c^+} - \overline{c^-}\right) = b(\zeta)\sigma_w \Delta c_w, \quad (5)$$

where σ_w is the standard deviation of w, and $b(\zeta)$ is a coefficient, which is a function of stability, characterized by $\zeta = z/L$, where z is the height above the surface and L is the Obukhov length. Often, a threshold value for w (a dead-band, centered at $w=0$, where no air sampling takes place) is used to increase Δc_w (Businger and Oncley 1990; Pattey et al. 1993). In HREA, the concept of a threshold value is continued in order to measure isotope fluxes by an REA system. The threshold for HREA sampling is defined by

$$H = \left|\frac{w'}{\sigma_w}\frac{c'}{\sigma_c}\right|, \quad (6)$$

where σ_c is the standard deviation of c, and w' and c' are the turbulent fluctuations of w and c, respectively. Sampling takes place in cases of $|(w'/\sigma_w)(c'/\sigma_c)| > H$; $w' < 0$ and $w' > 0$ are used to decide for downdraft and updraft sampling, respectively. Thus, only those turbulent exchange events that contribute strongly to the flux of *c* are collected. Additional information about the scalar c (c' and σ_c) or a similar scalar (i.e., similar probability density distribution of the scalars, cf. Bowling et al. 1999) are needed for HREA application. Fast response sensors for CO_2 are available in the case of HREA measurements for $^{13}CO_2$; thus the HREA method can be applied consistently.

9.3.1 Measuring System

The measurements were carried out at the 'Waldstein/Weidenbrunnen' site of the University of Bayreuth in the hilly region of the Fichtelgebirge, 50°09′N, 11°52′E, 775 m a.s.l. at a height of 30 m a.g.l., i.e., 11 m above a 19-m-high spruce forest. This site was the EUROFLUX site GE1 (Valentini et al. 2000) and has been systematically investigated in respect to data quality of turbulent fluxes (Rebmann et al., this Vol.).

Based on the experiences of first HREA $^{13}CO_2$ flux measurements (Bowling et al. 1999), the design of our HREA measurement system was quite similar (Fig. 9.1). Figure 9.2 shows details of the valve control box of the HREA system. We added an additional third line in both sampling paths (up- and downdraft) to solve the problem of simultaneous mixing ratio measurements for CO_2, a precondition to calculate a consistent $^{13}CO_2$ flux (Bowling et al. 1999). An additional infrared gas analyzer (IRGA) (Li-Cor 6262, LI-COR Inc.) in differential mode was used to measure the CO_2 mixing ratio of either the updraft or the downdraft air. A threshold value of H=1.1 was chosen, in accordance with earlier measurements of $^{13}CO_2$ fluxes (Bowling et al. 1999) and with HREA simulation runs that were performed using eddy covariance measurements available for the site from previous field campaigns.

The eddy covariance (EC) measurement system consisted of a sonic Solent (Gill Instruments, 1012 R2) and an IRGA Li-Cor 6262, measuring in absolute mode. The system was operated as one of the EUROFLUX network stations and a full description of the layout and operation mode of the system can be

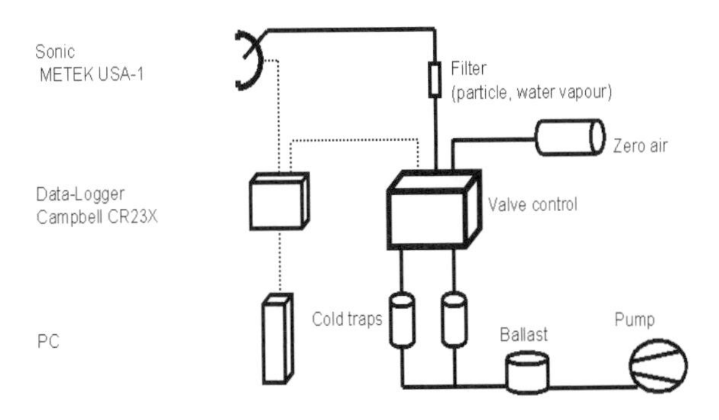

Fig. 9.1. Schematic diagram of the HREA system layout

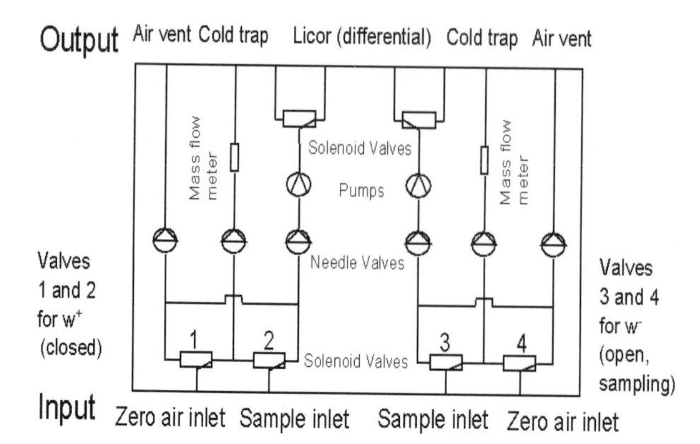

Fig. 9.2. Schematic diagram of the HREA valve control box

Table 9.1. Instrumentation for CO_2 and ^{13}C flux measurements

Method Year	CO_2 flux EC 1999	CO_2 flux EC 2000	^{13}C flux HREA 1999 and 2000
Wind components and sonic temperature	Solent 1012 R2	METEK USA-1	METEK USA-1
CO_2 concentration	Li-Cor 6262	Li-Cor 7500	Li-Cor 6262 (control)
$^{13}C/^{12}C$ isotope ratio			Mass spectrometer

found elsewhere (Aubinet et al. 2000). The system was changed in 2000 by using an open-path gas analyser (Li-Cor 7500; Table 9.1).

Profile measurements of wind velocity, temperature and humidity, measurements of the radiation budget, the soil heat flux and the soil temperature were available at the site in addition to HREA and EC measurements.

9.3.2 Methodological Problems

Methods based on relaxed eddy accumulation (REA) are indirect methods for flux measurements, because they rely on a parameterization in which the coefficient b in Eq. (5) is determined. The coefficient b is well defined with a value of 0.627 for an ideal Gaussian joint frequency distribution (JFD) of the vertical wind velocity and a scalar (Wyngaard and Moeng 1992). However, measured JFDs tend to show a high degree of skewness, so that b must be determined from measurements of a second scalar with high temporal resolution (e.g., 10 Hz). Similar turbulent exchange characteristics (scalar similarity) for the scalar of interest and the second scalar must be assumed. Simulations of REA with a data set from the EBEX 2000 experiment (San Joaquin Valley, California, USA, 20 August 2000, Oncley et al. 2002) showed that errors related to the methodological assumptions are relatively small. Fixed *b* coefficients produced larger uncertainty in flux results of REA simulations than *b* coefficients derived from a second scalar for each measuring period individually (Ruppert et al. 2002).

The validity of scalar similarity is even more critical for HREA. It must also be assumed for the second scalar when determining the hyperbolic deadband from online measurements with high temporal resolution. This might introduce an additional source of error, highly dependent on scalar similarity. HREA simulations showed relative flux errors up to 26% underestimation due to poor scalar similarity for water vapor and CO_2 before noon and sonic temperature and CO_2 after noon. The observed errors increased with the size of the hyperbolic deadband H (Fig. 9.3). The large errors originate from incorrect weighting of samples close to the borders of the deadband compared to the weighting in eddy covariance. These errors cancel out, as long as the JFDs of vertical wind velocity and the scalars match, which means a high degree of scalar–scalar cross correlation. *b* coefficients are determined incorrectly as soon as the shapes of the JFDs differ. In addition to this source of error, large hyperbolic deadbands reduce the number of samples and thereby the statistical representativeness of the measurement becomes crucial. A theoretic optimum hyperbolic deadband size of H=0.8 was found only after excluding periods with poor scalar similarity.

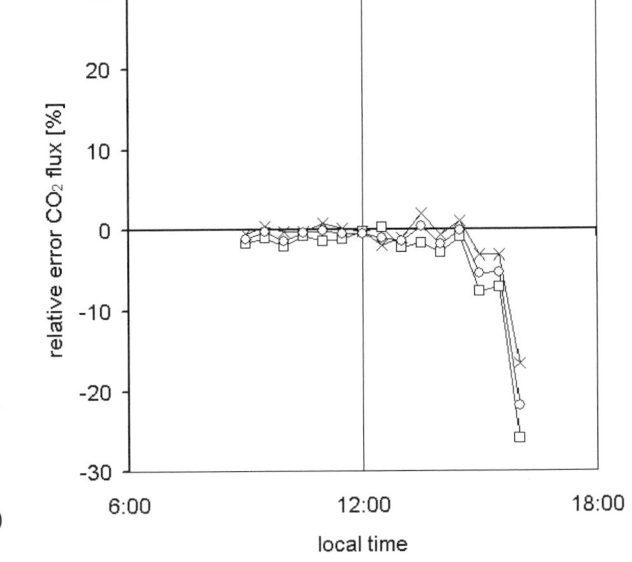

Fig. 9.3. Relative CO_2 flux errors observed in HREA simulations Using *b* coefficients calculated from buoyancy flux. Hyperbolic deadband sizes are H=0.5 (*crosses*), H=0.8 (*circles*) and H=1.1 (*squares*)

The use of HREA is required us to increase the concentration differences Δc_w, e.g., for the detection of differences in $\delta^{13}C$ with enough accuracy (Bowling et al. 1999). Limitation in sensor resolution may require hyperbolic deadbands even larger than H=0.8. Therefore, HREA must be applied with great care to avoid systematic errors, particularly in regard to the underlying assumption of scalar similarity under changing meteorological conditions. The relationship between ^{13}C isotope ratios and CO_2 concentrations found in Keeling plots (e.g., Buchmann et al. 1998) suggests that CO_2 may show the highest degree of scalar similarity to ^{13}C isotopes. However, this assumption can only be tested by simultaneous fast measurements of ^{13}C and CO_2 with high accuracy. A perfect similarity of both scalars would mean that no additional information is present in the isotopic signature so that $^{13}CO_2$ fluxes could be directly deducted from CO_2 fluxes.

A complete analysis of the advantages of HREA over REA for measuring certain species fluxes will therefore need a careful assessment of the error introduced by measuring very close to the mass spectrometer detection limits. Simulation results for the concentration difference might help to estimate errors related to sensor accuracy.

9.4 Results

9.4.1 Carbon Flux Measurements

HREA measurements were carried out in August 1999 and in June 2000. We present here a case study for 19 August 1999, to point out essential results. The daily courses of mean meteorological parameters, in particular for water vapor pressure, indicate the decoupling of canopy air from the atmosphere until a strong increase of down-welling radiation at about 10 a.m. (Fig. 9.4). A wavelet transform was used to account for the development of coherent structures in turbulent data time series (Heinz et al. 1999). A formation of coherent structures at about 10:20 a.m. was found (Fig. 9.5), coinciding with the time when the canopy air is well mixed with air from the atmosphere above.

The CO_2 flux, measured by eddy covariance technique, responds to the variations in photosynthetically active radiation (PAR; see Fig. 9.4), i.e., it shows a strong increase after 10 a.m. and a strong decrease at about 12:30 a.m. (Fig. 9.6). HREA CO_2 up- and downdraft results reproduce the trend for the CO_2 flux (cf. Fig. 9.6) and show differences large enough to resolve measurable differences in $\delta^{13}C$ (Bowling et al. 1999). The statement is valid, even though we used CO_2 differences calculated by HREA algorithm from the EC measurements, because of problems with the differential operation of the HREA Li-Cor.

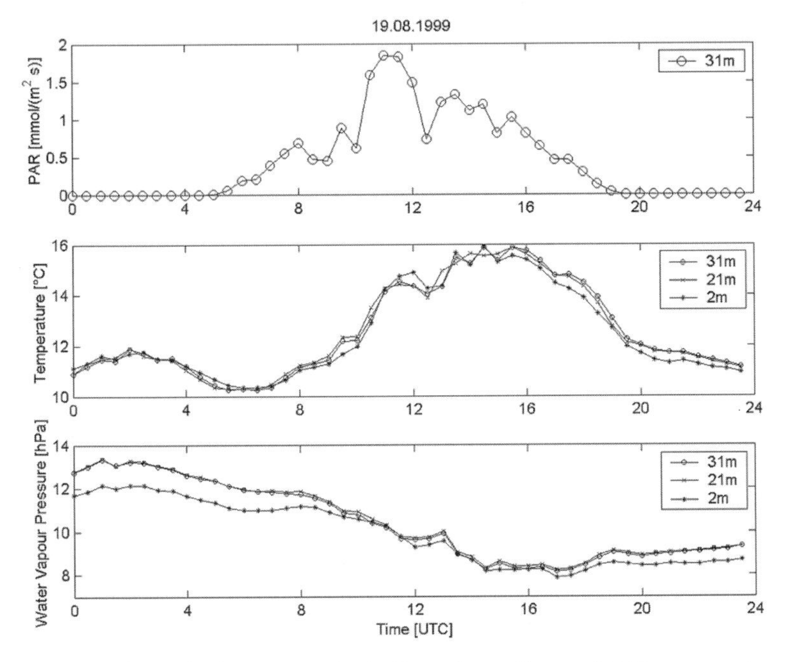

Fig. 9.4. Diurnal trend of photosynthetically active radiation (*PAR*), temperature and water vapor pressure on 19 August 1999

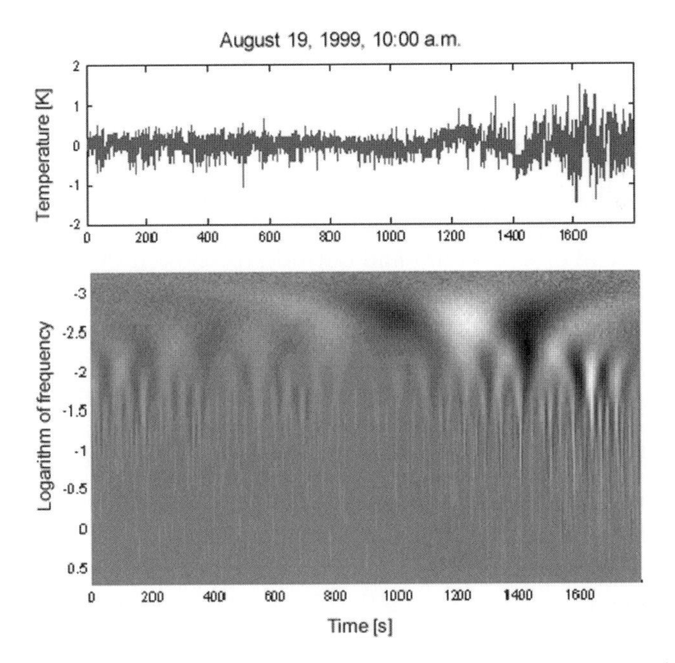

Fig. 9.5. Gray-scale image of the wavelet coefficients $T(a,b)$ as a function of translation time a (abscissa) and dilation b (ordinate), and time trace of temperature fluctuations ΔT vs. time t

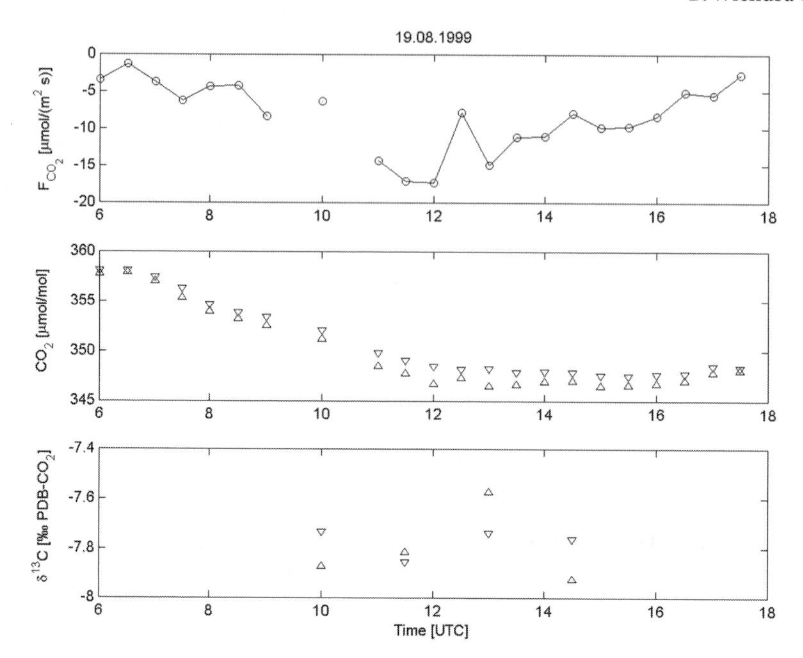

Fig. 9.6. Diurnal trends in eddy covariance CO_2 flux, CO_2 mixing ratios and $\delta^{13}C$ values for HREA updrafts (*triangles*) and downdrafts (*inverted triangles*)

The results for $\delta^{13}C$ up- and downdrafts show significant differences for most of the HREA measurements (Fig. 9.6), taking into account the accuracy of the mass spectrometer analysis (0.01‰ vs. PDB, analyzed by the Max-Planck-Institute for Biogeochemistry, Jena) and the results for control samples (maximum difference of 0.04‰ vs. PDB), for which the same air was sampled simultaneously in the up- and downdraft line of the HREA system. The diurnal trend of $\delta^{13}C$ in up- and downdrafts showed $^{13}CO_2$-enriched air for downdrafts (i.e., $\delta^{13}C$ is less negative for down- than for updrafts) until about 11:00 a.m., whereas after about 11:30 a.m. the updraft air is more enriched in $^{13}CO_2$ than the downdraft air. This result is consistent with our understanding of ecosystem processes. In the morning, the canopy air (updrafts) is less enriched in $^{13}CO_2$ ($\delta^{13}C$ is more negative) because of respiratory processes in the canopy during the night. At noon and in the afternoon, the updraft air is more enriched in $^{13}CO_2$ because of discrimination against $^{13}CO_2$ during photosynthetic processes in the ecosystem. Note the reverse in $\delta^{13}C$ differences already at about 2:30 p.m., indicating an increase of respiration in the canopy due to a strong decrease in PAR at about 12:30 a.m. (cf. Fig. 9.4). The results show CO_2 fluxes that are already (still) directed to the canopy, whereas the $\delta^{13}C$ differences indicate fluxes that are still (already) directed out of the canopy in the morning (afternoon), a result that was also found for measurements above a deciduous forest (Bowling et al. 1999). The effect seems to be caused by a complex interaction of turbulent exchange processes with processes of ecosystem activity (assimilation, respiration). The findings indicate a time shift between both processes, a result found by other measurements as well (Foken et al. 2000). Additional measurements will be necessary in future to understand and explain these processes in detail. Similar results as for 19 August 1999 were obtained for other daily courses in HREA $\delta^{13}C$ up- and downdraft measurements. The $\delta^{13}C$ up- and downdraft values ranged between –8.29 and –7.51‰ (vs. PDB). The differences in $\delta^{13}C$ up- and downdrafts varied between –0.16 and 0.17‰.

9.4.2 Structure of the Exchange Process Conclusions

The results of the flux measurements on 22 June 2000 are shown in Fig. 9.7 with the typical daily cycle, with downward CO_2 fluxes corresponding to upward ^{13}C fluxes caused by assimilation. The analysis of $^{13}C/^{12}C$ isotope ratios for up- and downdraft measurements corresponds partly to respiration instead of assimilation during the day. Therefore, the structure of turbulence was investigated using the mixing layer analogy for exchange processes above a forest canopy according to Raupach et al. (1986).

The shear scale L_S is one of the characteristic scales of a mixing layer. It depends on the wind velocity $u(h)$ and the wind gradient du/dz near the canopy height h, with the vorticity thickness δ_w:

$$L_s = \delta_w / 2 = u(h) / (du / dz)_{z=h} \quad (7)$$

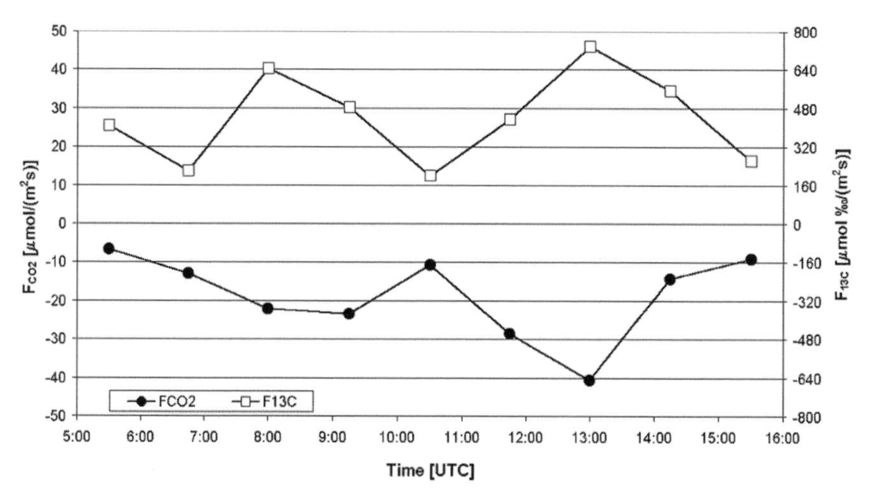

Fig. 9.7. Corresponding CO_2 and ^{13}C flux measurements for 22 June 2000 at Waldstein/Weidenbrunnen site

The second characteristic length scale is the wavelength of coherent structures Λ_x that corresponds to the wavelength of the initial Kelvin-Helmholtz instability of a mixing layer (Finnigan 2000). Λ_x can be estimated by analyzing a time series with wavelet technique (Brunet and Irvine 2000). According to Raupach et al. (1986), both length scales are related linearly by a factor *m*, even for non-neutral stratification (Brunet and Irvine 2000), with *m*=7...10 estimated by several experimental studies:

$$\Lambda_x = m \cdot L_s \tag{8}$$

Comparing the length scales for the vertical wind velocity and CO_2 exchange, four different situations can be found:

1. The length scale of the vertical wind component corresponds to the conditions of the mixing layer but is not similar to that of the carbon dioxide exchange. The atmosphere and the canopy are coupled for strong vertical wind events only.
2. The length scales of the vertical wind component and the carbon dioxide exchange are similar, and correspond to the conditions of the mixing layer. The atmosphere and the canopy are well coupled.
3. The length scales of the vertical wind component and the carbon dioxide exchange are different to each other. They do not correspond to the conditions of the mixing layer. Often, no significant carbon dioxide scale L_s and Λ_x exists. The atmosphere and the canopy are not coupled.
4. The length scales of the vertical wind component and the carbon dioxide exchange are similar, but they do not correspond to the conditions of the mixing layer (long waves). The atmosphere and the canopy are not coupled.

These four different exchange situations were found on 22 June 2000 and are illustrated in Fig. 9.8.

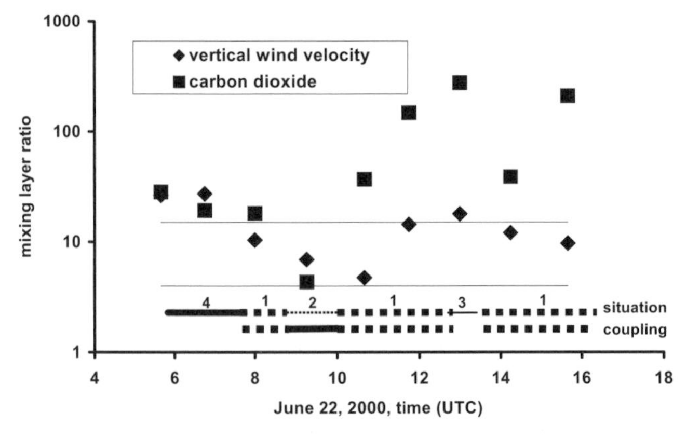

Fig. 9.8. Parameter *m* according to Eq. (8) for the vertical wind velocity and CO_2 exchange and classification of exchange conditions for 22 June 2000

9.5. Conclusions

The purpose of employing $^{13}CO_2$ fluxes in addition to commonly used CO_2 fluxes was to gather supplementary information about the carbon budget of an ecosystem. Stable isotopes can, in principle, provide information about either the sources/sinks or the processes. Thus, $^{13}CO_2$ fluxes could be used to identify CO_2 sources/sinks or to study the interaction of (turbulent) exchange and metabolic processes with both an ecosystem and the atmosphere. Recent studies (e.g., Valentini et al. 2000; Matzner et al., this Vol.; Rebmann et al., this Vol.) clearly show the need for both sources/sinks and process-information in order to explain and interpret results of studies with varying measurements and/or modelling techniques, the latter used for differing scales.

The study shows the methodological background for the combined use of $^{13}CO_2$ and CO_2 fluxes and a technology to measure $^{13}CO_2$ fluxes. It marks the first step in the use and the potential that is inherent in stable carbon isotope information for the understanding and interpretation of ecosystem carbon budgets. The main findings, which would be particularly useful when conducting further examinations, are:

- Fluxes of CO_2 and the stable isotope ^{13}C can be used in principle to partition the NEE into assimilation and respiration. The proposed partition approach is based on simplified mass balances for CO_2 and ^{13}C (Lloyd et al. 1996). When using simplified balance equations, the partition method is only valid if the canopy layer is well coupled with the atmosphere above. The development (see case 1, above) or the existence (case 2) of a mixing layer, along with the measurement of ^{13}C fluxes within this layer, are required to separately quantify assimilation and respiration of an ecosystem.
- Nevertheless, the HREA $\delta^{13}C$ results show, in conjunction with an analysis of turbulent exchange processes by means of wavelet analysis, the complex interaction of turbulent exchange and ecosystem activity. The results indicate that simple bulk and storage assumptions, i.e., simplified mass balance assumptions for the carbon budget, are often not valid. This finding has far-reaching consequences for the examination and interpretation of ecosystem carbon budgets using other measurement and/or modeling techniques. It may, for example, influence the NEE measurements and its interpretation with respect to the measuring height and the height of the mixing layer (Rebmann et al., this Vol.). As modeling approaches fail, for example, when a gradient approach is used to describe fluxes, more adequate models could be applied, for instance, using non-local transilient closure (Berger et al., this Vol.).
- It was shown that $^{13}CO_2$ flux measurements are feasible when using an HREA system. However, the use of an HREA system requires not only highly accurate isotope analyses, but also detailed micrometeorological

tests. Both are necessary in order to use the adequate mean vertical wind velocity for the measuring site and to control the scalar similarity between the measured scalars and proxy scalars which are used to determine an optimal deadband. Because of the high sensitivity of HREA to the deadband, a more robust REA method should be achieved by increasing the accuracy of the isotope analyses. Nevertheless, up to now, only a few HREA data sets of ^{13}C isotopes are available worldwide, and the data set presented is the first one obtained for a coniferous forest.

- It has to be assumed that a changing environment, with higher temperatures in the vegetation period, may influence the exchange processes significantly. Presently, due to the decoupling in the morning, the assimilation starts in the trunk space with respired carbon dioxide that was accumulated during the night. This process may change in the future. On the other hand, hot situations with advection of warm air in the boundary layer and stable stratification can, even during the daytime, reduce assimilation. To study these effects, adequate measuring methods and a better micrometeorological understanding must be developed. This chapter reveals some of the options to study these special processes. Only a combination of long-term carbon flux measurements and detailed studies can help us to understand the changing conditions of ecosystem exchange.

Acknowledgements. We are grateful to Dave Bowling for useful advise and discussions regarding the HREA system and isotope flux measurements. The project was funded by the Federal Ministry of Education, Science, Research and Technology (PT BEO-0339476 C and D).

References

Aubinet M, Grelle A, Ibrom A, Rannik Ü, Moncrieff J, Foken T, Kowalski AS, Martin PH, Berbigier P, Bernhofer Ch, Clement R, Elbers J, Granier A, Grünwald T, Morgenstern K, Pilegaard K, Rebmann C, Snijders W, Valentini R, Vesala T (2000) Estimates of the annual net carbon and water exchange of forests: the EUROFLUX methodology. Adv Ecol Res 30:113–175

Baldocchi D, Meyers T (1998) On using eco-physiological, micrometeorological and biogeochemical theory to evaluate carbon dioxide, water vapor and trace gas fluxes over vegetation: a perspective. Agric For Meteorol 90:1–25

Baldocchi D, Falge EH, Gu L, Olson R, Hollinger D, Running S, Anthoni P, Bernhofer C, Davis K, Evans R, Fuentes J, Goldstein A, Katul G, Law B, Lee XH, Malhi Y, Meyers T, Munger W, Oechel W, Paw UKT, Pilegaard K, Schmid HP, Valentini R, Verma S, Vesala T (2001) FLUXNET: a new tool to study the temporal and spatial variability of ecosystem-scale carbon dioxide, water vapor and energy flux densities. Bull Am Meteorol Soc 82(11):2415–2434

Bowling DR, Delany AC, Turnispseed AA, Baldocchi DD, Monson RK (1999) Modification of the relaxed eddy accumulation technique to maximize measured scalar mixing ratio differences in updrafts and downdrafts. J Geophys Res 104:9121–9133

Brunet Y, Irvine MR (2000) The control of coherent eddies in vegetation canopies: streamwise structure spacing, canopy shear scale and atmospheric stability. Boundary-Layer Meteorol 94:139–163

Buchmann N (2002) Plant ecophysiology and forest response to global change. Tree Physiol 22:1177–1184

Buchmann N, Hinckley TM, Ehleringer JR (1998) Carbon isotope dynamics in *Abies amabilis* stands in the Cascades. Can J For Res 28:808–819

Businger JA, Oncley SP (1990) Flux measurement with conditional sampling. J Atmos Ocean Tech 7:349–352

Finnigan J (2000) Turbulence in plant canopies. Annu Rev Fluid Mech 32:519–571

Foken T, Kartschall T, Badeck F, Waloszczyk K, Wichura B, Gerchau J (2000) Time response characteristics for the atmosphere–plant-interaction, measured during the total solar eclipses in southern Germany on August 11, 1999. In: Proc 14th Symp on Boundary Layer and Turbulence, 7–11 Aug, American Meteorological Society, pp 159–160

Heinz G, Handorf D, Foken T (1999) Direct visualization of the energy transfer from coherent structures to turbulence via wavelet analysis. In: Proc 13th Symp on Boundary Layer and Turbulence, Dallas, Texas, 10–15 Jan, American Meteorological Society, pp 664–665

Keeling CD (1958) The concentration and isotopic abundances of atmospheric carbon dioxide in rural areas. Geochim Cosmochim Acta 13:322–334

Lloyd J, Kruijt B, Hollinger DY, Grace J, Francey RJ, Wong S-C, Kelliher FM, Miranda AC, Farquhar GD, Gash JHC, Vygodskaya NN, Wright IR, Miranda HS, Schulze E-D (1996) Vegetation effects on the iso-topic composition of atmospheric CO_2 at local and regional scales: theoretical aspects and a comparison between rain forest in Amazonia and a boreal forest in Siberia. Aust J Plant Physiol 23:371–399

O'Leary MH (1988) Carbon isotopes in photosynthesis. BioScience 38:328–336

Oncley SP, Foken T, Vogt R, Bernhofer C, Kohsiek W, Liu H, Pitacco A, Grantz D, Ribeiro L, Weidinger T (2002) The energy balance experiment EBEX-2000. In: Proc 15th Symp on Boundary Layer and Turbulence, Wageningen, 15–19 July, American Meterological Society, pp 1–4

Pattey E, Desjardins RL, Rochette P (1993) Accuracy of the relaxed eddy accumulation technique. Boundary-Layer Meteorol 66:341–355

Raupach MR, Coppin PA, Legg BJ (1986) Experiments on scalar dispersion within a model plant canopy, part I. The turbulence structure. Boundary-Layer Meteorol 35:21–52

Raupach MR, Finnigan JJ, Brunet Y (1996) Coherent eddies and turbulence in vegetation canopies: the mixing-layer analogy. Boundary-Layer Meteorol 78:351–382

Ruppert J, Wichura B, Delany AC, Foken T (2002) Eddy sampling methods, a comparison using simulation results. In: Proc 15th Symp on Boundary Layer and Turbulence, Wageningen, 15–19 July, American Meteorological Society, pp 27–30

Valentini R, Matteucci G, Dolman AJ, Schulze E-D, Rebmann C, Moors EJ, Granier A, Gross P, Jensen NO, Pilegaard K, Lindroth A, Grelle A, Bernhofer C, Grünwald T, Aubinet M, Ceulemans R, Kowalski AS, Vesala T, Rannik Ü, Bergigier P, Loustau D, Guomundsson J, Thorgeirsson H, Ibrom A, Morgenstern K, Clement R, Moncrieff J, Montagnani L, Minerbi S, Jarvis PG (2000) Respiration as the main determinant of carbon balance in European forests. Nature 404:861–865

Wyngaard JC, Moeng C-H (1992) Parameterizing turbulent diffusion through the joint probability density. Boundary-Layer Meteorol 60:1–13

Yakir D, Wang X-F (1996) Fluxes of CO_2 and water between terrestrial vegetation and the atmosphere estimated from isotope measurements. Nature 380:515–517

10 Modeling the Vegetation Atmospheric Exchange with a Transilient Model

M. Berger, R. Dlugi, and T. Foken

10.1 Introduction

The interaction between the atmosphere and the soil–vegetation system in ecological models is often modeled with the Penman-Monteith (PM) approach (Monteith 1980) which is based on the energy balance equation. This method describes the interaction with the surface by a resistance approach and relies on the constant flux layer approach which is equivalent to the classical K approach (see, e.g., Stull 1988; Kramm 1995). The PM method is mainly forced by radiation and includes some simplification which does not generally allow the use of short averaging periods in the order of one hour. The constant flux layer assumption is not valid to describe the exchange processes for a forest canopy (Shaw et al. 1974; Raupach et al. 1991; Kaimal and Finnigan 1994), with typical effects like counter gradients (Denmead and Bradley 1985), coherent structures (Amiro 1990) or mixing layers (Raupach et al. 1996). The big leaf approach or the PM method cannot describe the effects of physical processes behind such features. Models using more suitable descriptions of energy transfer, thermodynamics and turbulence (e.g. higher order closure formulations, see Kurata (1982); Meyers and PawU 1986, 1987) yield results that better agree with measured data from the field. Another possible method of atmosphere–forest modeling may be the transilient non-local turbulence theory (TTT) according to Stull (1984, 1993), which allows us to consider also transports between different layers and in different directions and not only between neighbouring model layers (Inclán 1996). Inclán et al. (1996, 1998, 1999) further developed the FLAME model (Forest–Land–Atmosphere-ModEl) to study various interactions in the system soil–vegetation–atmosphere with special emphasis on forest canopies. Up to now, the model has been applied to mainly flat regions (Inclán et al. 1998).

Extensive studies of the atmosphere–forest exchange at the Weidenbrunnen site (50.08°N, 11.87°E, 775 m a.s.l.) also need adequate modeling with a high time resolution below one hour. The aim of this investigation is to apply the FLAME model in a complex terrain with limited access to boundary

Ecological Studies, Vol. 172
E. Matzner (Ed.), Biogeochemistry of Forested Catchments in a Changing Environment

layer data such as from radiosonde station Meiningen (50.57°N, 10.38°E) and of the DWD DM4 Model (14 July 1998). Therefore, also advection is considered. Furthermore, it is tested if the available data and information of the Weidenbrunnen site (Gerstberger 2001) site allow an adequate model initialization. For validation of the modeled sensible and latent heat fluxes above the 19-m-tall spruce forest at Weidenbrunnen, a data set of measured turbulent fluxes between 14 July and 8 August 1998 was available (Mangold 1999; Foken 2003).

This modeling study including validation is the basis to add more sophisticated atmosphere–forest models to the long-term flux measurements for gap filling procedures and for modeling of exchange processes and budgets for different forests of the region in the future.

10.2 Short Description of FLAME

The FLAME (Forest–Land–Atmosphere-ModEl) according to Inclán (1996) and Inclán et al. (1996, 1998) applied the TTT (transilient turbulence theory) to study interactions in the system soil–vegetation–atmosphere, with special emphasis on forest canopies. The mean flow dynamic and thermodynamic conditions are determined from prognostic budget equations for horizontal wind components, potential temperature and specific humidity including the vertical divergences of the corresponding turbulent fluxes of momentum, sensible and latent heat and Coriolis effect. Further on, budget equations for trace substances are solved numerically (Inclán et al. 1999).

The model considers sinks for momentum in the canopy and at the soil surface, as well as sinks and sources of sensible and latent heat and trace substances at the surfaces of plant elements and into the soil. Within this framework, advection, for example of sensible or latent heat, acts like a volume source or sink.

This prognostic one-dimensional, first-order, non-local turbulence closure model is driven by the flow field and the radiation budget and therefore also solves the radiation budget:

$$Q^{*} = \left(K{\downarrow} - K{\uparrow}\right) + \left(L{\uparrow} - L{\downarrow}\right) \quad (1)$$

and the energy budget

$$Q^{*} = Q_H + Q_E + \Delta S \quad (2)$$

with Q^{*} the net radiation, K the downward (↓) and upward (↑) short-wave radiation fluxes, L the downward (↓) and upward (↑) long-wave radiation fluxes, Q_H and Q_E the turbulent sensible and latent heat fluxes and ΔS the storage term. ΔS includes all possible storages like the soil heat flux as well as the heat storage in the plant material and the photosynthesis.

FLAME consists of three interactively coupled submodels describing the atmospheric boundary layer (ABL), the vegetation and the soil. The vertical domain of FLAME reaches from the soil-saturated zone up to 3,000 m height above the ABL and is normally divided into a maximum of 85 unequally spaced layers. In the case discussed in this study, the idealized tree is divided into five layers within the canopy crown and five in the trunk space. The layer height within the canopy is a function of leaf area index (LAI) increment Δ_{LAI} as discussed by Inclán et al. (1996).

Above the canopy the layers are unequally spaced up to 1,000 m height in the atmosphere. Between 1,000 m and the top of the atmospheric model the layer heights are nearly constant with steps of about 200 m. In the soil, 15 layers are used. They are also unequally spaced and are mounted between 0.001 m at surface to 1.20 m depth.

FLAME has been currently applied to model exchange processes for different tall vegetation canopies like spruce and pine (Inclán 1996; Constantin et al. 1998; Inclán et al. 1998) as well as for a mixed forest mainly composed of oaks and pines (Inclán et al. 1999). In this study we compare FLAME results with measured fluxes of sensible and latent heat and evaluate effects of local advection which may be of significant influence on budgets of water, carbon dioxide and energy.

10.3 Model Initialization for the Case Study

For model initialization various input parameters are required, such as profiles of air temperature (t in degrees Centigrade), humidity (q in grams per kilogram) and horizontal wind velocity components (u, v and u_g, v_g; all in meters per second) from the surface to the top of the atmospheric model, as well as profiles of temperature and liquid water content in the soil. Also information on soil type and various morphological and physiological parameters is needed.

For the present study, the model run starts corresponding to the initialization of atmospheric boundary layer conditions for 14 July 1998 (DOY 195):

1. *Profile A*: run starts at 7 MEZ with a combined atmospheric profile of DWD DM4 Model (grid box 78, 111; 50.11°N, 11.93°E; elevation: 637 m) for the ABL conditions and for the atmosphere within the canopy with measurements at the Weidenbrunnen test site (50.08°N, 11.87°E; elevation: 775 m).
2. At 13 MEZ, run starts with two different atmospheric profiles: *profile B*, a combined atmospheric profile of DWD DM4 Model and measurements at the Weidenbrunnen test site and *profile C*, a combined atmospheric profile of radiosonde profile from station Meiningen (50.57°N, 10.38°E; elevation:

450 m) for the ABL conditions and for the atmosphere within the canopy with measurements at Weidenbrunnen.

Profiles A, B and C finish at 20 MEZ. For 8 August 1998 (DOY 220) only a combined atmospheric profile of radiosonde Meiningen and measurements at Weidenbrunnen at 13 MEZ could be used (*profile D*). The model run finishes also at 20 MEZ.

Figure 10.1 shows the atmospheric initialization profiles B and C of FLAME for 14 July 1998 at 13 MEZ with the two possible profiles for ABL conditions using the radiosonde profile (76–3,700 m) and the profile from DWD DM4 Model (71–2,800 m). All heights and the pressure were reduced to sea level height because the altitude of test sites is not yet included in FLAME.

The radiosonde data include pressure, geopotential height, temperature, wind velocity, wind direction and relative humidity up to a height of 500 hPa. So the humidity profile (q) was calculated from temperature, relative humidity and pressure (Buck 1981; Ross and Elliott 1996). The profiles of horizontal wind components (u and v) were computed with wind velocity U and wind direction using the geometrical knowledge of the two-dimensional wind vector U.

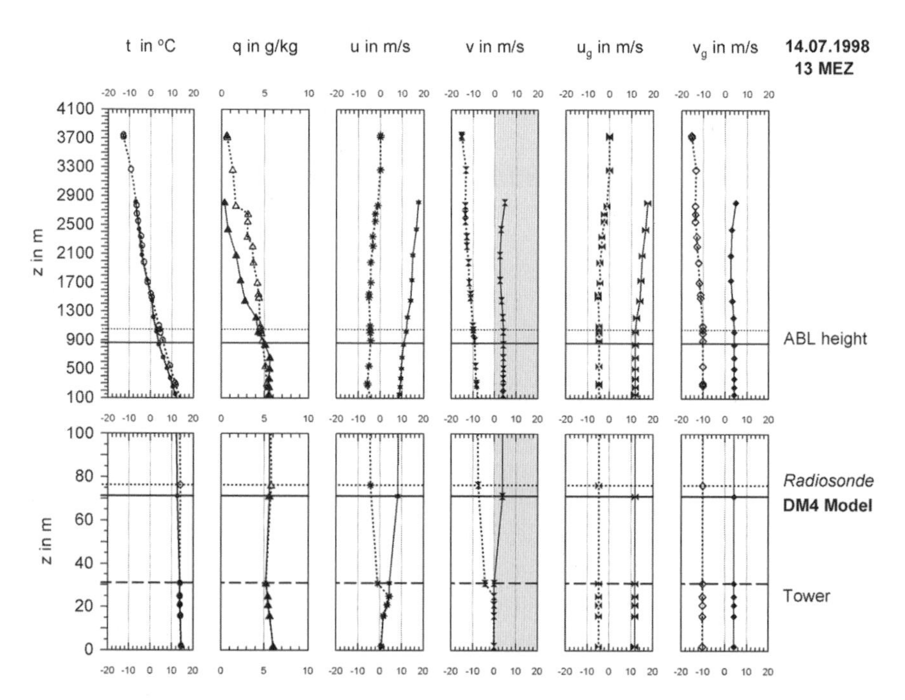

Fig. 10.1. Atmospheric initialization profile B and C for FLAME, 14 July 1998 (DOY 195) at 13 MEZ. *Dotted lines* Radiosonde; *solid lines* DWD DM4 Model or atmospheric boundary layer (*ABL*) height; *dashed line* tower height; *symbols on lines* measuring or modeling points. Field of positive horizontal wind components v is marked *gray*

Form DWD DM4 Model output profiles of temperature, humidity, horizontal wind velocity components and geopotential height exist up to a height of about 3,000 m in the atmosphere.

For lower levels up to 31 m, measurements were taken from the Weidenbrunnen site (Mangold 1999). At 31 m, wind velocity and wind direction are measured. At 25, 21, 16 and 2 m the wind velocity is observed. To calculate the horizontal wind velocity components, u was set equal to the measured wind velocity and v=0 m s^{-1}. Excluding the height of 16 m, also temperature and humidity profiles are measured at the same levels as the wind velocity. Therefore, applying a linear interpolation procedure, temperature and humidity at 16 m are calculated.

The profiles of geostrophic wind components (u_g and v_g) cannot be derived directly from radiosonde data and DM4 Model output. As a first approach the observed wind velocity correspondence with the geostrophic wind above the ABL can be taken if the pressure gradient is small. Such conditions are found at middle and high latitudes and so the profiles of geostrophic wind components are set equal to profiles of horizontal wind components above the ABL height. Below ABL height (1,050 m radiosonde; 863 m DM4 Model) the profiles of geostrophic wind components are constant.

Figure 10.1 shows the similarity of the temperature gradient within both initialization profiles, while the humidity profiles vary much more above the ABL. The DM4 Model humidity profile is drier (6 g kg^{-1} <q <0.4 g kg^{-1}) than the radiosonde profile (6 g kg^{-1} <q <1.3 g kg^{-1}). Larger differences can be seen for the profiles of horizontal wind components u and v and corresponding to this for profiles of geostrophic wind components u_g and v_g. These differences are caused by the difference in wind direction, and therefore the field of the horizontal wind components v is marked gray in Fig. 10.1. For radiosonde data, wind direction was found to be between SW (DD=235°) and W (DD=270°) and for DM4 Model between N (DD=9°) and NE to N (DD=27°).

Initialization of the soil model profiles of soil temperature and liquid water content of the soil is necessary and measured at the Weidenbrunnen site. For 14 July 1998at 13 MEZ the initialization profile for soil is shown in Fig. 10.2. The observed soil temperature t_G will be reduced from t_G=11.5 °C near the surface to t_G=9.4 °C at a depth of 64 cm. For the soil model, also some other soil properties such as the heat capacity, moisture potential and porosity are required. This information can be surveyed in the soil type. At Weidenbrunnen a texture analysis of the first 30 cm of soil found 42.4 % sand, 29.46 % silt and 28.07 % clay. Using the USDA (US Department of Agriculture) soil texture triangle after FitzPatrick (1980), the soil type is characterized at the borderline between clay-loam and loam. Therefore *loam* is the used soil type in the model.

For the vegetation model various morphological and physiological parameters are determined (E. Falge 2000, pers. comm.; Gerstberger 2001) and

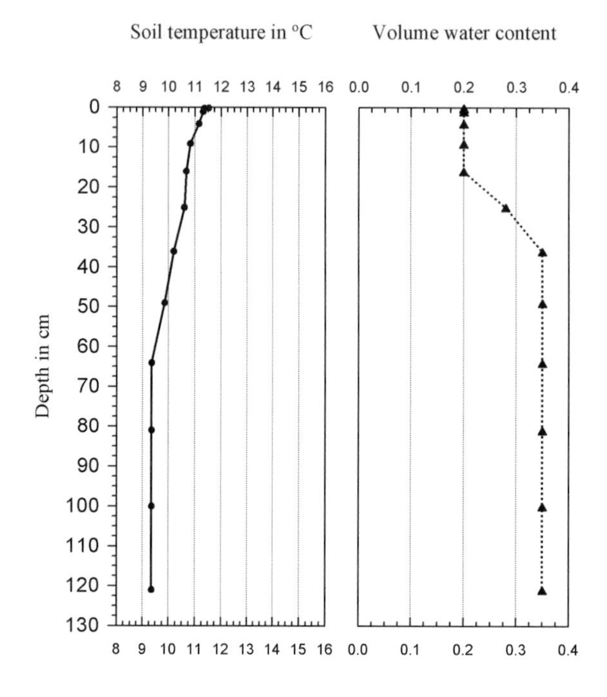

Fig. 10.2. Soil initialization profile of FLAME for 14 July 1998 (DOY 195) at 13 MEZ for temperature (*circles*) and volume water content (*triangles*)

Table 10.1. Model input parameters

Symbol	Description	Value
d_x, d_y	Horizontal length in x , y direction	3.5 m
D_t	Diameter of tree trunk	0.20 m
LAI	Leaf area index	4 m^2 m^{-2}
S_a	Index of area occupied by stems and trunk	0.13
r_{min}	Minimum stomata resistance	150 s m^{-1}
r_{vis}, r_{nir}, r_{ir}	Leaf reflectivity in VIS, NIR and IR bands	0.1, 0.4, 0.04
$r_{s,vis}$, $r_{s,nir}$, $r_{s,ir}$	Soil reflectivity in VIS, NIR and IR bands	0.1, 0.2, 0.1
τ_{vis}, τ_{nir}, τ_{ir}	Leaf transmissivity in VIS, NIR and IR bands	0.03, 0.43, 0.0
Ψ_1, Ψ_2	Leaf water potential for stomata response	−10, −25 bar
R_M	Diameter of canopy at z_M	1.6 m
R_T	Diameter of canopy at z_T	0.8 m
z_L	Trunk height	6 m
z_M	Height of maximum LAI	13.5 m
z_T	Vegetation height	19 m
Δ_{LAI}	Leaf area index increment	0.8

listed in Table 10.1, such as height of the canopy, height of trunk or height of maximum LAI, and also parameters for the stomata resistance, as well as radiative properties of the canopy elements. LAI=4 $m^2\ m^{-2}$ was chosen for the Weidenbrunnen test site based on the correspondence between measured and calculated momentum transfer.

Once the initialization is completed, a loop over time is carried out to make the forecast. The time step is set to 3 min. For the present study, the incoming short-wave radiation above the canopy is not predicted at each time step, but is taken for each time step from recalculated 10-min radiation measurements at 30 m height at the Weidenbrunnen site and applied in Eq. (1) to run the model.

10.4 Comparison of Measured and Modeled Fluxes

In this section a comparison between modeled and measured fluxes above the coniferous forest at Weidenbrunnen is presented mainly for 14 July 1998. The measurements of sensible Q_H and latent Q_E heat fluxes were performed by an eddy covariance technique at 32 m height above the canopy providing half-hourly means (Mangold 1999). For the comparison the modeled fluxes have been averaged arithmetically for the same 30-min time interval. $Q_{E,M}$ and $Q_{H,M}$ are the fluxes modeled with initialization profiles A and B (DWD DM4 Model). For the initialization profiles C and D (radiosonde) the symbols used are $Q_{E,R}$ and $Q_{H,R}$.

Due to the different initialization profiles at 13 MEZ (Fig. 10.1) the modeled fluxes should be compared at first. The right graph in Fig. 10.3 points out the small differences between the simulated sensible and latent fluxes $Q_{H,M}$, $Q_{H,R}$ and $Q_{E,R}$, $Q_{E,M}$. The main discrepancy is found during the first hour due to the time required to adapt the model to boundary and initial conditions. If this first hour is neglected the correlation coefficient is r=0.999 and the differences in modeled fluxes become about $\pm$10 W m^{-2}.

The left graph in Fig. 10.3 shows the initialization at 7 MEZ. For about the first 90 min $Q_{E,M}$ is smaller than Q_E because dew evaporation enhances water vapour transfer from the canopy. This process is not considered by FLAME.

In Fig. 10.4 (left), the transilient matrix at 15:30 MEZ for 14 July 1998 applying the initialization profile B at 13 MEZ is displayed. The magnitude of matrix elements indicates how much air is involved in a mixing process between a source and a destination grid box. Larger matrix elements indicate more mixing. The matrix elements have been contoured and multiplied by 1,000 for display purposes. The height of the source and the destination grid points are indicated along the ordinate and abscissa, respectively. The local turbulent domain is characterized by the matrix elements on the diagonal showing the presence of small-scale turbulent mixing up to ca. 530 m. Non-

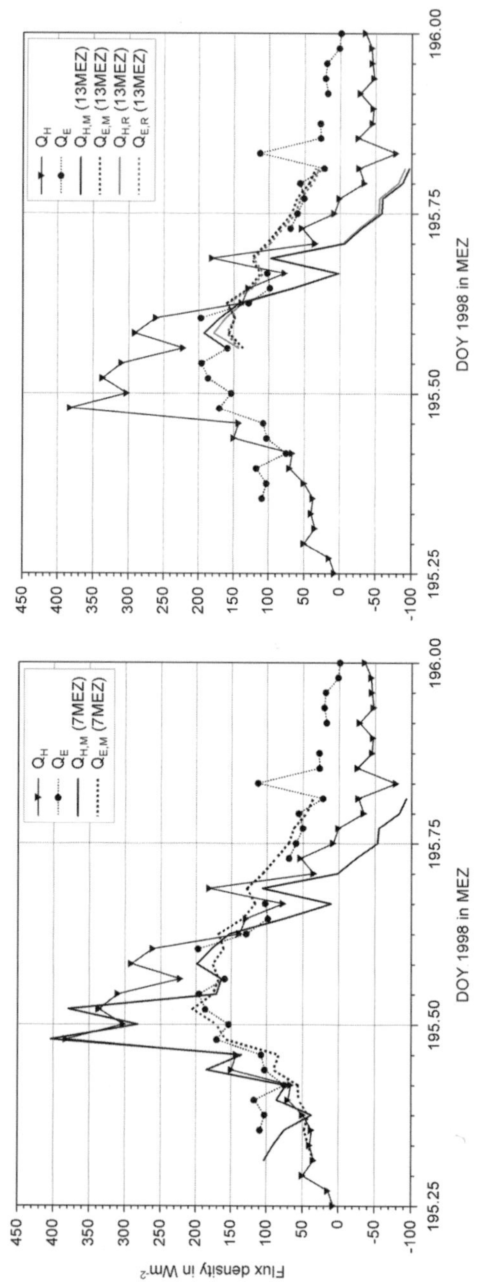

Fig. 10.3. Measured (Q_H, Q_E) and modeled sensible and latent heat flux densities at top of canopy for 14 July 1998 (DOY 195) using initialization at 7 MEZ (*left* $Q_{H,M}$, $Q_{E,M}$, profile A) and at 13 MEZ (*right* $Q_{H,M}$, $Q_{E,M}$; $Q_{H,R}$, $Q_{E,R}$, profiles B and C)

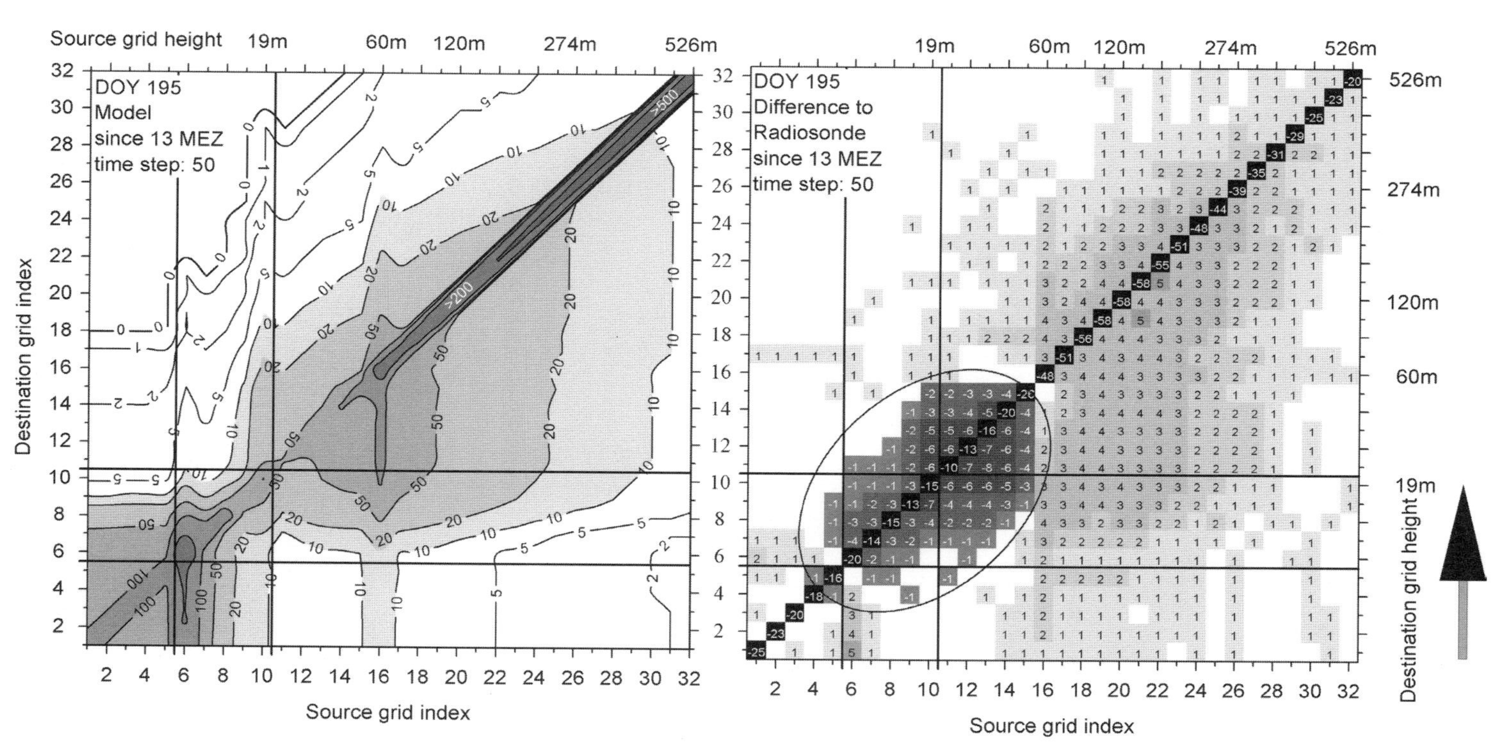

Fig. 10.4. Transilient matrix at 15:30 MEZ for 14 July 1998 calculated with initialization profile B (*left*) and difference to transilient matrix at 15:30 MEZ calculated with initialization profile C (*right*). Positive differences (DWD-Model radiosonde) are marked in *light gray* and negative differences are in *darker gray*. To display the fraction of air reaching any destination location that came from each source location, transilient elements have been multiplied by 1,000

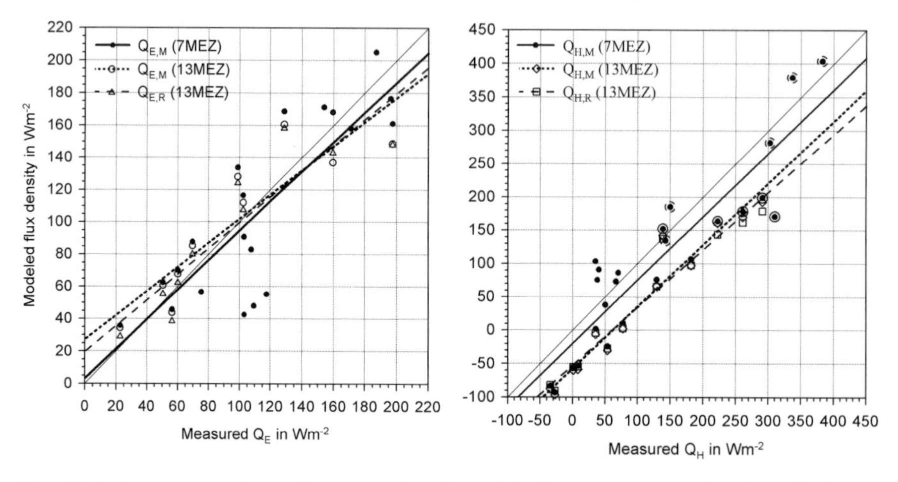

Fig. 10.5. Intercomparison of measured and modeled latent heat flux densities (*left*) and sensible heat flux densities (*right*) at top of canopy for 14 July 1998 (DOY 195) using initialization at 7 and 13 MEZ with their calculated linear regression line; *dot with broken circle* (10:30–12:30 MEZ); *dot with solid circle* (13–15 MEZ)

Table 10.2. Statistical analysis with linear regression. Predictor indicates flux density; predictand indicates Q_H sensible heat flux or Q_E latent heat flux; *r* correlation coefficient; *s* standard deviation of residuum; *a* and *b* linear regression coefficients

Flux density	*n*	*r*	*s*	*a*	*b*
$Q_{E,M}$ (7 MEZ)	20	0.837	29.7	3.1	0.92
$Q_{E,M}$ (13 MEZ)	10	0.893	19.4	27.2	0.75
$Q_{E,R}$(13 MEZ)	10	0.907	19.1	19.5	0.80
$Q_{H,M}$ (7 MEZ)	25	0.912	51.7	−19.9	0.95
$Q_{H,M}$ (13 MEZ)	13	0.972	23.6	−57.5	0.93
$Q_{H,R}$ (13 MEZ)	13	0.971	22.7	−52.7	0.87

local transport is described by the other elements (e.g. Stull 1993). At this time period a turbulent transport from layers located within and above the canopy, up to 120 m, down to layers in the trunk space can be noted.

The difference to the transilient matrix at 15:30 MEZ calculated for the initialization profile C can be seen in Fig. 10.4 (right). Positive differences (DWD-Model radiosonde) are indicated in light gray and negative differences are displayed as darker gray.

Next, the comparison between the measured and simulated sensible and latent heat fluxes will be presented and discussed. Figure 10.5 shows simulated (line) and measured (line with marker) sensible and latent heat fluxes using the initialization at 7 and at 13 MEZ. Corresponding to Fig. 10.3, the intercomparison of measured and modeled latent heat fluxes together with their calculated linear regression lines are shown in Fig. 10.5 and the statistical analysis is listed in Table 10.2.

After dew evaporation, the agreement between the measured and with FLAME modeled fluxes appears to be quite satisfactory, as the modeled sensible flux $Q_{H,M}$ (7 MEZ) is strongly correlated with the incoming short-wave radiation/net radiation during 10:30 and 15 MEZ. In Fig. 10.5, these values are marked for the time period 10:30–12:30 MEZ by a dot with a broken circle (good agreement) and for the time period 13–15 MEZ by a dot with a solid circle where $Q_H > Q_{H,M}$ is observed for wind direction change to a sector DD >260°. The modeled sensible heat fluxes $Q_{H,M}$ and $Q_{H,R}$ with the initialization at 13 MEZ show the same behavior. Looking at the linear regression lines ($Q_H/Q_{H,M;}\ Q_{H,R}$) there is a linear shift to the 1:1 line of about 50 W m^{-2}. For 8 August 1998 this linear shift to the 1:1 line is about 100 W m^{-2} for the same wind sector, where low vegetation and drier soils dominate. Therefore, an analysis of the influence of local advection on the budgets and fluxes of sensible and latent heat was performed.

10.5 Conclusions

For comparison with data from field studies the local advection of sensible Q_H and latent Q_E heat is considered in such a way that these terms are added to the budget equation for potential temperature and humidity and the energy budget equation. Entrainment into the ABL, as discussed, for example, by McNaughton and Spriggs (1986), Jacobs (1994) or Dlugi et al. (2002), shows a comparable influence on Q_H and Q_E in such a way that terms (index ADV) are added or subtracted from the energy budget term, e.g.

$$Q_H = Q_H \pm Q_{H,ADV} \text{ or } Q_E = Q_E \pm Q_{E,ADV}. \quad (3)$$

First intercomparisons were performed by Inclán et al. (1998) for the LINEX-97/2 experiment from measurements above a pine forest and by Inclán (1996) and Constantin et al. (1998) for a spruce forest. A recent re-analysis showed that during some days of the LINEX-97/2 study, vertical and/or horizontal advection of sensible and latent heat may have influenced the fluxes measured above the pine forest. Therefore, the analysis of the Weidenbrunnen data and the comparison with the results obtained by FLAME including advection of heat or moisture for other sites especially point out the important role of local dynamic processes on the magnitude of sensible and latent heat fluxes (and fluxes of other quantities) measured above the canopy.

In general, the comparison of modeled and measured data of the sensible and latent heat flux is satisfactory even for the non-optimal boundary layer forcing and the hilly terrain. Also the available data set of plant ecological and micrometeorological data allows a good initialization of the model. For further investigations at the Weidenbrunnen site the FLAME model appears to be an adequate model approach. The results also underline the possible usage of the model for non-ideal FLUXNET sites.

More specifically, we can draw the following conclusions:

- The exchange processes between the forest canopy and the atmosphere are characterized by phenomena like counter gradient transport inside the canopy and intermittent transport of momentum and scalar quantities in the mixing layer near the top of, as well as inside, the canopy (e.g. Raupach et al. 1991, 1996; Wichura et al., this Vol.). Therefore, transfer processes cannot be described by models based on local closure techniques like the K approach. Two methods of model formulations have been developed: (1) prognostic or forward and (2) diagnostic or inverse (e.g. Katul 2001). When compared with field data, these approaches can partially reproduce measured turbulent fluxes of scalar quantities; however, they need further refinement as many results show only qualitative agreement. Non-local, or higher order, closure techniques for modeling exchange processes can be applied for prognostic processes. The FLAME model by Inclán et al. (1996) is applied in a prognostic version and is shown to be an adequate approach, not only for rather homogeneous sites (e.g. Inclán et al. 1998), but also for the orographically structured Weidenbrunnen site if advection is considered.
- The non-linear and non-local exchange in the system soil–vegetation–atmospheric boundary layer requires time-dependent descriptions of dynamic and thermodynamic processes in both the canopy and ABL flow, coupled with the energy, water and nutrient transfers in the soil–plant system. Such models are essential to realistically describe the potential influence of environmental change at global and regional scales as well as at the scale of ecosystems.
- In the present phase of climate change, with an increase of the mean temperature and a higher variability of meteorological parameters, models must also be able to reproduce these factors. If the matrix is parameterized for such conditions, the transilient theory is able to handle these non-linear processes. Higher-order closure models need an accurate description of both the second moments of velocity statistics and the scalar quantities inside the canopy, such as humidity, temperature and trace gases. Such research issues are necessary to model exchange processes between high vegetation and the atmosphere in a changing environment.

Acknowledgements. This research was supported by the Federal Ministry of Education and Research (contract PT BEO51-0339476C).

References

Amiro BD (1990) Comparison of turbulence statistics within three boreal forest canopies. Boundary-Layer Meteorol 51:99–121

Buck AL (1981) New equations for computing vapour pressure and enhancement factor. J Appl Meteorol 20:1527–1532

Constantin J, Inclán MG, Raschendorfer M (1998) The energy budget of a spruce forest: field measurements and comparison with the forest-land- atmosphere model (FLAME). J Hydrol 212/213:22–35

Denmead DT, Bradley EF (1985) Flux-gradient relationships in a forest canopy. In: Hutchison BA, Hicks BB (eds) The forest-atmosphere interaction. Reidel, Dordrecht, pp 421–442

Dlugi R, Berger M, Kramm G, Rube S (2002) Modellierung von Austauschprozessen und chemischen Reaktionen im Vergleich mit Messdaten. Bericht zum 6. Teil des BMBF Vorhabens ECHO, FZ-Jülich. http://www.fe-juelich.de/icg/icg-ii/ECHO/echo.ger.html

FitzPatrick EA (1980) Soils: their formation, classification and distribution. Longman, New York

Foken T (2003) Angewandte Meteorologie, mikrometeorologische Methoden. Springer, Berlin Heidelberg New York

Gerstberger P (2001) Die BITÖK-Untersuchungsflächen im Fichtelgebirge und Steigerwald. Bayreuther Forum Ökol 90:193

Inclán MG (1996) Modellierung nicht lokaler Austauschprozesse in und über hohen Pflanzenbeständen. PhD Thesis, University (LMU) Munich, Germany

Inclán MG, Stull RB, Dlugi R (1996) Application of transilient turbulence theory of forest canopies. Boundary-Layer Meteorol 79:315–344

Inclán MG, Dlugi R, Zelger M (1998) Vorbereitung und Validierungsläufe des Waldmodelles FLAME mit Daten des Experimentes LINEX-97/2, Teil 2: Vergleich zwischen gemessenen und modellierten turbulenten Flüssen über dem Kiefernbestand. Abschlußbericht des DWD-Werkvertrages, 60 pp. (Available from R. Dlugi at rdlugi@gmx.de)

Inclán MG, Schween J, Dlugi R (1999) Estimation of volatile organic compound fluxes using the forest-land-atmosphere model (FLAME). J Appl Meteorol 38:913–921

Jacobs CMJ (1994) Direct impact of atmospheric CO_2 enrichment on regional transpiration. PhD Thesis, Department of Meteorology, Wageningen Agricultural University, The Netherlands

Kaimal JC, Finnigan JJ (1994) Atmospheric boundary layer flows: their structure and measurement. Oxford University Press, New York

Katul GG, Lai C-T, Siqueira M, Schäfer K, Albertson JD, Wesson KH, Ellswoth D, Oren R (2001) Inferring scalar sources and sinks within canopies using forward and inverse methods. In: Lakshmi V, Albertson J, Schaake J (eds) Land surface hydrology, meteorology and climate: observations and modeling. Water science and application, vol 3. American Geophysical Union, Washington, DC, pp 13–15

Kramm G (1995) Zum Austausch von Ozon und reaktiven Stickstoffverbindungen zwischen Atmosphäre und Biosphäre. Maraun Verlag, Frankfurt

Kurata K (1982) Theoretische Untersuchung der Turbulenz innerhalb eines Pflanzenbestandes. Ber Inst Meteorol Climatol 20. University Hannover

Mangold A (1999) Untersuchung der lokalen Einflüsse auf Turbulenzmessungen der Station Weidenbrunnen. Diploma Thesis, University Bayreuth

McNaughton KG, Spriggs TW (1986) A mixed layer model for regional evaporation. Boundary-Layer Meteorol 34:243–262

Meyers TP, PawU KT (1986) Testing a higher-order closure model for modelling airflow within and above plant canopies. Boundary-Layer Meteorol 37:297–311

Meyers TP, PawU KT (1987) Modelling the plant canopy micrometeorology with higher-order closure principles. Agric For Meteorol 41:143–163

Monteith JL (1980) Principles of environmental physics. Arnold, London

Raupach MR, Antonia RA, Rajagopalan S (1991) Rough-wall turbulent boundary layers. Appl Mech Rev 44:1–25

Raupach MR, Finnigan JJ, Brunet Y (1996) Coherent eddies and turbulence in vegetation canopies: the mixing-layer analogy. Boundary-Layer Meteorol 78:351–382

Ross RJ, Elliott WP (1996) Tropospheric water vapor climatology and trends over North America: 1973–1993. J Climatol 9:3561–3574

Shaw RH, Siversides RH, Thurtell GW (1974) Some observation of turbulence and turbulent transport within and above plant canopies. Boundary-Layer Meteorol 5:429–449

Stull RB (1984) Transilient turbulence theory, part 1. The concept of eddy mixing across finite distances. J Atmos Sci 41:3351–3367

Stull RB (1988) An introduction to boundary layer meteorology. Kluwer, Dordrecht

Stull RB (1993) Review of non local mixing in turbulent atmospheres: transilient turbulence theory. Boundary-Layer Meteorol 62:21–96

11 Fog Deposition and its Role in Biogeochemical Cycles of Nutrients and Pollutants

T. WRZESINSKY, C. SCHEER, and O. KLEMM

11.1 Introduction

It has been recognized for about 100 years now that the deposition of fog (*occult deposition*) can play an important role in the hydrological cycles of various mountainous ecosystems (Marloth 1906; Linke 1916; Grunow 1955; Baumgartner 1958, 1959). Considering the fact that the concentrations of trace substances in fog water are typically higher than those of comparable rain water, it becomes evident that the deposition of nutrients and pollutants through fog deposition may be as high as, or even higher than, the deposition through rainwater (e.g., Saxena and Lin 1990). There is a large variability in physical and chemical conditions of fog events, so that any estimates of their potential roles in biogeochemical cycles must be studied individually for each site and each time period of interest.

Many studies of fog deposition refer to modelling efforts. The most prominent model was first described by Lovett (1984) and has been widely used, partly in modified form (Lovett and Reiners 1986; Joslin et al. 1990; Mueller 1991; Mueller et al. 1991; Pahl and Winkler 1995; Pahl 1996; Baumgardner et al. 2003). Some recent studies introduced the eddy-covariance technique, a direct experimental approach to measure fog deposition (Beswick et al. 1991; Vong and Kowalski 1995; Kowalski et al. 1997). Comparisons of such experimental results with the deposition model showed general agreement of the deposition fluxes; however, the time periods for these intercomparisons were short.

In this chapter, we quantify the deposition of fog water with the experimental and the model approach. Together with information about the chemical composition, we will quantify the deposition of nutrients and pollutants through occult deposition. The importance of these deposition fluxes for the biogeochemical cycles of various elements is discussed. The results are further evaluated together with results from the canopy balance method by Matzner et al. (this Vol.).

Ecological Studies, Vol. 172
E. Matzner (Ed.), Biogeochemistry of Forested Catchments in a Changing Environment

11.2 Site

All experiments with fog chemistry and physics were performed on a meteorological walk-up tower at the Weidenbrunnen site at 50°08′32″ N, 11°52′04″ E, 775 m a.s.l., in a Norway spruce plantation, about 18 m high, planted around 1945. It is located within the Lehstenbach catchment, close to its water divide. The area is often called the Waldstein site, according to the nearby mountain peak. The eddy-covariance setup was installed at 31 m above ground. The cloud water collector was installed at the 24-m platform of the tower. Rainwater was collected at the nearby Pflanzgarten site.

11.3 Fog Chemistry

A cloud is an aerosol with liquid particles (droplets) dispersed in air. Fog is a cloud with direct contact to the earth surface. For this definition it is unimportant what the exact cloud-forming mechanisms are: isobaric cooling through radiation cooling, mixing of water-saturated air masses of different temperatures, adiabatic cooling during a lifting process, etc. The size of fog droplets range typically between 2 and 50 µm diameter. The collection of fog water for the purpose of chemical analysis is established through impaction of droplets on collection surfaces. The required acceleration is established either through exposure of a collection surface into the wind (passive collectors) or through pumping of air around a collection surface with high speed (active collectors). We used the Bayreuth heatable active strand cloud water collector (BCC) as described by Wrzesinsky (2003). It is a modified version of the Caltech active strand cloud water collector (CASCC; Daube et al. 1987). In contrast to the original CASCC, the collection strands of the BCC are heatable to enable collection during freezing conditions. The cut-off diameter and the theoretical maximum collection efficiency are 7.1 µm and 88 %, respectively. For an event-based sampling, an automatic switching system, employing the visibility data, was established. More details about the sampling strategy and routine are given in Wrzesinsky and Klemm (2000) and Wrzesinsky (2003).

For sample periods of about 2–30 h, we yielded volumes of 60 to more than 1,000 ml. Samples were taken shortly after event endings and stored in precleaned polyethylene bottles at −18 °C. For analyses, filtered aliquots were used (0.45 µm). Cations were analyzed using ICP-AES, anions using IC, and ammonia using FIA at BITÖK's central laboratory Zentrale Analytik (ZAN).

Figure 11.1 shows that for the period 27 June to 5 December 2000 the inter-event variability of the fog water composition is high. The total ion sum ranges from about 300 to over 13,000 µeq l^{-1}. The median composition is dominated by NH_4^+ (43 %), NO_3^- (27 %), and SO_4^{2-} (20 %).

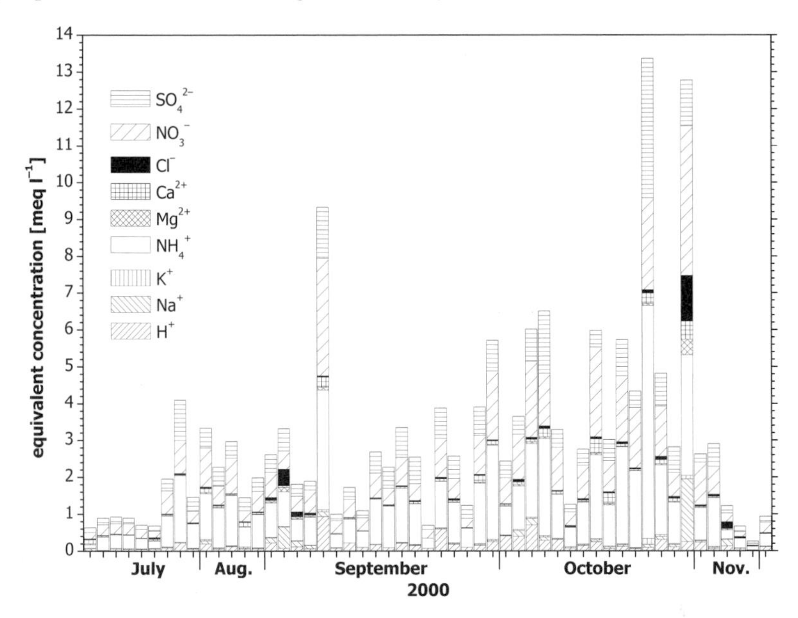

Fig. 11.1. Chemical composition (inorganic ions) of fog water samples from the Waldstein site

The concentrations of ions in fog water are generally higher than those of rainwater that was collected at the same site, and within a short time period of the respective fog event. Median concentration ratios are 16 for H^+ (calculated as 10^{-pH}), 3.1 for Cl^-, 22 for NO_3^-, 17 for SO_4^{2-}, 24 for NH_4^+, 6.1 for Ca^{2+}, 1.7 for K^+, 9.4 for Mg^{2+}, and 2.3 for Na^+, respectively. During the formation of precipitation within a cloud, droplets grow to the size of raindrops. One important droplet growth process is condensation of water from the gas phase into existing droplets, thus leading to a dilution of the liquid water concentrations of substances. However, the extreme concentration ratios of ions that are representative for boundary layer air, namely nitrate, ammonium, and sulfate, indicate that fog represents the boundary layer air, whereas rain falls through a deeper layer of the atmosphere and therefore exhibits generally lower concentrations of pollutants than fog. This reasoning is certainly valid for our site in the Fichtelgebirge mountain range.

It should be noted that the pHs of fog events since 1997 are between 3.0 and 5.8, with a median of 4.2, which is 0.70 pH units higher than during the late 1980s (Klemm, this Vol.). Due to reductions in the emissions of SO_2 in Europe, the acidity in fog water was significantly reduced.

Other trace substances in fog water such as heavy metals, organo-metals, aldehydes, peroxides, N-phenols, haloacetates, toxicity parameters, organic and inorganic carbon have been analyzed in selected samples (Herterich 1987; Trautner 1988; Richartz 1989; Trautner et al. 1989; Richartz et al. 1990;

Wrzesinsky 1998; Römpp 1999; Bopp 2000; Hottenroth 2001; Weigl 2001; Grießbaum 2002). Due to the episodic nature of the data sets, these results are not discussed here.

11.4 Direct Measurement of Fog Deposition

The turbulent deposition flux of fog water was measured with the eddy-covariance method. Wind and droplet size distributions were measured with a Young 81000 ultrasonic anemometer and a fast droplet spectrometer FM-100 (Droplet Measurement Technologies, Inc.). The data collection rate was 8.6 Hz. Droplet size distributions were measured in 40 size channels for diameters up to 50 μm. For each time step, the liquid water content (LWC) was computed from the droplet size spectrum, and the turbulent deposition flux $F_{t,fog}$ was computed for 30-min intervals:

$$F_{t,fog} = \overline{w' \cdot LWC'} \tag{1}$$

with w' the deviation from the mean of the vertical wind component w, and LWC' the deviation of the LWC. Sedimentation, i.e. gravitational fluxes, $F_{S,fog}$, was calculated from Stokes' law, and the total deposition flux, $F_{tot,fog}$, was computed as the sum of $F_{t,fog}$ and $F_{S,fog}$. More details about the applied techniques are described in Burkard et al. (2002).

For the quantification of the deposition of ions or other solutes in fog water through fog deposition, F_c, the deposition flux $F_{tot,fog}$ is multiplied by the respective fog water concentration c on an event-basis:

$$F_c = F_{tot,fog} \cdot c \tag{2}$$

for example for nitrate:

$$F_{NO_3^-} = F_{tot,fog} \cdot c_{NO_3^-} \tag{2.1}$$

Table 11.1 shows results of the field campaign from 27 June to 5 December 2000. Until 18 October 2000, an experimental setup of the University of Bern was employed, and from 19 October 2000 on, the almost identical setup of the University of Bayreuth was used (see Burkard et al. 2002 for details). During the entire period, the research site was in fog for 24 % of the time. The total fog water deposition flux was F_{tot}=20.4 mm. Compared to 433 mm deposition through rain and snow during the same time period, the fog water deposition accounts for about 4.5 % of the total wet deposition.

A contribution of 4.5 % of fog water deposition to the water balance in this particular ecosystem may be regarded as a minor contribution. However, for

Table 11.1. Minimum, maximum, mean, and sum of mean fog water flux rate, duration, mean liquid water content (*LWC*), and fog water deposition per event for the field campaign 27 June to 5 December 2000

	Minimum	Maximum	Mean	Sum
Mean flux rate (mg m^{-2} s^{-1})	0.85	21.6	5.99	
Event duration (h)	1.5	175		923
Mean LWC (g m^{-3})	0.013	0.263	0.07	
Fog water deposition (mm)	0.005	4.23		20.4

Table 11.2. Comparison of ion deposition by rain and fog at the Waldstein site as a mean for the field campaign 27 June to 5 December 2000. (After Thalmann et al. 2002)

	Rain (kg ha^{-1} $year^{-1}$)	Fog (%)	Fog/rain
Na^+	2.2	0.47	21
Mg^{2+}	0.15	0.07	50
K^+	0.69	0.18	26
Ca^{2+}	1.5	0.47	32
NH_4^+	6.5	7.1	110
Cl^-	1.9	0.77	40
NO_3^-	16.2	17.2	106
SO_4^{2-}	11.5	9.0	79

the deposition of solutes in fog water, the deposition through fog water is large because the liquid concentrations are much higher in fog as compared to rain (see previous section). According to the results of the experimental period in 2000, as presented in Table 11.2, the contribution of fog deposition to total deposition through fog and rain is between 21 and 50 % for the ions Mg^{2+}, Na^+, K^+, Ca^{2+}, and Cl^-. For the more surface-related NO_3^-, NH_4^+, and SO_4^{2-} with higher enhancements of concentrations of fog over rain water, the contribution to the total wet and occult deposition is between 79 and 110 %.

11.5 Fog Deposition Model

A one-dimensional cloud water deposition model was applied to predict the deposition of fog water. The model was developed by Lovett (1984) and has been widely applied since. We used the parameterization by Pahl and Winkler (1995) and Pahl (1996) for a mountainous spruce forest in central Europe. For the model application, the forest stand is considered to consist of several (in our case 17) layers of 1 m depth each. The turbulent transport of fog droplets from one layer to a neighboring layer underneath is estimated with an inferential model, in which the transfer velocity is parameterized from the measured wind speed above the forest canopy. In each layer, deposition of fog droplets to the vegetation surfaces is allowed and estimated. The sedimentation is estimated by use of Stokes' law. The model is driven by static parameters such as a vertical distribution of the vegetation surface area index (SAI) and by input data of the horizontal wind speed. As the model is highly sensitive to the horizontal wind, measured wind data are essential. Other important parameters are cloud physical data, namely liquid water content (LWC) and droplet size distribution. A direct measurement or sophisticated estimate seems very important. In our application, these parameters were either directly measured or parameterized from visibility data. For times when the turbulent fluxes were directly measured (see previous section), the actual droplet size distributions and LWC data could be used to drive the model. For most of the 5-year time period from 1998–2002, however, these parameters had to be estimated from the visibility data.

First, the LWC had to be estimated from the visibility (vis, in meters) data. We used a potential function of the form:

$$\mathrm{LWC} = 171.4 \cdot \mathrm{vis}^{-1.45} \tag{3}$$

to predict LWC from vis. A parameterization of the droplet size distribution from LWC is more difficult. We classified the LWC data into 0.1 g m^{-3} classes and parameterized a function to each size class. Figure 11.2 shows an example of the measured and adapted droplet size distributions for LWCs between 0.3 and 0.4 g m^{-3}. Although the measured data seem to exhibit three modes at 10-, 16-, and 23-μm droplet diameters (*D*), the predominance of the middle mode and large variability of the data seem to justify the use of a unimodal approximation (with maximum at D=15 μm in the presented case). The adapted equation for this size class is:

$$\mathrm{LWC_D} \approx 0.050 \cdot \exp\left(-\frac{\left(\log_{10}\left(D/2\right) - 0.857\right)^2}{0.0346}\right) \tag{4}$$

$\mathrm{LWC_D}$ is the liquid water content in the droplet size class with diameter D. The parameterizations after Best (1951) and Deirmendjian (1969), which lead to relatively good results in the Pahl (1996) study, are also shown in Fig. 11.2. It appears that parameterizations are not easily transferable from one forest or mountain range to another. The atmospheric processes leading to typical droplet size distributions for various ecosystems are systematically different from each other. For the quantification of the deposition flux, using all size

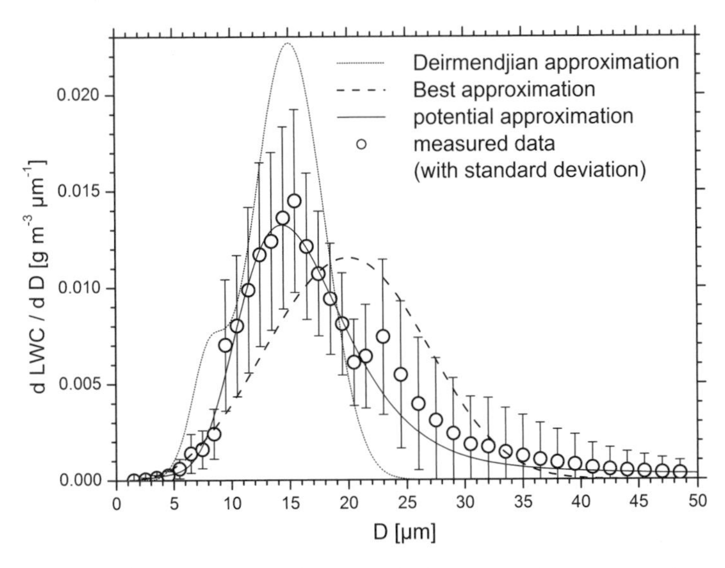

Fig. 11.2. Measured droplet spectra for the period 27 June to 5 December 2000 and LWC class between 0.3 and 0.4 g m^{-3} (with standard deviation), potential approximation (Eq. 3), and approximations after Best (1951) and after Deirmendjian (1969). *d* Derivative; *D* droplet diameter

classes and the entire time series, the new approximation and the approximation after Best (1951) lead, on average, to the best agreement between measured and modelled fog deposition, with maximum deviations of +12 % (Best distribution) and –15 % (new approximation).

Due to obvious drawbacks of the parameterization of LWC and droplet size distributions from visibility data, we first operated the deposition model by use of the measured drop size distribution. Figure 11.3 shows the calculated results for two periods: 27 September to 5 December 2000 (period 1) and 4 April 2001 to 18 March 2002 (period 2). The modelled deposition fluxes are clearly higher than the measured ones. For period 2, the deviation is +28 %, which may be regarded as a relatively large discrepancy. For period 1, however, the deviation is in the order of 100 %, which cannot be regarded as acceptable agreement. For single months (not shown), the deviations between model and measurements vary highly. In most cases, the model overestimates the measured deposition (by up to +900 %); however, negative deviations up to –39 % also occurred. The largest deviations between model and measurement occurred in the droplet size range with diameters 5–10 µm.

If the deposition was not computed by use of the measured droplet size spectra, but with spectra that were calculated from the visibility, the agreement between modelled and measured deposition was better. Considering that the parameterization was problematic itself, we cannot consider this a

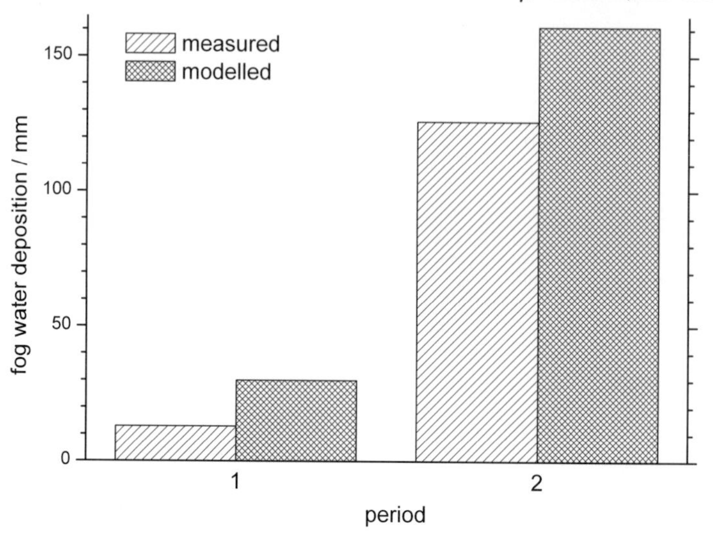

Fig. 11.3. Measured and modelled fog deposition for the experimental period 1 (27 June to 5 December 2000) and 2 (4 April 2001 to 18 March 2002)

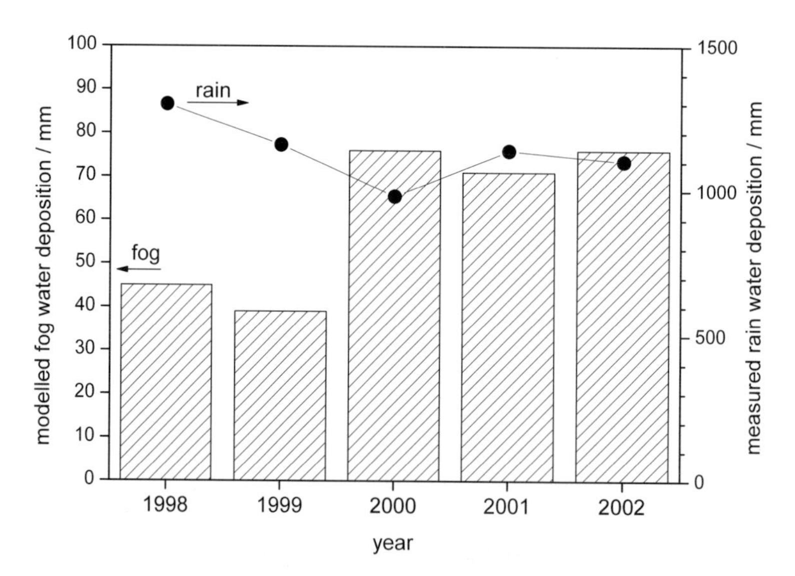

Fig. 11.4. Modelled fog water deposition (*bars, left y-axis*) and measured rain and snow deposition (*circles, right y-axis*) for the years 1998–2002 at the Waldstein site in NE Bavaria

true improvement of the modelling of fog deposition. Overall, the modelled deposition of fog water was higher than the measured one.

The advantage of the model is doubtless its ability to estimate the fog deposition with a rather simple input data set. Figure 11.4 shows the modelled fog water deposition for the time period 1998–2002. Given the tendency of the model to overestimate the fog deposition, deposition rates between 39 and 76 mm should be regarded as a maximum deposition of fog water at the Waldstein site in the Fichtelgebirge mountains, NE Bavaria, at 800 m a.s.l. Compared to the total deposition through rain, snow, and fog, the fog deposition contributes between 3.3 % (year 1999) and 7.2 % (year 2000).

11.6 Discussion

Fog is a frequent phenomenon in mountainous ranges of central Europe. Depending on the local conditions, various processes may lead to the formation of clouds and fog with varying relative importance. In NE Bavaria (central Europe), advection of clouds is the most common phenomenon leading to foggy conditions in the forest ecosystems on the slopes of the Fichtelgebirge mountain range. The higher the altitude above sea level, the higher the fog frequency and typical density. At the Waldstein ecosystem research site at approximately 800 m a.s.l., fog occurred about 20 % of the time (for the period 1997–2002).

Fog deposition makes up a small yet significant term in the ecosystem water balance. The interannual variability of the absolute and relative contribution is high. The one-dimensional model after Lovett (1984) and Pahl and Winkler (1995) and Pahl (1996) estimates the annual contribution to be between 3.3 and 7.2 %. However, we found clear evidence that the model overestimates the true deposition by up to 100 % for time periods extending over several months. Direct measurements of turbulent exchange of fog droplets lead to an estimated contribution to the water balance of 4.5 % for a 6-month period. As this period in 2000 yielded the highest relative contribution of fog water to the water balance, we estimate the annual contribution of fog water to water balance to be between 2 and 4.5 %.

The importance of fog deposition lies mainly in the input of nutrients and pollutants through fog deposition, because the solute concentrations in fog are much higher than those in rain water. For some ions, the enrichment in fog water overcompensates the small contribution in the water balance, meaning that more deposition takes place through fog than through rain and snow. For the 6-month intensive experiment period in 2000, the deposition of NH_4^+ and NO_3^- through fog was 5.55 and 3.91 kg N ha^{-1} $year^{-1}$, respectively. During the same time period, the depositions through rain and snow was 5.04 and 3.65 kg N ha^{-1} $year^{-1}$, respectively. The deposition of ionic nitrogen through fog was thus about 10 % larger than that through rain and snow. Therefore, fog deposition plays a very important role in the biogeochemical cycles of nutrients and forest fertilization through atmospheric deposition.

In our opinion, the deposition of pollutants and nutrients through fog to various ecosystems deserves further study. In particular, for mountain ecosystems far too little is known about the magnitude, temporal and spatial variability of fog deposition and its driving forces. Both, models and experimental approaches need to be further developed in order to arrive at more accurate and precise estimates.

For a more direct quantification of turbulent deposition through the eddy covariance approach, the development of single droplet analysis bears the potential to develop further understanding of the fog deposition process. With knowledge about the size and chemical composition of single droplets, a disjunct eddy-covariance approach (Rinne et al. 2000) could be directly applied and yield information about the deposition (or emission) fluxes of size-segregated droplets.

11.7 Conclusions

From 6 months of fog water flux measurements in the year 2000 at the Waldstein site we found the following:

- Fog water deposition is about 4.5 % of total water input at the site for the experimental period. Additional measurements in 2001 and 2002 were performed to further examine these numbers but have not yet been completely evaluated. The direct flux measurements show that common models after Lovett (1984) and Pahl (1996) may overestimate the fog water input at the site.
- Fog deposition of major ions is about the same as rain deposition. This is due to the enrichment of these ions in fog water compared to rain water. For the major ions NH_4^+, NO_3^-, and SO_4^{2-} fog deposition varies between 79 and 110 % of the rain deposition.
- Fog deposits 9.5 kg N ha^{-1} $year^{-1}$ (NH_4^+-N and NO_3^--N) which is slightly higher than that deposited by rain (8.7 kg N ha^{-1} $year^{-1}$).
- Fog water acidity has declined throughout recent years at the Waldstein site.

Acknowledgements. We thank J. Gerchau for help during the field campaigns and BITÖK's ZAN for chemical analyses. We are grateful to W. Eugster, R. Burkard, and E. Thalmann for fruitful discussions. The cooperation of P. Winkler (Deutscher Wetterdienst, Hohenpeißenberg) sharing an operational version of the fog deposition model with us is gratefully acknowledged. The studies were supported by the Deutsche Forschungsgemeinschaft (DFG) through grant Kl 623/4.

References

Baumgardner RE, Selma SI, Lavery TF, Rogers CM, Mohnen VM (2003) Estimates of cloud water deposition at mountain acid deposition program sites in the Appalachian Mountains. J Air Waste Manage Assoc 53:291–308

Baumgartner A (1958) Nebel und Niederschlag als Standortfaktor am Großen Falkenstein (Bayrischer Wald). Forstwiss Centralbl 13:257–272

Baumgartner A (1959) Das Wasserangebot aus Regen und Nebel sowie die Schneeverteilung in den Wäldern am Großen Falkenstein (Bayrischer Wald). Wald Wasser 3:45–54

Best AC (1951) Drop-size distribution in cloud and fog. Q J R Meteorol Soc 77:418–426

Beswick KM, Hargreaves K, Gallagher MW, Choularton TW, Fowler D (1991) Size-resolved measurements of cloud droplet deposition velocity to a forest canopy using an eddy correlation technique. Q J R Meteorol Soc 117:623–645

Bopp S (2000) Entwicklung eines Biotestverfahrens zur Toxizitätsabschätzung von Stoffgemischen in Nebelproben. Diploma Thesis, University of Bayreuth

Burkard R, Eugster W, Wrzesinsky T, Klemm O (2002) Vertical divergences of fogwater fluxes above a spruce forest. Atmos Res 64:133–145

Daube BC, Flagan RC, Hoffmann MR (1987) Active cloudwater collector. Patent no 4697462:1–13. United States Patent and Trademark Office, Arlington

Deirmendjian D (1969) Electromagnetic scattering on spherical polydispersions. Elsevier, New York

Grießbaum F (2002) Chemistry of cloud interstitial particles: sample collection for ion chromatography (IC), scanning electron microscopy (SEM), and carbon (TC, EC, BC, and OC) analysis. Diploma Thesis, University of Bayreuth

Grunow J (1955) Der Nebelniederschlag im Bergwald. Forstwiss Centralbl 74:21–36

Herterich R (1987) Eignung und Anwendung von Rotating Arm Collectoren zur Bestimmung von Nebeleigenschaften – Flüssigwassergehalt, ionische Spurenstoffe, Wasserstoffperoxid. Diploma Thesis, University of Bayreuth

Hottenroth S (2001) Nitrophenole im Nebel: Analytik und Interpretation atmosphärischer Parameter. Diploma Thesis, University of Bayreuth

Joslin JD, Mueller SF, Wolfe MH (1990) Tests of models of cloudwater deposition to forest canopies using artificial and living collectors. Atmos Environ 24A:3007–3019

Kowalski AS, Anthoni PM, Vong RJ, Delany AC, Maclean GD (1997) Development and evaluation of a system for ground-based measurement of cloud liquid water turbulent fluxes. J Atmos Ocean Technol 14:468–479

Linke F (1916) Niederschlagsmessungen unter Bäumen. Meteorol Z 33:140–141

Lovett G (1984) Rates and mechanisms of cloud water deposition to a subalpine balsam fir forest. Atmos Environ 18:361–371

Lovett G, Reiners W (1986) Canopy structure and cloud water deposition in a subalpine coniferous forest. Tellus 38B:319–327

Marloth H (1906) Über Wassermengen welche Sträucher uns Bäume aus treibendem Nebel und Wolken Auffangen. Meteorol Z 23:547–553

Mueller S (1991) Estimating cloud water deposition to subalpine spruce-fir forests I. Modifications to an existing model. Atmos Environ 25A:1093–1104

Mueller S, Joslin L, Wolfe M (1991) Estimating cloud water deposition to subalpine spruce-fir forests I. Model testing. Atmos Environ 25A:1105–1122

Pahl S (1996) Feuchte Deposition auf Nadelwälder in den Hochlagen der Mittelgebirge. Berichte des Deutschen Wetterdienstes, Offenbach

Pahl S, Winkler P (1995) Höhenabhängigkeit der Spurenstoffdeposition durch Wolken auf Wälder. Abschlussbericht, Deutscher Wetterdienst, Meteorologischen Observatorium Hohenpeißenberg

Richartz H (1989) Nitrierte Phenole im Nebel. Diploma Thesis, University of Bayreuth
Richartz H, Reischl A, Trautner F, Hutzinger O (1990) Nitrated phenols in fog. Atmos Environ 24A:3067–3071
Rinne HJI, Delany JP, Greenberg JP, Guenther AB (2000) A true eddy accumulation system for trace gas fluxes using disjunct eddy sampling method. J Geophys Res 105:24791–24798
Römpp A (1999) Haloacetate und Nitrophenole im Nebel. Diploma Thesis, University of Bayreuth
Saxena VK, Lin N-H (1990) Cloud chemistry measurements and estimates of acidic deposition on an above cloudbase coniferous Forest. Atmos Environ 24A:329–253
Thalmann E, Burkard R, Wrzesinsky T, Eugster W, Klemm O (2002) Ion fluxes from fog and rain to an agricultural and a forest ecosystem in Europe. Atmos Res 64:147–158
Trautner F (1988) Entwicklung und Anwendung von Meßsystemen zur Untersuchung der chemischen und physikalischen Eigenschaften von Nebelwasser und dessen Deposition auf Fichten. Dissertation, University of Bayreuth
Trautner F, Reischl A, Hutzinger O (1989) Nitrierte Phenole im Nebelwasser – Beitrag ur Waldschadensforschung. USSF-Z Umweltchem Ökotox 3:10–11
Vong RJ, Kowalski AS (1995) Eddy correlation measurements of size dependent cloud droplet turbulent fluxes to complex terrain. Tellus 47B:331–332
Weigl W (2001) Toxizitätsgeleitete Identifizierung von organischen Substanzen und Bestimmung des Anteils an Metalltoxizität im Nebel. Diploma Thesis, University of Bayreuth
Wrzesinsky T (1998) Sommerlicher Nebel im Fichtelgebirge: Häufigkeit und chemische Zusammensetzung. Diploma Thesis, University of Bayreuth
Wrzesinsky T (2003) Direkte Messung und Bewertung des nebelgebundenen Eintrags von Wasser und Spurenstoffen in ein montanes Ökosystem. Dissertation, University of Bayreuth
Wrzesinsky T, Klemm O (2000) Summertime fog chemistry at a mountainous site in central Europe. Atmos Environ 34:1487–1496

12 Turbulent Deposition of Ozone to a Mountainous Forest Ecosystem

O. KLEMM, A. MANGOLD, and A. HELD

12.1 Introduction

Intensive research activities within the past 15 years have yielded a good understanding of the processes within the troposphere that lead to photochemical episodes with increased ozone (O_3) and other secondary trace gas concentrations (e.g., Finlayson-Pitts and Pitts 2000). Atmospheric ozone may act phytotoxicologically (e.g., Sandermann et al. 1997). The evaluation and quantification of the potential damage to vegetation through high O_3 concentrations are not straightforward because O_3 (like any other gas) can act on the plant metabolism only if it is deposited onto the surfaces or taken up into the mesophyll. The AOT (accumulated exposure over a threshold) concept (Fuhrer et al. 1997) attempts to quantify toxic O_3 levels. For example, AOT40 is the sum of all hourly O_3 mixing ratios over 40 ppb during the daylight hours in one vegetation period. For a Norway spruce forest, an AOT40 value of 10,000 ppb $\times$ h (=10 ppm $\times$ h) may be regarded as a critical level. This concept implies that, once the plant metabolism is active, high O_3 concentrations increase the phytotoxic potential. With the AOT40 concept, the possible effect of reduced O_3 mixing ratios since 1988 (as observed for our mountainous ecosystem research site in central Europe; Klemm, this Vol.) may be quantified.

However, a high concentration of O_3 in ambient air does not necessarily mean that a high amount of O_3 is taken up by the plants. Transport through the boundary layer and subsequent deposition to plant surfaces and uptake through the stomata must be carefully considered in this context. Grünhage et al. (1999) argue that results of routine ambient air quality monitoring stations, typically measured within the turbulent boundary layer at 3 m above ground, should not be used to calculate AOTs because the concentrations at the laminar boundary layer of the plant surfaces are lower than that within the turbulent boundary layer. They propose the use of a simple deposition model to parameterize the deposition velocity. However, the uncertainties

Ecological Studies, Vol. 172
E. Matzner (Ed.), Biogeochemistry of Forested Catchments in a Changing Environment

concerning the precision and accuracy of this concept when applied to various types of vegetation remain large.

We measured and modelled the deposition of ozone at the Waldstein forest ecosystem research site in the Fichtelgebirge mountains (NE Bavaria) over extended time periods between 1999 and 2002. We identified large discrepancies between modelled and measured fluxes (Klemm and Mangold 2001), particularly during the nights, and more pronounced during the early and late phases of the growing seasons, rather than during the months of June, July and August. Detailed micrometeorological analyses confirmed these findings and ruled out the roles of artefacts or biases, such as drainage flow, in the measured turbulent deposition of ozone.

This chapter summarizes findings from several growing seasons. Results from more detailed experiments are shown, when two parallel flux setups were operated in 2002, one above the canopy and the other one below the canopy. These investigations aimed to study the contribution of O_3 deposited on the soil surface to the overall deposition flux.

12.2 Site

The Bayreuth Institute for Terrestrial Ecosystem Research (BITÖK) operates an ecosystem research site in the Fichtelgebirge mountain range, NE Bavaria. Most investigations focus on the Lehstenbach catchment (Gerstberger et al., this Vol.) which is often called the Waldstein area, according to the nearby mountain peak. Most ozone deposition measurements were performed at a meteorological walk-up tower at the Weidenbrunnen site, at 50°08′32″N, 11°52′04″E, 775 m a.s.l., a Norway spruce plantation (about 18 m high) planted in about 1945. A parallel system was operated in summer 2002 within the Coulissenhieb site (Gerstberger et al., this Vol.). This site is about 200 m to the NNW from the tower site and lies within its main footprint area during westerly winds (Mangold 1999). The Norway spruce trees of the Coulissenhieb site are older (about 120 years) and taller (about 30 m) than those at the Weidenbrunnen site.

12.3 Experimental Setup

We measured the deposition flux of ozone with the eddy covariance method (e.g., Foken et al. 1995; Güsten and Heinrich 1996; Aubinet et al. 2000). At the tower site, we measured turbulence parameters between 1999 and 2002 with a Solent-R2 (Gill, UK) sonic anemometer at 32 m above ground, operating at 20.8 Hz. The ozone concentration was determined with a setup as described by Güsten and Heinrich (1996) and Klemm and Mangold (2001), using the fast response GFAS ozone sonde OS-G-2. The raw signals were calibrated in a post-processing routine with the routine O_3 data from the nearby Pflanzgarten site (Klemm 2004). We computed the fluxes over 30-min averaging periods. More details about data processing and quality control are given in Klemm and Mangold (2001). A similar setup was operated from 18 July to 5 August 2002, at 3 m above ground level within the Coulissenhieb site. A Young model 81000 ultrasonic anemometer and a second GFAS ozone sonde OS-G-2 were employed. Sample acquisition frequency was 10 Hz. A standard UV ozone monitor (MLU M-400, inlet at 3 m above ground level) was used for measurement of below-canopy O_3 concentrations and calibration of the fast ozone sonde in 30-min intervals. The quality control routines for the flux computation were identical to those employed at the nearby tower site insofar as the flux data records were filtered by application of stationarity tests and exclusion of data records with too low friction velocity (<0.1 m s^{-1}). The integral turbulence parameters were not used for quality control of these data sets. Both UV ozone monitors were calibrated in June 2002 applying an externally supervised quality control procedure.

12.4 Big Leaf Model

We predicted the deposition of O_3 by use of a simple model. The deposition flux F_{O3} (in units nmol m^{-2} s^{-1}) is computed as:

$$F_{O_3} = -\frac{1000}{MV} \cdot q_{O_3} \cdot v_{d,O_3} \tag{1}$$

with q_{O3} being the air mixing ratio of O_3 (in units ppb, which is nmol mol^{-1}), MV the mol volume (e.g., 24.45 l mol^{-1} under standard conditions 25 °C, 1,013 hPa), and $v_{d,O3}$ the deposition velocity of O_3 (in units m s^{-1}). This inferential model approach is parameterized as the *big leaf model*, which means that only one receptor surface is present in the model and that turbulent exchange within the forest canopy is neglected. The deposition velocity is parameterized as a series of resistances against transport (Fig. 12.1). Before deposition can occur, atmospheric O_3 has to overcome the turbulent atmospheric resistance, R_a, and the resistance of the quasi laminar boundary layer R_b (all resistances are in units s m^{-1}). Finally, either it deposits to a cuticle or it diffuses into a stoma, against R_{cut} or R_{stom}, respectively. The deposition velocity can now be computed as:

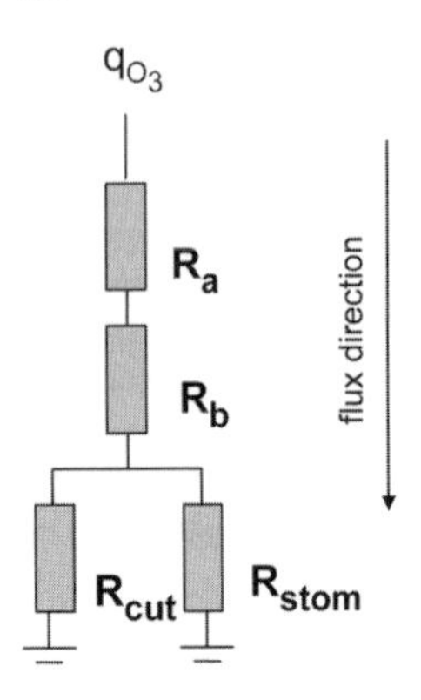

Fig. 12.1. Schematics of the simple big leaf approach to model the deposition of O_3. R_a and R_b are the turbulent and quasi laminar boundary layer resistances, respectively. R_{cut} and R_{stom} represent cuticular and stomatal resistances

$$v_{d,O_3} = \frac{1}{R_a + R_b + \left(\frac{1}{R_{cut}} + \frac{1}{R_{stom}}\right)^{-1}} \quad (2)$$

</equation>

The parameterization of the resistances is described in detail in Klemm and Mangold (2001). Briefly, R_a and R_b are calculated from measured data of the ultrasonic anemometer above the tree canopy following formulations given by Hicks et al. (1987), R_{cut} is parameterized to values between 1,000 and 2,000 s m^{-1} as a function of the measured leaf surface wetness (Klemm et al. 2002), and the stomatal resistances R_{stom} are derived from a plant physiological model (Falge et al. 2000). Obviously, this model allows only deposition of O_3, but no emission.

12.5 Modelled Fluxes

The big leaf model is a simple tool to predict long time series of O_3 deposition. It requires standard meteorological data driving the plant physiology model, ultrasonic wind data for the parameterizations of R_a and R_b, and leaf wetness data to estimate R_{cut}. An example time series is shown for 29 June 2000 in Fig. 12.2 (upper graph). During the nights, the magnitude of the deposition flux is limited to −0.4...−1.0 nmol m^{-2} s^{-1} (depending on the time of year), because the stomata are closed and high R_{cut} values allow for no significant deposition to the outer plant surfaces. During daytime, however, the deposition flux is high (up to −9 nmol m^{-2} s^{-1} in May and June), mainly due to the high stomatal conductance (low R_{stom}) during photosynthetic activity of the spruce trees. The pronounced diurnal behavior is established throughout the entire growing season. Figure 12.3 shows modelled deposition fluxes over the 6-month vegetation period in 2000. For each month, a median diur-

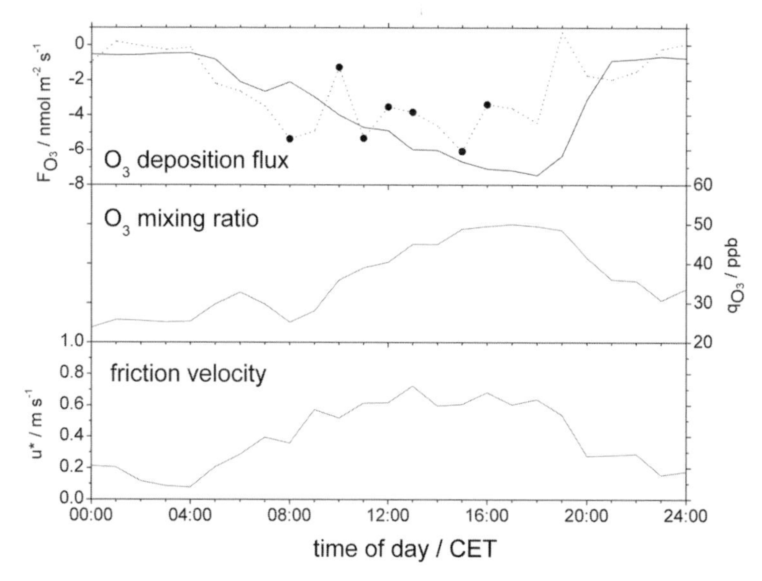

Fig. 12.2. *Above* Modelled (*solid line*) and measured (*dotted line*) O_3 deposition flux at 32 m above ground level (above canopy) at the tower site Weidenbrunnen on 29 June 2000. *Bullets* indicate the high-quality data. *Middle* O_3 mixing ratio at the nearby Pflanzgarten site. *Below* Friction velocity at the tower site. *CET* Central European time, which is 1 h ahead of universal time (UT)

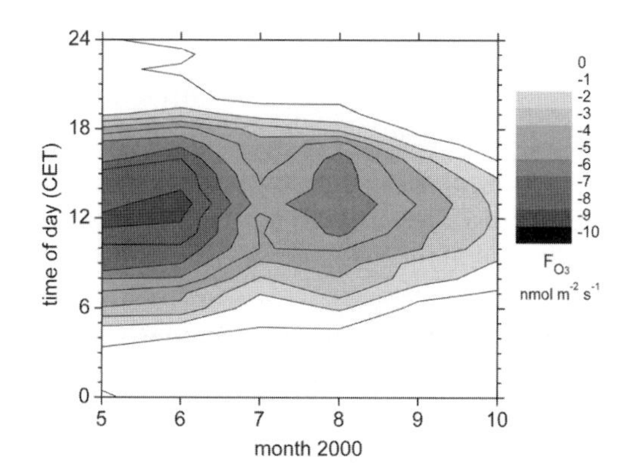

Fig. 12.3. Interpolated contour of monthly medians of the hourly deposition fluxes of O_3 during the year 2000, modelled data

nal cycle was calculated from the medians for each hour of the day. It is evident from Fig. 12.3 that the highest modelled deposition fluxes occurred at 12 h during the months of May and June. The peaks around noon-time persist throughout October; however, the maxima decrease from almost -10 nmol m^{-2} s^{-1} to values lower than -3 nmol m^{-2} s^{-1}. This is due to the reduction in the intensity of incoming solar radiation and the shortening of daylight duration. Overall, stomatal conductivity (which equals $1/R_{stom}$) is the main driver for O_3 deposition.

12.6 Deposition to the Forest Canopy

For the measured fluxes at the tower site, micrometeorological quality control reduced the data sets to about one half of the available flux raw data. Data losses are due to not only instrumental malfunctions (resulting, for example, from power failures), but also low friction velocities, non-stationarities, and poorly developed turbulence (see Klemm and Mangold (2001) for more details). Nevertheless, large data pools remain for the growing seasons of 1999 and 2000, and further experiments of several weeks' duration each were successfully conducted in 2001 and 2002. Figure 12.2 (upper graph) shows measured O_3 fluxes on 29 June 2000. Several periods of significant deposition fluxes between −1.3 and −6.0 nmol m^{-2} s^{-1} were observed between 08:00 and 16:00 h CET (central European time, which is 1 h ahead of UT, universal time). Other fluxes were smaller in magnitude, some slightly positive, but did not pass quality control. If we accept that these fluxes were small in magnitude, a clear diurnal behavior of the measured deposition flux is evident from Fig. 12.2 (upper graph), although the measured fluxes exhibit a much larger variability than the modelled ones. The results shown in Fig. 12.2 suggest a correlation of the measured deposition flux with the ozone mixing ratio (middle graph) and with the friction velocity (lower graph), respectively. However, a statistical analysis over the entire vegetation period shows that there is no correlation whatsoever between F_{O3} and q_{O3} (Klemm and Mangold 2001), nor between F_{O3} and u^* (data not shown). In fact, the diurnal behavior as presented in Fig. 12.2 does not reflect typical conditions. More frequently, large (negative) deposition fluxes were observed during the nights, particularly later in the seasons.

In Fig. 12.4, the monthly medians of measured hourly fluxes are plotted in a similar fashion as the modelled ones in Fig. 12.3. Although the general magnitude of the measured median fluxes (−0.5...−8.6 nmol m^{-2} s^{-1}) agrees with the modelled ones, the differences between these two plots are striking. Typically, diurnal cycles with maximum (negative) fluxes during the daylight hours are absent in the data sets. In addition, the nighttime fluxes are generally much higher than predicted in the model. Thorough data analysis confirms the validity of these nighttime fluxes. The variability of the measured fluxes is high (for an example see Fig. 12.2). Some periods even exhibited upward fluxes.

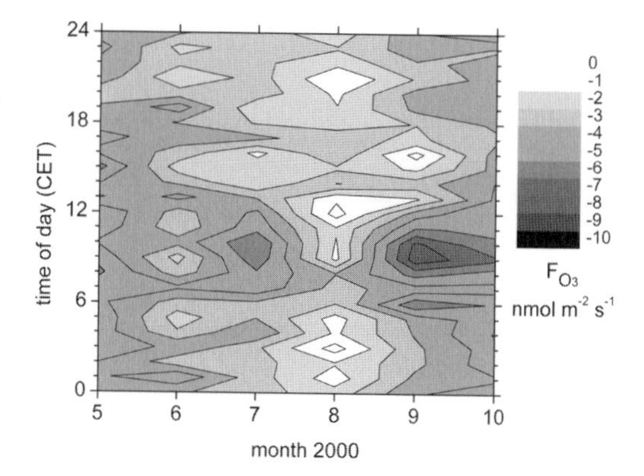

Fig. 12.4. Interpolated contour of monthly medians of the hourly deposition fluxes of O_3 during the year 2000, measured data. Scale and color coding are identical to those in Fig. 12.3

12.7 Deposition to the Forest Ground

The below-canopy measurements were conducted for a limited time period during summer 2002. Technical problems led to large data losses, and quality control led to a further reduction of the data sets by about 85 %. In many cases, the measured friction velocity u* within the forest stand was lower than the required 0.1 m s^{-1} which led to data rejection.

In Figs. 12.5 and 12.6, diurnal cycles of vertical fluxes as measured above and below the canopy are shown. On 29 July 2002 (Fig. 12.5), the vertical flux above the canopy is mostly oriented downward, i.e. from the atmosphere to the vegetation. No clear diurnal cycling is evident from the presented data. Most of the above-canopy fluxes are of high quality and lead to a reliable estimate of the average deposition flux of F_{O3}=–4.2 nmol m^{-2} s^{-1} for the time period shown. The below-canopy fluxes exhibit a lower quality before 08:00 and after 21:00 h, mostly due to limited turbulence and resulting low friction velocities within the forest stand. Therefore it is not clear if the upward fluxes during the early morning hours should be interpreted as fluxes from the forest stem space upward into the canopy, or as erroneous data. If we accept all data from Fig. 12.5 for the below-canopy site (including the ones with limited quality), the average flux is oriented downward (F_{O3}=–0.5 nmol m^{-2} s^{-1}) and thus by a factor of 9 smaller than the deposition flux above the canopy.

The fluxes as measured on 3 August 2002 (Fig. 12.6) show a strong diurnal cycle, with high deposition fluxes during the day, and low, but mostly downward, fluxes during the night-time hours. If we accept, again, the entire data sets by including the low-quality data (implicitly accepting that these fluxes are low), the ratio of the above-canopy flux (–5.7 nmol m^{-2} s^{-1}) to below-canopy flux (–1.7 nmol m^{-2} s^{-1}) is 3.3 in this case.

Diurnal cycles of the deposition fluxes of ozone could be calculated for four individual days. The results are summarized in Table 12.1 and indicate that for

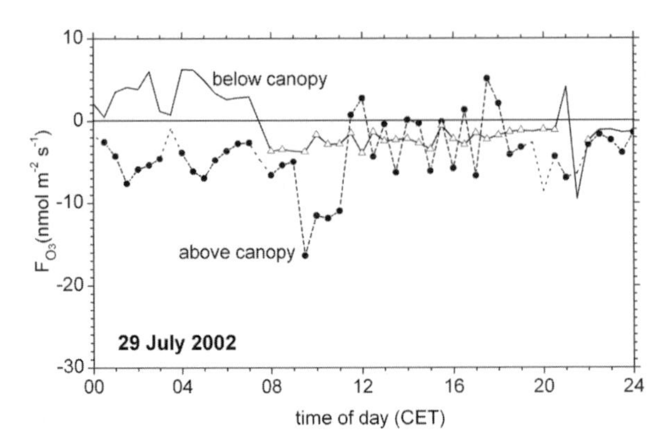

Fig. 12.5. Ozone deposition fluxes F_{O3} at the Weidenbrunnen tower site at 32 m above ground level (*above canopy, dotted line*) and within the Coulissenhieb forest site at 3 m above ground level (*below canopy, solid line*), respectively, as measured on 29 July 2002. *Bullets* indicate the high-quality data

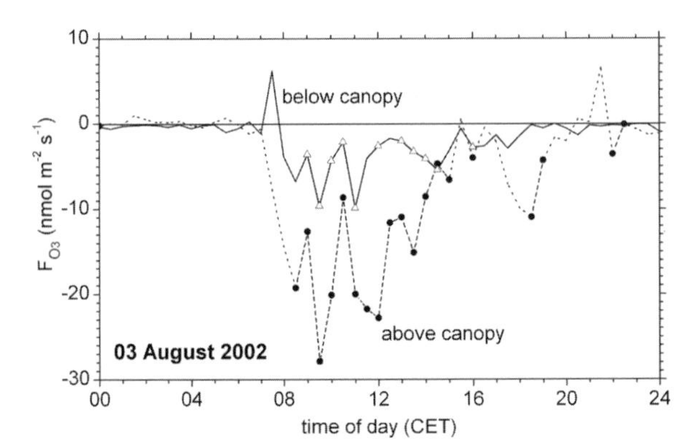

Fig. 12.6. Same as Fig. 12.5, but for 3 August 2002

Table 12.1. Average diurnal deposition fluxes of ozone as measured at the tower site Weidenbrunnen and within the Coulissenhieb forest, respectively. All measured flux data, including those with limited quality, were considered in this computation

Date	Average ozone deposition flux F_{O3} (nmol m^{-2} s^{-1})		
	Above canopy	Below canopy	Flux ratio
20 July 2002	−7.4	−0.9	8.0
29 July 2002	−4.2	−0.5	8.9
31 July 2002	−9.1	−1.4	6.6
03 August 2002	−5.7	−1.7	3.3

these days the flux ratio above/below canopy was between 3.3 and 8.9, respectively, indicating that a portion between 9 and 23 % of the flux that deposits above the forest penetrates through the canopy and passes a virtual balance level at 3 m above ground.

12.8 Synthesis and Conclusions

Modelling the dry deposition of ozone to a mountainous spruce forest ecosystem showed that the O_3 deposition is supposed to be largely driven by the ozone concentration in the atmospheric boundary layer, and by the stomatal conductivity of the vegetation. According to this modelling framework, most of the deposited ozone enters the stomata of Norway spruce. At our central European site, the modelled deposition fluxes were highest during daytime hours. Seasonal maxima occurred during June.

The measurement of O_3 deposition over more than two vegetation periods exhibited results that were different in various aspects. The variability of the measured deposition fluxes was generally much larger than that of the modelled ones. A correlation with O_3 mixing ratios or atmospheric turbulence (u^*) could not be found. Although the magnitude of daytime fluxes was comparable to those modelled with the big leaf model, the diurnal behavior was often different. Maximum deposition fluxes during the day did occur during the months of May to July, but hardly between September and October. The night-time fluxes were in many cases much higher than the modelled ones, often higher than the corresponding daytime fluxes. This indicates that during the night, stomatal fluxes did not drive the deposition of O_3. If we still accept the general validity of the big leaf model concept (Fig. 12.1), the atmospheric resistances R_a and R_b are small enough to allow high fluxes. Consequently, either R_{stom} or R_{cut} are generally too high in the initial parameterization. If we accept that the parameterization of R_{stom} is appropriate because it is based on a rich base of experiments and models developed for this experimental site, R_{cut} has to be adapted, in particular to allow high nocturnal fluxes. A recalculation of R_{cut} by use of the measured fluxes (for details see Klemm and Mangold 2001) leads to a mean cuticular resistance of $R_{cut} \approx 247$ s m^{-1}, which is four to eight times smaller than in the initial parameterization.

The absence of any significant correlation between the deposition flux on the one hand and the ozone concentration or mixing ratio on the other hand puts the AOT40 concept further into question. This concept implies that a higher mixing ratio of O_3 bears a higher phytotoxic potential. However, if a higher concentration fails to lead to a higher deposition flux, it is questionable if it may lead to a more intense or more direct damage to the plants. Further, it is unclear if a reduction of the O_3 mixing ratios, as observed at the Waldstein site since 1988 (Klemm, this Vol.) leads to a reduction of the phytotoxic potential of ozone at this forest site.

A further step towards understanding the dynamics of the O_3 deposition is the comparison of above-canopy vertical fluxes with below-canopy vertical fluxes. This implies that the model concept as displayed in Fig. 12.1 is given up, or expanded, and allows for partial flux of ozone through the canopy. Analysis of four diurnal cycles in July and August 2002 showed that between 9 and 23 % of the vertical ozone flux above the canopy could be found below the canopy. Either this flux will reach the ground and eventually deposit to the soil surface or another possibility is reaction with nitric oxide (NO) that may have been emitted from the soil. The reaction of NO with O_3 is fast and has been described to be important in other forest ecosystems (Rummel et al. 2002). Our results (from only 4 days of data coverage) indicate that an emission flux of NO between 0.5 and 1.7 nmol m^{-2} s^{-1} (daily averages, Table 12.1) would compensate for the measured deposition fluxes of ozone. This is in the same range as estimated NO emissions from spruce forest soils, as estimated with a process-oriented model (Stange et al. 2000).

Any extension of the big leaf model with the new flux pathway would require the introduction of further resistances (for example, in-canopy aerodynamic resistance), which would be hard to parameterize and would therefore introduce further uncertainties. In our view, this approach promises no further insight into ozone deposition pathways or quantification of vertical fluxes of ozone. A more complete experimental coverage of the deposition and emission processes near the forest soil surface, however, would lead to a better understanding of the processes involved. The absence of correlations between the measured O_3 deposition flux, on the one hand, and either the O_3 mixing ratio in air or the friction velocity, on the other hand, supports the conclusion that simplified models are not appropriate. Further experimental data are needed to better quantify the phytotoxic potential of atmospheric ozone.

Acknowledgements. We are grateful to G. Spindler (Institut für Troposphärenforschung, Leipzig) for lending us a second ozone sonde for this experiment. E. Falge (Universität Bayreuth) computed the STANDFLUX data for the estimate of stomatal conductivity. H. Kanter (Institut für Meteorologie und Klimaforschung, Garmisch-Partenkirchen) performed the external calibration of the ozone analyzers. Most of the 2002 experiments were conducted within the BEWA2000 experiment which was funded by the German Federal Ministry of Education and Research (BMBF).

References

Aubinet M, Grelle A, Ibrom A, Rannik Ü, Moncrieff J, Foken Th, Kowalski AS, Martin PH, Berbigier P, Bernhofer Ch, Clement R, Elbers J, Granier A, Grünwald T, Morgenstern K, Pilegaard K, Rebmann C, Snijders W, Valentini R, Vesala T (2000) Estimates of the annual net carbon and water exchange of forests: the EUROFLUX methodology. Adv Ecol Res 30:113–175

Falge E, Tenhunen JD, Ryel R, Alsheimer M, Köstner B (2000) Modelling age- and density-related gas exchange of *Picea abies* canopies in the Fichtelgebirge, Germany. Ann For Sci 57:229–243

Finlayson-Pitts BJ, Pitts JN Jr (2000) Chemistry of the upper and lower atmosphere. Academic Press, San Diego, 969 pp

Foken T, Dlugi R, Kramm G (1995) On the determination of dry deposition and emission of gaseous compounds of the biosphere–atmosphere-interface. Meteorol Z NF 4:91–118

Fuhrer J, Skärby L, Ashmore MR (1997) Critical levels for ozone effects on vegetation in Europe. Environ Pollut 97:91–106

Grünhage L, Jäger H-J, Haenel H-D, Löpmeier F-J, Hanewald K (1999) The European critical levels for ozone: improving their usage. Environ Pollut 105:163–173

Güsten H, Heinrich G (1996) On-line measurements of ozone surface fluxes, part I. Methodology and instrumentation. Atmos Environ 30:897–909

Hicks BB, Baldocchi DD, Meyers TP, Hosker RP, Matt DR (1987) A preliminary multiple resistance routine for deriving dry deposition velocities from measured quantities. Water Air Soil Pollut 36:311–330

Klemm O, Mangold A (2001) Ozone deposition at a forest site in NE Bavaria. Water Air Soil Pollut Focus 1(5):223–232

Klemm O, Milford C, Sutton MA, van Putten E, Spindler G (2002) A climatology of leaf surface wetness. Theor Appl Clim 71:107–117

Mangold A (1999) Untersuchung der lokalen Einflüsse auf die Turbulenzmessungen an der Station Weidenbrunnen. Diploma Thesis, Department of Micrometeorology, University of Bayreuth

Rummel U, Ammann C, Gut A, Meixner FX, Andreae MO (2002) Eddy covariance measurements of nitric oxide flux within an Amazonian rain forest. J Geophys Res 107: 8050–8058

Sandermann H, Wellburn AR, Heath RL (1997) Forest decline and ozone: a comparison of controlled chamber and field experiments. Ecological studies 127. Springer, Berlin Heidelberg New York

Stange F, Butterbach-Bahl K, Papen H (2000) A process-oriented model of N_2O and NO emissions from forest soils. 2. Sensitivity analysis and validation. J Geophys Res 105:4385–4398

13 The Emissions of Biogenic Volatile Organic Compounds (BVOC) and Their Relevance to Atmospheric Particle Dynamics

R. Steinbrecher, B. Rappenglück, A. Hansel, M. Graus, O. Klemm, A. Held, A. Wiedensohler, and A. Nowak

13.1 Introduction

The exchange of carbon compounds between the earth's surface and the atmosphere is a key process in global change. The sources and sinks of CO_2 and CH_4 are major triggers for the greenhouse gas budget of the atmosphere. The importance of carbon sequestration in terrestrial ecosystems is still one of the open questions in efforts to quantify key processes within the global biogeochemical carbon cycle. Vice versa, the exchange of C between land surface and the atmosphere is highly sensitive to global change. Changes of solar insulation, temperature, humidity and precipitation will affect the C fluxes in ecosystems. Secondary effects such as changed species interaction through land-use change and the deposition of toxic substances into ecosystems are also important. In the short term, major feedback loops include effects of stomatal control on the uptake of CO_2 and the loss of water by the vegetation as a result of the direct coupling of gas exchange to the energy input into ecosystems. On longer time scales (years to decades and centuries), additional feedbacks on biogeochemical cycles play an important role due to the time lag between, e.g., vegetation growth (C storage) and C loss via organic matter degradation, or vegetation extinction through extreme events (fire, herbivory), resulting in changed dynamics of the vegetation cover including its management.

Besides CO_2 and CH_4, other, more complex volatile organic compounds (VOC) are emitted from the soil/vegetation complex in ecosystems. These compounds play important roles in the carbon balance of the vegetation, as well as in the carbon chemistry of the atmosphere. Although the carbon loss of the biosphere by biogenic VOC emission amounts to only 2 % of the annual net-C uptake on the global scale (Guenther et al. 1995; Prentice et al. 2001), Mediterranean ecosystems, e.g. *Quercus ilex* forests, may emit all C assimilated in the form of monoterpenes during the 'midday depression' (Stein-

Ecological Studies, Vol. 172
E. Matzner (Ed.), Biogeochemistry of Forested Catchments in a Changing Environment

brecher et al. 1997). Considering the reactivity of monoterpenes and their particle formation potential in the atmosphere, the interaction with aerosol particle dynamics is high, closing another, yet more complex feedback loop between biosphere and atmosphere.

VOC released from anthropogenic as well as from biogenic sources comprise a large variety of different hydrocarbon groups. The biogenic volatile organic compounds (BVOC) include the quantitatively dominant group of the isoprenoids [isoprene, monoterpenes (e.g. α-pinene), sesquiterpenes] and oxygen-containing species (OVOC), such as alcohols, carbonyls and organic acids. In general, methane is separated from the VOC mainly due to the fact that CH_4 is produced (or oxidized) through soil microbial processes, while VOCs are predominantly produced by plants. In addition, due to its low reactivity, CH_4 plays only a minor role in air chemistra in regions where anthropogenic or biogenic hydrocarbons are abundant. The global anthropogenic VOC flux is estimated as 1×10^{14} g C year^{-1}, while biogenic sources may be higher by a factor of ten (1.2×10^{15} g C year^{-1}) (Steinbrecher 1994; Guenther et al. 1995). The magnitude of emissions varies significantly between ecosystems and between regions (e.g. Guenther et al. 1995; Simpson et al. 1999). Due to the high reactivity of BVOC towards atmospheric oxidants such as ozone and the HO and NO_3 radicals, BVOC contribute significantly to the regional and global atmospheric photo-oxidant budget.

Within the plant kingdom only certain plant species emit isoprenoids. Generally, isoprene emission is more frequently reported for deciduous plants whereas monoterpene emission is common for evergreen leaves (Helas et al. 1997). Both isoprene and monoterpene emission rates of some evergreen oak species (e.g. *Quercus ilex* L.) are strongly regulated by light and temperature via their direct impacts on leaf biochemistry (Steinbrecher et al. 1997). The monoterpene emission from resin-containing plants is strongly temperature controlled via increasing the vapour pressure of the target compounds with increasing temperature and thus increasing the diffusion rates through the cell walls of the resin ducts. Emission of all biogenic VOC is also constrained by water and nitrogen availability in the soil and the development stage of the leaves (via gene as well as enzyme activities) (Kreuzwieser et al. 1999). The complex regulation processes of the emission of an ecosystem as a result of the interaction of molecular biology, biochemistry, ecology, air chemistry and physics still remain to be elucidated.

The huge amount of chemical compounds emitted into the atmosphere is constantly removed from the atmosphere through oxidation followed by dry and wet deposition processes. The chemistry involved is complex (for review see, e.g., Atkinson 2000). The major initial step is the reaction with the HO radical, also called the atmospheric 'detergent', which in turn is formed during the photodissociation of O_3 and subsequent reaction with water. The relative amounts of hydrocarbons and nitrogen oxides (NO_x=NO+NO_2) also play crucial and interrelated roles in the mechanism of the chemical oxidation of hydrocarbons. At very low levels of NO_x, hydrocarbon oxidation removes O_3 and consumes HO, while at higher NO_x levels, more O_3 and reactive radicals are produced. In the case of O_3 addition to C=C double bounds, first reaction products are energy-rich biradicals that follow a number of chemical reaction pathways. Of significant importance is a set of chemical reactions that lead to formation of HO radicals and substituted alkyl radicals via formation of energy-rich hydroperoxide from hot esters. This formation of HO may be an important tropospheric HO source particularly at night (Atkinson 2000). During the night, or at low light conditions, the NO_3-radical chemistry may be important in the plant canopy, an issue that has not yet been studied very well. Due to the highly non-linear character of the chemical reactions involved in the degradation of VOC in the atmosphere, only detailed chemistry schemes (e.g. Derwent et al. 2001) in combination with good estimates of the initial conditions are able to describe and forecast, e.g. the ozone distributions in the atmosphere.

The oxidation of biogenic mono- and sesquiterpenes may contribute not only to the formation of ozone, but also to the production of secondary organic aerosol (SOA) as a result of the condensation of bifunctional highly condensable carboxylic acids, and thus affect the radiation balance of the earth (Hoffmann et al. 1997). Particle nucleation from organic vapours may increase the total particle number concentration, while condensation of low-volatile VOC on existing particles may increase their size and make them more efficient as cloud condensation nuclei (CCN). The ozone concentration in the atmosphere seems to play a crucial role in that processes. Under low ozone conditions, the aerosol yield from the photo-oxidation of monoterpenes is quite low, because the prevailing reaction chains lead to compounds that do not readily condense into particles. However, at higher O_3 concentrations, more low-volatile products are formed, which can then form particles (Kanakidou et al. 2000). The formation of aerosol particles and their subsequent growth to CCN size has been observed at remote continental sites, e.g. in Finland, and their relation to biospheric processes has been identified (Kulmala et al. 2001). However, the formation mechanisms and sources of these aerosols are poorly known, so their global source strength is still uncertain by an order of magnitude (Penner et al. 2001).

In summary, land surfaces emit and absorb a variety of non-CO_2 and non-CH_4 VOC of significant importance for the regional and global air chemistry and particle distributions, with implications for air quality as well as for the radiative forcing potential of the atmosphere and cloud microphysical processes. Atmospheric oxidant chemistry is directly or indirectly mediated by soil moisture patterns through the surface energy balance, trace gas emissions, fire frequency/severity, boundary layer development, and convective activity. Therefore, the investigations on the cycling of reactive carbon compounds in forests discussed in this section contribute significantly to the understanding of the role of the biosphere within the earth system.

A large national experiment was conducted in north-east Bavaria, Germany, in 2001 and 2002, focusing on the *Regional biogenic emissions of reactive volatile organic compounds (BVOC) from forests: Process studies, modelling and validation experiments* (BEWA 2000). The experiments were performed on a meteorological walk-up tower at the Weidenbrunnen site, located within the Lehstenbach catchment, close to its water divide (Matzner et al., this Vol.). The area is often called the Waldstein site, according to the nearby mountain peak. The Waldstein site is a well-suited ecosystem for this type of study because the dominant tree species Norway spruce [*Picea abies* (L.) Karst., with a canopy height of about 18 m] characterizes one of the most representative forest types in Germany, the fetch conditions for micrometeorological flux studies at this site are very good (Mangold 1999), and long-term records of ecosystem as well as atmospheric parameters exist. The long-term data sets make it possible to evaluate the short-term integrated field experiment data in the light of ongoing changes in ecosystem processes possibly as a result of global change.

13.2 Flux Measurement Methods

Micrometeorological flux measurement systems are widely used to investigate the trace compound exchange between the earth surface and the atmosphere. Currently two approaches are favoured: (1) the eddy covariance (EC) technique, and (2) the relaxed eddy accumulation (REA) technique (Dabberdt et al. 1993; Oncley et al. 1993; Steinbrecher et al. 2000).

The basic method for determining turbulent energy and mass transport is the EC technique. This technique directly determines the turbulent flux of, e.g. trace gases by determining the covariance of the mixing ratio between the vertical wind speed. However, EC-flux calculations are only possible if (1) the sensor has a continuous fast response time (sampling frequency $\geq$ 10 Hz) with sufficient sensitivity and accuracy and (2) the sensor is specific for the compound of interest. These two constraints have limited the application of EC within the group of BVOC to isoprene by using an ozone chemiluminescence detector (Hills and Zimmerman 1990) and to acetone and formic acid by using a tandem mass spectrometer (Shaw et al. 1998).

In order to investigate the turbulent matter exchange between the biosphere and the atmosphere without having fast sensors available, adaptations of the EC technique are necessary. One approach is the eddy accumulation technique. This technique separates air according to the direction of the vertical wind speed w, measured in the vicinity of a sonic anemometer. The updraft and downdraft eddies, respectively, are pumped into two reservoirs at a rate proportional to the magnitude of w. The mixing ratios of the target compounds in the two reservoirs are determined by slow-response sensors.

Fig. 13.1. Relaxed eddy accumulation (REA) system at the Waldstein site at 50°08′32″N, 11°52′04″E, 775 m a.s.l., and 32 m above ground level. In the vicinity of the sonic anemometer inlets for the up- and down-draft eddies are shown. The control unit for the sonic anemometer as well as the fast switching solenoid valves for the REA system are mounted directly on the mast

Relaxed eddy accumulation (REA) is based on the suggestion of Businger and Oncley (1990) that the demands of the eddy accumulation may be relaxed by sampling air at a constant rate for updrafts and downdrafts, rather than proportionally. Recent setup of REA systems can be found, e.g., in Steinbrecher et al. (2000) and Schade and Goldstein (2001) and is shown in Fig. 13.1. Based on the REA system, the vertical flux is determined as:

$$F = b(\xi) * \sigma_w * \left(\overline{c^+} - \overline{c^-}\right) \quad (1)$$

with σ_w being the standard deviation of the vertical wind velocity, c^+ the concentration in the updraft reservoir, and c^- the concentration in the downdraft reservoir. The Businger–Oncley parameter b is a dimensionless factor. Empirical values are around 0.6 (e.g. Businger and Oncley 1990). The parameter b depends on the atmospheric stability ξ; however, if a threshold $|w_0|$ is applied, i.e. the 'up'-sample is collected in the case $w>w_0$, and the 'down'-sample is taken in the case $w<-w_0$, this dependence is minimized. Instead, b is dependent on the width of the dead-band. Introducing dynamic dead bands (Christensen et al. 2000), where the threshold $|w_0|$ is a function of σ_w, leads to an almost constant value of b. Introducing a threshold also leads to a better signal-to-noise ratio of (c^+-c^-). The value of b can be calculated based on concurrent REA and EC measurements of water vapour, CO_2, and air temperature, by inverting Eq. (1) (e.g. Pattey et al. 1993, 1999).

13.3 Fluxes of Primary and Secondary BVOC

During the BEWA 2000 field experiments, fluxes of VOC were determined with proton transfer reaction mass spectrometry (PTR-MS) coupled to a REA system. The PTR-MS is capable of measuring simultaneously a variety of organic trace gases, including oxygenated compounds. The PTR-MS technique is described in detail in Lindinger et al. (1998). Over the last few years the PTR-MS technique has gained in popularity in environmental sciences as a fast VOC sensor. Questions related to the fate of biogenic hydrocarbons both in physiological processes and in the atmosphere have been addressed (e.g. Fall et al. 1999; Warneke et al. 1999; Crutzen et al. 2000; Kreuzwieser et al. 2002). Recently, PTR-MS has been used to measure BVOC fluxes including oxygenated compounds from fields before and after cutting by direct EC or disjunct EC methods (Karl et al. 2001; Warneke et al. 2002), as well as VOC fluxes from a subalpine forest using virtual disjunct EC techniques (Karl et al. 2002). In our studies we connected the PTR-MS to a REA system that was used in previous studies as described in Steinbrecher et al. (2000). Air samples were taken at the top of the meteorological tower at the height of 32 m above ground, close to the Gill Sonic anemometer that controlled the REA sampling. Micrometeorological data were routinely screened for unfavourable effects such as rain/fog events and turbulence disturbances caused by the tower construction itself. For REA measurements the Businger–Oncley parameter was calculated using the vertical wind w and the virtual temperature T_v, i.e. parameters that are measured concurrently by the sonic anemometer. Rearranging Eq. (1) leads to:

$$b = \frac{F_{T_v}}{\sigma_w \cdot \Delta T_v} \qquad (2)$$

REA valves were activated based on σ_w results. For this canopy a b value of 0.39 (threshold velocity w_0=0.6) was determined.

Figure 13.2 shows some results of VOC fluxes obtained on 28/29 July 2001. The results clearly show diurnal variation of primary biogenic compounds such as isoprene and monoterpenes (Fig. 13.2A, B). Canopy fluxes of isoprene reached up to 7 nmol m^{-2} s^{-1} during daytime. The fluxes of the sum of monoterpenes were in the same range. On 28 July, the meteorological conditions were slightly more favourable for biogenic emissions due to enhanced temperatures and almost clear sky (see Fig. 13.3). In Fig. 13.3 the photolysis rate $J_{(O1D)}$ is plotted as an indicator of the primary production rate of the HO radical, the main scavenger for VOC, given by Poppe et al. (1995):

$$P(HO) \propto J_{(O^1D)} [O_3][H_2O] \qquad (3)$$

</equation>

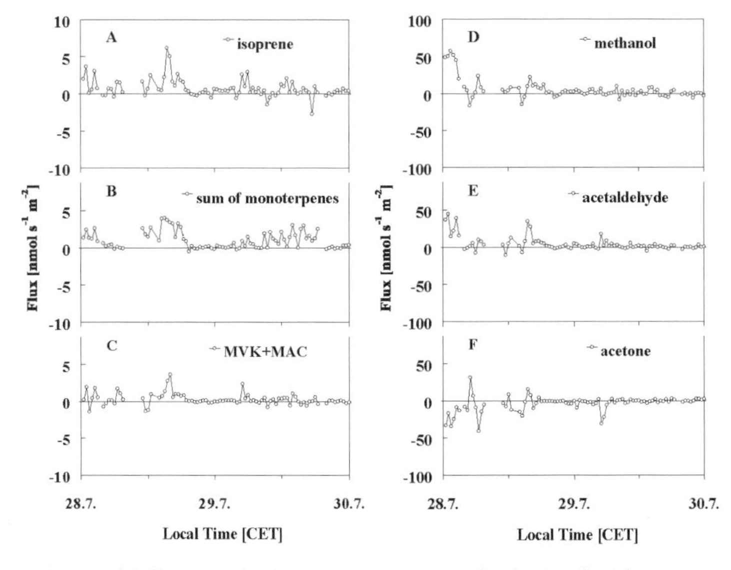

Fig. 13.2. Fluxes of different volatile organic compounds obtained with a REA-PTR-MS system over a Norway spruce at the Waldstein site on 28/29 July 2001. **A** Isoprene; **B** the sum of monoterpenes; **C** the sum of methyl-vinyl-ketone (*MVK*) and methacrolein (*MAC*); **D** methanol; **E** acetaldehyde; **F** acetone

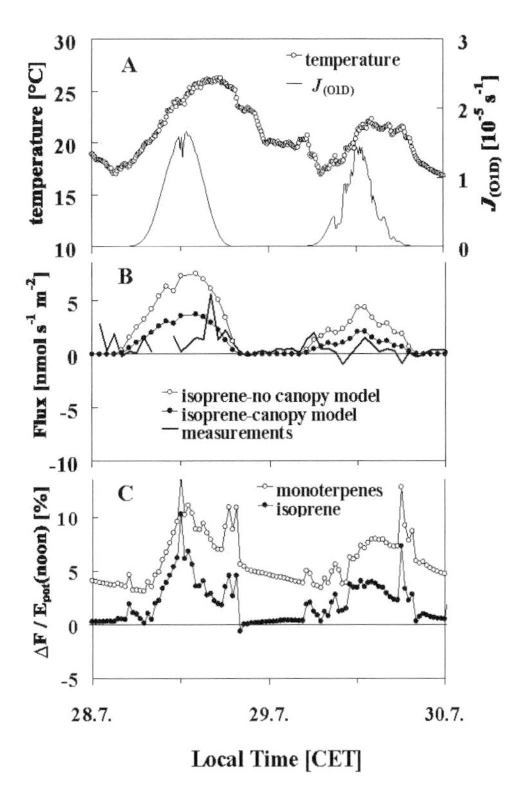

Fig. 13.3. **A** Air temperature and photolysis rate of ozone [$J_{(O1D)}$] at top of the Waldstein meteorological tower at height 31 m on 28/29 July 2001. **B** Sensitivity studies with calculations for isoprene emissions based on a BVOC model addressing tree canopy effects. **C** Calculations with the canopy atmospheric chemistry emission model (CACHE) for the difference between potential and net emission ΔF at the top of the Norway spruce canopy relative to average noon-time value of potential emission E_{pot} (noon). Measured BVOC fluxes are shown in **B** (Gerhard Smiatek, Renate Forkel, IMK-IFU, unpubl. results)

The PTR-MS technique also allowed retrieving data of oxygenated VOC such as methyl-vinyl-ketone (MVK) and methacrolein (MAC) (Fig. 13.2C), methanol (Fig. 13.2D), and carbonyls (Fig. 13.2E, F). MVK and MAC are secondary products from isoprene degradation (Tuazon and Atkinson 1990; Stroud et al. 2001). The BEWA 2000 data confirm this relationship and reveal a better correlation of MVK+MAC with isoprene (r^2=0.78) than with the sum of monoterpenes (r^2=0.30). In our study, MVK+MAC fluxes were about 30 % lower than isoprene fluxes. Both observations indicate active photochemical degradation of isoprene in this area contrary to a study of Lamanna and Goldstein (1999), who reported both time lags between the time series of isoprene and MVK+MAC concentrations and MVK+MAC concentrations that were twice the isoprene values, and related this behaviour to transport processes. Among the oxidation products of isoprene, MAC is of particular interest since it is the only precursor of peroxymethacryloyl nitrate (MPAN), a compound of the peroxyacyl nitrates group (PANs) that may be used as a specific biogenic marker in transport studies (Nouaime et al. 1993; Williams et al. 1997).

Figure 13.2D–F shows results for some oxygenated compounds, for which flux data have scarcely been reported to date. Figure 13.2D shows the range of observed methanol fluxes which reached a value of 22 nmol s^{-1} m^{-2} on 28 July in the afternoon. This is comparable with previously reported values of 1–2 mg h^{-1} m^{-2} from a lodgepole pine plantation (Schade and Goldstein 2001) and a subalpine coniferous forest (Baker et al. 2001; Karl et al. 2002). Daytime methanol mixing ratios reach peak values of 15 ppbv comparable to values reported by Goldan et al. (1995) above a pine forest plantation in western Alabama during the Southern Oxidant Study (SOS). These high methanol mixing ratios are consistent with its relatively long atmospheric lifetime of approximately 10 days (Heikes et al. 2002), making advection an important issue. According to present understanding, methanol in the biosphere is produced primarily due to cell expansion through demethylation of pectine (Galbally and Kirstine 2002, and references therein). This would infer enhanced methanol flux values during the growing season. Warneke et al. (1999) reported methanol production from decaying plant matter which also could contribute to methanol fluxes above forest canopies. However, enclosure measurements of a Norway spruce twig during the summer 2002 intensive field experiment at the Waldstein site did not reveal significant direct methanol fluxes, possibly due to lower temperatures compared to 2001 (data not shown). In summary, methanol fluxes were observed in summer 2001 at the Waldstein site, which are comparable to previous findings above pine forest sites. The data set is very limited and does not allow a quantification of the source strength of individual contributions such as cell wall expansion, plant decay, or direct temperature-dependent emissions.

Figure 13.2E,F shows fluxes of acetaldehyde and acetone, two compounds that are typical intermediate compounds in the photochemical degradation of both anthropogenic and biogenic VOC. They may also be emitted directly in combustion processes. Growing evidence shows that these species have significant direct biogenic sources (Fall 1999). Plants are likely to emit acetaldehyde under a variety of stress conditions, especially when their roots are flooded (Fall 1999; Kreuzwieser et al. 1999), e.g. after intensive rainfalls. Acetone may be released directly from coniferous forests from either buds (MacDonald and Fall 1993) or decaying matter (Warneke et al. 1999). The results of the BEWA campaign reveal only a poor correlation between acetaldehyde and acetone fluxes (r^2=0.02), indicating different origins of both species. The observations show that acetone deposition prevails, suggesting that biogenic sources of acetone play a minor role, at least at this site. Acetaldehyde fluxes usually better match with isoprene fluxes (r^2=0.55). However, they correlate best with MVK+MAC fluxes (r^2=0.81). This is an indication that acetaldehyde transported from the forest canopy into the boundary layer is primarily of photochemical origin by oxidation of VOC rather than a result of direct plant emission.

13.4 Modelled Versus Measured Isoprenoid Emission

For modelling BVOC emissions on ecosystem, regional, and global scales, three different modules are necessary (Smiatek 2001; Steinbrecher and Smiatek 2004): (1) databases storing all required tabular and spatial geo-data, (2) a GIS/RDBMS interface processing the input data according to specific model domains and subdomains and (3) the BVOC emission module.

Norway spruce is considered to be the major BVOC emitter at the Waldstein site. Therefore, the biochemical, ecological, and biophysical data of this tree species are used in the modelling activities. However, the BVOC modelling tool includes 128 different major plant species of Europe emitting up to 33 different chemical compounds. For modelling regional BVOC emissions, 29 land-use data sets are available covering large parts of Europe.

The implemented semi-empirical BVOC emission module is based on approaches presented by Guenther et al. (1995), Simpson et al. (1999), and Steinbrecher et al. (1999) and discussed in detail in Stewart et al. (2003). The emission E of the chemical compound i by the plant species k is described as:

$$E_{ik} = A_k * EF * SLW_k * C(T,PAR) \tag{4}$$

In Eq. (4), A_k describes the area of the emitting plant species k (in m^2), EF is the average emission factor (in $\mu g\ g^{-1}\ h^{-1}$), SLW_k is the foliar biomass density (in $g\ m^{-2}$), and C is a dimensionless environmental correction factor. It takes into account the effects of temperature (T) and solar radiation, given as the photosynthetically active radiation (PAR), on the emission factor.

The canopy/chemistry model CACHE (canopy atmospheric chemistry emission model; Forkel et al. 1999) predicts diurnal courses of temperature and concentrations of water vapour and chemical constituents by solving the prognostic equations for heat and mass. Vertical transport of heat and of 77 trace species in the atmosphere is calculated between the soil surface and 3 km above the ground. Presently, the turbulent fluxes between the model layers are parameterized by a gradient description applying an exchange coefficient that is dependent on thermal stratification and the wind profile (Forkel et al. 1990). Soil temperature and water content are also predicted (Forkel et al. 1984). Chemical transformation rates are computed with the RACM (regional atmospheric chemistry mechanism) gas phase chemistry mechanism (Stockwell et al. 1997). The RACM mechanism includes oxidation schemes for isoprene, monoterpenes with one double bond, which are represented by α-pinene (API), and monoterpenes with two double bonds, which are represented by d-limonene (LIM). The required photolysis frequencies above the canopy are also modeled (Madronich 1987). Inside the canopy the photolysis frequencies are parameterized on the basis of the computed decrease of the PAR. Leaf surface temperatures and heat and water vapour fluxes between the leaves or needles and the ambient air are calculated for each model layer (Norman 1979). The short-wave and long-wave radiative fluxes within the canopy are also considered as well as the stomatal resistances (Jarvis 1976).

Emissions of isoprene, monoterpenes, and other VOC by the leaves are modeled, taking into account the predicted leaf temperatures and the PAR fluxes in the canopy. The emission is given by

$$E = EF(pool) * C_T + EF(syn) * C_T * C_L \tag{5}$$

C_T and C_L are correction terms accounting for the temperature as well as PAR dependency of the emission factors EF of VOC pools and actual synthesis, respectively (Guenther et al. 1995; Steinbrecher et al. 1999). NO emissions at the soil surface are considered to be dependent on temperature. Corresponding calculations refer to the results obtained within the EU project FOREXNOX (Emeis et al. 1998). Deposition of trace gases at the soil and leaf surfaces are also considered (Wesely 1989). The total number of model layers, the number of layers in the canopy, and the trunk space, the trunk height, and the canopy height can be varied by the user within reasonable limits. For the model applications shown below, 40 layers ranging from subground up to a height of 3,000 m above ground were taken. The trunk space height and the layers were according to the Waldstein canopy structure. The soil is described by 15 layers. CACHE has been validated for several boreal forest types in Europe, e.g. a *Pinus sylvestris* (L.) forest in the River Rhine valley (Forkel et al. 2001).

In Fig. 13.3 meteorological parameters, along with measured and modelled isoprenoid fluxes for the Waldstein site, are shown for two selected days in

July 2001. The first day is characterized by more or less cloudless sky with temperatures up to 26 °C. The second day is partly cloudy with temperatures not exceeding 23 °C. In this presentation, only the ozone photolysis rate is plotted as an indicator for PAR as $J_{(O1D)}$ is directly linked to the HO reactivity of the atmosphere [see Eq. (3)]. In Fig. 13.3B, the impact of the canopy structure on the potential isoprene emission is shown by performing a sensitivity study. Without taking into account canopy effects (e.g. shading, transpiration cooling, tree density), the potential emission of the forest stand is overestimated by up to a factor of two. In the case where canopy effects are considered, the modelled fluxes match better the measured fluxes; however, still some systematic deviations exist. The measured fluxes are smaller compared with the modelled potential fluxes. This may be a result of the degradation of isoprene during its transport from the emission source to the measurement receptor site as well as the transport properties of the atmosphere. As pointed out in previous sections, isoprene reacts fast with oxidants that are present in the atmosphere and thus secondary products are formed already inside the canopy. The chemical degradation of the primary emitted compounds will likely lead to a smaller flux for these species. For isoprene, the chemical degradation occurs mostly during the day whereas monoterpenes are also degraded by night-time chemistry. Therefore, the emission is quite low at night. Apart from chemical transformations, the net emission of compounds from a forest may further be lowered through limited turbulent transport as shown in Figs. 13.3B and C. Around noon (28 July 2001), the model run by CACHE indicates a strong decrease in the net emission source strength. The decrease in the net emission is also visible in the low measured flux at the same time. An opposite event occurs in the afternoon of the same day, when a sharply increasing measured flux occurs along with reduced chemistry but increased turbulent transport. In general, the differences between modelled potential flux and modelled net emission are in the range between 5 and 12 % for these specific days. The results presented in this chapter show that net emission of compounds from a forest stand is a complex interaction between physical, chemical, and biological processes. Only parts of it are currently known. Therefore, present efforts in modelling the trace compound exchange between the biosphere and atmosphere are still at their early beginnings and not able to simulate adequately the effect of complex situations like cloudy weather on the net emission of a forest.

13.5 Particle Size Distribution and Fluxes

Determination of vertical particle fluxes through eddy covariance (EC) is still a challenging task. In recent years, however, much effort has been spent on measuring particle number fluxes above heterogeneous terrain such as forests (Gallagher et al. 1997; Buzorius et al. 1998, 2000) and urban areas (Dorsey et al. 2002). A combination of particle flux and size distribution measurements allows deeper insight into the aerosol dynamics at the Waldstein site. During the BEWA campaigns, an EC system comprising a sonic anemometer and two condensation particle counters (Held and Klemm 2002) was set up to measure particle number fluxes at 22 m above ground level, above the Norway spruce forest. Particle size distributions (PSD) were determined in the size range from 3–900 nm particle diameter at the same height as particle flux measurements using a differential mobility particle sizer (DMPS, e.g. Birmili et al. 1999).

To our knowledge, BEWA provided the first framework, where concurrent particle flux, PSD, and BVOC flux measurements were conducted in central Europe.

Figure 13.4 displays a typical evolution of the PSD (upper graph) and the corresponding particle number fluxes (lower graph) measured on 2 August 2001. On this day, a diurnal pattern of the turbulent particle number fluxes is found with small or even no fluxes during night-time and deposition fluxes in the order of 10^7–10^8 particles m^{-2} s^{-1} during daytime. Deposition fluxes clearly dominate over emission; however, the occurrence of short-term emission flux events shows a great impact on the PSD of the ambient particle population. For example, on 2 August a short-term particle emission episode can be observed around noon in conjunction with a steep drop of the particle concentration. The sudden occurrence of ultra-fine particles (<10 nm particle diameter) coincides with the onset of significant deposition fluxes at 08:00 h CET. Overall, the PSD exhibits a characteristic 'banana-shaped' evolution, indicating particle formation as well as particle growth through coagulation and condensation of low-volatile precursor gases. To a certain extent, these particle precursor compounds may result from BVOC reaction chains leading to low-volatile products especially at high ozone concentrations.

From these considerations one would expect particle emission fluxes from the forest stand. However, during episodes of particle formation the strongest particle deposition fluxes have been observed. Therefore, if particles have been formed within the forest stand, i.e. below the measuring level of the particle flux system, they must have been emitted from the stand as particles smaller than the minimum detectable particle size of 3 nm. The PSD evolution measured above the forest suggests that particle growth continues above the canopy. Thus, particles grown to sizes larger than 3 nm may be detected by the flux system as depositing particles (cf. Fig. 13.4).

Quantification of the aerosol yield from possible condensation processes of BVOC through EC particle flux measurements is limited in various ways. First, without knowledge of the chemical composition of the aerosol particles one can only speculate about the condensing species. Second, it is not straightforward to obtain particle mass fluxes from the measured particle number fluxes. Particles of different sizes are exchanged between the bios-

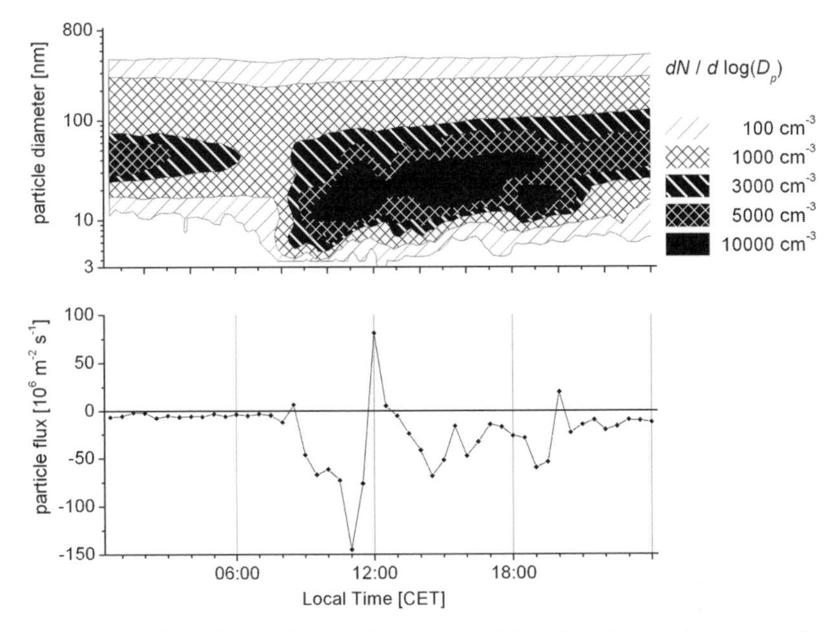

Fig. 13.4. Evolution of particle size distribution (*above*) and particle number fluxes (*below*) over Norway spruce forest at the Waldstein site on 2 August 2001. Shading indicates particle number concentration N normalised by the logarithmmic diameter Dp, dN/dlog (Dp)

phere and the atmosphere with different transfer velocities. To obtain mass fluxes, particle number fluxes have to be determined individually for various size ranges. This measurement challenge is currently the issue of intense research efforts. A modelling approach to convert number fluxes to mass fluxes using particle deposition models may be considered as semiquantitative at the most. Nevertheless, direct measurements of particle fluxes in combination with particle size distributions will help to evaluate the links between particle formation and particle growth processes, on the one hand, and biospheric processes such as BVOC emissions, on the other hand.

13.6 Conclusions

- The comprehensive understanding of biosphere–atmosphere interactions, in general, explicitly the contribution of VOC and particles to the carbon cycle, is still at its very beginning. A large national experiment was conducted at the Waldstein site in 2001 and 2002 (BEWA 2000) which contributed to both the understanding of matter exchange between forests and the atmospheric boundary layer, including further development of methods for quantifying trace compound exchange on an ecosystem scale, and the establishment of advanced modelling tools for calculating primary and net emission of biogenic VOC.

- The results presented demonstrate that new techniques in micrometeorological determination of matter exchange between the ecosystem and the atmosphere significantly enhance our understanding of emission and deposition processes of gases and particles. The application of new analytical techniques (PTR-MS) further elucidated the complexity of BVOC exchange of Norway spruce forests. It is concluded that in the summertime not only the C budget of a coniferous forest is governed by CO_2/CH_4 sources and sinks, but also an additional loss of carbon through biogenic carbon compound emission has to be considered. There is sound evidence that global change (higher temperature, longer drought periods, changed nutrient conditions, etc) is putting more and more stress on vegetation, likely favouring the emission of reactive VOCs, with negative effects on the C balance of ecosystems.
- The Waldstein experiments provided some answers about the fate of primary emitted substances from forest into the atmosphere but also gave rise to new questions: What is the contribution of BVOC to the C balance of ecosystems? How do other VOC apart from isoprenoids contribute to the reactivity of the forest atmosphere? How important is NO_3-radical chemistry inside the canopy? What is the contribution of different chemical degradation pathways of BVOC to particle formation? Answering these questions will be a major challenge for the future and will contribute to our understanding of the earth-system, which is needed for ensuring the sustainable development of our environment.

Acknowledgements. The authors gratefully acknowledge the funding of the German Department of Education and Research (BMBF) in the framework of the Atmosphere Research Program (AFO 2000). The model contributions and technical assistance from IMK-IFU by Renate Forkel, Gerhard Smiatek, and Dominik Steigner, and from the Institute of Ionphysics by Christian Lindinger are also gratefully acknowledged.

References

Atkinson R (2000) Atmospheric chemistry of VOCs and NO_x. Atmos Environ 34:2063–2101

Baker B, Guenther A, Greenberg J, Fall R (2001) Canopy level fluxes of 2-methyl-3-buten-2-ol, acetone and methanol by a portable relaxed eddy accumulation system. Environ Sci Technol 35:1701–1708

Birmili W, Stratmann F, Wiedensohler A (1999) Design of a DMA-based size spectrometer for a large particle size range and stable operation. J Aerosol Sci 30:549–553

Businger JA, Oncley SP (1990) Flux measurement with conditional sampling. J Atmos Ocean Technol 7:349–352

Buzorius G, Rannik Ü, Mäkelä JM, Vesala T, Kulmala M (1998) Vertical aerosol particle fluxes measured by eddy covariance technique using condensational particle counter. J Aerosol Sci 29:157–171

Buzorius G, Rannik Ü, Mäkelä JM, Keronen P, Vesala T, Kulmala M (2000) Vertical aerosol fluxes measured by the eddy covariance method and deposition of nucleation mode particles above a Scots pine forest in southern Finland. J Geophys Res 105:19905–19916

Christensen CS, Hummelshøj P, Jensen NO, Larsen B, Lohse C, Pilegaard K, Skov H (2000) Determination of the terpene flux from orange species and Norway spruce by relaxed eddy accumulation. Atmos Environ 34:3057–3067

Crutzen PJ, Williams J, Pöschl U, Hoor P, Fischer H, Warneke C, Holzinger R, Hansel A, Lindinger W, Scheeren B, Lelieveld J (2000) High spatial and temporal resolution measurements of primary organics and their oxidation products over the tropical forests of Surinam. Atmos Environ 34:1161–1165

Dabberdt WF, Lenschow DH, Horst TH, Zimmerman PR, Oncley SP, Delany AC (1993) Atmosphere–surface exchange measurements. Science 260:1472–1482

Derwent RG, Jenkin EM, Saunders SM, Pilling MJ (2001) Characterization of the reactivities of volatile organic compounds using a master chemical mechanism. J Air Waste Manage Assoc 51:699–707

Dorsey JR, Nemitz E, Gallagher MW, Fowler D, Williams PI, Bower KN, Beswick KM (2002) Direct measurements and parameterisation of aerosol flux, concentration and emission velocity above a city. Atmos Environ 36:791–800

Emeis S, Forkel R, Schoenemeyer T (1998) European forests as a source of atmospheric nitrogen oxides; exchange of oxidants at the forest–atmosphere interface. Final report of EU project ENV-CT95-0023 TNO-MEP-R 98/365. TNO, Apeldoorn, The Netherlands

Fall R (1999) Biogenic emissions of volatile organic compounds from higher plants. In: Hewitt CN (ed) Reactive hydrocarbons in the atmosphere. Academic Press, London, pp 43–96

Fall R, Karl T, Hansel A, Jordan A, Lindinger W (1999) Volatile organic compounds emitted after leaf wounding: on-line analysis by proton-transfer-reaction mass spectrometry. J Geophys Res 104:15963–15974

Forkel R, Panhans WG, Welch R, Zdunkowski W (1984) A one-dimensional numerical study to simulate the influence of soil moisture, pollution and vertical exchange on the evolution of radiation fog. Beitr Phys Atmos 57:72–91

Forkel R, Seidl W, Dlugi R, Deigele R (1990) A one-dimensional numerical model to simulate formation and balance of sulfate during radiation fog events. J Geophys Res 95D:18501–18515

Forkel R, Stockwell WR, Steinbrecher R (1999) Multilayer canopy/chemistry model to simulate the effect of in-canopy processes on the emission rates of biogenic VOCs. In: Borell PM, Borell P (eds) Proc EUROTRAC Symp 1998. Witpress, Southampton, pp 45–49

Forkel R, Steinbrecher R, Stockwell WR (2001) Modelling of the effect of in-canopy chemical reactions on the emission rates of biogenic VOC. In: Midgley PM, Reuther M, Williams M (eds) Proc EUROTRAC Symposium 2000. Springer, Berlin Heidelberg New York, CD-ROM, GENEMIS-2, pp 1–4

Gallagher MW, Beswick KM, Duyzer J, Westrate H, Choularton TW, Hummelshoj P (1997) Measurements of aerosol fluxes to Speulder forest using a micrometeorological technique. Atmos Environ 31:359–373

Galbally IE, Kirstine W (2002) The production of methanol by flowering plants and the global cycle of methanol. J Atmos Chem 43:195–229

Goldan PD, Kuster WC, Fehsenfeld FC, Montzka SA (1995) Hydrocarbon measurements in the southeastern United States: the Rural Oxidants in the Southern Environment (ROSE) program 1990. J Geophys Res 100:35945–35963

Guenther A, Hewitt NC, Erickson D, Fall R, Geron C, Graedel T, Harley P, Klinger L, Lerdau M, McKay WA, Pierce T, Scholes B, Steinbrecher R, Tallamraju R, Taylor J, Zimmerman PA (1995) Global model of natural volatile organic compound emissions. J Geophys Res 100:8873–8892

Helas G, Slanina J, Steinbrecher R (1997) Biogenic volatile organic compounds in the atmosphere. SPB, The Hague

Heikes BG, Chang W, Pilson MEQ, Swift E, Singh HB, Guenther A, Jacob DJ, Field BD, Fall R, Riemer D, Brand L (2002) Atmospheric methanol budget and ocean implication. Global Biogeochem Cycles 16:80/1–80/13

Held A, Klemm O (2002) Vertical particle fluxes over a coniferous forest. In: Midgley PM, Reuther M (eds) Proc EUROTRAC-2 Symp 2002, Transport and Chemical Transformation in the Troposphere.. Backhuys, Leiden, CD-ROM, GUEST-39, pp 1–4

Hills AJ, Zimmerman PR (1990) Isoprene measurement by ozone-induced chemiluminescence. Anal Chem 62:1055–1060

Hoffmann T, Odum JR, Bowman F, Collins D, Klokow D, Flagan RC, Seinfeld JH (1997) Formation of organic aerosols from the oxidation of biogenic hydrocarbons. J Atmos Chem 26:189–222

Jarvis PG (1976) The interpretation of the variations in leaf water potential and stomatal conductance found in canopies in the field. Philos Trans R Soc Lond Ser B 273:593–610

Kanakidou M, Tsigaridis K, Dentener FJ, Crutzen PJ (2000) Human-activity-enhanced formation of organic aerosols by biogenic hydrocarbon oxidation. J Geophys Res 105:9243–9254

Karl T, Guenther A, Lindinger C, Jordan A, Fall R, Lindinger W (2001) Eddy covariance measurements of oxygenated VOC fluxes from crop harvesting using a redesigned proton-transfer-reaction-mass-spectrometer. J Geophys Res 106:24157–24167

Karl T, Spirig C, Prevost P, Stroud C, Rinne J, Greenberg J, Fall R, Guenther A (2002) Virtual disjunct eddy covariance measurements of organic trace compound fluxes from a subalpine forest using proton transfer reaction mass spectrometry. Atmos Chem Phys 2:279–291

Kreuzwieser J., Schnitzler JP, Steinbrecher R (1999) Biosynthesis of organic compounds emitted by plants. Plant Biol 1:149–159

Kreuzwieser J, Graus M, Wisthaler A, Hansel A, Rennenberg H, Schnitzler J-P (2002) Xylem-transported glucose as an additional carbon source for leaf isoprene formation in *Quercus robur*. New Phytol 156:171–178

Kulmala M, Hämeri K, Aalto PP, Mäkelä JM, Pirjola L, Douglas Nilsson E, Buzorius G, Rannik Ü, Dal Maso M, Seidl W, Hoffman T, Janson R, Hansson HC, Viisanen Y, Laaksonen A, O´Dowd CD (2001) Overview of the international project on biogenic aerosol formation in the boreal forest (BIOFOR). Tellus 53:324–343

Lamanna MS, Goldstein AH (1999) In situ measurements of C_2-C_{10} volatile organic compounds above a Sierra Nevada ponderosa pine plantation. J Geophys Res 104:21247–21262

Lindinger W, Hansel A, Jordan A (1998) On-line monitoring of volatile organic compounds at pptv levels by means of proton-transfer-reaction mass spectrometry (PTR-MS). Medical applications, food control and environmental research. Int J Mass Spectrom Ion Processes 173:191–241

MacDonald RC, Fall R (1993) Acetone emissions from conifer buds. Phytochemistry 34:991–994

Madronich S (1987) Photodissociation in the atmosphere. 1. Actinic flux and the effects of ground reflections and clouds. J Geophys Res 92:9740–9752

Mangold A (1999) Untersuchung der lokalen Einflüsse auf Turbulenzmessungen der Station Weidenbrunnen. Diploma Thesis, Abt Mikrometeorologie, Universität Bayreuth

Norman J (1979) Modelling the complete canopy. In: Barfield BJ, Gerber JF (eds) Modification of the aerial environment of plants. Am Soc Agric Eng Monogr 2:249–277

Nouaime G, Bertman SB, Seaver C, Elyea D, Huang H, Shepson PB, Starn TK, Riemer DD, Zika RG, Olszyna K (1998) Sequential oxidation products from tropospheric isoprene chemistry: MACR and MPAN at a NO_x-rich forest environment in the southeastern United States. J. Geophys Res 103:22463–22471

Oncley SP, Delany AC, Horst TW, Tans PP (1993) Verification of flux measurement using relaxed eddy accumulation. Atmos Environ 27A:2417–2426

Pattey E, Desjardins RL, Rochette P (1993) Accuracy of relaxed eddy-accumulation technique, evaluated using CO_2 flux measurements. Boundary-Layer Meteorol 66:341–355

Pattey E, Desjardins RL, Westberg H, Lamb B, Zhu T (1999) Measurement of isoprene emissions over a black spruce stand using a tower-based relaxed eddy-accumulation system. J Appl Meteorol 38:870–877

Penner JE, Andreae MO, Annegarn H, Barrie L, Feichter J, Hegg D, Jayaraman A, Leaithc R, Murphy D, Nganga J, Pitari G (2001) Aerosols, their direct and indirect effects. Climate change 2001: the scientific basis. Cambridge University Press, New York, pp 289–348

Poppe D, Zimmermann J, Dorn HP (1995) Field data and model calculations for the hydroxyl radical. J Atmos Sci 52:3402–3407

Prentice IC, Farquhar GD, Fasham MJR, Gouldan ML, Heimann M, Jaramillio VJ, Kheshgi HS, Le Quéré C, Scholes RJ, Wallace DWR (2001) The carbon cycle and atmospheric carbon dioxide. Climate change 2001: the scientific basis. http://www.grida.no/climate/ipcc_tar/wg1/095.htm

Schade WG, Goldstein AH (2001) Fluxes of oxygenated volatile organic compounds from a ponderosa pine plantation. J Geophys Res 106:3111–3123

Shaw WJ, Spicer CW, Kenny D (1998) Eddy correlation fluxes of trace gases using a tandem mass spectrometer. Atmos Environ 32:2887–2898

Simpson D, Winiwarter W, Börjesson G, Cinderby S, Ferreiro A, Guenther A, Hewitt N, Janson R, Khalil MAK, Owen S, Pierce T, Puxbaum H, Shearer M, Skiba U, Steinbrecher R, Tarrason L, Öquist MG (1999) Inventorying emissions from nature in Europe. J Geophys Res 104:8113–8152

Smiatek G (2001) GIS and RDBMS class system in support of BVOC emission inventories. In: Keller J (ed) Proc 5th GLOREAM Worksh. Paul Scherrer Institut, Villingen, pp 24–26

Steinbrecher R (1994) Emission of VOCs from selected European ecosystems: the state of the art. In: Borrell PM, Borrell P, Cvitas T, Seiler W (eds) Proc EUROTRAC Symp 1994. SPB, The Hague, pp 448–454

Steinbrecher R, Smiatek G (2004) VOC emissions from biogenic sources. In: Friedrich R, Reis S (eds) Emissions of air pollutants – measurements, calculation, uncertainties. Springer, Berlin Heidelberg New York, pp 16–24

Steinbrecher R, Hauff K, Rabong R, Steinbrecher J (1997) Isoprenoid emission of oak species typical for the Mediterranean area: source strength and controlling variables. Atmos Environ 31/S1:79–88

Steinbrecher R, Hauff K, Hakola H, Rössler T (1999) A revised parameterisation for emission modelling of isoprenoids for boreal plants. In: Laurila T, Lindfors V (eds) Bio-

genic VOC emission and photochemistry in the boreal regions of Europe. EUR 18910 EN. EC, Brussels, pp 29–43
Steinbrecher R, Klauer M, Hauff K, Mayer H (2000) Canopy fluxes of reactive volatile organic compounds over a Scots pine plantation in southern Germany. http://zdb-imk.physik.uni-karlsruhe.de/cgi-bin/fetch.pl/poster/lt2-c5.pdf
Stewart EH, Hewitt CN, Bunce RGH, Steinbrecher R, Smiatek G, Schoenemeyer T (2003) A highly spatially and temporally resolved inventory for biogenic isoprene and monoterpene emissions – model description and application to Great Britain. J Geophys Res 108(D20):4644
Stockwell WR, Kirchner F, Kuhn M, Seefeld F (1997) A new mechanism for regional atmospheric chemistry modeling. J Geophys Res 102D:25847–25879
Stroud CA, Roberts JM, Goldan PD, Kuster WC, Murphy PC, Williams EJ, Hereid D, Parrish D, Sueper D, Trainer M, Fehsenfeld FC, Apel EC, Riemer D, Wert B, Henry B, Fried A. Mertinez-Harder M, Harder H, Brune WH, Li G, Xie H, Young VL (2001) Isoprene and its oxidation products, methacrolein and methylvinyl ketone, at an urban forested site during the 1999 Southern Oxidants study. J Geophys Res 106:8035–8046
Tuazon EC, Atkinson R (1990) A product study of the gas-phase reaction of isoprene with the OH radical in the presence of NO_x. Int J Chem Kinet 22:1221–1236
Warneke C, Karl T, Judmaier H, Hansel A, Jordan A, Lindinger W, Crutzen PJ (1999) Acetone, methanol, and other partially oxidized volatile organic emissions from dead plant matter by abiological processes: significance for atmospheric HO_x chemistry. Global Biogeochem Cycles 13:9–17
Warneke C, Luxembourg SL, de Gouw JA, Rinne HJI, Guenther AB, Fall R (2002) Disjunct eddy covariance measurements of oxygenated VOC fluxes from an alfalfa field before and after cutting. J Geophys Res 107, D8, 10.1029/2001JD000594
Wesely ML (1989) Parameterization of surface resistances to gaseous dry deposition in regional-scale numerical models. Atmos Environ 23:1293–1304
Williams J, Roberts JM, Fehsenfeld FC, Bertman SB, Buhr MP, Goldan PD, Hübler G, Kuster WC, Ryerson TB, Trainer M, Young V (1997) Regional ozone from biogenic hydrocarbons deduced from airborne measurements of PAN, PPN, and MPAN. Geophys Res Lett 24:1099–1102

14 Trends in Deposition and Canopy Leaching of Mineral Elements as Indicated by Bulk Deposition and Throughfall Measurements

E. Matzner, T. Zuber, C. Alewell, G. Lischeid, and K. Moritz

14.1 Introduction

In the past three decades, numerous studies on the biogeochemistry of forested ecosystems in Europe and North America have shown that the deposition of mineral elements from the atmosphere strongly influences their functioning. Acidification of soils, surface- and groundwaters, N saturation and forest decline are key processes that change with rates of deposition of mineral elements (Ulrich 1994; Fenn et al. 1998; Evans et al. 2001). As an example of ecosystem functioning, the losses of elements from the ecosystem by seepage and runoff can be considered. On a European-wide, scale the deposition of S and N was shown to determine the Al losses from seepage and runoff in acid forest soils (Dise et al. 2001), the N deposition to determine the NO_3 losses (MacDonald et al. 2002) and the Mg deposition to largely determine the Mg losses (Armbruster et al. 2002).

The deposition rates of mineral elements depend strongly on anthropogenic emissions of dust, SO_2, NH_3 and NO_x and thus were subjected to changes in the past that are related to the development of emission rates. A number of long-term studies have documented trends in the deposition of mineral elements in forest ecosystems, partly reaching back to the 1960s (Likens et al. 1990; Matzner and Meiwes 1994; Stoddard et al. 1999; Alewell et al. 2000). These studies have shown a significant reduction in the deposition of S, H^+, Ca and Mg in the 1980s and early 1990s. The observed trends were attributed to the development of dust and SO_2 emissions. Only minor trends were found in the case of N deposition, while the emissions of N should have decreased in Europe from 1990 onwards (Skeffington 2002).

The atmospheric deposition of mineral elements also influences processes in the phyllosphere, namely the leaching of cations from foliage and the uptake of nutrients by foliage, especially of N. The importance of cation leaching from canopies as a pathway of cations into forest soils has increased because the measures to reduce the dust emissions in the last decades have led

Ecological Studies, Vol. 172
E. Matzner (Ed.), Biogeochemistry of Forested Catchments in a Changing Environment

to a decrease in Ca and Mg deposition in central European forest ecosystems. Among other factors, the amount of precipitation and the buffering of H^+ in the canopy were shown to influence the canopy leaching rates. In laboratory and field irrigation experiments, acid mist caused increased leaching of Ca and Mg from the foliage, particularly at low pH (Wood and Bormann 1975; Puckett 1990; Lovett and Hubbell 1991; Cappellato et al. 1993; Liechty et al. 1993; Neary and Gizyn 1994; Sayre and Fahey 1999). Besides the H^+ deposition, the NH_4 deposition was also shown to trigger the leaching of cations from needles (Roelofs et al. 1985). The present state of knowledge on the rates of cation leaching from canopies and the driving forces is based mostly on local field observations and controlled laboratory experiments that have been carried out at relatively short-term temporal scales. Local properties such as climate, tree species and chemical status of forest soils may affect the leaching of cations from forest canopies. Whether the factors influencing canopy leaching identified in laboratory experiments and local field studies can be generalized for assessments on longer time scales and on regional scales is a matter of debate. Lovett et al. (1996) compared throughfall measurements from eight forested sites in the northeastern United States and reported that leaching of Ca and Mg was higher in forests with more acidic rain events. Langusch et al. (2003) analysed regional data on throughfall fluxes in European forests and found a positive correlation between Ca leaching and H^+ deposition, but the influence of H^+ deposition on Mg and K leaching was less.

As the influencing factors change with time, one can also expect long-term trends in cation leaching from foliage. However, this has not been reported yet.

The uptake of NH_4 and NO_3 in the canopy by needles and leaves has been known for a long time. This specific behaviour of N makes it difficult to estimate rates of total deposition in forest ecosystems and to establish N budgets. Processes and rates of canopy uptake of N have recently been reviewed by Harrison et al. (2000). They conclude that the uptake of N from the solution phase exceeds that from the gas phase and that NH_4 is taken up preferentially over NO_3. Uptake from the liquid phase seems to be linearly concentration dependent and is thus changing with N deposition. Estimated rates of canopy uptake for European forests differ strongly and may exceed 10 kg N ha^{-1} $year^{-1}$ by a long way (Harrison et al. 2000).

In summary, changes in the deposition rate and the chemical composition of rain not only might affect the cycling of elements in the soil, but also may have significant effects on the element turnover in the canopy. Here, we report data on long-term development of element fluxes with bulk deposition and throughfall for the period 1988–2001 for two coniferous sites in the Lehstenbach catchment and for the deciduous Steinkreuz site. We focus on H^+, K, Ca, Mg, S, NH_4 and NO_3. Our aims are to document the actual deposition rates, to investigate trends in deposition and canopy leaching and to investigate differences between the spruce and beech forest.

14.2 Trends of Concentrations

Bulk deposition at both catchments was sampled with open rain samplers of 326 cm^2 each as described in Gerstberger (2001) with five, three and three replicates for Lehstenbach, LFW01 and Steinkreuz, respectively. Throughfall at the Coulissenhieb site was sampled at four measurement fields each installed with five samplers (distance 2 m), yielding a total number of 20 samplers. Throughfall at the site LFW01, established by the Bavarian Water Authorities, was collected with the same sampler type as in Coulissenhieb and Steinkreuz. The number of replicates here was 15. For throughfall measurements at Steinkreuz, 12 samplers were installed in a regular grid (distance between grid points was 5 m). Stemflow at the beech site was measured at five stems. In the following, the term 'throughfall' is used for the sum of stemflow and canopy drip for the Steinkreuz site. Stemflow represented $<10\%$ of the element input to the soil at the Steinkreuz site (Chang and Matzner 2000) and is negligible at the Coulissenhieb and LFW01 sites.

Sampling of bulk deposition and throughfall started in autumn 1987 at the site LFW01, in summer 1992 at Coulissenhieb and in January 1995 at Steinkreuz. Bulk deposition and throughfall samples were collected fortnightly. To determine the variability within the site, throughfall samples at the site Coulissenhieb were analysed separately for each of the 20 samplers. At Steinkreuz, three throughfall samplers were mixed to give four replicates for chemical analysis. At the site LFW01 only one mixed sample from all 15 samplers was analysed. After 0.45 μm filtration of the samples, the analysis covered pH (glass electrode), Na, K, Ca, Mg, Al, Mn, Fe (ICP-AES), NH_4 (colorimetrically), NO_3, SO_4, PO_4 and Cl (ion chromatography). Trend analysis in the concentrations was performed by simple linear regression vs. time based on the single measurements. For reasons of clarity we present only figures showing volume weighted annual average concentrations. In the case of the Steinkreuz site, the volume weighted concentrations in throughfall reflect canopy drip as well as stemflow. Deposition data were summed up for calendar years in the case of sites Coulissenhieb and Steinkreuz, whereas for site LFW01 the hydrological year is used. The differences seemed to be negligible.

The already well-documented decrease in deposition of sulfate and H^+ throughout Europe and North America (EMEP 1999; Stoddard et al. 1999) is clearly reflected in the concentration trends in bulk deposition and throughfall at both catchments (Fig. 14.1). At Coulissenhieb, a significant decrease in sulfate and proton concentrations was found in throughfall. In the last 4 years (1998–2001), the average annual concentrations were around 1.5 mg SO_4-S l^{-1}, while in the years 1993–1997, the average concentrations were 3–4 mg SO_4-S l^{-1}. Sulfate in bulk deposition decreased in a similar way, but on a much lower concentration level. The decrease in sulfate coincides with decreasing proton concentrations. The pH of throughfall at the Coulissenhieb site

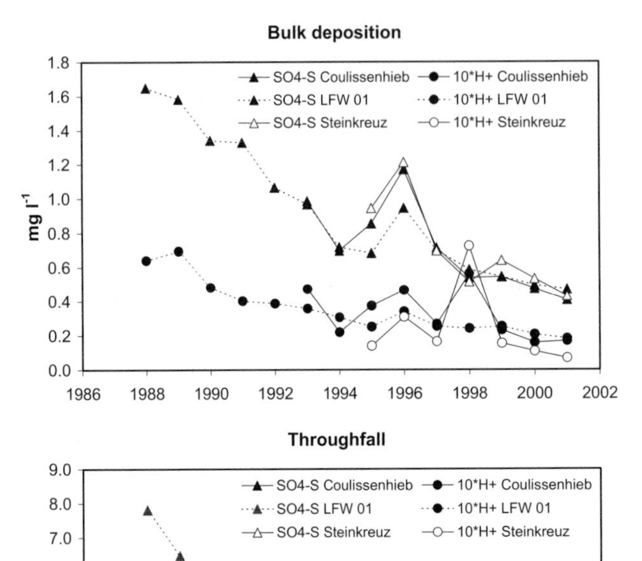

Fig. 14.1. Volume weighted concentrations of SO_4-S and H^+ in bulk deposition (*above*) and throughfall (*below*)

increased from about 4.0 in 1993 to 5.0 in 2001. The time series of the site LFW01 match the data from Coulissenhieb well and indicate that the decrease in sulfate and H^+ concentrations in bulk deposition and throughfall extend back to 1988, where the highest concentrations (8 mg SO_4-S l^{-1} in throughfall) were found.

The pH of throughfall at Steinkreuz was already around 5.0 in 1994 and no trend was found for the annual averages. However, pH had a clear seasonality with high values in summer and low values in winter. The winter values tend to increase (data not shown). The sulfate concentrations in throughfall decreased. Following 1998, the average annual concentration was around 1.2 mg SO_4-S l^{-1} while from 1995–1997 about 1.8 mg l^{-1} was observed. The development of sulfate concentrations in bulk deposition at Steinkreuz was similar to throughfall with concentrations between 1.2 and 0.5 mg SO_4-S l^{-1}. Sulfate concentrations in bulk deposition were also similar when comparing Steinkreuz and Coulissenhieb.

There was no clear trend in the concentrations of K, Ca and Mg in bulk deposition of Steinkreuz and Lehstenbach in the time period from 1993–2001 (Fig. 14.2). Note that the Mg concentrations were extremely low in bulk depo-

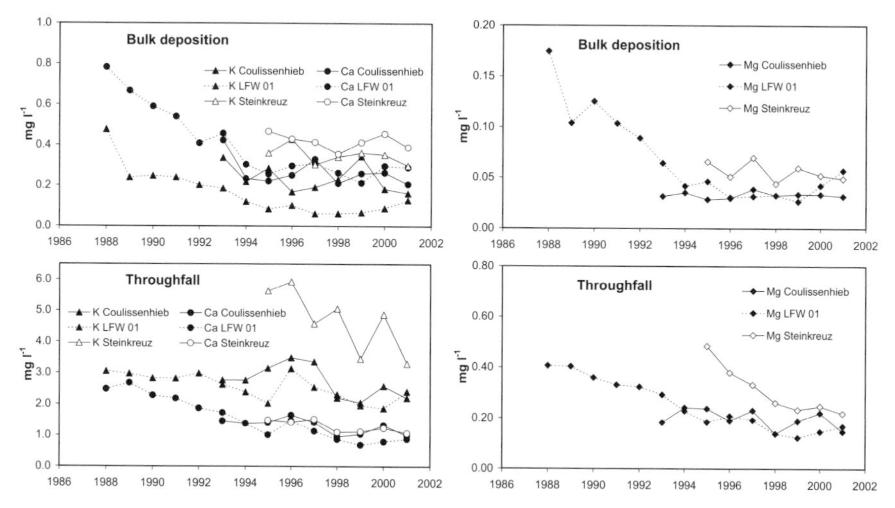

Fig. 14.2. Volume weighted concentrations of Ca, K and Mg in bulk deposition (*above*) and throughfall (*below*)

sition, often under the detection limit of 0.05 mg l^{-1}. In these cases, a value of 0.025 mg l^{-1} was taken for the calculation of fluxes and the average annual concentrations. The decrease of Ca and Mg deposition at the Lehstenbach catchment dates back to the period before 1993, which can also be clearly seen from the data of site LFW01. Concentrations decreased substantially from 1988–1994 similar to the development in the German Solling region (Matzner and Meiwes 1994).

There were consistent trends in the concentrations of these cations in throughfall. A decrease in the average annual concentrations of Ca in throughfall at all sites was observed. From 1993–1997, the average Ca concentrations in throughfall were about 1.5 mg l^{-1} at the Coulissenhieb site and decreased by almost 30 % to about 1.1 mg l^{-1} from 1998–2001. Similar patterns of Ca concentrations were observed at the Steinkreuz site. No trend was found for the Mg concentrations in throughfall at Coulissenhieb, but at Steinkreuz the Mg concentrations decreased from 1995–2001 from about 0.40 to 0.23 mg l^{-1}. Concentrations of Ca and Mg in throughfall were surprisingly similar at both sites, given the different tree species soil and environmental conditions. However, at the beech site the concentrations in the vegetation period are higher than those in winter (not shown) and the similarity of the concentrations will decrease if seasonal patterns are considered.

The average K concentrations in throughfall also decreased substantially at the Steinkreuz site and – to a lesser extent – at the Coulissenhieb and LFW01 sites.

Ammonium and nitrate concentrations were in a similar range at all sites, both in bulk deposition and throughfall. In the period 1993–2001, the concen-

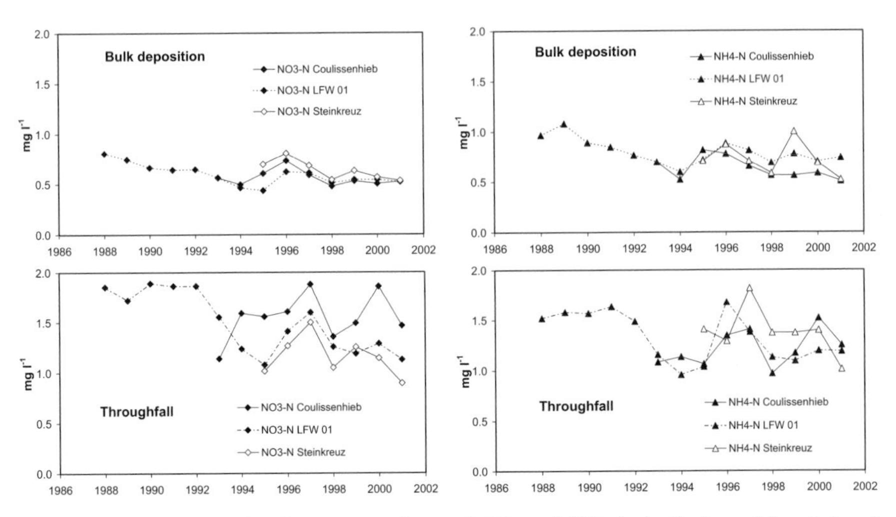

Fig. 14.3. Volume weighted concentrations of NH_4 and NO_3 in bulk deposition (*above*) and throughfall (*below*)

trations of NH_4 and NO_3 in bulk deposition and throughfall had no trend at all three sites.(Fig. 14.3). However, the longer time series of the LFW01 site indicated a slight decrease of the NH_4 and NO_3 concentrations in bulk deposition and throughfall from 1988–1993. Nitrogen content of fog samples did not show a trend (Klemm, this Vol.).

14.3 Trends of Fluxes and Total Deposition

Fluxes with bulk deposition and throughfall were calculated as concentration times measured precipitation amount. Element fluxes with throughfall are the result of dry and wet deposition from the atmosphere, canopy uptake and canopy leaching. The separation of the different processes in a quantitative way is notoriously difficult. Beside micrometeorological methods, annual fluxes with bulk deposition and throughfall are used to estimate canopy leaching and total deposition rates. Here, we apply the method developed by Ulrich (1983) to the sites Coulissenhieb and Steinkreuz (see also Bredemeier et al. 1988; Draaijers et al. 1997). According to Ulrich (1983) total deposition of an element 'x' is calculated as the sum from bulk deposition (BD_x) and interception deposition (ID_x):

$$TD_x = BD_x + ID_x \qquad x = Ca,\ Mg,\ K,\ H^+,\ NH_4\text{-N},\ NO_3\text{-N},\ SO_4\text{-S} \qquad (1)$$

Several authors argue that SO_4, Cl and Na deposition with throughfall adequately describe patterns of total deposition, because of the low exchange between the forest canopy and SO_4, Na and Cl (Ulrich 1983; Matzner 1988; Lovett et al. 1997). Thus, for Na, Cl and SO_4-S interception deposition is defined as the difference between throughfall (TF) and bulk deposition.

Total interception deposition comprises particles (part) and gases (gas). In order to calculate the particle deposition of Ca, Mg, K, S and H^+, the 'Na factor' ($f_{Na}=ID_{Na}/BD_{Na}$) was used, assuming that Na behaves inert in the canopy and represents the particle deposition of other major elements:

$$ID_{part,x} = f_{Na} \cdot BD_x \qquad x = Ca,\ Mg,\ K,\ H^+,\ NH_4\text{-}N,\ NO_3\text{-}N,\ SO_4\text{-}S \tag{2}$$

The source or sink of the canopy for a certain cation was calculated as follows:

$$Q_x = TF_x - TD_x \quad x = Ca,\ Mg,\ K,\ H^+ \tag{3}$$

For the calculation of total H^+ deposition it was assumed that the gaseous interception of H^+ is equivalent to the gaseous interception of SO_2, which was calculated as follows:

$$ID_{gas,H} = ID_{gas,SO2\text{-}S} = TF_{SO4\text{-}S} - BD_{SO4\text{-}S} - ID_{part,SO4\text{-}S} \tag{4}$$

Because of the substantial uptake of N in the canopy, ID_{gas} and total deposition of NH_3 and NO_x cannot be calculated with this method. Potential effects of the deposition of gaseous N compounds on the H^+ deposition are thus also not considered.

The amount of precipitation did not change in a systematic way throughout the observation period and the trends in the average annual concentrations are thus reflected to a large extent in the fluxes. There was a strong decrease in the H^+ and S fluxes with bulk deposition and throughfall at all sites. At Coulissenhieb, the total deposition of S decreased from about 30 kg S ha^{-1} $year^{-1}$ in 1993 to around 12 kg ha^{-1} $year^{-1}$ at the end of the observation period (Table 14.1). Similarly, the calculated total deposition of H^+ at Coulissenhieb decreased from a maximum of 1.8 kg ha^{-1} $year^{-1}$ in 1993 to about 0.3 kg ha^{-1} $year^{-1}$ in recent years. At Steinkreuz, the total deposition of S was generally much lower than that at Coulissenhieb, but still decreased with time (Table 14.2). In recent years, the S deposition at Steinkreuz was only about 7 kg ha^{-1} $year^{-1}$ and the H^+ deposition became very low (<0.14 kg ha^{-1} $year^{-1}$). The largest decrease is obvious from the time series of the LWF01 site; here, the fluxes of S with throughfall decreased constantly from 58 kg S ha^{-1} $year^{-1}$ in 1988 to around 12 in recent years and the H^+ deposition decreased from 3.1 to 0.2 kg ha^{-1} $year^{-1}$ (Table 14.3). At two other sites in the

Table 14.1. Fluxes with bulk deposition, throughfall and calculated total deposition of mineral elements at Coulissenhieb

Year	H_2O (mm)	H	Na	K	Ca	Mg	NH_4-N	NO_3-N	SO_4-S	Cl
Bulk deposition (kg ha^{-1} year^{-1})										
1993	1,047	0.49	6.3	3.5	4.4	0.3	7.3	5.9	10.3	9.4
1994	1,143	0.25	4.8	2.5	2.7	0.4	6.0	5.7	8.0	6.2
1995	1,312	0.49	6.3	3.7	2.9	0.4	10.7	7.9	11.2	6.1
1996	901	0.42	2.3	1.5	2.2	0.3	7.0	6.6	10.5	5.8
1997	822	0.22	2.5	1.6	2.7	0.3	5.4	4.8	5.9	4.0
1998	1,270	0.71	5.0	2.9	2.7	0.4	7.2	6.1	6.9	7.6
1999	1,038	0.24	3.1	3.6	2.7	0.3	5.8	5.5	5.6	8.1
2000	994	0.16	2.2	1.8	2.6	0.3	5.8	5.0	4.7	4.2
2001	1,069	0.18	2.6	1.7	2.2	0.3	5.4	5.6	4.4	5.0
Mean	1,066	0.35	3.9	2.5	2.8	0.3	6.7	5.9	7.5	6.3
Throughfall (kg ha^{-1} year^{-1})										
1993	774	1.13	6.1	21.4	11.3	1.4	8.4	8.8	31.0	12.0
1994	870	0.80	8.0	24.1	12.0	2.1	9.9	13.8	25.9	14.8
1995	1,015	0.77	6.3	32.0	14.3	2.4	10.8	15.8	26.1	12.5
1996	701	1.23	4.7	24.5	11.6	1.3	9.4	11.3	31.8	10.2
1997	591	0.63	5.4	19.8	8.4	1.4	8.3	11.1	17.3	10.0
1998	951	0.82	8.8	21.0	9.0	1.3	9.2	12.9	15.8	13.5
1999	817	0.38	6.2	16.8	8.5	1.5	9.6	12.2	13.2	11.1
2000	761	0.27	5.5	19.7	10.1	1.7	11.6	14.2	12.5	10.4
2001	825	0.24	5.8	18.2	8.2	1.2	10.3	12.1	11.8	10.2
Mean	812	0.70	6.3	21.9	10.4	1.6	9.7	12.5	20.6	11.6
Total deposition (kg ha^{-1} year^{-1})										
1993		1.80	6.3	3.5	4.4	0.3			31.0	12.0
1994		1.06	8.0	4.1	4.4	0.7			25.9	14.8
1995		1.42	6.3	3.7	2.9	0.4			26.1	12.5
1996		1.05	4.7	3.1	4.6	0.6			31.8	10.2
1997		0.51	5.4	3.4	5.9	0.7			17.3	10.0
1998		0.95	8.8	5.1	4.7	0.7			15.8	13.5
1999		0.36	6.2	7.1	5.4	0.7			13.2	11.1
2000		0.22	5.5	4.4	6.4	0.8			12.5	10.4
2001		0.32	5.8	3.8	4.9	0.7			11.8	10.2
Mean		0.85	6.3	4.3	4.8	0.6			20.6	11.6
Total deposition – throughfall (kg ha^{-1} year^{-1})										
1993		0.67		−17.9	−6.8	−1.1				
1994		0.26		−20.0	−7.7	−1.4				
1995		0.65		−28.2	−11.4	−2.0				
1996		−0.18		−21.3	−6.9	−0.8				
1997		−0.13		−16.4	−2.5	−0.7				
1998		0.14		−15.9	−4.4	−0.6				
1999		−0.02		−9.6	−3.1	−0.8				
2000		−0.05		−15.2	−3.7	−0.9				
2001		0.05		−14.4	−3.3	−0.5				
Mean		0.16		−17.7	−5.5	−1.0				

Table 14.2. Fluxes with bulk deposition, throughfall and calculated total deposition of mineral elements at Steinkreuz

Year	H_2O (mm)	H	Na	K	Ca	Mg	NH_4-N	NO_3-N	SO_4-S	Cl
Bulk deposition (kg ha^{-1} year^{-1})										
1995	920	0.13	6.4	3.3	4.3	0.6	6.5	6.4	8.7	5.2
1996	724	0.22	3.4	3.1	3.1	0.4	6.3	5.8	8.8	5.8
1997	665	0.11	2.0	2.0	2.8	0.5	4.7	4.6	4.6	3.5
1998	838	0.61	2.3	2.8	3.0	0.4	4.9	4.5	4.3	4.8
1999	786	0.12	2.6	2.8	3.3	0.5	7.9	5.0	5.0	4.8
2000	767	0.08	2.0	2.7	3.5	0.4	5.3	4.4	4.1	3.5
2001	1,001	0.07	2.2	3.0	3.9	0.5	5.2	5.3	4.3	4.4
Mean	814	0.19	3.0	2.8	3.4	0.5	5.8	5.1	5.7	4.6
Throughfall (kg ha^{-1} year^{-1})										
1995	713	0.11	7.3	40.1	10.6	3.5	10.0	7.3	11.8	9.1
1996	576	0.12	3.1	34.1	8.2	2.2	7.4	7.3	11.3	7.2
1997	484	0.08	3.1	22.1	7.3	1.6	8.8	7.3	8.0	6.7
1998	615	0.07	4.5	31.1	6.8	1.6	8.5	6.5	7.4	8.5
1999	589	0.08	3.6	20.4	6.6	1.4	8.1	7.4	7.3	7.6
2000	581	0.05	3.4	28.3	7.2	1.4	8.1	6.7	6.6	6.8
2001	779	0.04	3.7	25.7	8.4	1.7	7.9	7.0	6.7	7.1
Mean	620	0.08	4.1	28.9	7.9	1.9	8.4	7.0	8.4	7.6
Total deposition (kg ha^{-1} year^{-1})										
1995		0.24	7.3	3.8	5.0	0.7			11.8	9.1
1996		0.38	3.1	3.1	3.1	0.4			11.3	7.2
1997		0.16	3.1	3.1	4.3	0.7			8.0	6.7
1998		0.54	4.5	5.6	5.8	0.7			7.4	8.5
1999		0.14	3.6	4.0	4.5	0.7			7.3	7.6
2000		0.07	3.4	4.5	5.9	0.7			6.6	6.8
2001		0.04	3.7	5.0	6.5	0.8			6.7	7.1
Mean		0.22	4.1	4.1	5.0	0.7			8.4	7.6
Total deposition – throughfall (kg ha^{-1} year^{-1})										
1995		0.13		−36.3	−5.6	−2.8				
1996		0.26		−31.0	−5.1	−1.8				
1997		0.08		−19.0	−3.0	−0.9				
1998		0.47		−25.6	−1.0	−0.9				
1999		0.05		−16.4	−2.1	−0.7				
2000		0.02		−23.8	−1.3	−0.8				
2001		0.00		−20.8	−1.9	−0.9				
Mean		0.14		−24.7	−2.9	−1.2				

Table 14.3. Fluxes with bulk deposition, throughfall and calculated total deposition at the LFW01 site

Year	H_2O (mm)	H	Na	K	Ca	Mg	NH_4-N	NO_3-N	SO_4-S	Cl
Bulk deposition (kg ha^{-1} year^{-1})										
1988	922	0.59	9.8	4.4	7.2	1.6	8.9	7.4	15.2	6.6
1989	950	0.66	5.3	2.3	6.3	1.0	10.3	7.1	15.0	4.4
1990	811	0.39	4.8	2.0	4.8	1.0	7.2	5.4	10.9	3.8
1991	821	0.33	3.4	2.0	4.4	0.9	7.0	5.3	10.9	1.7
1992	930	0.36	4.6	1.9	3.8	0.8	7.1	6.0	9.9	1.6
1993	1,009	0.36	5.4	1.9	4.6	0.6	7.1	5.7	9.8	4.3
1994	1,275	0.39	4.7	1.5	3.9	0.5	7.6	5.9	9.1	3.4
1995	1,198	0.30	3.8	1.0	3.1	0.6	8.6	5.2	8.2	4.2
1996	883	0.30	2.0	0.9	2.6	0.3	7.8	5.5	8.3	2.8
1997	870	0.22	2.5[a]	0.5	2.7	0.3	7.1	5.3	6.1	3.7
1998	1,162	0.28	3.2	0.7	3.0	0.4	8.0	5.9	6.8	5.4
1999	1,109	0.28	3.3	0.7	2.4	0.3	8.7	6.0	6.0	5.8
2000	1,079	0.22	2.9	1.0	3.2	0.5	7,6	5.8	5.3	5.3
2001	993	0.18	2.4	1.3	2.9	0.6	7.4	5.2	4.7	4.5
Mean	1,001	0.3	4.3	1.6	3.9	0.7	7.9	5.8	9.0	4.1
Throughfall (kg ha^{-1} year^{-1})										
1988	747	2.7	12.1	22.8	18.6	3.0	11.3	13.9	58.4	13.6
1989	815	1.99	10.0	24.3	21.9	3.3	12.9	14.0	52.8	14.1
1990	623	1.36	10.0	17.6	14.2	2.2	9.8	11.8	32.4	15.2
1991	655	1.57	6.4	18.5	14.3	2.2	10.7	12.2	39.8	9.6
1992	713	1.53	7.9	21.3	13.4	2.3	10.6	13.3	36.9	11.5
1993	772	1.52	9.5	20.3	13.3	2.3	8.9	12.0	34.1	11.8
1994	972	1.58	8.7	23.3	13.5	2.2	9.3	12.0	33.9	13.6
1995	1,060	0.94	7.2	21.5	10.8	2.0	11.0	11.4	25.8	12.6
1996	773	1.03	3.8	24.3	11.3	1.6	13.0	10.9	32.2	9.4
1997	640	0.67	4.4	16.3	7.3	1.2	8.8	10.2	17.7	9.4
1998	941	0.59	6.5	21.8	8.2	1.3	10.7	11.8	17.5	11.6
1999	943	0.57	6.0	18.5	6.5	1.2	10.3	11.2	13.7	12.4
2000	879	0.46	6.2	16.4	7.0	1.3	10.5	11.3	12.3	11.2
2001	768	0.27	5.8[a]	18.5	6.8	1.3	9.1	8.7	11.4	9.9
Mean	807	1.2	7.6	20.4	11.9	2.0	10.5	11.8	29.9	11.8
Total deposition (kg ha^{-1} year^{-1})										
1988		3.1	12.1	5.4	8.9	2.0			58.4	13.6
1989		2.2	10.0	4.3	11.9	1.9			52.8	14.1
1990		1.0	10.0	4.1	9.9	2.1			32.4	15.2
1991		1.5	6.4	3.8	8.5	1.6			39.8	9.6
1992		1.6	7.9	3.2	6.5	1.4			36.9	11.5
1993		1.4	9.5	3.3	8.2	1.1			34.1	11.8
1994		1.5	8.7	2.8	7.1	1.0			33.9	13.6
1995		1.0	7.2	1.9	5.8	1.0			25.8	12.6
1996		1.3	3.8	1.7	5.1	0.5			32.2	9.4
1997		0.6	4.4	0.9	4.8	0.5			17.4	9.4
1998		0.5	6.5	1.5	6.2	0.8			17.5	11.6
1999		0.4	6.0	1.3	4.3	0.5			13.7	12.4
2000		0.3	6.2	2.1	6.9	1.8			12.3	11.2
2001		0.2	5.8	3.1	7.0	1.4			11.4	9.9
Mean		1.2	7.5	2.8	7.2	1.2			29.9	11.8

Table 14.3. (*Continued*)

Year	H_2O (mm)	H	K	Ca	Mg
Total deposition – throughfall (kg ha^{-1} year^{-1})					
1988		0.4	–17.4	–9.7	–1.1
1989		0.2	–20.0	–9.9	–1.4
1990		–0.4	–13.5	–4.3	–0.1
1991		–0.1	–14.7	–5.8	–0.5
1992		0.1	–18.1	–6.9	–0.9
1993		–0.1	–17.0	–5.1	–1.1
1994		–0.1	–20.4	–6.3	–1.2
1995		0.0	–19.6	–5.0	–0.9
1996		0.3	–22.5	–6.3	–1.1
1997		0.0	–15.3	–2.5	–0.7
1998		–0.1	–20.3	–2.0	–0.5
1999		–0.1	–17.2	–2.2	–0.6
2000		–0.2	–14.4	–0.1	–0.3
2001		–0.1	–15.4	–0.2	0.1
Mean		0.0	–17.6	–4.7	–0.8

[a] Value taken from the Coulissenhieb site.

Fichtelgebirge, Horn et al. (1989) measured S fluxes with throughfall of 35 and 54 kg S ha^{-1} year^{-1} in 1985.

The significant decline of sulfate and proton deposition in both catchments is clearly a result of the reduction of SO_2 emissions throughout Europe. According to the reports of the European Monitoring and Evaluation Programme (EMEP 1999), the total sulfur emissions in Europe fell between 1980 and 1997 by more than 60 % (EMEP 1999). Some countries, such as Austria, Finland, pre-unification West Germany, Norway, Sweden and Switzerland, have cut down their sulfur emissions by as much as 75–80 % between 1980 and 2000 (Ferrier et al. 2001). The SO_2 concentration measurements in the Fichtelgebirge also showed a strong decline (Klemm, this Vol.), coinciding with the decreasing S deposition rates.

The measured fluxes with bulk deposition did not indicate a decrease in the Ca and Mg input at Coulissenhieb and Steinkreuz from 1993–2001 (Tables 14.1 and 14.2). Fluxes of Ca and Mg with bulk deposition were found to be very similar at both sites, with about 0.4 kg Mg ha^{-1} year^{-1} and 3 kg Ca ha^{-1} year^{-1}. The development of the calculated rates of total deposition of Ca and Mg supports the findings from the bulk deposition measurements; there was no further decline of deposition in the observation period from 1993–2001. As with bulk deposition, the total deposition rates of Ca and Mg at Coulissenhieb and Steinkreuz were very similar (Tables 14.1–14.3). On average, the Ca deposition was about 4.8 kg Ca ha^{-1} year^{-1} and the Mg deposition amounted to 0.6 kg ha^{-1} year^{-1}.

The time series of the LFW01 site, reaching back to 1988, however, indicated clearly a decrease in the deposition of Ca and Mg. Total deposition rates of both elements were highest from 1988–1991. Similar to the other sites, there was no further decrease from 1993 onwards. In 1985, the fluxes of Ca and Mg with bulk deposition in the Fichtelgebirge area were 10 and 1 kg ha^{-1} $year^{-1}$, respectively (Horn et al. 1989), supporting the higher rates of Ca and Mg deposition in the 1980s.

The Mg deposition depends largely on the distance to the sea, since sea spray is a significant source of atmospheric Mg (Armbruster et al. 2002). Both sites located in southern Germany are around 500 km away from the North and Baltic Seas and the deposition rates of Mg are thus low. The distance effect to the sea can be observed up to 300 km (Armbruster et al. 2002). Close to the sea, deposition of Mg may exceed 10 kg ha^{-1} $year^{-1}$.

In the case of K, no trend in the total deposition rates was observed for Coulissenhieb and Steinkreuz given the time from 1993–2001. At the LWF01 site, K and Na deposition rates decreased from 1988–1992.

Fluxes of NH_4 and NO_3 with bulk deposition were similar in the Lehstenbach catchment and at Steinkreuz. On average, the bulk deposition of mineral N (NH_4+NO_3) amounted to about 13 and 11 kg N ha^{-1} $year^{-1}$ at the two coniferous sites (Coulissenhieb and LFW01) and the Steinkreuz site, respectively. In contrast, the fluxes of mineral N with throughfall were significantly higher at Coulissenhieb and LWF01, reaching 22 kg N ha^{-1} $year^{-1}$ on average, vs. 15 kg N ha^{-1} $year^{-1}$ at Steinkreuz. The observed N deposition rates at these sites seem to be typical for many central European forests. In a review of N fluxes in throughfall from 181 European forest ecosystems, MacDonald et al. (2002) report average mineral N fluxes of 16.8 kg N ha^{-1} $year^{-1}$. The spatial variation was found to be enormous, reaching from <5 kg N ha^{-1} $year^{-1}$ in northern Scandinavia to >50 kg N ha^{-1} $year^{-1}$ in the Netherlands.

Nitrogen fluxes in throughfall do not represent the total deposition, because canopy uptake of deposited N has to be accounted for. Nevertheless, the higher N fluxes in throughfall at Coulissenhieb can be attributed to higher rates of total deposition, since the canopy uptake of N in spruce exceeds the one of beech trees (Brumme et al. 1992; Eilers et al. 1992). This interpretation is supported by the high rates of N deposition through fog deposition at the Coulissenhieb site (Wrzesinski et al., this Vol.). Rates of total N deposition are estimated in Chapter 25.

There was no trend observed in bulk deposition and throughfall for N fluxes at all sites. Thus, there is no indication from our data that national and international emission reduction measures for NH_3 and NO_x, as reported by EMEP (1999), Evans et al. (2001), Skeffington (2002) and Klemm (this Vol.), have already successfully reduced the deposition of N to these forests.

The decrease in the Ca concentration in throughfall at the Coulissenhieb and LWF01 sites is reflected in the temporal development of the throughfall fluxes. As an average of the years 1993–1995, the throughfall flux of Ca amounted to 12.5 kg ha^{-1} $year^{-1}$, while in the last 3 years on average only 9 kg ha^{-1} $year^{-1}$ was measured. At the LWF01 site the average Ca flux with throughfall was as high as 18 kg ha^{-1} $year^{-1}$ from 1988–1990, decreasing to 6.8 kg ha^{-1} $year^{-1}$ in the last 3 years. The Mg fluxes with throughfall at Coulissenhieb were highest in 1994 and 1995, but the trend is not as clear as for Ca. The Ca fluxes at Coulissenhieb were slightly higher than those at Steinkreuz. In contrast, the Mg fluxes at Steinkreuz exceeded those at Coulissenhieb. For Ca and Mg there was no clear trend in throughfall fluxes at Steinkreuz. However, for both elements the highest fluxes were observed in the first year of observation, in 1995.

The calculated total deposition of K at the sites was surprisingly similar with 3–4 kg K ha^{-1} $year^{-1}$. Throughfall fluxes of K at the Coulissenhieb and Steinkreuz sites seem to have decreased. Comparing the average K fluxes in the first 3 years of the measurement with recent years' results in fluxes of 26 vs. 18 kg ha^{-1} $year^{-1}$ in the case of Coulissenhieb and 32 vs. 24 kg ha^{-1} $year^{-1}$ in the case of Steinkreuz. The average K fluxes with throughfall at the Steinkreuz beech site were somewhat higher than those at the Coulissenhieb spruce site (29 vs. 22 kg K ha^{-1} $year^{-1}$). The time series of the LFW01 site did not show a trend of the K fluxes with throughfall, since in 1990 and 1991 very low fluxes were found, similar to recent years. Thus, the decrease in K fluxes at Steinkreuz and Coulissenhieb should be viewed with caution.

14.4 Rates and Trends of Canopy Leaching

Leaching rates of K, Mg and Ca are calculated by subtracting total deposition from throughfall fluxes (Tables 14.1–14.3). The method used to calculate the total deposition of cations certainly has shortcomings and the errors involved cannot be exactly quantified. Major problems arise by assuming inert behavior of Na in the canopy and that Na interception deposition reflects the interception deposition of K, Ca and Mg. However, the calculated leaching rates are plausible in terms of the amount and relative contribution of the different elements. For Coulissenhieb, the average leaching rates of K, Ca and Mg are 18, 5.5 and 1.0 kg ha^{-1} $year^{-1}$, respectively. Similar rates were found for the LWF01 spruce site (Table 14.3). For Steinkreuz these rates amount to 25, 2.9 and 1.2. The difference in leaching between the two sites was thus most pronounced for Ca, being two times higher in the coniferous sites as compared to Steinkreuz. Langusch et al. (2003), using the same method to quantify total deposition in 37 European forests, report average annual canopy leaching rates of 15, 6 and 1.3 kg ha^{-1} $year^{-1}$ for K, Ca and Mg, respectively. The variation among the sites was very high. The average rates observed at the Lehstenbach and Steinkreuz sites are well within that range.

The temporal development of the leaching rates as well as the H^+ deposition as a potentially driving factor of canopy leaching are given in Fig. 14.4. For Ca and Mg, a decrease in the leaching rates was observed at both sites. For example, the Ca leaching at Coulissenhieb decreased from roughly 8.1 kg ha^{-1} $year^{-1}$ (first 4 years) to 3.6 kg ha^{-1} $year^{-1}$ (last 4 years); the Ca leaching also decreased at Steinkreuz by about 50 %. By far the largest rates were observed in the first 2 years and Ca leaching was relatively low from 1998 onwards. The decrease in the calculated Mg leaching rates at both sites was in a similar range to the Ca leaching (50 %), while the decrease in the K leaching amounted to about 30 %. These developments are confirmed by the long-term data from the LFW01 site in the case of Ca and Mg, but not for K.

Decreasing leaching of Ca, Mg and K might be a result of decreasing H^+ deposition, since H^+ buffering in the canopy triggers the leaching of cations (Hansen 1996; Lovett et al. 1996). Langusch et al. (2003), in a regional assessment, found Ca to be the dominant cation in buffering H^+ ions in the canopy.

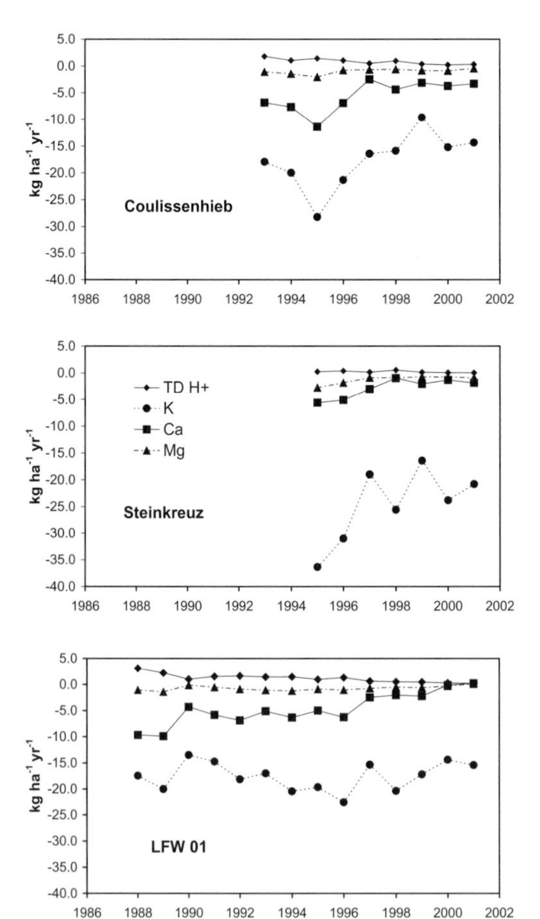

Fig. 14.4. Temporal development of calculated cation leaching in the canopy (values less than 0 indicate leaching; TD = Total Deposition)

The canopy leaching of Ca showed a higher correlation to the total H^+ deposition and to the amount of H^+ buffered as compared to Mg and K. Leaching of K was positively, but weakly, related to the amount of precipitation and SO_4-S deposition. In the case of Coulissenhieb and LFW01, the reduction of H^+ deposition might thus be involved in the trend of Ca and Mg leaching, but at Steinkreuz the H^+ deposition was very low throughout the whole period and cannot explain the trends in leaching. An alternative explanation of changing leaching rates of cations might be the development of foliage cation contents. At Coulissenhieb, the Ca and Mg concentrations of needles in different age classes declined in the past decade (Alewell et al. 2000), which may also result in decreasing rates of cation leaching. Such effects of the Ca content in leaves on Ca leaching were shown in the case of oak trees (Berger and Glatzel 1998). The decline in needle Ca is most likely related to the depletion of Ca in the soil solution (Alewell et al. 2000; Matzner et al., this Vol.).

14.5 Conclusions

The data presented clearly indicate changes in the environmental conditions of both forest ecosystems that cause changes in ecosystem functioning.

- The decrease in S and H^+ deposition is most obvious and has led to actual deposition rates that are in the range of estimated critical loads for the deposition of acidity. In recent years, the calculated deposition of protons at Steinkreuz was <0.1 $kmol_c$ ha^{-1} $year^{-1}$ and at Coulissenhieb around 0.3 $kmol_c$ ha^{-1} $year^{-1}$. Accounting for the potential acidity resulting from the NH_4 deposition (and neglecting the alkalinity production by NO_3 turnover), these numbers increase by about 0.6 $kmol_c$ ha^{-1} $year^{-1}$. Van der Salm and de Vries (2001) gave average values for critical load of acidity in the case of poor forest soils of around 0.4 $kmol_c$ ha^{-1} $year^{-1}$, while Becker et al. (2000) calculated for a large number of German forest sites critical loads of acidity mostly around 1 $kmol_c$ ha^{-1} $year^{-1}$. According to the critical load concept, the present deposition rates of acidity found in our ecosystems should – in the long run – result in conditions where negative effect on soil functions and tree growth should be avoided.
- The changing leaching rates of cations from the canopy, likely caused by the change in H^+ deposition, are an example of how ecosystem functioning is affected by changing deposition. The overall Ca, Mg and K cycling through the trees seems to decline with time, since not only leaching decreased, but also – in case of the Coulissenhieb site – the Ca contents of needles. Both will result in a strong decrease in cation uptake by roots.
- The trend of declining Ca and Mg deposition seems to have leveled out since 1994. The present rates of deposition are much lower than in the past,

causing the mitigating effects of Ca and Mg deposition on soil acidity to decline.

- There was no trend in N deposition as indicated by the throughfall fluxes. Taking the results from other studies into account, it seems that N depositions in central Europe were quite stable for more than 30 years until now (Matzner and Meiwes 1994; Wright et al. 2001). The N fluxes with throughfall presently observed at the Coulissenhieb and Steinkreuz sites are higher than the critical loads derived from empirical studies. For temperate forests, Bobbink et al. (2002) gave a range of 10–20 kg N ha^{-1} $year^{-1}$ (measured as throughfall N) for the critical load, depending on the effect considered. If 10 kg N ha^{-1} $year^{-1}$ is considered as safe to avoid NO_3 losses, nutrient imbalances and destabilization of trees, then the reduction of emission and deposition needed to achieve the critical load is about 50 % of the present rates for Coulissenhieb and LWF01 and about 30 % for Steinkreuz.

Acknowledgements. We would like to thank the members of the Central Analytic Department of the Bayreuth Institute of Terrestrial Ecosystem Research (BITÖK) for the chemical analysis of numerous water samples and for their help in the field sampling. Thanks are also due to Uwe Hell, Andreas Kolb, Gerhard Müller, Roland Blasek and Uwe Wunderlich for their help in the field sampling and site maintenance. Financial support was given by the Federal Ministry of Education and Research (grant no. PT BEO51-0339476 C+D).

References

Alewell C, Manderscheid B, Gerstberger P, Matzner E (2000) Effects of reduced atmospheric deposition on soil solution chemistry and elemental contents of spruce needles in NE-Bavaria, Germany. J Plant Nutr Soil Sci 163:509–516

Armbruster M, MacDonald J, Dise NB Matzner E (2002) Throughfall and output fluxes of Mg in European forest ecosystems: a regional assessment. For Ecol Manage 164:137–147

Becker R, Block J, Schimming CG, Spranger T, Wellbrock N (2000) Critical Loads für Waldökosysteme-Methoden und Ergebnisse für Standorte des Level II-Programms. BML-Report 2001. Bundesministerium für Ernährung und Verbraucherschutz, Bonn

Berger TW, Glatzel G (1998) Canopy leaching, dry deposition, and cycling of calcium in Austrian oak stand as a function of calcium availability and distance from a lime quarry. Can J For Res 28:1388–1397

Bobbink R, Ashmore M, Braun S, Flückiger W, van den Wyngaert IJJ (2002) Empirical nitrogen critical loads for natural and semi-natural ecosystems: 2002 update. Background document for the expert workshop on empirical critical loads for nitrogen on (semi)natural ecosystems. Swiss Agency for the Environment, Forest and Landscape, Berne, Switzerland

Bredemeier M, Matzner E, Ulrich B (1988) A simple and appropriate method for the determination of total atmospheric deposition in forest ecosystem monitoring. In:

Unsworth MH, Fowler E (eds) Acid deposition processes at high elevation sites. Reidel, Dordrecht, pp 607–614

Brumme R, Leimcke U, Matzner E (1992) The uptake of NH_4 and NO_3 from wet deposition by above ground parts of young beech (*Fagus silvatica* L.) trees. Plant Soil 142:273–279

Cappellato R, Peters NE, Ragsdale HL (1993) Acidic atmospheric deposition and canopy interactions of adjacent deciduous and coniferous forests in the Georgia Piedmont. Can J For Res 23:114–124

Chang SC, Matzner E (2000) The effect of beech stemflow on spatial patterns of soil solution chemistry and seepage fluxes in a mixed beech/oak stand. Hydrol Proc 14:135–144

Dise NB, Matzner E, Armbruster M, MacDonald J (2001) Aluminium output fluxes from forest ecosystems in Europe: a regional assessment. J Environ Qual 30:1747–1755

Draaijers GPJ, VanLeeuwen EP, DeJong PGH, Erisman JW (1997) Base cation deposition in Europe, part II. Acid neutralization capacity and contribution to forest nutrition. Atmos Environ 31:4159–4168

Eilers G, Brumme R, Matzner E (1992) Above ground N-uptake from wet deposition by Norway spruce (*Picea abies* Karst.). For Ecol Manage 51:239–249

EMEP (1999) Transboundary acid deposition in Europe. EMEP emission data. Status report 1999 of the European Monitoring and Evaluation Programme. EMEP/MSC-W Report 1/1999. http://www.emep.int/ladm.html

Evans CD, Cullen JM, Alewell C, Marchetto A, Moldan F, Kopáèek J, Prechtel A, Rogora M, Vesely J, Wright R (2001) Recovery from acidification in European surface waters. Hydrol Earth Syst Sci 5:283–298

Fenn ME, Poth MA, Aber JD, Baron JS, Bormann BT, Johnson DW, Lemley AD, McNulty SG, Ryan DF, Stottlemeyer R (1998) Nitrogen excess in North American ecosystems: predisposing factors, ecosystem responses, and management strategies. Ecol Appl 8:706–733

Ferrier RC, Jenkins A, Wright RF, Schöpp W, Barth H (2001) Assessment of recovery of European surface waters from acidification 1970–2000: an introduction to the Special Issue. Hydrol Earth Syst Sci 5:274–282

Gerstberger P (2001) Waldökosystemforschung in Nordbayern: die BITÖK-Untersuchungsflächen im Fichtelgebirge und Steigerwald. Bayreuther Forum Ökol 90:1–186

Hansen K (1996) In-canopy throughfall measurements of ion fluxes in Norway spruce. Atmos Environ 30:4065–4076

Harrison AF, Schulze ED, Gebauer G, Bruckner G (2000) Canopy uptake and utilization of atmospheric nitrogen. In: Schulze ED (ed) Carbon and nitrogen cycling in European forest ecosystems. Ecological studies 142. Springer, Berlin Heidelberg New York, pp 171–188

Horn R, Schulze, ED, Hantschel R (1989) Nutrient balance and element cycling in healthy and declining Norway spruce stands. Ecological studies 77. Springer, Berlin Heidelberg New York, pp 444–455

Langusch JJ, Borken W, Armbruster M, Dise NB, Matzner E (2003) Canopy leaching of cations in central European forest ecosystems, a regional assessment. J Plant Nutr Soil Sci 166:168–174

Liechty HO, Mroz GD, Reed DD (1993) Cation and anion fluxes in northern hardwood throughfall along an acidic deposition gradient. Can J For Res 23:457–467

Likens GE, Bormann FH, Hedin LO, Driscoll CT, Eaton JS (1990) Dry deposition of sulfur: a 23-year record for the Hubbard Brook Forest ecosystem. Tellus 42B:319–329

Lovett GM, Hubbell JG (1991) Effects of ozone and acid mist on foliar leaching from eastern white pine and sugar maple. Can J For Res 21:794–802

Lovett GJ, Nolan SS, Driscoll CT, Fahey TJ (1996) Factors regulating throughfall flux in a New-Hampshire forested landscape. Can J For Res 26:2134–2144
Lovett GM, Bowser JJ, Edgerton ES (1997) Atmospheric deposition to catchments in complex terrain. Hydrol Proc 11:645–654
MacDonald JA, Dise NB, Matzner E, Armbruster M, Gundersen P, Forsius M (2002) Nitrogen input together with ecosystem nitrogen enrichment predict nitrate leaching from European forests. Global Change Biol 8:1028–1033
Matzner E (1988) Der Stoffumsatz zweier Waldökosysteme im Solling. Ber Forschungszentrums Waldökosysteme/Waldsterben Reihe A40:1–217
Matzner E, Meiwes KJ (1994) Long-term development of element fluxes with bulk precipitation and throughfall in two forested ecosystems of the German Solling area. J Environ Qual 23:162–166
Neary AJ, Gizyn WI (1994) Throughfall and stemflow chemistry under deciduous and coniferous forest canopies in south-central Ontario. Can J For Res 24:1089–1100
Puckett LJ (1990) Time- and pH-dependent leaching of ions from deciduous and coniferous foliage. Can J For Res 20:1779–1785
Roelofs JGM, Kempers AJ, Houdijk AL, Jansen J (1985) The effect of air-borne ammonium sulfate on *Pinus nigra* var. *maritima* in the Netherlands. Plant Soil 84:45–56
Sayre RG, Fahey TJ (1999) Effects of rainfall acidity and ozone on foliar leaching in red spruce (*Picea rubens*). Can J For Res 29:487–496
Skeffington R (2002) European nitrogen policies, nitrate in rivers and the use of the INCA model. Hydrol Earth Syst Sci 6:315–324
Stoddard JL, Jeffries DS, Lükewille A, Clair TA, Dillon PJ, Driscoll CT, Forsius M, Johannessen M, Kahl JS, Kellogg JH, Kemp A, Mannio J, Monteith DT, Murdoch PS, Patrick S, Rebsdorf A, Skjelkvale BL, Stainton MP, Traaen T, van Dam H, Webster KE, Wieting J, Wilander A (1999) Regional trends in aquatic recovery from acidification in North America and Europe. Nature 401:575–578
Ulrich B (1983) Interaction of forest canopies with atmospheric constituents: SO_2, alkali and earth alkali cations and chloride. In: Ulrich B, Pankrath J (eds) Effects of accumulation of air pollutants in forest ecosystems. Reidel, Dordrecht, pp 33–45
Ulrich B (1994) Nutrient and acid-base budget of central European forest ecosystems. In: Godbold DL, Hüttermann A (eds) Effects of acid rain on forest processes. Wiley-Liss, New York, pp 1–50
Van der Salm C, de Vries W (2001) A review of the calculation procedure for critical acid loads for terrestrial ecosystems. Sci Total Environ 271:11–25
Wood T, Bormann FH (1975) Increases in foliar leaching caused by acidification on artificial mist. Ambio 4:169–171
Wright RF, Alewell C, Cullen JM, Evans CD, Marchetto A, Moldan F, Prechtel A, Rogara M (2001) Trends in nitrogen deposition and leaching in acid sensitive streams in Europe. Hydrol Earth Syst Sci 5:299–310

15 Phyllosphere Ecology in a Changing Environment: The Role of Insects in Forest Ecosystems

B. STADLER and B. MICHALZIK

15.1 Introduction

A major challenge to understanding the role of species in ecosystems (Lawton 1994) is to collect and filter those pieces of information from different levels of organization that are likely to affect ecosystem functioning and maintenance of key processes; for example, the approaches that link ecosystem and population ecology, evolution and system ecology (Loehle and Pechmann 1988) or different scales from leaves to landscapes (Holling 1992; Levin 1992; Wiens 1995). However, integrated approaches are notoriously difficult to pursue. In addition, the approaches of conventional research have different aims. For example, ecosystem ecologists are often more concerned with average rates of flow for a particular period of time, viewing an ecosystem as a gigantic 'black box' (Grimm 1995) whose behaviour is independent of its history (Higashi and Burns 1991) (*engineer approach*). Ecologists, in contrast, working with organisms, populations or communities, are concerned with specific adaptations, species interactions, community structures and population dynamics in an evolutionary context (*organismic approach*).

Implicit in these contrasting views of the world is a fundamentally different perception of temporal and spatial scales. For example, a catchment is a spatial scale unimportant to organismic biologists, because most organisms are unlikely to appreciate this boundary. Ecosystem ecologists, however, need to estimate inputs and outputs from ecosystems, which make catchments a useful tool. When trying to understand the behaviour of ecosystems in a changing environment, each of these approaches has merits and shortcomings, because they reflect only part of the real world. Usually, it is a combination of effects, physical, chemical, climatic and biotic, that influence ecosystem functions. For example, predictions on inorganic nitrogen losses from ecosystems now start to recognize the full spectrum of disturbance events and changes in environmental conditions (Aber et al. 2001). However, canopy insect herbivores are usually inconspicuous and not considered except when population outbreaks cause severe defoliation (Swank et al. 1981). As a consequence, their

Ecological Studies, Vol. 172
E. Matzner (Ed.), Biogeochemistry of Forested Catchments in a Changing Environment

impacts on ecosystem processes remain unnoted even though there are early indicators like an increased potassium flux in throughfall at nominal herbivory (Seastedt et al. 1983; Seastedt and Crossley 1984; Stadler et al. 2001b). To understand the direction of change in ecosystem processes under changing environmental conditions, an important question is when are perturbations amplified and under which conditions are they dampened? It is likely that there are many compensatory dynamics in forest ecosystems, especially in fast-growing, rapidly multiplying organisms, which are connected via trophic interactions. To disentangle these relationships and evaluate their role for ecosystem processes is the aim of this chapter.

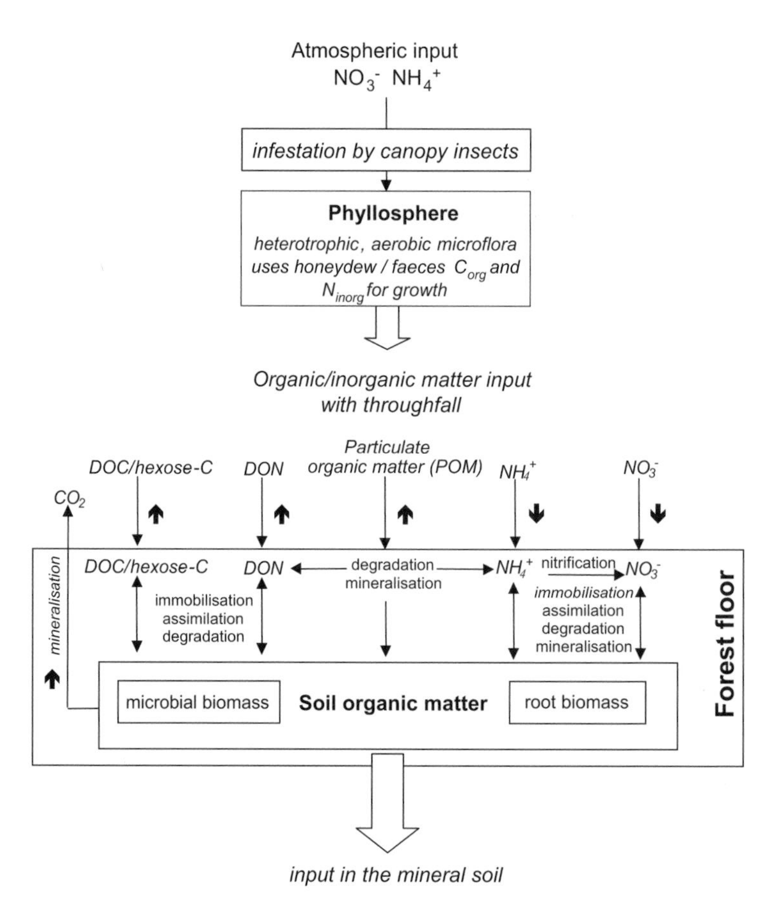

Fig. 15.1. Conceptual approach to the study of the role of phytophagous insects in forest ecosystems. Excreta of aphids and lepidopterous larvae had an effect on the epiphytic microorganisms and nutrient cycling in the canopy of Norway spruce and beech/oak. Eventually, input of energy and nutrients affects soil processes. *Italics* indicate those aspects that are addressed in this chapter. *Bold arrows* indicate that the phytophagous insects significantly increased or decreased nutrient concentrations/fluxes in throughfall

This synopsis of our field studies in coniferous and deciduous forests offers an integrated approach by recognizing the importance of details (identifying mechanisms), while painting a larger picture (identifying patterns). In addition, an attempt is made to connect the biological processes occurring above- and below ground, for example, in the canopy and forest soil. We only present results for two functional groups of phytophagous insects, aphids and lepidopterous larvae (sap and leaf feeders), feeding in the canopies of coniferous and deciduous trees at the two main study sites (Gerstberger et al., this Vol.) and their effects on the nutrient and energy fluxes through the canopy (Fig. 15.1). In this approach, we link the population dynamics and trophic interactions between herbivores and microorganisms with throughfall and soil solution chemistry. We do not address the indirect effects, e.g. herbivore-induced changes in plant chemistry, community or canopy structure, although we acknowledge their importance. Instead, we focus on environmental change that may affect insect populations, which subsequently affect energy and nutrient input into different compartments of forested ecosystems.

Given that forest ecosystems are complex adaptive systems (Levin 1998, 1999), we argue that, in order to understand ecosystem functioning, it is necessary to arrive at an understanding of processes based on feedback loops, which link organisms with biogeochemistry. We identify gaps in our understanding of the driving forces of key ecosystem processes and suggest avenues for future research, which better link the engineer and organismic approach, in order to identify the direction and magnitude of changes in ecosystem processes.

15.2 Ecology of Canopy Insects

15.2.1 Aphids

Aphids are most diverse and abundant in the temperate regions (Dixon et al. 1987). Due to parthenogenetic reproduction they are able to exploit ephemeral resources and become abundant in a single season (Scheurer 1964; Kidd 1990). An outstanding characteristic of aphids is that they often produce copious amounts of honeydew (sugary excreta). In forest ecosystems, wet weight of the honeydew produced may amount to 400–700 kg ha^{-1} $year^{-1}$ (Zoebelein 1954; Table 15.1). The quantities of honeydew may not be comparable because of the different methods and units of measurement used. Nevertheless, the drain of energy from trees by aphids is substantial. For example, Llewellyn (1975) reported that the average loss of energy from mature lime trees is 19 % of the annual net production due to infestation with lime aphids,

Table 15.1. Quantities of honeydew produced by different species of aphids recorded in the literature. Methods used and amounts of honeydew recorded differ substantially between studies (direct sampling with aluminium foil, washing of honeydew from leaves, determining differences in weight of attending ants before and after collecting honeydew), making it difficult to compare results directly (*Fm* fresh mass; *Dm* dry mass)

Author	Aphid species	Quantities of honeydew produced
Herzig (1937)	*Aphis* sp. on different plants	1.4–6.1 kg Fm/100 days/nest (*Lasius* sp.)
Zwölfer (1952)	*Cinara* on spruce	400 kg Fm ha^{-1} year^{-1}
Zoebelein (1954)	*Cinara* on spruce	4 kg Dm tree^{-1} year^{-1} $\rightarrow\sum$ 700 kg Fm ha^{-1} year^{-1}
	Aphids on beech	6–7 kg Dm tree^{-1} year^{-1} (collected by ants)
Müller (1956)	Aphids on pine + beech	4,800 l/100 trees (collected by three ant colonies)
Wellenstein (1961)	Lachnid on pine + Lachnid on oak	10,000 l ha^{-1} year^{-1}
Eckloff (1972)	*Cinara* spp. on spruce	5–70 kg Fm tree^{-1} year^{-1} (400–700 kg ha^{-1} year^{-1})
Llewellyn (1972)	*Eucallipterus tiliae*	8–10 kg Dm tree^{-1} year^{-1}
Heimbach (1986)	*E. tiliae*	7 mg Dm during larval development
	Tuberolachnus annulatus	2–50 kg Dm tree^{-1} year^{-1} (*Tilia* spp. 40–50 years old) 5 mg Dm during larval development
Spiller and Llewellyn (1987)	*Rhopalosiphum padi*	0.14–0.23 μl/24 h
	Metopolophium dirhodum	0.11–0.40 μl/24 h (different cereal crops)
Novak (1989)	*Hyalopterus pruni*	100–300 ml m^{-2} hedge (e.g. *Prunus spinosa*)
Stadler et al. (1998)	*Cinara pilicornis*	0.165 mg Dm/24 h (on Norway spruce)
Stadler et al. (2001b)	*E. abietinum*	0.0157 mg Dm/24 h (on Sitka spruce)

and 26 aphids/leaf would drain completely the net primary production of a tree. When the aphid *Drepanosiphum plantanoides* (Schr.) is present on newly developing leaves of *Acer pseudoplantanus* (L.), leaf area reduction is proportional to the number of aphids feeding on a leaf during leaf development and growth of stem wood of infested trees per year may be reduced by 280% (Dixon 1971).

At our study sites in the Waldstein area, population size of *Cinara pilicornis* (Hartig) on Norway spruce varied considerably both between and within years and usually showed a single peak in abundance (Fig. 15.2). Aphids were

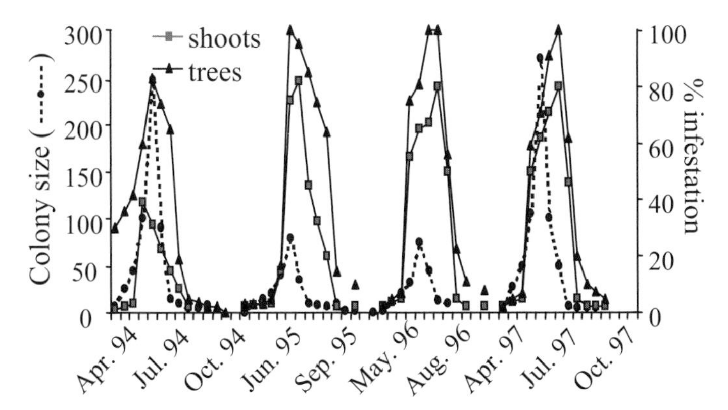

Fig. 15.2. Mean percentage of infested trees and infested shoots (*solid lines*), as well as number of aphids on shoots (colony size, *dotted lines*) on Norway spruce at the Waldstein site from 1994–1997

abundant in June in 1994 and 1997, and relatively uncommon in 1995 and 1996. After the dispersal of alates, colony size declined. As a consequence, the number of infested shoots and trees increased with time and often reached 100% in July/August. Therefore, population dynamics of conifer aphids is characterized by a clumped distribution early in the season and a more even distribution later in the year. This pattern has consequences for the spatial distribution of honeydew within the canopy of trees and forest stands. Seemingly, a multigeneration life cycle of opportunistic organisms such as aphids produces population fluctuations, which are highly erratic and may depend heavily on plant quality, current weather conditions and migrating behaviour (Kindlmann and Dixon 1996).

15.2.2 Lepidopterous Larvae

Several species of lepidoptera become pests in forests under favourable conditions. Among the best-understood examples are *Lymantria dispar* (L.), *L. monacha* (L.), *Operophtera brumata* (L.) and *Euproctis chrysorrhoea* (L.) (Varley et al. 1973; Barbosa and Schaefer 1997; Hassell 2000; Liebhold et al. 2000; Allen and Humble 2002). Detailed studies of these species have provided good insights into host–parasitoid/predator ecology as well as competitive interactions with other species in the same guild-influencing population cycles and outbreaks (e.g. Schowalter et al. 1986; Royama 1996). Almost all of these species have a single generation per year. Thus, a numerical response to favourable environmental conditions is likely to result in population cycles with periods of repeated outbreaks and crashes, which usually span several years to decades. Changes in abundance are usually less erratic and the numerical response to environmental change less pronounced than in aphids. The most visible direct damage is destructive feeding on the foliage of the host plants sometimes leading to considerable leaf area loss (Schowalter et al. 1986; Lowman 1995).

At the Steigerwald site, beech and oak suffered considerable leaf area loss due to feeding by *O. brumata* (Fig. 15.3). Loss in leaf area was visually classified as either 0, 1–5, 6–25, 26–50 or more than 50 % and monitored at 2-week intervals. Lepidopterous larvae began to feed when the leaves started to develop in mid-May (Fig. 15.3a). At that time 70–80 % of the leaves on beech and oak were damaged and this increased to 90 % by mid-October. On the uninfested control sites, few leaves were damaged initially, but later in the season up to 40 % of the terminal leaves showed signs of feeding damage. The loss of leaf area, however, was considerably greater in the infested stands, where leaf area losses ranged between 25 and 40 % on beech and 20 and 35 % on oak. In contrast, lepidopterous larvae only consumed 1–5 % of the leaf area of the control trees on an adjacent site (Fig. 15.3b).

A major problem in evaluating the effects of these canopy insects is the difficulty of sampling the upper strata of mature trees (Lowman and Wittman 1996). Therefore, much of our knowledge on trophic interactions and links to

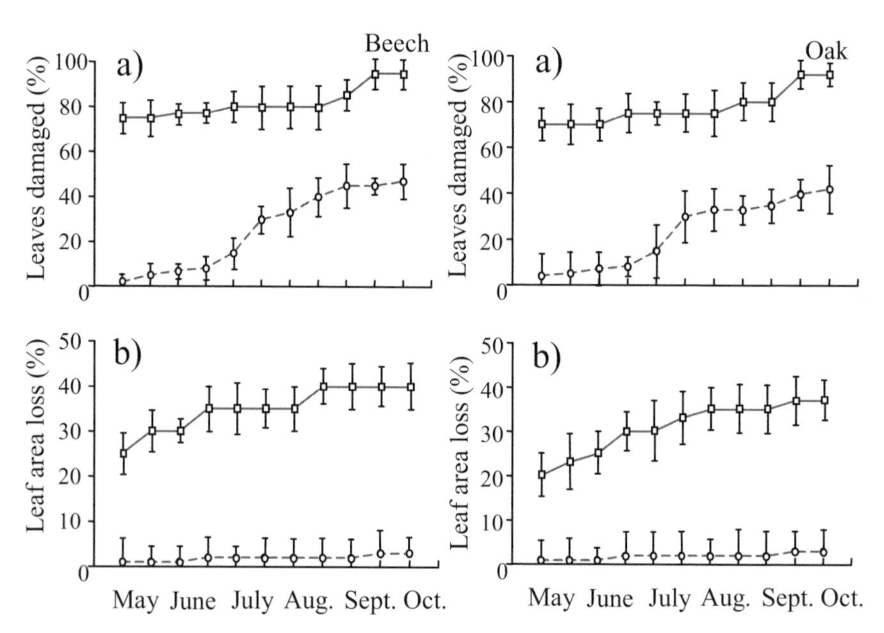

Fig. 15.3. Percentage of leaves damaged (a) and reduction in leaf area (b) of beech and oak during the growing season at the Steigerwald site. *Solid lines with squares* Trees that were infested with leaf-feeding lepidopterous larvae; *dashed lines with circles* uninfested controls. Distance between trees in the infested and uninfested plots was 300 m. Values are mean±SE. (Stadler et al. 2001a)

ecosystem processes is derived from small trees and from studies of the peripheral leaves in the lower strata of large trees. Nevertheless, these studies provide a starting point for the qualitative and quantitative assessment of trophic interactions.

In summary, the available information suggests that aphids and larvae of lepidoptera may initiate: (1) direct effects such as (a) increased deposition of honeydew/faeces onto needles and leaves, (b) transfer of energy and nutrients from the plants to insects and (c) on foliage growth, chemistry and quantity; and (2) indirect effects by (a) altering decomposition processes, (b) changing plant community structure, (c) increasing root respiration and exudation and (d) especially during outbreaks, changing abiotic conditions such as light penetration through the canopy, plant transpiration and soil hydrology. These effects are likely to vary with species, insect abundance and environmental conditions, and have different rates.

15.3 Interactions Between Insects and Microorganisms

Insects and epiphytic microorganisms are two of the most abundant groups of organisms, and they likely affect the distribution and abundance of each other. In areas in many parts of central Europe with a high atmospheric nitrogen deposition (Gundersen et al. 1998; Schulze 2000), the limiting resource for microorganisms is energy (Tsai et al. 1997). Therefore, any canopy process that increases the availability of energy should favour the abundance of microorganisms. We determined the number of colony-forming units (CFU) by collecting five shoots of beech and five shoots of oak, each with about 10–15 leaves. Shoots were collected in spring, summer and early autumn from the lower strata at the periphery of the canopy. In all our studies on the abundance of epiphytic microorganisms, we found a 2–3 order of magnitude increase in the number of bacteria, yeasts and filamentous fungi on needles and leaves of infested trees (Stadler et al. 1998; Stadler and Müller 2000). The results were the same irrespective of the cultivation media used for growing microbes. For example, specialist microbes that are able to grow by utilizing inorganic nitrogen, such as ammonia-N or nitrate-N as the sole N source, thrived significantly better when honeydew was available. In addition to changes in abundance, there was also a change in the species composition on leaves of infested trees. For example, on uninfested leaves of oak, 10 species of bacteria and, on damaged leaves, 9 species were identified, while on uninfested and damaged leaves of beech 9 species and 16 species were identified, respectively (Müller et al. 2004). The rank abundance plot (Fig. 15.4) shows that it was mainly three to four species of bacteria on beech that were abundant, while on oak it was only one species. Many species of bacteria were rare. Therefore, the effect of herbivores on species richness and diversity of epi-

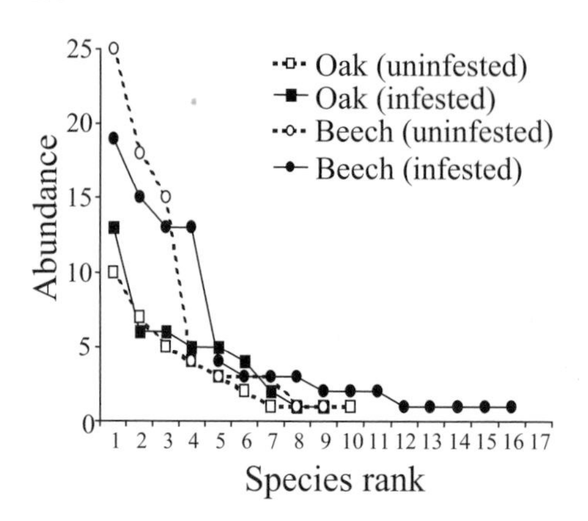

Fig. 15.4. Rank abundance plot for species of bacteria living on leaves of beech and oak either infested (*solid lines*) or uninfested (*dotted lines*) with lepidopterous larvae

phytic bacteria was much greater on beech than on oak, probably due to differences in surface micromorphology (Neinhuis and Barthlott 1998). For example, leaves of beech are smooth, lack wax crystals and are wettable, while those of oak contain higher concentrations of hydrolyzable tannin (Shure et al. 1998), which might suppress growth of some species of microorganisms. The different species of bacteria had different abilities to utilize inorganic nitrogen, especially on oak, which we believe affects nutrient cycling in the canopy of infested trees, and subsequently alters nutrient flows from the canopy to the forest floor.

15.4 Biotic Interactions and Throughfall Chemistry

Less well understood are the effects of the trophic interactions, which were described above, on throughfall chemistry. This is, however, an important aspect, because processes mediated by canopy insects determine the input of nutrients and ions into the forest floor. The effects of herbivores may be largely underestimated if only direct effects, such as the amount of lost leaf area due to feeding of caterpillars, are considered. Indirect effects such as an increased nitrogen content of infested leaves (Dixon 1971), effects on mycorrhiza (Gange et al. 2002), changes in litter quality (Findlay et al. 1996) or changes in root exudation (Bardgett et al. 1998; Wardle 2002) might be equally important.

In a field experiment, we studied the effects of aphids feeding on Norway spruce and lepidopterous larvae feeding on beech and oak on throughfall chemistry (Fig. 15.5). To do so, five throughfall samplers were placed beneath trees close to the periphery to be able to relate aphid abundance and leaf area loss with throughfall chemistry. Solutions were collected at 2-week intervals and, after returning to the laboratory, immediately filtered (0.45-μm, cellu-

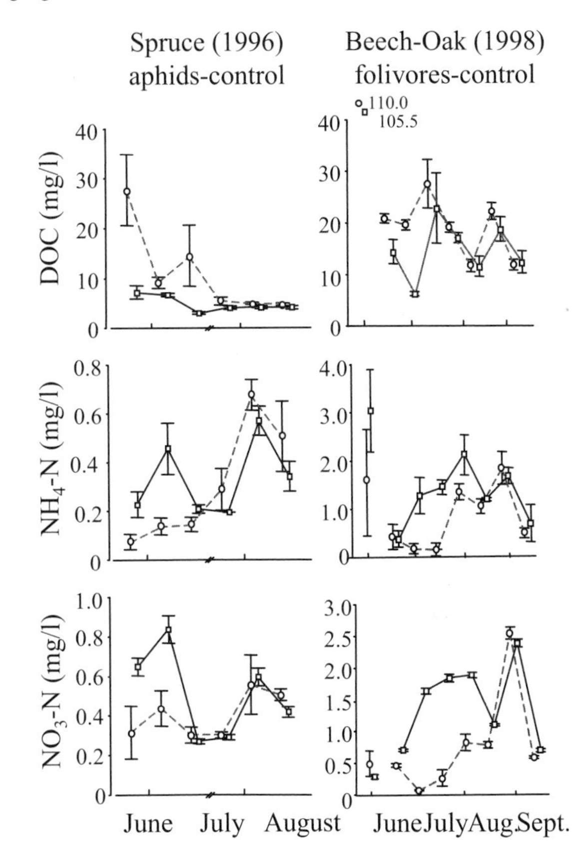

Fig. 15.5. Concentrations (mean±SE) of different chemical compounds in throughfall collected beneath infested (*dashed lines with circles*) and uninfested (*solid lines with squares*) trees during the growing season at the Waldstein (Norway spruce, aphid infested; *left*) and Steigerwald (beech–oak, lepidoptera infested; *right*) sites

lose-acetate membrane) and analyzed. In June, at the time when aphids were most abundant (Fig. 15.2), dissolved organic carbon (DOC) concentrations in throughfall collected beneath infested spruce were highest. With declining aphid numbers DOC concentrations also declined (Fig. 15.5, left). The correlation between DOC concentrations and colony size above throughfall samples was highly significant (r=0.72, P <0.001). Beneath trees attacked by lepidopterous larvae, the concentrations of DOC were higher at the beginning of the sampling period (Fig. 15.5, right) when the lepidopterous larvae were actively feeding. The high value at the first collection in June was due to an unusually high input of pollen from pine in that year.

Dissolved organic nitrogen (DON) concentrations in throughfall collected beneath spruce increased after the aphid population (and DOC concentrations, respectively) declined, while no significant differences were recorded in the throughfall collected beneath beech/oak attacked by lepidopterous larvae. Concentrations of inorganic nitrogen were lower in the throughfall collected beneath spruce as long as aphids were present on the trees, and differences disappeared with declining aphid numbers (DOC concentrations, respectively). Averaged over the complete sampling period, NH_4-N and NO_3-N concentrations in throughfall, collected beneath damaged beech/oak trees, were 25–45 % lower than beneath undamaged trees (Fig. 15.5, right). These differences were especially pronounced during June and July when the caterpillars were actively feeding.

15.5 Beyond the Canopy: Influences on Soil Processes and Ecosystems

It is interesting to know whether the increased input of DOC and DON and reduced input of inorganic nitrogen beneath infested Norway spruce affect processes in the forest floor.

Field experiments showed that the passage of DOC from the canopy to the floor led to significantly higher DOC concentrations (Mann-Whitney U-test, P=0.003) in forest floor solutions, with the initial response slightly delayed and longer lasting than the aboveground infestation maxima (Fig. 15.6). Significantly higher concentrations of DON (P <0.001) and NO_3-N (P=0.006) were observed right from the beginning of the experiment in the floor solu-

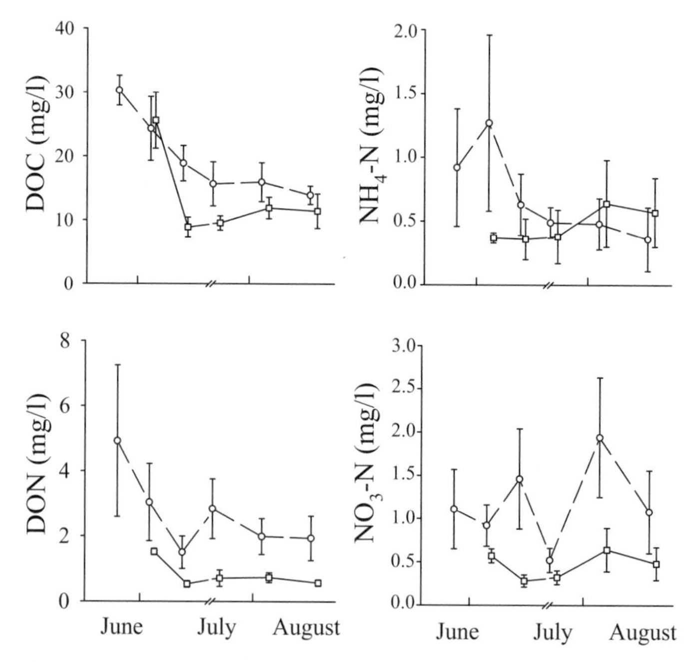

Fig. 15.6. Concentrations (mean±SE) of different compounds in forest floor solution collected from under infested (*dashed lines with circles*) and uninfested (*solid lines with squares*) Norway spruce trees during the growing season at the Waldstein site

tions collected beneath infested trees, whereas no statistically significant trend was observed for NH_4-N concentrations. While the concentrations in the forest floor solution were affected by aphid infestation, differences in fluxes of organic matter and inorganic nitrogen species between infested and uninfested sites showed no consistent trend (data not shown).

These results suggest that the less pronounced effects of phytophagous insects on fluxes in soil solution chemistry might be due to (1) a high variability in rainfall and (2) the temporal frequency of measurements (e.g. weekly sample collection), which is too coarse to adequately record biological soil processes. Therefore, to study the effects mediated by aphids on forest floor solution chemistry under controlled conditions, column irrigation experiments were carried out (Michalzik and Stadler 2000).

The different levels of aphid infestation (uninfested, moderately infested, highly infested) were simulated in these experiments by applying different amounts (0, 16 and 80 mg hexose-C day^{-1}) of honeydew-DOC to forest floor cores. The amounts corresponded to the amount of honeydew produced by *C. pilicornis* on Norway spruce. Honeydew was applied with simulated precipitation at t=24, 48 and 72 h. Sampling was carried out hourly and daily in order to determine the effects on the CO_2 production as well as on DOC, DON and N_{inorg} dynamics in forest floor solutions.

The simulation of a moderate level of aphid infestation, i.e. a low input of honeydew-DOC, significantly enhanced base respiration within 1 h of honeydew addition (Mann-Whitney U-test, P=0.001) to an average of 109 mg CO_2-C h^{-1} m^{-2} (Fig. 15.7). The simulation of the heavy infestation levels observed in

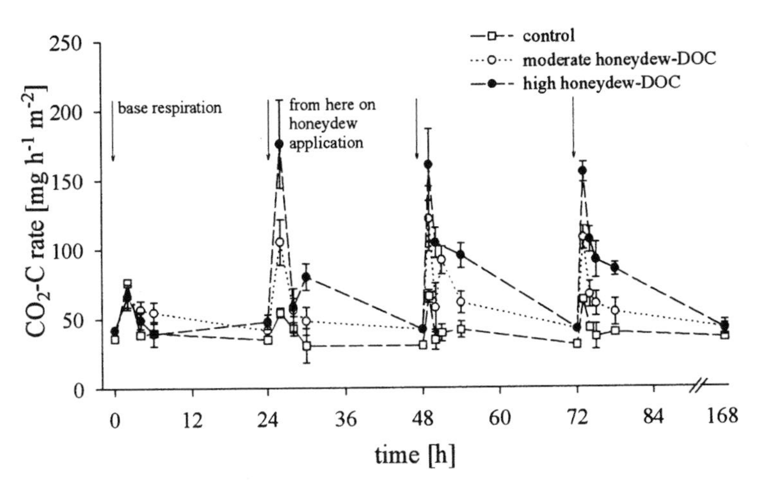

Fig. 15.7. CO_2 emitted (mean±SE) from the forest floor. Columns of soil either received high (*solid circles*) or moderate (*open circles*) amounts of honeydew-DOC. To the control treatment (*squares*) equal amounts of water, but no DOC were applied. (Michalzik and Stadler 2000, reprinted with permission from Urban and Fischer Verlag)

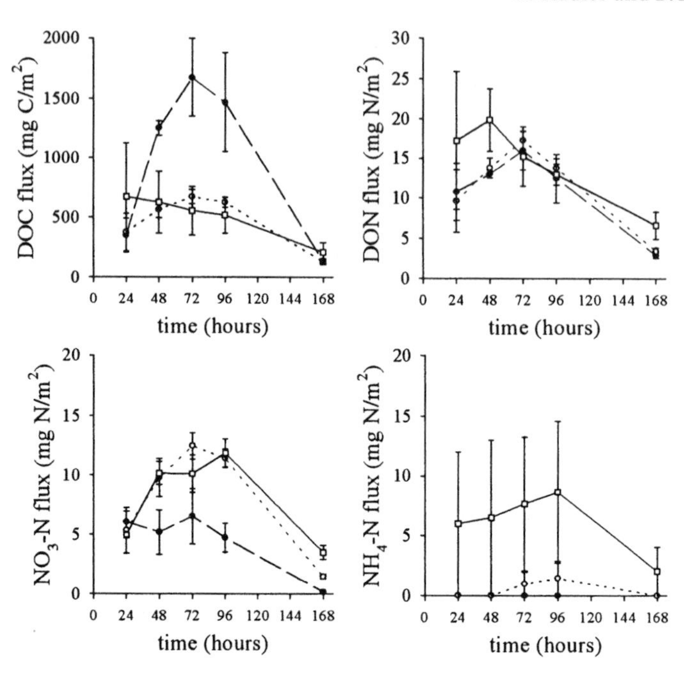

Fig. 15.8. Fluxes (mean±SE) of compounds in forest floor percolates irrigated with high (*solid circles*) and moderate (*open circles*) amounts of honeydew-DOC, compared to the control treatment (*squares*) that received no DOC

the field in 1994 and 1997 (see Sect. 15.2.1 and Fig. 15.2) further increased CO_2 efflux (P=0.0003) to 175 mg CO_2-C h^{-1} m^{-2}.

The addition of moderate amounts of honeydew caused a significant decline in NH_4-N fluxes (P=0.043) in forest floor solution, but compared to the control treatment had no effect on NO_3-N, DON or DOC fluxes. In contrast, a large input of honeydew increased the immobilization of both NH_4-N and NO_3-N (P=0.019 and 0.014, respectively) and slightly reduced DON fluxes (Fig. 15.8). DOC fluxes increased considerably but nevertheless declined to the base level within 72 h of applying honeydew (Fig. 15.8).

These results indicate the importance of inputs of organic carbon for forest soil CO_2 respiration and solution chemistry over short periods of time. Over larger temporal and greater spatial scales (e.g. annual fluxes in a forest stand), the variability in the fluxes caused by canopy insects is buffered when passing through the other ecosystem compartments, such as the forest floor. Therefore, annual averages obscure the effect of short-term biological processes responsible for variation in ecosystem key variables. It appears unlikely that soil processes such as mineralization, mobilization and transport of organic matter will be understood without an adequate knowledge of the factors that affect these processes. Biological processes appear to be more affected by variability in the availability of resources (*organismic approach*) than by annual mean values (*engineer approach*).

15.6 Synthesis and Future Prospects

In order to achieve a good understanding of how species and communities affect ecosystem processes, it is necessary to combine life history, population ecology and evolutionary ecology with ecosystem ecology. This is not a simple task. A prerequisite is that the different research approaches operate in concert and are aware that some patterns may only become apparent as a result of the interdisciplinary co-operation. Our approach (Fig. 15.1) combined inputs from zoology, microbiology and soil ecology and generated a number of robust results. For example, the two functional groups of insects initiated similar events. While feeding they produced excreta, which supplied energy and nutrients for epiphytic microorganisms, which thrived better on infested trees (Fig.15.4; Stadler and Müller 2000). These biotic interactions affected throughfall chemistry by increasing energy and reducing nutrient flows (Fig. 15.5). As not all of the energy-rich compounds are consumed in the canopy, they reach the forest floor (Fig. 15.6) where the same processes affect the soil solution. Soil microorganisms utilize the labile carbon and immobilize nitrogen (Figs. 15.7, 15.8). The magnitudes of these reactions initiated by aphids and lepidopterous larvae differ, but are qualitatively identical. Thus, in general, excreta seem to be the main driving force for a chain of processes cascading from the canopy down to the forest floor.

If the excreta of insects are an important factor driving short-term nutrient cycles in the canopy (Fig. 15.5) and forest floor (Figs. 15.6, 15.7), then there is a need to combine information on the life history response of insects, their population dynamics, migration behaviour and evolutionary response to changing environmental conditions. For example, *C. pilicornis* produces more honeydew on experimentally stressed trees (Stadler et al. 1998) and *O. brumata* produces more faeces (Buse et al. 1998) on trees exposed to higher levels of CO_2. In particular, sap-feeding insects seem to do better on stressed trees, but not necessarily leaf feeders (Koricheva et al. 1998).

15.6.1 Linking Above- and Below-Ground Processes

Another important aspect is the linking of above- and below-ground processes (Bardgett et al. 1998; Bonkowski et al. 2001; Hunter 2001; Scheu 2001). In temperate forest ecosystems, annual fluxes of DOC and DON in the forest floor are positively associated with fluxes of DOC and DON in throughfall (Michalzik et al. 2001). Variation in throughfall fluxes explained about 46 and 65 % of the variation in DOC and DON fluxes in the forest floor, respectively. An intercorrelation with water fluxes was not observed. The results suggest that dissolved organic matter (DOM) fluxes in throughfall positively affect the fluxes of DOM in the forest floor. It is unlikely that the DOM in throughfall solution passes through the canopy and forest floor without changes in its chemical composition. As reported by Qualls and Haines (1992), up to 60 % of the DOM in throughfall is decomposed, with the highest rates of decomposition occurring in the first 3 weeks. Guggenberger and Zech (1994) found that 50 % (35–50 kg C ha^{-1} $year^{-1}$) of the DOC in throughfall consists of carbohydrates, mainly microbial metabolites washed out of the canopy. Thus, throughfall is loaded with easily decomposable C and N compounds, which probably act as co-substrates or promoters accelerating decomposition and mineralization processes in the forest floor. Eventually, these processes might lead to increased fluxes of DOC and DON in forest floor leachates. However, it needs to be shown that this is not just a consequence of certain environmental factors, such as high temperatures and optimal moisture conditions, enhancing the production of DOM in both the canopy and the forest floor at the same time.

15.6.2 The 0.45-µm Barrier

The link between above- and below-ground processes should become even more apparent if DOM fractions larger than 0.45 µm are studied in more detail. So far, the effects of organic matter larger than 0.45 µm, but smaller than 1 mm, the size of fine particulate organic matter (FPOM), in throughfall on soil processes have not been assessed. The only study we are aware of is that of Carlisle et al. (1966) in which field data on DOM fractions larger than 0.45 µm were collected in an oak stand. For the size class up to 200 µm (0.2 mm) these authors observed an annual carbohydrate input in throughfall of 89.2 kg ha^{-1} and a (particulate) organic carbon input of 226.6 kg C ha^{-1}. Highest amounts of glucose, fructose and melizitose were reported during August. This is considerably greater than that reported by Guggenberger and Zech (1994) for dissolved organic carbon at a spruce site (see above) or by Currie et al. (1996) with 117 kg C ha^{-1} $year^{-1}$ in a hardwood forest. At our study site in the Steigerwald, we determined the fluxes of hexose-C in unfiltered throughfall solutions in order to assess the importance of organic matter transport via the particulate pathway compared to the dissolved fraction (Fig. 15.9). We found that particulate hexose-C fluxes were 30 % higher at the infested than at the control sites, whereas the fluxes (variation within 10 %) were nearly identical in the filtered samples. It is not known how these particulates influence soil processes such as litter decomposition and microbial activity. From a quantitative point of view, the preliminary data suggest that particulate organic matter is an important component of the throughfall

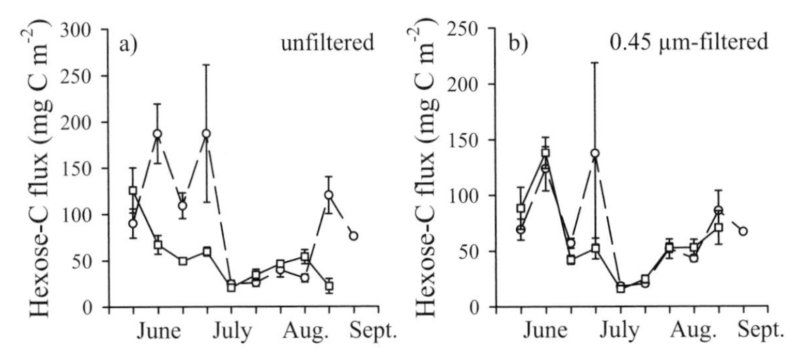

Fig. 15.9. Fluxes (mean±SE) of hexose-C in a unfiltered and b 0.45-μm-filtered throughfall solutions collected from beneath infested (*dashed lines with circles*) and control (*solid lines with squares*) trees at the Steigerwald site

input of organic substances into the forest floor. Thus, given the fact that this pathway is rarely considered in ecosystem studies, future research should focus on the amounts, forms and function of canopy-derived organic matter produced under different climates and environmental conditions.

15.6.3 Indirect Effects

A further drawback limiting our understanding of the links between food-web interactions and ecosystem processes is that studies on the *indirect effects* of herbivores on plants and ecosystems only describe processes qualitatively and do not develop methods to quantify the impacts on vital ecosystem functions. For example, it would be interesting to know how leaves affected by excreta or damaged by herbivores influence decomposition rates (per unit area) and what this means for soil-building processes. Field experiments on indirect effects are needed. Similarly, quantifying the effects of herbivores on root exudation is virtually impossible at the moment. Therefore, in order to complete the picture on the effects of herbivory on soil nutrient dynamics, more effort should be directed at obtaining estimates of the magnitude of these processes.

15.7 Conclusions

The results reported here may be considered as conservative, because the climatic conditions at our study sites did not promote any major insect outbreaks. However, assuming there has been an increase in annual average temperature, precipitation and, more likely, more variability in these factors within and between years (Schneider 1993; Foken, this Vol.), we propose the following hypotheses for forest insects, which are all testable. These hypotheses are based on our knowledge of the ecology of these two groups of canopy insects and their trophic links with microorganisms and ecosystem process.

- Large fluctuations in abiotic conditions will favour fast-reproducing insects with a high phenotypic plasticity, which are able to respond numerically to changes in environmental quality. Opportunistic species like many sap feeders are likely to respond most markedly.
- If host plants tend to become less favourable for phytophagous insects, e.g. with increase in atmospheric CO_2 concentrations, then the expectation is that even if the frequency of outbreaks of forest pest species does not increase, more energy and nutrients tend to become available in the canopies of forest trees. This is due to longer developmental times, extended periods of infestation, increasing consumption of leaves or sap or larger amounts of faeces produced.
- The abundance of epiphytic microorganisms in the phyllosphere will increase considerably if more energy is made available due to the compensatory feeding of herbivores. A larger standing crop of microorganisms will immobilize atmospheric N by a buildup of biomass, transforming inorganic into organic N. Thus, both the quality and amount of the N input into the forest ecosystem will change.
- A higher input of labile carbon compounds and organic nitrogen (e.g. DOC, DON, POM) into the forest floor may speed up decomposition processes and initiate a positive feedback loop, increasing the rate of CO_2 emission from the forest floor.

These hypotheses are interconnected and an experimental test of the underlying mechanisms will need a multidisciplinary approach. The challenge will be to filter the important details obtained from the organismic ecological approach and combine them with those from the large-scale patterns and processes. Especially in climate change scenarios, it is important to understand the link between biotic interactions and ecosystem processes, because the rates of response might be very different. There have been several attempts to predict the response of herbivores to climate change (Kareiva et al. 1993; Docherty et al. 1997; Bale et al. 2002; Watt and McFarlane 2002; Dixon 2003). However, there is a shortage of integrated studies dealing with more than a single trophic level (Schowalter 2000). Even studies on multitrophic interactions or bottom-up and top-down forces are on a specific set of organisms and population ecology is poorly integrated into ecosystem ecology. The effect of phytophagous insects on ecosystems extends further than one or two trophic levels up and down the food chain. Growing evidence suggests that they also affect ecosystem functioning (e.g., Swank et al. 1981; Lovett and

Ruesink 1995) even at endemic levels of infestation (Stadler et al. 2001a). Therefore, the elaboration of links between organismic ecology, with its many separate disciplines, and ecosystem science is mandatory if threats to many ecosystems are to be better anticipated and a mechanistic understanding of ecosystem functioning is to be achieved.

Acknowledgements. We are indebted to Tony Dixon for helpful comments on an earlier draft of the manuscript. Financial support was given by the German Ministry for Research and Technology (grant no. BMBF no. PT BEO 51-0339476).

References

Aber JD, Ollinger SV, Driscoll CT, Likens GE, Holmes RT, Freuder RJ, Goodale CL (2001) Inorganic nitrogen losses from a forested ecosystem in response to physical, chemical, biotic, and climatic perturbations. Ecosystems 5:648–658

Allen EA, Humble LM (2002) Nonindigenous species introductions: a threat to Canada's forests and forest economy. Can J Plant Pathol 24:103–110

Barbosa P, Schaefer PW (1997) Comparative analysis of patterns of invasion and spread of related lymantriids. In: Watt AD, Stork NE, Hunter MD (eds) Forests and insects. Chapman and Hall, London, pp 153–175

Bale JS, Masters GJ, Hodkinson ID, Awmack C, Bezemer TM, Brown VK, Butterfield J, Buse A, Coulson JC, Farrar J, Good JE, Harrington R, Hartley S, Jones TH, Lindroth RL, Press MC, Symrnioudis I, Watt AD, Whittaker JB (2002) Herbivory in global climate change research: direct effects of rising temperature on insect herbivores. Global Change Biol 8:1–16

Bardgett RD, Wardle DA, Yeates GW (1998) Linking above-ground and below-ground interactions: how plant responses to foliar herbivory influence soil organisms. Soil Biol Biochem 30:1867–1878

Bonkowski M, Geoghegan IE, Birch ANE, Griffiths BS (2001) Effects of soil decomposer invertebrates (protozoa and earthworms) on an above-ground phytophagous insect (cereal aphid) mediated through changes in the host plant. Oikos 95:441–450

Buse AJ, Good EG, Dury S, Perrins CM (1998) Effects of elevated temperature and carbon dioxide on the nutritional quality of leaves of oak (*Quercus robur* L.) as food of the winter moth (*Operophtera brumata* L.). Funct Ecol 12:742–749

Carlisle A, Brown AHF, White EJ (1966) The organic matter and nutrient elements in the precipitation beneath a sessile oak (*Quercus petraea*) canopy. J Ecol 54:87–98

Currie WS, Aber JD, McDowell WH, Boone RD, Magill AH (1996) Vertical transport of dissolved organic C and N under long-term N amendments in pine and hardwood forests. Biogeochemistry 35:471–505

Dixon AFG (1971) The role of aphids in wood formation. I. The effect of the sycamore aphid *Drepanosiphum plantanoides* (Schr.) (Aphididae), on the growth of sycamore, *Acer pseudoplantanus* (L.). J Appl Ecol 8:165–179

Dixon AFG (2003) Climate change and phenological asynchrony. Ecol Entomol 28:380–381

Dixon AFG, Kindlmann P, Lepš J, Holman J (1987) Why there are so few species of aphids, especially in the tropics. Am Nat 129:580–592

Docherty M, Salt DT, Holopainen JK (1997) The impacts of climate change and pollution on forest pests. In: Watt AD, Stork NE, Hunter MD (eds) Forests and insects. Chapman and Hall, London, pp 229–247

Eckloff W (1972) Beitrag zur Ökologie und forstlichen Bedeutung bienenwirtschaftlich wichtiger Rindenläuse. Z Angew Entomol 70:134–157

Findlay S, Carreiro M, Krischik V, Jones CG (1996) Effects of damage to living plants on leaf litter quality. Ecol Appl 6:269–275

Gange AC, Bower E, Brown VK (2002) Differential effects of insect herbivory on arbuscular mycorrhizal colonization. Oecologia 131:103–112

Grimm NB (1995) Why link species and ecosystems? A perspective from ecosystem ecology. In: Jones CG, Lawton JH (eds) Linking species and ecosystems. Chapman and Hall, London, pp 5–15

Guggenberger G, Zech W (1994) Composition and dynamics of dissolved organic carbohydrates and lignin-degradation products in two coniferous forests, N.E. Bavaria, Germany. Soil Biol Biochem 26:19–27

Gundersen P, Emmett BA, Kjonaas OJ, Koopmans CJ, Tietema A (1998) Impact of nitrogen deposition on nitrogen cycling in forests: a synthesis of NITREX data. For Ecol Manage 101:37–55

Hassell MP (2000) The spatial and temporal dynamics of host–parasitoid interactions. Oxford University Press, Oxford

Heimbach U (1986) Freilanduntersuchungen zur Honigtauabgabe zweier Zierlausarten (Aphidina). J Appl Entomol 101:396–413

Herzig J (1937) Ameisen und Blattläuse. Z Angew Entomol 24:367–435

Higashi M, Burns TP (1991) Theoretical studies of ecosystems: the network approach. Cambridge University Press, Cambridge

Holling CS (1992) Cross-scale morphology, geometry, and dynamics of ecosystems. Ecol Monogr 62:447–502

Hunter MD (2001) Insect population dynamics meets ecosystem ecology: effects of herbivory on soil nutrient dynamics. Agric Forest Entomol 3:77–84

Kareiva PM, Kingsolver JG, Huey RB (1993) Biotic interactions and global change. Sinauer Associates, Sunderland, Massachusetts

Kidd NAC (1990) The population dynamics of the large pine aphid, *Cinara pinea* (Mordv.). II. Simulation of field populations. Res Popul Ecol 32:209–226

Kindlmann P, Dixon AFG (1996) Population dynamics of tree-dwelling aphid: individuals to populations. Ecol Model 89:23–30

Koricheva J, Larsson S, Haukioja E (1998) Insect performance on experimentally stressed woody plants: a meta-analysis. Annu Rev Entomol 43:195–216

Lawton JH (1994) What do species do in ecosystems? Oikos 71:364–374

Levin SA (1992) The problem of pattern and scale in ecology. Ecology 73:1943–1967

Levin SA (1998) Ecosystems and the biosphere as complex adaptive systems. Ecosystems 1:431–436

Levin SA (1999) Fragile dominion. Perseus, Boulder

Liebhold AM, Elkinton JS, Williams D, Muzika RM (2000) What causes outbreaks of gypsy moth in North America? Popul Ecol 42:257–266

Llewellyn M (1972) The effects of the lime aphid *Eucallipterus tiliae* L. (Aphididae), on the growth of the lime *Tilia* × *vulgaris* Hayne I. Energy requirements of the aphid population. J Appl Ecol 9:261–282

Llewellyn M (1975) The effects of the lime aphid *Eucallipterus tiliae* L. (Aphididae) on the growth of the lime (*Tilia* × *vulgaris* Hayne). J Appl Ecol 12:15–23

Loehle C, Pechmann JK (1988) Evolution: the missing ingredient in system ecology. Am Nat 132:884–899

Lovett G, Ruesink A (1995) Carbon and nitrogen mineralization from decomposing gypsy moth frass. Oecologia 104:133–138

Lowman MD (1995) Forest canopies. Academic Press, San Diego
Lowman MD, Wittman PK (1996) Forest canopies: methods, hypotheses, and future directions. Annu Rev Ecol Syst 27:55–81
Michalzik B, Stadler B (2000) Effects of phytophagous insects on soil solution chemistry: herbivores as switches for nutrient dynamics in the soil. Basic Appl Ecol 1:117–123
Michalzik B, Kalbitz K, Park J-H, Solinger S, Matzner E (2001) Fluxes and concentrations of dissolved organic matter – a synthesis for temperate forests. Biogeochemistry 52:173–205
Müller H (1956) Können Honigtau liefernde Baumläuse (Lachnidae) ihre Wirtspflanzen schädigen? Z Angew Entomol 39:168–177
Müller T, Müller M, Behrendt U, Stadler B (2004) Diversity of culturable phyllosphere bacteria on beech and oak: effects of lepidopterous larvae. Microb Res 158:291–297
Neinhuis C, Barthlott W (1998) Seasonal changes of leaf surface contamination in beech, oak, and gingko in relation to leaf micromorphology and wettability. New Phytol 138:91–98
Novak H (1989) Untersuchungen über Produktion und Konsum von Honigtau in ausgewählten Hecken Oberfrankens. Diploma Thesis, Universität Bayreuth, Germany
Qualls RG, Haines BL (1992) Biodegradability of dissolved organic matter in forest throughfall, soil solution, and stream water. Soil Sci Soc Am J 56:578–586
Royama T (1996) Analytical population dynamics. Chapman and Hall, London
Scheu S (2001) Plants and generalist predators as links between the below-ground and above-ground system. Basic Appl Ecol 2:3–13
Scheurer S (1964) Zur Biologie einiger Fichten bewohnender Lachnidenarten (Homoptera, Aphidina). Z Angew Entomol 53:153–178
Schneider SH (1993) Scenarios of global warming. In: Kareiva PM, Kingsolver JG, Huey RB (eds) Biotic interactions and global change. Sinauer Associates, Sunderland, Massachusetts, pp 9–23
Schowalter TD (2000) Insect ecology. An ecosystem approach. Academic Press, New York
Schowalter TD, Hargrove WW, Crossley DA (1986) Herbivory in forested ecosystems. Annu Rev Entomol 31:177–196
Schulze E-D (2000) The carbon and nitrogen cycle of forest ecosystems. In: Schulze E-D (ed) Carbon and nitrogen cycling in European forest ecosystems, vol 142. Springer, Berlin Heidelberg New York, pp 3–13
Seastedt TR, Crossley DA (1984) The influence of arthropods on ecosystems. BioScience 34:157–161
Seastedt TR, Crossley DA, Hargrove WW (1983) The effects of low-level consumption by canopy arthropods on the growth and nutrient dynamics of black locust and red maple trees in the southern Appalachians. Ecology 64:1040–1048
Shure DJ, Mooreside PD, Ogle SM (1998) Rainfall effects on plant–herbivore processes in an upland oak forest. Ecology 79:604–617
Spiller NJ, Llewellyn M (1987) Honeydew production and sap ingestion by the cereal aphids *Rhopalosiphum padi* and *Metopolophium dirhodum* on seedlings of resistant and susceptible wheat species. Ann Appl Biol 110:585–590
Stadler B, Müller T (2000) Effects of herbivores on epiphytic micro-organisms in canopies of forest trees. Can J For Res 30:631–638
Stadler B, Michalzik B, Müller T (1998) Linking aphid ecology with nutrient fluxes in a coniferous forest. Ecology 79:1514–1525
Stadler B, Solinger S, Michalzik B (2001a) Insect herbivores and the nutrient flow from the canopy to the soil. Oecologia 126:104–113
Stadler B, Müller T, Sheppard L, Crossley A (2001b) Effects of *Elatobium abietinum* on nutrient fluxes in Sitka spruce canopies receiving elevated nitrogen and sulphur deposition. Agric For Entomol 3:253–261

Swank WT, Waide JB, Crossley DA, Todd RL (1981) Insect defoliation enhances nitrate export from forest systems. Oecologia 51:297–299

Tsai CS, Killham K, Cresser MS (1997) Dynamic response of microbial biomass, respiration rate and ATP to glucose additions. Soil Biol Biochem 29:1249–1256

Varley GC, Gradwell GR, Hassell MP (1973) Insect population ecology. Blackwell, Oxford

Wardle A (2002) Communities and ecosystems: linking the aboveground and belowground components. Princeton University Press, Princeton

Watt AD, McFarlane AM (2002) Will climate change have a different impact on different trophic levels? Phenological development of winter moth *Operophtera brumata* and its host plant. Ecol Entomol 27:254–256

Wellenstein G (1961) Honigtaubildende Forstinsekten und ihre wirtschaftliche Bedeutung. In: Schwenke W (ed) Forstwissenschaft im Dienste der Praxis. BLV, Munich, pp 184–199

Wiens JA (1995) Landscape mosaics and ecological theory. In: Hansson L, Fahrig L, Merriam G (eds) Mosaik landscapes and ecological processes. Chapman and Hall, London, pp 1–26

Zoebelein G (1954) Versuche zur Feststellung des Honigtauertrages von Fichtenbeständen mit Hilfe von Waldameisen. Z Angew Entomol 36:358–362

Zwölfer W (1952) Die Waldbienenweide und ihre Nutzung als forstentomologisches Problem. Verh Dtsch Ges Angew Entomol 12:1–15

16 Element Fluxes with Litterfall in Mature Stands of Norway Spruce and European Beech in Bavaria, South Germany

B. BERG and P. GERSTBERGER

16.1 Introduction

Litterfall is the largest natural inflow of organic material and nutrients to the forest floor and in most European forests is dominated by that from the trees. The chemical composition of this material and the temperature and moisture content of the upper soil layers are considered to be the main factors controlling the turnover rates of the shed litter and the release of nutrients. Thus they also determine the quantity of nutrients released and the accumulation of humus and nutrients. Few larger studies have been made on litterfall on a European basis. For needles of Norway spruce and different pine species, Berg and Meentemeyer (2001) made a study covering the main part of western Europe, whereas for European beech, for example, few values have been published.

Nutrient concentrations in foliar litter vary with the litter species, climate, and pollution level (e.g. Berg and McClaugherty 2003). In general, it appears that in unpolluted stands beech is richer in the main nutrients than spruce and beech leaf litter often has a concentration of around 9 mg N g^{-1}, whereas spruce may have only half that concentration (Staaf 1982; Berg and McClaugherty 2003). In stands with N pollution or in stands subject to N fertilization, the concentration of N in litter will increase above the level considered to be normal (e.g. Berg and Tamm 1991). Further, in such polluted stands, the difference in between-species N concentration in foliar litter may be less pronounced, and at the N-polluted site Solling in central Germany the average N concentration in a stand of European beech was 12.7 mg g^{-1} and in one with Norway spruce it was 10.7 mg g^{-1} (Matzner et al. 1982).

An increased N level in litter often means an increased or changed level of other nutrients, at least for those that structurally are connected to N, such as P and S but also K (Berg and McClaugherty 2003). Norway spruce needle litter appeared to differ from a more general pattern, and Berg and Tamm (1991) investigating the variation between sets of nutrients in spruce needle litter after N fertilization found that the concentrations of N were significantly related to those of S, Ca, Mn, Al, and Zn.

Ecological Studies, Vol. 172
E. Matzner (Ed.), Biogeochemistry of Forested Catchments in a Changing Environment

It has been found that a changed chemical composition of foliar litter may change the accumulation rate of humus in the forest floor and thus also the storage rate of nutrients, for example N (e.g. Berg and McClaugherty 2003; Berg, this Vol.). This means that although there is a return of nutrients to the forest floor with litterfall, not all nutrients are available for plant uptake. Such far-going influences of N additions, either in the form of deposition or as N fertilization, clearly motivate a closer investigation of nutrient budgets with the intention of finding deviations from the normal pattern. With a changed chemical composition of foliar litter the flow of nutrients back to the soil system is also changed. A higher concentration of N in foliar litter could thus result in changed levels of other nutrients in the humus leading to a changed pattern in the nutrient supply.

The aim of this chapter is to present and evaluate the return of nutrients in two N-polluted stands of Norway spruce and European beech. For this purpose, litterfall was collected at both stands and analyzed for the nutrients N, P, S, K, Ca, Mg, Mn, Al, and Fe. We compared the returned amounts at the two sites and used data from an additional site at Solling (Matzner 1988).

16.2 Discussion

16.2.1 Litterfall Amounts

We used two sites for the budget studies, site Steinkreuz and site Coulissenhieb (Gerstberger et al. 2004).

Site *Steinkreuz*, located at 49°52′21″N, 10 27′45″E, at an altitude of 400–460 m, in the mountain range Steigerwald in northwestern Bavaria, has an average annual temperature of 7.9 °C, and an annual precipitation of 700–800 mm. The forest is dominated by European beech (*Fagus sylvatica* L.) and sessile oak (*Quercus petraea*) with a few hornbeam (*Carpinus betulus* L.) trees.

At the Steinkreuz site litter was collected from the beginning of the vegetation period in March/April. Litterfall was collected at 4-week intervals until mid December using ten replicate terylene net collectors with 0.5-mm mesh, each of 1.0 m^2 placed in a straight line at 8-m distance. After drying at 60 °C for 48 h leaves of beech and oak were separated as well as the nuts of the two species and the woody material, resulting in six litter components.

Over the period from 1996 until 2001 (Table 16.1) the beech foliar litterfall varied from 2,871–3,528 kg ha^{-1}, thus with a factor of 1.23 between the highest and lowest values. For oak the leaf litterfall ranged from 683 to 786 kg ha^{-1}, thus with a factor of 1.15 over the same period. Average fall of beech leaves was 3,118.2 kg ha^{-1} and oak leaves was 729.7 kg ha^{-1}. The average fall of total

Table 16.1. Amounts of biomass and concentrations of nutrients over time in newly shed foliar litter of European beech and sessile oak at the site Steinkreuz and in needle litter from Norway spruce at the site Coulissenhieb

Year	Foliar litter (kg ha^{-1})	C (mg g^{-1})	N	Ca	K	P	S	Mn	Mg	Al	Fe
Steinkreuz – European beech											
1996	2,923.2	477	12.4	6.5	5.7	–	–	1.8	0.8	0.11	0.11
1997	3,444.1	470	15.7	6.2	4.5	0.9	1.1	1.4	0.9	0.11	0.10
1998	3,528.3	469	14.1	7.1	3.7	1.0	1.1	1.7	1.0	0.14	0.12
1999	2,869.4	472	13.3	6.3	5.1	0.8	1.0	1.9	0.8	0.14	0.12
2000	3,073.1	474	14.6	7.9	5.5	0.8	1.0	2.4	1.1	0.20	0.10
2001	2,870.9	438	9.50	8.9	3.3	0.7	0.8	1.8	0.8	0.15	0.12
Average	3,118.2	467	13.3	7.0	5.0	0.9	1.0	1.8	0.9	0.13	0.11
SD	270.0	13.1	2.0	1.0	0.9	0.1	0.1	0.3	0.1	0.03	0.01
Steinkreuz – sessile oak											
1996	732.3	478	9.93	7.9	3.1	–	–	2.6	1.1	0.12	0.11
1997	682.9	464	16.0	7.7	4.1	1.2	1.1	1.6	1.5	0.11	0.12
1998	786.5	465	17.4	7.2	3.5	1.1	1.2	1.8	1.1	0.10	0.12
1999	702.8	472	12.2	6.6	3.6	0.8	0.9	1.7	1.0	0.12	0.13
2000	683.8	467	11.0	9.7	3.4	0.8	0.8	3.5	1.4	0.14	0.14
2001	790.1	470	10.6	9.5	3.5	0.9	0.9	2.1	1.3	0.14	0.14
Average	729.7	469	12.8	8.1	3.5	1.0	1.0	2.2	1.2	0.12	0.13
SD	44.5	4.7	2.8	1.1	0.3	0.1	0.1	0.7	0.2	0.01	0.01
Coulissenhieb – Norway spruce											
1995	1,847.6	483	10.5	5.3	2.5	–	–	0.3	0.4	0.13	0.13
1996	2,411.1	487	11.7	4.3	2.0	–	–	0.2	0.4	0.12	0.11
1997	1,826.4	489	14.2	7.2	2.3	0.7	1.0	0.3	0.3	0.14	0.06
1998	2,642.4	480	9.7	7.5	1.8	0.8	1.0	0.3	0.4	0.13	0.08
1999	1,975.9	491	14.6	–	–	–	–	–	–	–	–
Average	2,140.7	486	112.1	6.1	2.1	0.7	1.0	0.3	0.4	0.13	0.10
SD	32.8	4.0	2.0	1.3	0.3	0.04	0.02	0.04	0.03	0.01	0.03

litter, including all components of both species, was 5,445.56 kg ha^{-1}. In this time period no loss of trees took place.

The foliar litterfall corresponded to ca 70 % of the total one. This is a figure similar to that estimated by Mälkönen (1974) from measurements in a mature boreal Scots pine forest. The relative size of the foliar litter fraction varies with stand age, and Berg et al. (1993) found for a chronosequence that whereas in an 18- to 25-year-old Scots pine forest the fraction of needle litter was 83 % it decreased to 68 % in a 55- to 61-year-old stand and ended at 58 % in a 120- to 126-year-old stand that had stopped growing. In this latter case, the trapping equipment was extensive, to catch the different litter fractions (Flower-Ellis 1985).

Whereas foliar litterfall is regular and rather evenly distributed over the forest floor and with relatively small variations over years, the cruder woody litter, e.g. twigs and branches, falls in connection with weather events, such as storms or heavy snowfalls, and nuts/cones would have a cyclic fall over years. In this period the ratio between highest and lowest fall for woody litter was ca. 3 and for nuts ca. 10.

Our other site, *Coulissenhieb*, was located in the Lehstenbach catchment, a watershed (4.2 km^2) in the Fichtelgebirge area in northeastern Bavaria/Germany, close to the border with the Czech Republic, at 50°08′35″N, 11°52′08″E and an altitude of 765–785 m a.s.l. The average precipitation in the area is 1,156 mm and the annual average temperature is 5.0 °C. The Coulissenhieb research plot is a 2.5-ha-large part of the catchment with a 150-year-old Norway spruce [*Picea abies* (L.) Karst.] stand. The soils at the site are acidic and classified as Haplic Podzols (FAO). Soil chemical parameters indicate a high degree of acidification (Alewell et al. 2000).

Litterfall was collected using 20 litter traps of polypropylene each with a surface of 0.166 m^2. The traps were located ca 1 m above the ground and spread over an area of 20×20 m. In the period from April to November; the traps were emptied once every 14 days. After drying at 60 °C the litter was sorted into needles, cones, and woody materials. Needle litterfall ranged between 1,826 and 2,642 kg ha^{-1} over a period of 5 years from 1995–1999, with an annual average of 2,141 kg ha^{-1} and a range factor of 1.4 (Table 16.1).

We used additional data from the site Solling. This site in central Germany, close to the city of Göttingen, is located at an altitude of ca 500 m. The annual average temperature is 6.4 °C and the precipitation 1,090 mm. The site has two stands, which we used for comparison, one with Norway spruce and one with European beech. Further site descriptions are given by Ellenberg et al. (1986).

16.2.2 Return of Elements – European Beech and Sessile Oak at Site Steinkreuz

The total return of elements for the site Steinkreuz is given in Table 16.2. After weighing, the samples were ground in a zirconium-ball mill into a powder (<45 m) and analyzed. Carbon and N were analyzed as CO_2 and N_2, respectively, after combustion using thermoconductivity detection. The cations (Al, Ca, Cu, Fe, K, Mg, and Mn), as well as P and S were determined using ICP-AES after digestion of 100 mg in 1 ml 1 M HNO_3 at 170 °C for 6–8 h.

In this mature stand the total annual return of N was on average 68.3 kg ha^{-1} and for Ca and K it was 35.5 and 23.9 kg ha^{-1}, respectively. Nutrients such as Mg and Mn were returned in smaller amounts with annually 4.7 and 8.4 kg ha^{-1}, respectively. For P and S the annual returns were 4.2 and 4.3 kg ha^{-1}, respectively. Smaller amounts of Al (0.98 kg) and Fe (0.60 kg) were returned (Table 16.2).

Table 16.2. Annual amounts of nutrients (in kg ha^{-1} year^{-1}) returned to the forest floor in litterfall at sites Steinkreuz and Coulissenhieb. We distinguished two main fractions, foliar litter and a combined fraction of woody parts and nuts/cones

	N	Ca	K	Mn	Mg	S	P	Al	Fe
Steinkreuz – European beech and sessile oak									
Total return	68.3	35.5	23.9	8.4	4.74	4.29	4.16	0.98	0.60
In foliar litter	51.1	28.1	16.9	7.3	3.70	3.90	3.45	0.52	0.47
In wood	6.6	1.54	1.49	0.55	0.36	0.48	0.38	0.06	0.07
In nuts	11.6	3.4	7.2	0.8	0.86	0.93	1.38	0.97	0.09
Fraction in foliar litter (%)	75	79	71	87	78	91	83	53	78
Coulissenhieb – Norway spruce									
Total return	41.2	16.2	7.3	0.7	1.60	3.14	2.19	1.20	1.14
In foliar litter	25.6	13.3	4.6	0.6	0.82	2.18	1.66	0.28	0.21
In wood	5.1	1.5	0.48	0.033	0.11	0.41	0.20	0.39	0.43
In cones	9.3	0.8	1.8	0.1	0.58	0.07	0.05	0.19	0.20
Fraction in foliar litter (%)	62	82	63	80	51	69	76	23	18

The main part of all nutrients was returned in the leaf litter, but still the size of the fractions varied. Thus, at this site 91 % of the S, 87 % of the Mn, and 83 % of the P was returned in the beech and oak foliar litter, but only 53 % of the Al. For both N, Ca, K, Mg, and Fe, between 71 and 79 % was returned in the leaf litter.

The amounts returned in the combined woody beech and oak litter followed the same pattern as the foliar litterfall. Nitrogen was the largest component (6.6 kg ha^{-1}) followed by Ca (1.54 kg ha^{-1}), and K (1.49 kg ha^{-1}). Manganese was returned in an amount of 0.55 kg ha^{-1} followed by S with 0.48 kg ha^{-1}, and P and Mg at rates of 0.38 and 0.36 kg ha^{-1}, respectively. As in the foliar litter, the amounts of Al and Fe were low, in this case 0.06 and 0.07 kg ha^{-1}, respectively.

16.2.3 Return of Elements – Norway Spruce at the Site Coulissenhieb

The total amounts of elements returned are given in Table 16.2. The samples to be analyzed were prepared as for the Steinkreuz site and the analytical process was identical.

In this mature stand the total annual return of N was on average 41.2 kg ha^{-1} and for Ca and K, 16.2 and 7.25 kg ha^{-1}, respectively. Nutrients such as Mg and Mn were returned in smaller amounts with annually 1.6 and 0.7 kg ha^{-1}, respectively. For S and P the annual returns were 3.14 and

2.19 kg ha^{-1}, whereas for Al and Fe they were 1.20 and 1.14 kg ha^{-1}, respectively.

The main part of all nutrients was returned in the leaf litter, but still the size of the fractions varied. Thus, 82 % of the Ca and 80 % of the Mn was returned in the spruce foliar litter but only 18 % of the Fe. Of N, P, and K, 62, 76, and 63 %, respectively, was returned in the leaf litter and 69 % of the S and 51 % of the Mg.

The amounts returned in the woody spruce litter combined followed the same pattern as the foliar litterfall. Nitrogen was the largest component (5.1 kg ha^{-1}) followed by Ca (1.5 kg ha^{-1}) and K (0.48 kg ha^{-1}). Sulfur was returned in an amount of 0.41 kg ha^{-1} followed by P with 0.20 kg ha^{-1}, and Mg and Mn at rates of 0.11 and 0.033 kg ha^{-1}, respectively. Like in the foliar litter the amounts of Al and Fe were about equal.

16.2.4 Some Trends in the Return of Nutrients

We compared the amounts returned in foliar litter for beech at Steinkreuz with those for beech at the site Solling (1969–1976) as well as the amounts returned in the spruce stand at sites Coulissenhieb and Solling (Table 16.3).

For both beech and spruce we may distinguish two groups of nutrients. A first group of N, Ca, and K has relatively high return rates and amounts in the order N>Ca>K. A second group encompasses Mn, Mg, S, P, Al, and Fe with lower rates and no clear trends. We may note that the beech stands on average returned higher amounts of N, Ca, K, Mn, P, and Mg. We may compare this with a heavily polluted stand of Norway spruce in southern Sweden (Nilsson et al. 2001). Also; in that case; the order of nutrient return was N>Ca>K in a first group, with S, P, and Mg returned in about the same amounts. This picture remained constant also after heavy experimental additions of N and S.

Table 16.3. Comparison of return pattern of nutrients (in kg ha^{-1} year^{-1}) in foliar litter at sites Steinkreuz (European beech), Coulissenhieb (Norway spruce), and Solling (paired stands of Norway spruce and European beech)

	N	Ca	K	Mn	S	Mg	P	Al	Fe
European beech									
Steinkreuz[a]	51.1	28.1	16.9	7.3	3.9	3.70	3.5	0.52	0.47
Solling	37.6	16.2	15.8	5.3	5.4	1.30	2.9	0.40	1.00
Norway spruce									
Coulissenhieb	25.6	13.3	4.6	0.6	2.2	0.82	1.7	0.28	0.21
Solling	35.0	11.8	6.2	4.0	5.6	1.04	2.8	0.90	2.90

[a] Beech and oak

A comparison with unpolluted stands of Norway spruce in Scandinavia (Berg et al. 1999, 2000) gives a somewhat different picture. The order of nutrient return in needle litter ranked after mass was Ca>N>Mn>K>Mg>P. These values were means of 12 investigated stands. The nutrient amounts are of course dependent on the amount of litterfall, yet 27 kg Ca ha^{-1} year^{-1} was returned on average compared with ca 13.3 kg at the site Coulissenhieb.

16.3 Concluding Remarks

The average foliar litterfall at a site with European beech (Steinkreuz) was 3,848 kg ha^{-1} year^{-1} and with a Norway spruce forest (Coulissenhieb) it was 2,141 kg ha^{-1} year^{-1}. For the main nutrients such as N, Ca, and K the return rates were clearly higher in the beech forest.

Acknowledgements. The study was in part carried out within the framework of the European Union project CN-ter (contract number QLKS-2001-00596).

References

Alewell C, Manderscheid B, Gerstberger P, Matzner E (2000) Effects of reduced atmospheric deposition on soil solution chemistry and elemental contents of spruce needles in NE Bavaria, Germany. J Plant Nutr Soil Sci 163:509–516

Berg B, Meentemeyer V (2001) Litterfall in some European coniferous forests as dependent on climate - a synthesis. Can J For Res 31:292–301

Berg B, McClaugherty C (2003) Plant litter. Decomposition. Humus formation. Carbon sequestration. Springer, Berlin Heidelberg New York

Berg B, Tamm CO (1991) Decomposition and nutrient dynamics of litter in long-term optimum nutrition experiments. I. Organic matter decomposition in Norway spruce (*Picea abies*) needle litter. Scand J For Res 6:305–321

Berg B, Berg M, Bottner P, Box E, Breymeyer A, Calvo de Anta R, Couteaux M, Gallardo A, Escudero A, Kratz W, Madeira M, Meentemeyer V, Muñoz F, Piussi P, Remacle J, Virzo De Santo A (1993) Litter mass loss in pine forests versus actual evapotranspiration on a European scale. In: Breymeyer A (ed) Conference papers 18. Proceedings from Scope Seminar, Geography of Carbon Budget Processes in Terrestrial Ecosystems, Szymbark, 17–23 Aug 1991, pp 81–109. Institute of Geography an Spatial Organization, Polish Academy of Sciences, Warsaw

Berg B, Johansson M, Tjarve I, Gaitnieks T, Rokjanis B, Beier C, Rothe A, Bolger T, Göttlein A, Gerstberger P (1999) Needle litterfall in a north European spruce forest transect. Departments of Forest Ecology and Forest Soils, Swedish University of Agricultural Sciences; Rep 80, 54 pp

Berg B, Johansson M-B, Meentemeyer V (2000) Litter decomposition in a transect of Norway spruce forests: substrate quality and climate control. Can J For Res 30:1136–1147

Ellenberg H, Mayer R, Schauermann J (1986) Ökosystemforschung: Ergebnisse des Sollingprojekts 1966–1986. Ulmer, Stuttgart

Flower-Ellis J (1985) Litterfall in an age series of Scots pine stands: summary of results for the period 1973–1983. Department of Ecology and Environmental Research, Swedish University of Agricultural Sciences; Rep 19, pp 75–94

Mälkönen E (1974) Annual primary production and nutrient cycle in some Scots pine stands. Commun Inst Forest Fenn 84:5

Matzner E (1988) Der Stoffumsatz zweier Waldökosysteme im Solling. Ber Forschungszentr Waldökosyst/Waldsterben, Reihe A:40:1–217

Matzner E, Khanna PK, Meiwes KJ, Lindheim J, Prenzel J, Ulrich B. (1982) Elementflüsse in Waldökosystemen im Solling – Datendokumentation. Göttinger Bodenkundl Ber 71:1–267

Nilsson LO, Östergren M, Wiklund K (2001) Hur paverkades träden ovan mark? In: Persson T, Nilsson LO (eds) Skogabyförsöket – effekter av langvarig kväve- och svaveltillförsel till ett skogsekosystem. Naturvardsverket 5173:51–67

Staaf H (1982) Plant nutrient changes in beech leaves during senescence as influenced by site characteristics. Acta Oecol/Oecol Plant 3:161–170

17 The Role of Woody Roots in Water Uptake of Mature Spruce, Beech, and Oak Trees

J. LINDENMAIR, E. MATZNER, and R. ZIMMERMANN

17.1 Introduction

Most rhizosphere studies about water uptake of trees have only considered the young fine root system. Much less is known about water uptake by older, suberized roots, although it seems probable that water transpired during periods of low fine root elongation is replaced by uptake through the suberized surfaces of mature roots (Kramer and Bullock 1966; Atkinson and Wilson 1979). As tree root systems are predominantly composed of woody roots (van Rees and Comerford 1990), it is of major importance to determine the functioning of this root surface. Thus, this chapter focuses on the water uptake characteristics of a widely ignored soil–plant interface – the woody coarse root system.

The capacity of different root zones for water uptake has been discussed for several decades in rhizosphere research. According to the classical theory, water uptake is mainly restricted to the young, unsuberized parts of the root system – the white root tips and elongation zones (Häussling et al. 1988; Rennenberg et al. 1996). However, several studies with measurements at suberized pine roots showed significant water uptake of older root zones (for small pine trees of 2–5 m height: Kramer 1946; and for 1-year-old pines: Chung and Kramer 1975; van Rees and Comerford 1990).

Root anatomy (McKenzie and Peterson 1995) as well as root system architecture and the vertical distribution of different root diameter classes vary depending on tree species and site characteristics (Friedrich 1992; Puhe 1994; Lee 1998). Water uptake by woody roots was mostly studied with pine (see above) or sometimes with fruit trees (Atkinson and Wilson 1979). These results cannot simply be transferred to other tree species. Laboratory measurements of hydraulic conductivity for spruce, pine, oak, and beech roots indicated clearly lower permeabilities for deciduous trees compared to conifers (Steudle and Heydt 1997).

In this chapter, root chambers were used to quantify water uptake by woody coarse roots (roots with secondary growth and diameters >0.2 cm),

Ecological Studies, Vol. 172
E. Matzner (Ed.), Biogeochemistry of Forested Catchments in a Changing Environment

allowing measurements at intact root segments in the field. Besides plant-specific factors (tree species, root diameter, and root position), the influence of site-specific factors (meteorological and soil physical parameters, transpiration rates of the tree) on water uptake is investigated. To examine the relationship between water uptake and variable soil and atmospheric conditions, measurements over longer time periods at the same root segment are needed.

17.2 Root Chamber Method

The acrylic glass-root chambers consist of two halves that are screwed together and sealed with rubber strips and by injection of a special sealant at the sides (Fig. 17.1). Each chamber has two scaled acrylic glass-tubes for reading of water volume. To avoid anaerobic conditions, the root chambers are aerated continuously. Root chambers were installed in the field at different coarse root segments, free of lateral roots, of a predominant mature spruce at the Coulissenhieb experimental plot in the growing seasons of 1998–2000. Furthermore, roots of beech and oak trees at the Steinkreuz experimental plot were investigated in the growing season of 2001. Roots were located in the forest floor or the upper mineral soil. For site description see Gerstberger et al. (this Vol.). Stem diameters of the investigated trees at breast height were 51 cm for spruce, 47 cm for oak, and 45 cm for beech. The roots investigated

Fig. 17.1. Continuously aerated root chamber with variable side stamp for direct measurements of water uptake in the field (inner length 6 cm, outer length 13.2 cm)

Table 17.1. Diameter of root segments investigated with root chambers for different tree species and sites

Root diameter (cm)	Coulissenhieb site			Steinkreuz site	
	Spruce			Beech	Oak
	Aug–Sept 1998	April–Oct 1999	May–Oct 2000	Aug–Oct 2001	
0.2–0.9	3	7	3	4	3
1.0–1.9	3	8	2	2	2
2.0–2.9	1	3	2	1	0
3.0–4.9	1	2	0	1	2
Total number of roots		35		8	7

varied in diameter from 0.2–4.6 cm for spruce and 0.3–3.8 and 0.4–4.0 cm for beech and oak, respectively (Table 17.1). After filling the chambers with a nutrient solution adapted to the soil solution chemistry found in the field, changes in water volume were recorded in daily to weekly intervals for several months throughout the growing season (Table 17.1). In order to quantify evaporation losses of the chamber itself, chambers without roots were installed as controls.

During the measurements in 1999, xylem sap fluxes were monitored at eight coarse roots (root diameter 1.8–5.5 cm) at the same spruce tree used for the root chamber studies. The roots were located both in the upper humus layer and in mineral soil. The thermal constant energy input method according to Granier (1985, 1987) was applied. These measurements were performed in order to compare temporal dynamics of the xylem sap flux with water uptake determined by the root chambers. Sapflow measurements in the stem of the spruce tree were used to calculate transpiration loss. The sapflow measurements were conducted with thermal flow meters consisting of a cylindrical heating and sensing element inserted into the stem at breast height or into the coarse root respectively (distance of 10–15 cm between the two elements). The upper elements or the elements located towards the stem were heated with constant power. The usual sensor length of 2 cm for stem flux measurements was reduced to 1 cm for measurements in roots. Sap flux density is a function of temperature difference sensed between the two elements and can be estimated via appropriate calibration factors (McDonald et al. 2002). Total sapflow per tree or per root was obtained by multiplying sap flux density by the cross-sectional area of hydroactive sapwood of the stem or roots. Further details about the experimental setup of xylem sap flux and root chamber measurements are given in Lindenmair et al. (2001) and Lindenmair (2003).

17.3 Water Uptake Rates of Coarse Roots for Different Tree Species

Using the root chamber method, water uptake from nutrient solution was determined for all coarse roots studied. The mean daily rates for spruce varied from 1–20 μl cm^{-2} root surface (Fig. 17.2), which is much less than rates known from the literature. The water uptake rates of woody pine roots reported by Kramer (1946) and van Rees and Comerford (1990) were 3–4 to 10–20 times higher, respectively, than those of the spruce roots in this study.

Mean daily rates for beech and oak (Fig. 17.3) varied from 1–7 and 1–6 μl cm^{-2}, respectively, and were about two times lower than those for spruce in each diameter class (root diameter classes: <1, 1–2, >2 cm). The difference between the uptake rates of beech and oak was not significant.

For all tree species, water uptake rates decreased with increasing root diameter, indicating reduced permeability during secondary growth. There was no effect of root position (forest floor – mineral soil) on water uptake of coarse roots (not shown). There is contradictory evidence in the literature concerning the relationship between root diameter and water uptake. Studies by Kramer and Bullock (1966) and Chung and Kramer (1975) of woody pine roots showed greater permeability with increasing root diameter (measurements under negative pressure gradients). Investigations with magnetic resonance imaging (MacFall et al. 1991; MacFall 1999, pers. comm.) of woody pine roots growing in sand resulted in decreasing uptake rates at the transition

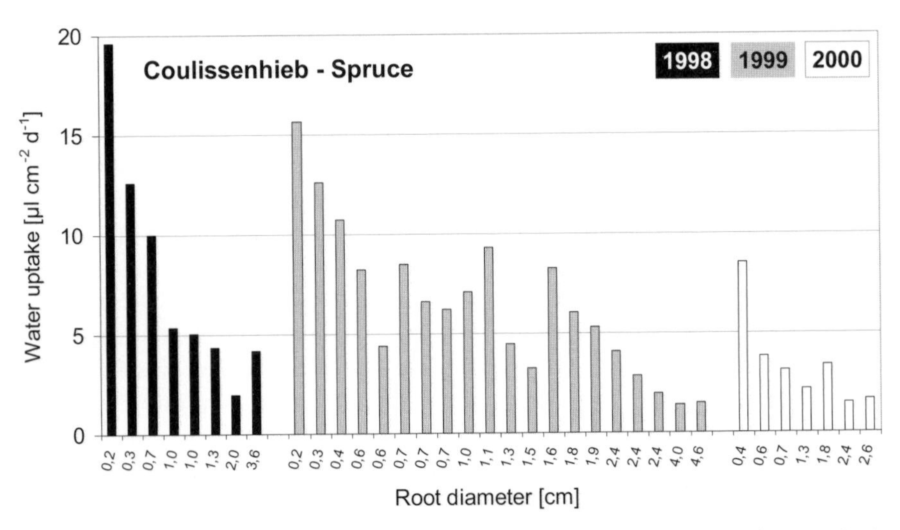

Fig. 17.2. Mean daily rates of water uptake of coarse spruce roots at the Coulissenhieb site for the years 1998–2000 (rates per unit root surface area, averaged over the experimental period August–September). Each *bar* represents one root

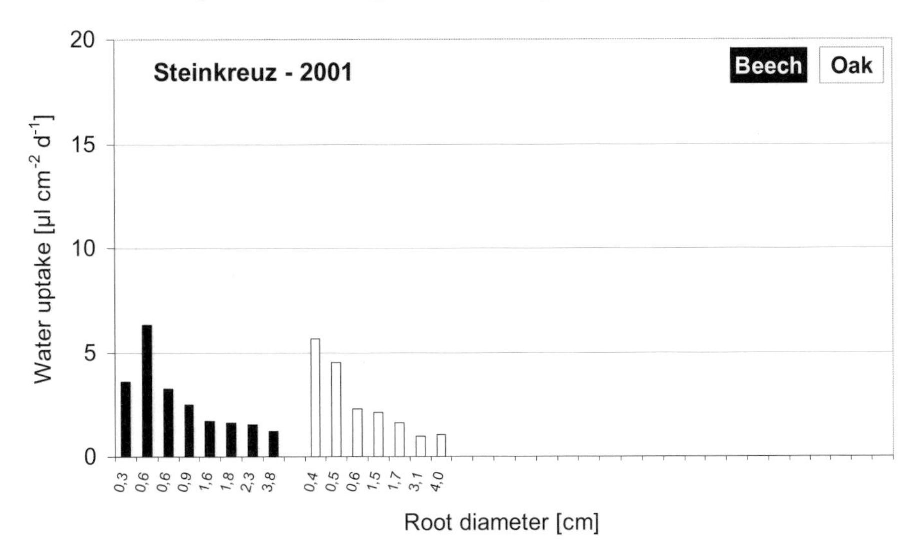

Fig. 17.3. Mean daily rates of water uptake of coarse beech and oak roots at the Steinkreuz site in the year 2001 (rates per unit root surface area, averaged over the experimental period August–September). Each *bar* represents one root

from fine to coarse roots. However, for very thick tap roots uptake rates increased again. The pathways whereby suberized roots take up water are still uncertain (van Rees and Comerford 1990; Steudle and Peterson 1998). Several pathways are discussed (Sanderson 1983; Moon et al. 1986; Enstone and Peterson 1992; Rüdinger et al. 1994; Sattelmacher et al. 1998). During differentiation and aging of young roots apoplastic transport barriers in endodermis and exodermis – like casparian bands and suberin lamellae – are formed (Jorns 1987; McCully and Canny 1988; Peterson et al. 1999), constricting water transport. However, there is new evidence that, for example, casparian bands do not totally restrict water flow, but stay permeable to some extent (Schreiber et al. 1999; Hose et al. 2001). During ongoing root browning and secondary growth, primary cortex and lastly endodermis are corrupted and sloughed away and a new root periderm consisting of cork layers is formed. In woody roots with advanced stages of secondary growth, the endodermis is no longer present to act as a water barrier. The cork tissue with a cork cambium may serves that function (MacFall et al. 1990; Comerford et al. 1994). Passive water entry across the periderm can occur through lenticels, around branch roots, and through wounds (Addoms 1946; Dumbroff and Peirson 1971; van Rees and Comerford 1990).

17.4 Time Course of Water Uptake for Spruce Roots in Relation to Climatic and Soil Physical Parameters and Xylem Sapflow Rates

Figure 17.4 presents the time course of surface-specific water uptake for selected coarse spruce roots with varying diameter in relation to climatic and soil physical parameters during the growing season of 1999. Water uptake

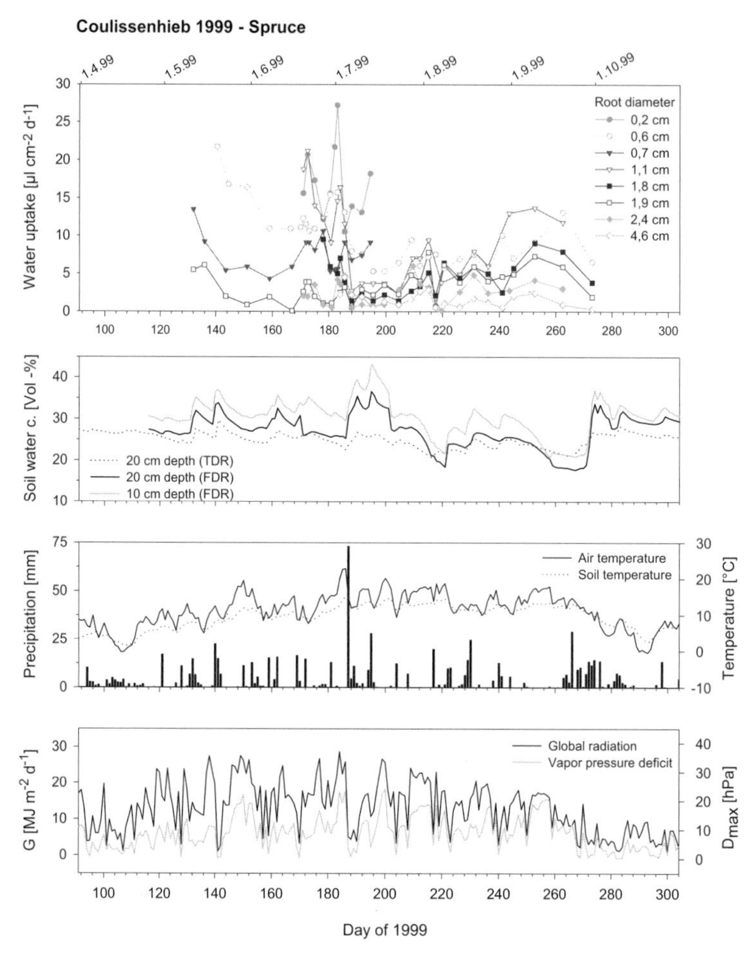

Fig. 17.4. Seasonal course of weather conditions and water uptake rates of coarse spruce roots during the growing season 1999 at the Coulissenhieb site. Graphs show soil water content in 10 and 20 cm depth measured with time domain reflectometry (*TDR*) (Coulissenhieb site) or frequency domain reflectometry (*FDR*) (nearby Weidenbrunnen site), mean air and soil temperature measured in 2 m height or the Oa horizon, respectively, precipitation, daily sum of global radiation (G), and daily maximum vapor pressure deficit (D_{max})

rates show clear temporal dynamics with some pronounced peaks. According to the mean uptake rates, thicker roots not only had a lower level of water uptake, but also showed a weaker response to environmental changes than the roots with smaller diameters (Fig. 17.2). Water uptake was, however, relatively synchronous for all roots investigated.

Water uptake of roots is influenced by various interdependent factors. Other studies (Alsheimer 1997; Köstner et al. 1999; Coners 2001) report highest correlations with global radiation, water pressure deficit, and soil humidity. Following the course of global radiation and water pressure deficit, there was an increase in water uptake at the beginning of the growing season in May and at the end in September (Fig. 17.4). The time course of water uptake runs opposite to soil humidity. In more humid periods with higher precipitation, water uptake of coarse roots decreased.

Diurnal course of sapflow in the stem and the roots of the upper humus layer corresponds well with global radiation (Fig. 17.5). Also, changes in vapor pressure deficit are clearly reflected in the pattern of sapflow. High rainfall with strong decrease in vapor pressure deficit can reduce sapflow to zero (day 240–241, Fig. 17.5). Compared to the roots in the humus layer (about

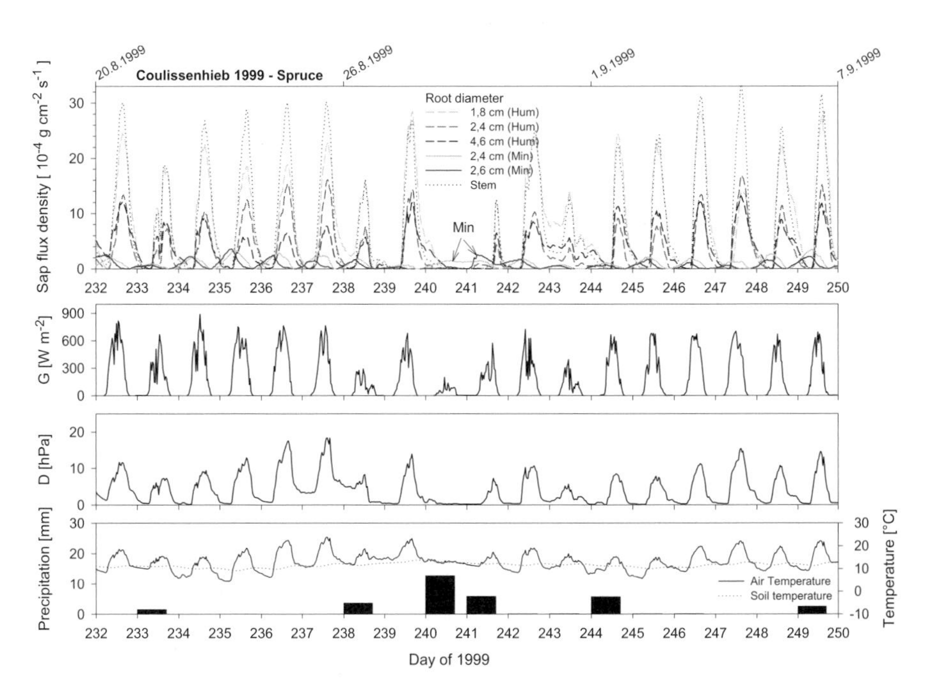

Fig. 17.5. Diurnal course of root and stem sapflow in one experimental spruce tree during the growing season 1999 at the Coulissenhieb site. Other *graphs* show the course of global radiation (*G*), vapor pressure deficit (*D*), air and soil temperature (measured in 2 m height or the Oa horizon, respectively), and daily precipitation. *Hum* and *Min* represent roots in the humus layer or mineral soil, respectively

50–100 g cm^{-2} day^{-1}), total daily sap flux in the roots of the deeper mineral horizons was only about 10 g cm^{-2} day^{-1}. Sapflow in the roots of the mineral layer seems more decoupled from climatic parameters and showed increasing flow even when stem sapflow decreases. This means that a considerable flux during night is observed in these roots.

Especially medium-term changes (on a seasonal scale) in water uptake from root chambers were related to sapflow dynamics (not shown). Water uptake of woody roots follows more strongly the total stem sapflow of the tree, representing the total transpiration loss, than the more variable sapflows measured in single roots. Thus high water uptake by the coarse roots in May and June as well as the increase in September (Fig. 17.4) correspond with increased transpiration rates of the tree.

17.5 Extrapolation from Root to Tree Scale

In order to compare water uptake rates of woody coarse roots with those of active (white) fine roots, additional measurements at six long roots of the spruce tree were carried out. Surface-specific uptake of woody coarse roots was in the case of the small coarse roots (<0.5 cm in diameter) about 20 times and in the case of the thickest roots (>2 cm in diameter) up to 70 times lower than for the young white roots. However, the relevance of woody roots for water uptake of the trees under field conditions depends on their amounts and distribution in the soil profile (Comerford et al. 1994). Based on the root biomass distribution at the Coulissenhieb site (according to Kummetz 1996) and on the measured surface-specific uptake rates of the different root diameter classes, water uptake by coarse roots contributes between 6 and 16 % to tree transpiration in the case of Norway spruce.

17.6 Conclusions

The results show that coarse roots of trees are able to take up considerable amounts of water. The findings confirm evidence from other studies (Chung and Kramer 1975; van Rees and Comerford 1990; MacFall et al. 1991; Escamilla and Comerford 1998) that woody roots are permeable to water. For all investigated tree species, water uptake rates decreased with increasing root diameter, indicating the reduction of permeability during secondary growth. Probably due to anatomical differences, spruce took up twice as much water compared to beech and oak per unit root surface.

As shown for spruce, the surface-specific uptake rates for woody coarse roots are much lower (factor 20–70) than for young, white, fine roots. However, during the year white roots with high surface-specific uptake rates often account only for less than 1 % of the entire root surface under field conditions (Kramer and Bullock 1966; Göttsche 1972; Comerford et al. 1994). Therefore their contribution to the overall water uptake of trees is strongly reduced. Non-suberized root surfaces alone seem not to be able to meet the whole water demand of perennial plants (Kramer and Boyer 1995). Calculations for spruce, based on the actual root distribution at the experimental plot Coulissenhieb, suggested water uptake through the coarse root system of about 6–16 % of the total tree transpiration. Water uptake of spruce roots measured in situ from moist soil (Lindenmair 2003) was similar to the experiments in nutrient solution, indicating that there was no general overestimation of uptake rates caused by the solution-containing chambers (see also Kramer 1946; MacFall et al. 1990, 1991).

The time course of water uptake rates during the year correlated positively with meteorological factors like global radiation and vapor pressure deficit and therefore with the measured transpiration rate of the tree. This indicates control of water uptake through coarse roots by gradients in water potential.

- The uptake through coarse roots might be physiologically important especially for time periods with low fine root growth, in particular cold and dry periods.
- Varying vertical distribution of fine and coarse roots in soil could result in different temporal and spatial uptake patterns, determining the adaptation of trees to changing environmental conditions.
- Thus, depending on tree species, the uptake through coarse roots could play an important role in drought resistance.
- The results may also be relevant for modeling of the water balance of woody plants since models are based only on data from the fine root system.

Acknowledgements. We would like to thank the engineering workshop of the University of Bayreuth for technical assistance and P. Gerstberger for the CAD drawing of the root chamber. We gratefully acknowledge the fruitful discussions with A. Göttlein (Forest Nutrition and Water Resources, Technical University of Munich) and A. Kuhn (Institute Phytosphere, Research Centre Jülich) concerning the progress of the root examinations. The studies were funded by the German Ministry of Education and Research under grant no. PT BEO 51-0339476 C+D.

References

Addoms RM (1946) Entrance of water into suberized roots of trees. Plant Physiol 21:109–111

Alsheimer M (1997) Xylemflußmessungen zur Charakterisierung raum-zeitlicher Heterogenitäten in der Transpiration montaner Fichtenbestände (*Picea abies* [L.] Karst.). Bayreuther Forum Ökol 49

Atkinson D, Wilson SA (1979) The root–soil interface and its significance for fruit tree roots of different ages. In: Harley JL, Russell RS (eds) The soil–root interface. Academic Press, London, pp 259–271
Chung H-H, Kramer PJ (1975) Absorption of water and ^{32}P through suberized and unsuberized roots of loblolly pine. Can J For Res 5:229–235
Comerford NB, Semthurst PJ, Escamilla JA (1994) Nutrient uptake by woody root systems. NZ J For Sci 24:195–212
Coners H (2001) Wasseraufnahme und artspezifische hydraulische Eigenschaften der Feinwurzeln von Buche, Eiche und Fichte: In situ-Messungen an Altbäumen. Dissertation Math-Naturwiss Fak, University of Göttingen, Germany
Dumbroff EB, Peirson DR (1971) Probable sites for passive movement of ions across the endodermis. Can J Bot 49:35–38
Enstone DE, Peterson CA (1992) A rapid fluorescence technique to probe the permeability of the root apoplast. Can J Bot 70:1493–1501
Escamilla JA, Comerford NB (1998) Measuring nutrient depletion by roots of mature trees in the field. Soil Sci Soc Am J 62:797–804
Friedrich J (1992) Räumliche Variation bodenchemischer und -physikalischer Merkmalsgrößen sowie der Wurzelverteilung in Buchen- und Fichtenökosystemen. Ber Forsch Zentr Waldökosyst A83
Göttsche D (1972) Verteilung von Feinwurzeln und Mykorrhizen im Bodenprofil eines Buchen- und Fichtenbestandes im Solling. Mitt Bundesforsch-Anst Forst-/Holzwirtsch 88
Granier A (1985) Une nouvelle méthode pour la mesure du flux de sève brute dans le tronc des arbres. Ann Sci For 42:81–88
Granier A (1987) Evaluation of transpiration in a Douglas-fir stand by means of sap flow measurements. Tree Physiol 3:309–320
Häussling M, Jorns CA, Lehmbecker G, Hecht-Buchholz C, Marschner H (1988) Ion and water uptake in relation to root development in Norway spruce (*Picea abies* [L.] Karst.). J Plant Physiol 133:486–491
Hose E, Clarkson DT, Steudle E, Schreiber L, Hartung W (2001) The exodermis: a variable apoplastic barrier. J Exp Bot 52:2245–2264
Jorns AC (1987) Presence and function of the Casparian band in roots of Norway spruce (*Picea abies* [L.] Karst.). J Plant Physiol 129:493–496
Köstner B, Tenhunen JD, Alsheimer M, Wedler M, Scharfenberg H-J, Zimmermann R, Falge E, Joss U (1999) Controls of evapotranspiration in spruce forest stands. In: Tenhunen JD, Lenz R, Hantschel R (eds) Ecosystem properties and landscape function in central Europe. Springer, Berlin Heidelberg New York
Kramer PJ (1946) Absorption of water through suberized roots of trees. Plant Physiol 21:37–41
Kramer PJ, Boyer JS (1995) Water relations of plants and soils. Academic Press, San Diego
Kramer PJ, Bullock HC (1966) Seasonal variations in the proportions of suberized and unsuberized roots of trees in relation to the absorption of water. Am J Bot 53:200–204
Kummetz E (1996) Die Wurzelentwicklung der Fichte (*Picea abies* [L.] Karst.). Untersuchungen zum C- und N-Haushalt unterschiedlich alter Bestände. Zulassungsarbeit Lehramt Gymnasien, University of Bayreuth, Germany
Lee D-H (1998) Architektur der Wurzelsysteme von Fichten (*Picea abies* [L.] Karst.) auf unterschiedlich versauerten Standorten. Ber Forsch-Zentr Waldökosyst A153
Lindenmair J (2003) Bedeutung von Grobwurzeln für die Wasser- und Ionenaufnahme von Altfichten – Freilandversuche mit Wurzelkammern. Dissertation, University of Bayreuth, Germany
Lindenmair J, Matzner E, Göttlein A (2001) Eine Wurzelkammer zur Quantifizierung der potentiellen Nährstoff- und Wasseraufnahme von Grobwurzeln an Bäumen im Frei-

land. In: Merbach W, Wittenmayer L, Augustin J (eds) Physiologie und Funktion von Pflanzenwurzeln. 11. Borkheider Seminar zur Ökophysiologie des Wurzelraumes. Teubner, Stuttgart, pp 124–130

MacFall JS, Johnson GA, Kramer PJ (1990) Observation of a water-depletion region surrounding loblolly pine roots by magnetic resonance imaging. Proc Natl Acad Sci USA 87:1203–1207

MacFall JS, Johnson GA, Kramer PJ (1991) Comparative water uptake by roots of different ages in seedlings of loblolly pine (*Pinus taeda* L.). New Phytol 119:551–560

McCully ME, Canny MJ (1988) Pathways and processes of water and nutrient movement in roots. Plant Soil 111:159–170

McDonald KC, Zimmermann R, Kimball JS (2002) Diurnal and spatial variation of xylem dielectric constant in Norway spruce (*Picea abies* [L.] Karst.) as related to micro-climate, xylem sap flow, and xylem chemistry. IEEE Transact Geosci Remote Sensing 40:2063–2082

McKenzie BE, Peterson CA (1995) Root browning in *Pinus banksiana* Lamb. and *Eucalyptus pilularis* Sm. 2. Anatomy and permeability of the cork zone. Bot Acta 108:138–143

Moon GJ, Clough BF, Peterson CA, Allaway WG (1986) Apoplastic and symplastic pathways in *Avicennia marina* (Forsk.) Vierh. roots revealed by fluorescent tracer dyes. Aust J Plant Physiol 13:637–648

Peterson CA, Enstone DE, Taylor JH (1999) Pine root structure and its potential significance for root function. Plant Soil 217:205–213

Puhe J (1994) Die Wurzelentwicklung der Fichte (*Picea abies* [L.] Karst.) bei unterschiedlichen chemischen Bodenbedingungen. Ber Forsch Zentr Waldökosyt A108

Rennenberg H, Schneider S, Weber P (1996) Analysis of uptake and allocation of nitrogen and sulphur compounds by trees in the field. J Exp Bot 47:1491–1498

Rüdinger M, Hallgren SW, Streudle E, Schulze E-D (1994) Hydraulic and osmotic properties of spruce roots. J Exp Bot 45:1413–1425

Sanderson J (1983) Water uptake by different regions of the barley root. Pathways of radial flow in relation to development of the endodermis. J Exp Bot 34:240–253

Sattelmacher B, Mühling K-H. Pennewiß K (1998) The apoplast – its significance for the nutrition of higher plants. Z Pflanzenernaehr Bodenkd 161:485–498

Schreiber L, Hartmann K, Skrabs M, Zeier J (1999) Apoplastic barriers in roots: chemical composition of endodermal and hypodermal cell walls. J Exp Bot 50:1267–1280

Steudle E, Heydt H (1997) Water transport across tree roots. In: Rennenberg H, Eschrich W, Ziegler H (eds) Trees – contributions to modern tree physiology. Backhuys, Leiden, pp 239–255

Steudle E, Peterson CA (1998) How does water get through roots? J Exp Bot 49:775–788

Van Rees KCJ, Comerford NB (1990) The role of woody roots of slash pine seedlings in water and potassium absorption. Can J For Res 20:1183–1191

18 Radial Growth of Norway Spruce [*Picea abies* (L.) Karst.] at the Coulissenhieb Site in Relation to Environmental Conditions and Comparison with Sites in the Fichtelgebirge and Erzgebirge

C. Dittmar and W. Elling

18.1 Introduction

Changing environmental conditions influence the evolution and growth of trees. With the building of annual increment layers, trees document their reaction to long-term as well as short-term changing growth conditions over a long period of time. Hence, for many years, tree-ring analyses have been used to detect growth and vitality variations in forest ecosystems.

Several studies have reported a remarkable increase of forest growth over the last 50 years (e.g. Becker et al. 1995; Röhle 1995; Spiecker et al. 1996; Pretzsch 1999; Mund et al. 2002). The main suggested causal factors are changes in forest management, anthropogenic N deposition, increased atmospheric CO_2 concentration and global warming (Spiecker et al. 1996).

Changes in management practices such as regeneration methods, tending, thinning and harvesting regimes have a positive effect on site productivity (Mund et al. 2002). In many regions, site conditions are improved because the intensive use of forests (e.g. by litter removal and pasturing in the past) has lost its relevance.

In many studies, the atmospheric deposition of N is regarded as the main growth-promoting factor over recent decades. This follows the assumption that N was, and is, the most limiting growth factor at many forest sites (Binkley and Högberg 1997). However, increasing growth trends were noted also in regions with low N deposition (Badeau et al. 1996), and, on the other hand, growth reductions under high N input have been found (Beck 2001; Elling and Dittmar 2003). Depending on species, soil and climatic conditions, N input in central Europe often exceeds critical levels (UN/ECE and EC 2000) and therefore endangers the stability and sustainability of forests. Hence, adverse effects of high atmospheric N input like eutrophication, acidification or nutrition imbalances on tree growth should be considered (Hofmann 1995; Beck 2001).

Ecological Studies, Vol. 172
E. Matzner (Ed.), Biogeochemistry of Forested Catchments in a Changing Environment

An improvement in growth conditions is also expected from the increased CO_2 content in the atmosphere. It can stimulate photosynthesis, reduce respiration and relieve water and low-light stress. Its effect on biomass production, however, is very difficult to quantify, because growth response to elevated CO_2 is strongly modified by species, site and climatic conditions. It is for these reasons that the potential of forests to mitigate the anthropogenic increase of atmospheric CO_2 concentrations is critically considered (WBGU 1998).

The uncontested rise in temperature during recent decades in the northern hemisphere and the resultant lengthening of the vegetation period is further given as an explanation of increased growth trends (Pretzsch 1999). A positive effect should be found especially at higher altitude sites, where growth is mainly controlled by temperature (Dittmar and Elling 1999). An increased frequency of drought and extreme weather events, however, which are also reported as consequences of global warming, may have more negative than positive effects on forest growth.

On very fertile soils in southern Bavaria, Röhle (1995) found distinct increased annual stem growth rates of Norway spruce. Compared to yield tables, radial growth exceeds more than 100 % of expected values. In the Lehstenbach watershed (Fichtelgebirge), Mund et al. (2002) showed that Norway spruce trees expressed strong growth increases at the tree level and modest growth increases at the stand level between 1985 and 1994. They investigated growth history and stand biomass along a chronosequence (graduated stand age) of six stands. Although parts of these stands had been limed and intensively thinned in the past, the authors assumed that increased N deposition and CO_2 concentration, as well as improved climatic conditions, are the main factors responsible for the observed changes of stem growth potential. An earlier investigation of more than 50 Norway spruce sites in northern and eastern Bavaria, however, revealed distinct growth reductions during recent decades. Strong relationships between needle loss and growth loss during the period from 1953–1987 were also found (Utschig 1989).

In the first half of the 1970s, conspicuous symptoms of decline in Norway spruce trees in the Fichtelgebirge were observed (Elling 1990). Initially, a yellowing of the older needles occurred, resulting later in complete discoloration. Finally, they died while showing a red-brown colour and fell from the trees (Schulze et al. 1989). This kind of needle discoloration (Chlorose) and loss is characteristic of the so-called mountainous yellowing (German: "*Montane Vergilbung*"). Trees at high altitudes (and on the higher parts of the slopes) were particularly affected by this decline phenomenon. These symptoms were more pronounced on windward sites (southwest to northwest) and slowly decreased in the second half of the 1980s. Several years before the symptoms of mountainous yellowing appeared in the high Fichtelgebirge, many trees at the Schneeberg, the highest mountain in the Fichtelgebirge, showed a dramatic reduction in tree-ring widths. This growth depression was accompanied by numerous missing rings on western slopes and a few individual missing rings on eastern slopes (Elling 1990). No comparable growth reductions and no occurrence of missing rings could be detected during the preceding 100 years. Hence, this phenomenon, undoubtedly, is to be defined as a new type of forest damage. Increasing growth trends only appear after 1986 as a post 1970s/early 1980s growth suppression recovery. An investigation of Norway spruce and silver fir chronologies from the Lower Bavarian Forest region, covering the last 500 years, revealed anomalous decline and recovery trends in the fir data during the second half of the twentieth century which are unique to the recent period and are not part of a long-term natural episodic phenomenon (Wilson and Elling 2004).

Several studies in the Fichtelgebirge and the Böhmerwald show a distinct higher degree of spruce damage on windward western slopes than on eastern leeward sites (Lochner and Schirbel 1991; Freilinger and Weber 1995; Picard et al. 1999). This is probably caused by a different deposition of acid compounds, which are considerably higher on windward sites compared to leeward sites (Katzensteiner and Glatzel 1997).

Changing environmental conditions affect short-term as well as long-term growth variations. Short-term growth variations are mainly controlled by weather fluctuations, although trees can show an increased sensitivity to drought and frost when experiencing environmental stress (Dittmar 1999). Long-term growth variations are a function of environmental influences, stand evolution and dynamics. Hence, in both cases, besides natural effects, human impacts must also be considered.

In order to study long-term and short-term growth variations and their relation to environmental changes, Norway spruce trees in the Lehstenbach watershed were investigated by dendroecological methods. At the Coulissenhieb site (forest department Weißenstadt, 775–785 m a.s.l., with a southwest aspect) 20 dominant trees were selected and sampled for tree-ring analyses. At each tree, two cores at breast height were taken using an increment borer.

In this chapter, the results of dendroecological analyses are presented and discussed in comparison with analogously investigated Norway spruce stands at similar altitudes in the Fichtelgebirge (Schneeberg), and in the western and eastern Erzgebirge. Thus, the interpretation of Norway spruce growth under different site and deposition conditions is based on temporal as well as regional comparisons. Experimental plots in the Fichtelgebirge and the western Erzgebirge are 130- to 150-year-old spruce stands, none of which were limed in the 1980s. In the eastern Erzgebirge, because of the intensive forest decline in this region, only a stand with 100-year-old trees was found. This site was also repeatedly limed in recent decades. Soil, climatic and deposition data were used for interpretation of tree-ring growth at these sites.

18.2 Status and Radial Growth of Norway Spruce at the Coulissenhieb Site

The Coulissenhieb site is located at an altitude of about 780 m a.s.l. on a southwest-orientated, wind-exposed site. Climatic and soil conditions are detailed by Gerstberger et al. (this Vol.). The investigated spruce trees are ca. 130 (120–140) years old. Due to the load of snow, numerous trees show crown breaks. The 20 sampled trees are generally dominant (tree class 2 according to Kraft 1884) and the average basal area per tree of the whole stand is represented in this group.

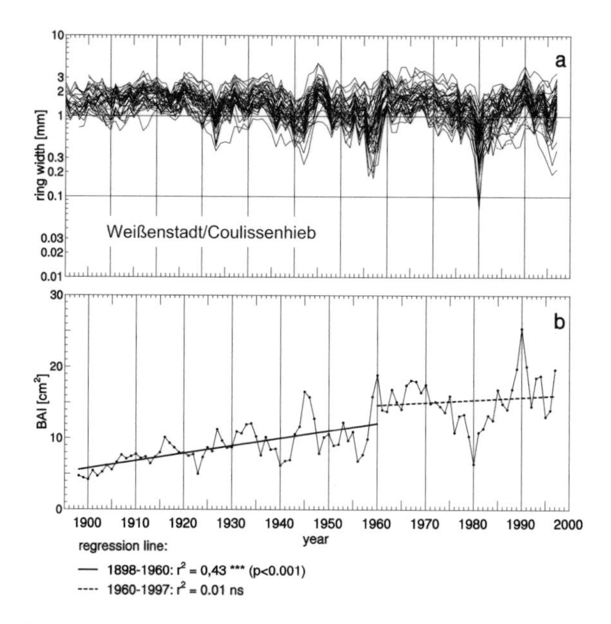

Fig. 18.1. **a** Radial growth curves of 40 increment cores from 20 spruce trees at site Weißenstadt/Coulissenhieb (780 m a.s.l., southwest aspect). Mean age of trees is ca. 130 (120–140) years. Due to a missing ring, one curve is asynchronous and has been highlighted as *dotted line* (data according to Schlecht 1998). For measurement of tree-ring widths, wood cores were taken at each tree reaching near the pith using an increment borer. Wood cores were taken at breast height (1.3 m) in two opposite wind directions, in 0.79 and 3.94 radial to widest stem diameter. Measurement of tree-ring widths, with a resolution of 0.01 mm, was made with the devices of Aniol and Johann in connection with the software program CATRAS (Aniol 1983, 1987). Each radial series was plotted and carefully synchronized to assign the correct calendar date. To plot growth of each sampled tree, arithmetic means of the two radial series for each tree were calculated. **b** Basal area increment (*BAI*) of mean basal area of 20 sampled trees at site Coulissenhieb. For calculation, diameter at breast height of the mean basal area tree, bark thickness according to Altherr et al. (1978) and mean radial increments of all measured radii without missing rings (39 radii) were used. Trends of BAI are illustrated as regression lines, significant between 1898 and 1960 (*solid lines*) and not significant between 1960 and 1997 (*dashed lines*)

The status of the crowns can be assigned a moderate damage degree. Using damage classes, which combine needle loss and discoloration, the crown condition of the sampled trees could be described in autumn 1997 as follows: 12 trees in damage class 0, five trees in damage class 1, one tree in damage class 2 and two trees in damage class 3. Needle analyses revealed a distinct Mg deficiency (1-year-old needles: 0.4–1.2 g kg^{-1} dry matter) while the supply of all other nutrients was sufficient (Schlecht 1998). With 1.4–1.7 % dry matter (1-year-old needles) N content is sufficient, but not high.

Tree-ring widths of Norway spruce at Coulissenhieb show an equable course with only small differences from year to year for several decades until around 1955 (Fig. 18.1a). Radial growth was also high until 1970 without showing an expected decrease in ring widths with increasing stem diameter (age trend). After 1970, incremental growth decreased, followed by a pronounced growth depression around 1980 accompanied by one missing ring at one tree. In the 1980s, a recovery of growth is apparent, but it did not continue after 1990.

The average basal area increment (BAI) per tree, calculated from the 20 sampled trees, increased significantly until 1960 (Fig. 18.1b). After 1960, how-

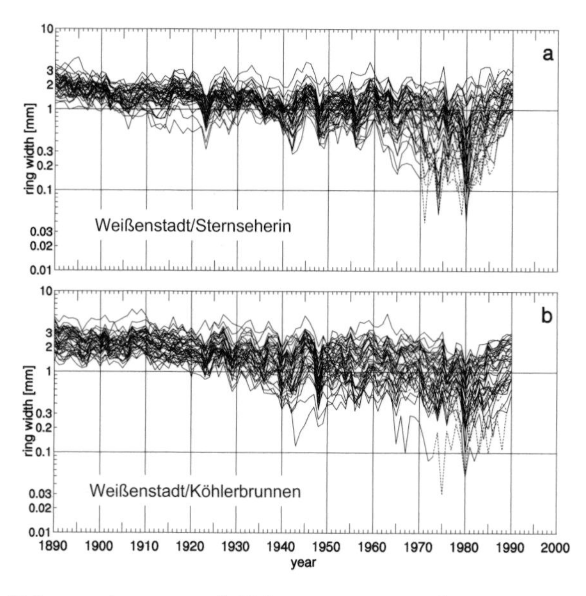

Fig. 18.2. a Radial growth curves of 40 increment cores from 20 spruce trees at site Weißenstadt/Sternseherin (800 m a.s.l., western aspect, Fichtelgebirge, Bavaria). Mean age of trees is ca. 140 (130–150) years. After first appearance of missing rings (34 missing rings in total) some curves become asynchronous and are highlighted as *dotted lines* (data according to Lochner and Schirbel 1991). **b** Radial growth curves of 40 increment cores from 20 spruce trees at site Weißenstadt/Köhlerbrunnen (770 m a.s.l., eastern aspect, Fichtelgebirge, Bavaria). Mean age of trees is ca. 145 (135–155) years. After first appearance of missing rings (six missing rings in total) some curves become asynchronous and are highlighted as *dotted lines* (data according to Lochner and Schirbel 1991)

ever, BAI only increased slightly. When considering site quality and tree class, this increase is in accordance with yield tables. Because of the strong reduction in the number of trees per hectare during the 1980s (Mund et al. 2002), this slight increase in BAI at the tree level does not correspond with a distinct growth increase at the stand level in recent decades.

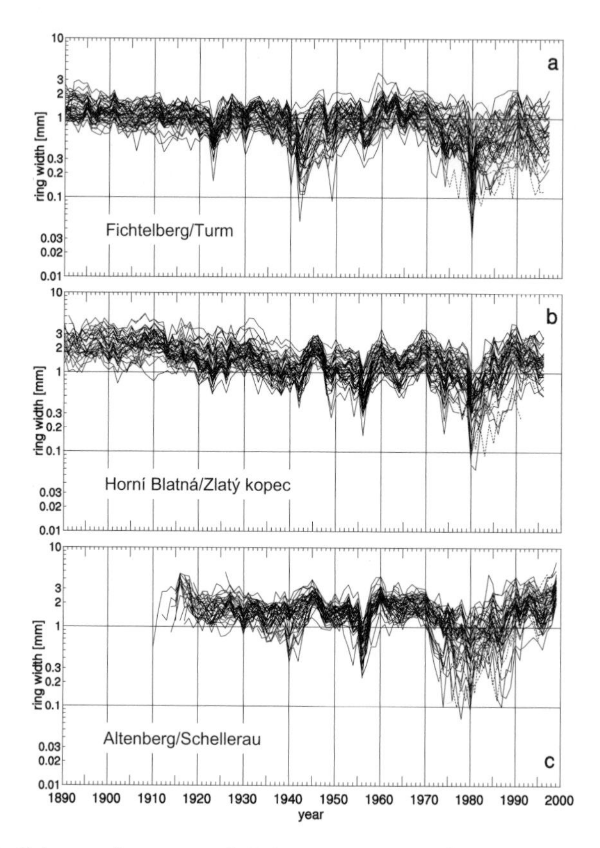

Fig. 18.3. **a** Radial growth curves of 40 increment cores from 20 spruce trees at site Fichtelberg/Turm (990 m a.s.l., Fichtelgebirge, Bavaria). Mean age of trees is ca. 175 (165–185) years. After first appearance of missing rings (eight missing rings in total) some curves become asynchronous and are highlighted as *dotted lines* (data according to Holzmann 1998). **b** Radial growth curves of 40 increment cores from 20 spruce trees at site Horní Blatná/Zlatý kopec (925 m a.s.l., western Erzgebirge, Czech Republic). Mean age of trees is ca. 130 (100–145) years. After first appearance of missing rings (seven missing rings at all) some curves become asynchronous and are highlighted as *dotted lines* (data according to Bendel and Preidt 1997). **c** Radial growth curves of 40 increment cores from 20 spruce trees at site Altenberg/Schellerhau (800 m a.s.l., eastern Erzgebirge, Saxony). Mean age of trees is ca. 100 (95–105) years. After first appearance of missing rings (six missing rings in total) some curves become asynchronous and are highlighted as *dotted lines* (data according to Triebenbacher 2001)

Decreasing radial growth between 1970 and the mid-1980s at the Coulissenhieb site is obvious, but not as pronounced as in higher and more western-located sites in the Fichtelgebirge and Erzgebirge (Figs. 18.2 and 18.3). There radial growth depression started earlier, was more intense and prolonged and was accompanied by many more missing rings, especially at the western-exposed site of the Schneeberg (Fig. 18.2a). Similar growth patterns are reported by Elling (1990) and Mund et al. (2002).

18.3 Growth Influences and Their Temporal and Regional Variations

18.3.1 Climate-Growth Relationships

For the interpretation of tree-ring variations, different dendroecological techniques are available. Single-factor analyses explore relationships between specific environmental factors and tree growth and enable the identification of the dominant growth controlling factors. Single-year analyses deal with the interpretation of extremely narrow or wide tree rings. Both techniques are described in detail by Dittmar and Elling (1999). Interpretations focus on climate influences which are known to be the main controlling factors of short-term growth variations. For this, regionalized climatic records or data from single suitable climatic stations, with daily resolution, were used and transformed by region-specific transfer functions to obtain the site-specific meteorological situation. Since tree-ring series are time series containing long-term growth variations (growth trends) and signal after-effects (autocorrelation), different steps of statistical processing are necessary to establish climate–growth relationships. Dittmar and Elling (1999) presented several of these methods and applied them in connection with linear models in order to establish climate–growth relationships for spruce and beech at different sites in southern Germany. Also for spruce, Dittmar and Elling (1999) obtained distinct altitude-dependent relationships. At cool–wet mountainous sites, temperature clearly is the most limiting growth factor. Wide tree rings are formed in years with warm and dry weather, whereas only small tree rings are possible in cold and wet weather conditions. For warm and dry sites in low altitudes the opposite connections are observed, where growth of spruce is strongly regulated by water supply. Therefore, at high altitude sites, positive correlations and in low altitude sites negative correlations between temperature and tree-ring widths are observed. In the case of precipitation during the vegetation period, the relationships are opposite: negative in high altitudes and positive in low altitudes. The altitude range 600–800 m a.s.l. is identified as a transition zone with weak relationships between weather and radial growth (Dittmar and Elling 1999). Spruce trees at these altitudes are growing, on average, in almost optimal climatic conditions and growth-limiting factors are changing from year to year.

According to its climate-growth relationships, the Coulissenhieb site can be assigned to this transition zone, although correlation analysis – calculated for a period of 53 years before growth depression (1922–1974) – identifies a stronger relationship with temperature than with precipitation for the vegetation period (May–August; see Table 18.1). This analysis demonstrates that temperature is the more frequent growth-limiting factor than water supply. In order to compare different growth patterns, additional spruce stands were investigated from exposed high altitudes:

- At the Schneeberg mountain in the Fichtelgebirge at 800 m a.s.l. (Weißenstadt/Sternseherin, west exposed) and at 770 m a.s.l. (Weißenstadt/Köhlerbrunnen, Schneeberg, east exposed; Lochner and Schirbel 1991).

Table 18.1. Single-factor analyses between radial growth, weather and water balance parameters. Regionalized climatic records are utilized for temperature and precipitation. Soil profile data, site-specific transformed meteorological data and the water balance model HyMo (Rötzer et al. 2004) were used for estimation of actual evapotranspiration and soil water content. After detrending and autoregressive modeling, the tree-ring indices of each tree were averaged together to develop site chronologies according to Dittmar and Elling (1999). *ns* Not significant; * significant with $p < 0.05$; *n.d.* not determined

Site	Temperature (mean May–Aug, 1922–1974)	Precipitation (sum May–Aug, 1922–1974)	Actual evapotranspiration (sum May–Aug, 1948–1974)	Soil water content (mean May–Aug, 1948–1974)
Coulissenhieb, 775–785 m a.s.l., southwest exposed	+0.31*	–0.09 ns	+0.48*	–0.37 ns
Sternseherin, 795–805 m a.s.l., Schneeberg, west exposed	+0.31*	–0.09 ns	+0.36 ns	–0.22 ns
Köhlerbrunnen, 765–780 m a.s.l., Schneeberg, east exposed	+0.27 ns	–0.05 ns	+0.43*	–0.22 ns
Turm, 985–995 m a.s.l. Ochsenkopf, southwest exposed	+0.26 ns	–0.12 ns	+0.43 ns (1960–1974)	+0.07 ns (1960–1974)
Zlatý kopez (western Erzgebirge), 920–935 m a.s.l., Horní Blatná, southwest exposed	+0.26 ns	–0.01 ns	n.d.	n.d.

- At the Ochsenkopf mountain at 990 m a.s.l. (Fichtelberg/Turm, Ochsenkopf; Holzmann 1998).
- In the Czech western Erzgebirge at 925 m a.s.l. (Horní Blatná/Zlatý kopez; Bendel and Preidt 1997).
- In the eastern Erzgebirge at 800 m a.s.l. (Altenberg/Schellerhau; Triebenbacher 2001).

Radial increment curves of spruce trees growing at these sites are presented in Figs. 18.2 and 18.3. Climate–growth relationships of the high altitude sites in the Fichtelgebirge and the western Erzgebirge are similar to that at the Coulissenhieb site (Table 18.1): positive correlations between ring width and temperature, with no correlation between ring width and precipitation. In addition, positive correlations between growth and actual evapotranspiration and no or even weak negative correlations between growth and soil water content confirm the described climate–growth relationships: actual evapotranspiration is mainly controlled by temperature and radiation and not by water supply, which can be assumed as sufficient in most years.

For the eastern Erzgebirge site, no significant climate–growth relationships were found. Single-year analyses according to Dittmar and Elling (1999), however, revealed that most of the positive pointer and event years are connected with warm and/or dry weather conditions. Wide ring widths are formed in the dry and warm years 1917, 1943, 1947, 1959, in the warm years 1945 and 1992 and in the dry year 1960.

18.3.1.1 Influence of Drought

Since the beginning of the forest decline debate, dry periods were often favoured as a dominant factor. Extreme weather events like drought and frost strongly influence radial growth and, therefore, can frequently be responsible for narrow ring widths in single years. In the literature, however, many cursory assessments about the impact of supposed or real dry years on the growth and vitality of spruce can be found. On dry sites, spruce can react sensitively to low precipitation in the current and following years (Elling 1990; Ellenberg 1996).

In the warm-dry region of Lower Franconia (Mainfranken), even on soils with high water storage capacity, a strong growth reduction in the dry year 1976 was detected. Missing rings were noted in individual trees in this year, while no growth depression took place before (Kaps and Stegmaier 2001). In all spruce trees, however, a strong and complete recovery of radial increment in the following year 1977 is obvious, needles showed no discoloration and their crowns only portrayed a slight transparency.

Important in this context are the observations and results of the Solling project. As a consequence of the dry period in 1983, no reduction of fine root biomass was found (Murach 1991). Both the duration and intensity of experimentally induced droughts in a roof experiment were much more severe than potential natural droughts during summer in central Europe (Bredemeier et al. 1998; Lamersdorf et al. 1998; Murach and Parth 1999). Nevertheless, ever under these dry conditions a decline of fine roots was not detected. In the following year, after the end of the drought experiment, the spruce growth recovered. Only with extreme water deficiency and only in the immediate following year, lag effects of dry years are to be expected. They are especially likely when dry periods occur in late summer or autumn. This was the case in 1947. At the Coulissenhieb site all known droughts (1893, 1911, 1921, 1934, 1947 with lag effect in 1948, 1976) caused only weak reactions in the tree-ring widths (Fig. 18.1). This is in line with the statistical confirmed climate–growth relationships described above.

18.3.1.2 Influence of Frost

Norway spruce is very resistant to strong winter frost. This is not the case, however, if it is affected by SO_2. This has been known for a long time, especially from observations in the heavily polluted Erzgebirge as well as from experimental findings (e.g. Wentzel 1956; Keller 1978, 1981, 1989; Däßler and Ranft 1986; Materna 1987; Bosch and Rehfuess 1988). Winter frost – especially in connection with a strong and rapid temperature decrease – played an important role in the stepwise decline of spruce forests in the Erzgebirge. The strongest frost event that took place in the last 100 years was likely during the winter 1955/1956.

After warm weather during December 1955 and January 1956 (in southern Germany *Daphne mezereum* L. already was flowering) temperatures dramatically decreased at the beginning of February. In the following days, temperatures below –30 °C were reached in the whole of Bavaria. Severe damage to fruit trees and silver fir were the consequence. In northeast Bavaria spruce also showed stress reactions in the 1956 growth ring. In southern Bavaria, no growth reduction was found (see also Ewald et al. 2000), even for trees at timber line, where the cool and wet vegetation period in 1956 reduced cambial activity. Investigations of temperature trends could not explain these differences between north and south Bavaria. However, a significant correlation was found between the S content in the needles and the extent of growth reduction in 1956 (Franz 1989). Although S data of needles are only available since 1977, we assume that regional differences in the distribution of pollution over Bavaria did not fundamentally change in the period concerned. Northeast Bavaria was especially affected by high SO_2 pollution over several decades. This, as well as the decrease of pollution in recent years, is well documented in the development of the S content of spruce needles over the years 1977–1995 (Elling and Pfaffelmoser 1997; Fig. 18.4).

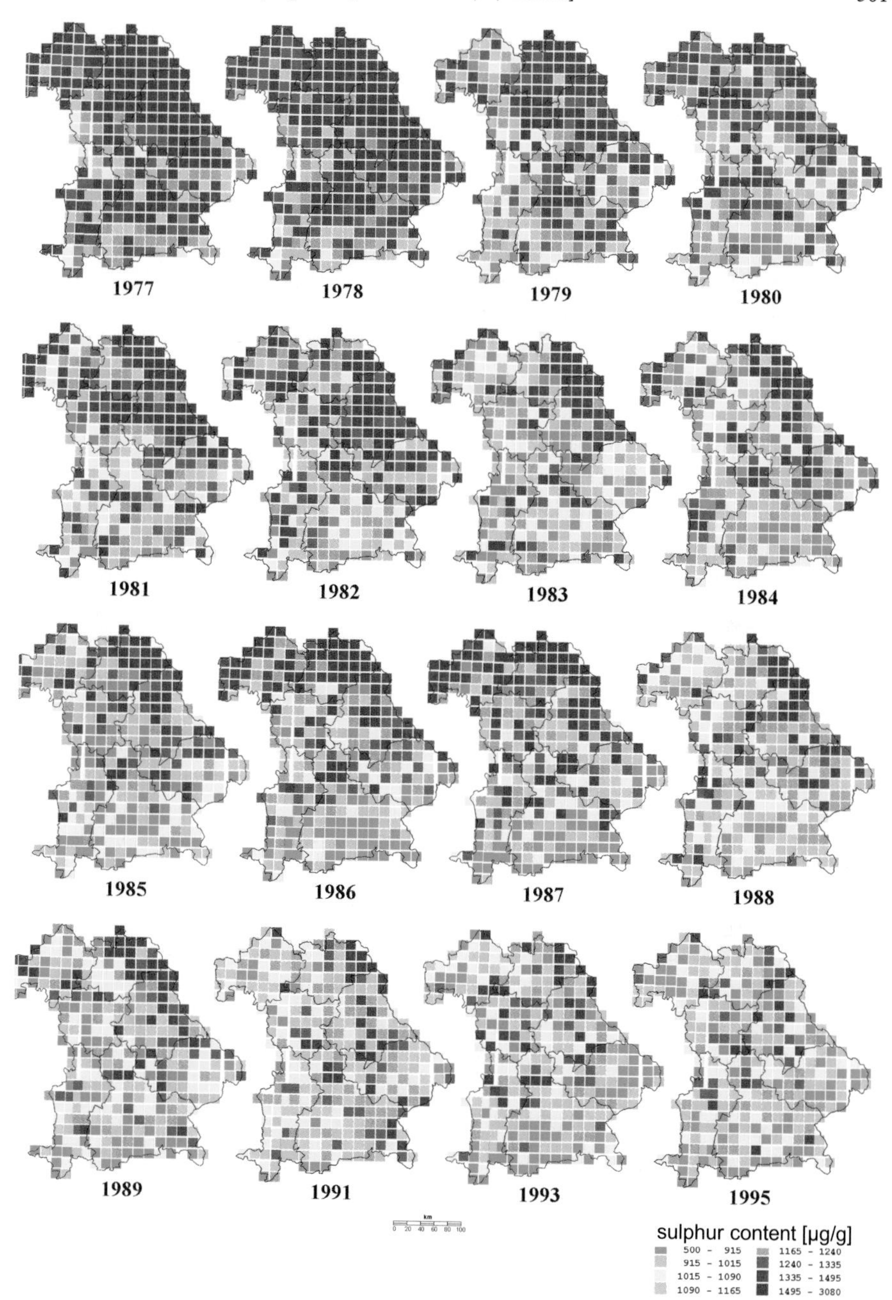

Fig. 18.4. Temporal trends of sulfur content (μg g^{-1} dry matter) in 6-month-old spruce needles in Bavaria over the period 1977–1995. Data come from the network of bio-indication for spruce developed by the Bayerisches Landesamt für Umweltschutz (according to Elling and Pfaffelmoser 1997)

Not only in the highly polluted Erzgebirge, but also in the Fichtelgebirge with lower acid deposition, spruce suffered from this extreme frost event. At the Coulissenhieb site especially, dominant trees reacted. After a distinct reduction of radial increment in 1956 some of the trees showed a distinct recovery of ring widths in 1958 or 1959. As all other weather conditions were favourable at this time, this suggests a clear lag effect from the frost event in February 1956.

The sudden drop in temperature during the New Year's Eve night of 1978/1979 often was suggested as a cause for forest decline in northeast Bavaria. However, this can be excluded, because yellowing and severe growth depression in the Fichtelgebirge had already taken place in the first half of the 1970s (Elling 1990). Observations in the Erzgebirge (Materna 1987), however, detail that special attention to this temperature decrease and its consequences is needed. At this location, with an SO_2 pollution of 20–30 μg m^{-3} (yearly average), spruce showed reddening, fall of needles and some evidence of bud decline. Even the decline of whole stands was observed during 1979.

As a consequence of severe frost damage, trees react with a striking decrease of radial growth in the current or following year as demonstrated for 1956 (see above). However, the strong growth minimum of 1980 (Fig. 18.1) cannot be explained by lag effects of the 1978/1979 sudden frost event. Further, the intensity of this growth minimum increases with altitude (see Figs. 18.1–18.3), while the probability of frost damage due to a drop in temperature during the New Year's Eve night 1978/1979 decreases with altitude. The small tree ring of 1980 is probably caused by the cool and wet weather conditions during the vegetation period of this year in connection with other, still unknown, impacts.

Frost resistance in spruce is controlled by day length and temperature (Kandler et al. 1979). In the middle of the winter, when days are short, a reduction of frost resistance is only possible if a long lasting and strong warming occurs. According to climate records of the German weather service this only occurs in lower altitudes (below approx. 700–800 m a.s.l.) and for a time span of ca. 1 week. Dendrochronological investigations from the whole of Bavaria have shown that damage of spruce due to the sudden drop in temperature during the New Year's Eve night 1978/1979 is only possible under specific conditions. Hence, a strong decrease of radial increment in 1979, which can be interpreted as a consequence of this sudden frost, is only to be expected under the following circumstances:

- Location in the northeast of Bavaria. This means a high load of S pollution and also a strong decrease of temperature between 31 December 1978 and 1 January 1979.
- Altitude below approx. 700–800 m a.s.l. Only at lower altitudes did a significant warming take place before the sudden frost.
- Potassium deficiency, e.g. on Cambisols of serpentinite with concurrent high Mg nutrition (Ehlich and Pfadenhauer 1992). Magnesium deficiency seems effective in the same way, but weaker.

Growth at lower altitudes in the Fichtelgebirge, on sites with sufficient K nutrition but deficient Mg, decreased noticeably according to tree rings from 1978–1979. This is also obvious at the Coulissenhieb site. The decrease is strongest in those stands that already show a sensitive reaction for 1956. Results from experiments under controlled conditions partly contradict this finding, however. For spruce saplings growing under K deficiency, no reduction in frost resistance was noticeable (Senser 1990) as substrate soil material from the Bavarian Alps was used, which is rich in lime and has an excellent supply of Mg and Ca. Deficiency of these elements in the strongly acidified soils from the Bavarian Forest, however, caused a distinct reduction in frost resistance.

At high altitude sites (e.g. at the site Fichtelberg/Turm, 990 m a.s.l, Fig. 18.3a) no reduction in radial growth in 1979 is noticeable. In fact, spruce trees growing at high altitude sites in the Bavarian Forest (approx. 1,300 m a.s.l) show the 1979 tree ring as distinctly wider than the 1978 ring (Bart 1987; Picard et al. 1999). This underlines the assumption that in higher altitudes no damage due to the sudden frost in the New Year's Eve night 1978/1979 occurred. According to temperature trends, it would not be expected either. Temperature data from the weather station Großer Falkenstein (1,307 m a.s.l.) in the Bavarian Forest show that minimum temperatures during the night exceeded 0 °C for only 1 day (29 December 1978) before the sharp drop occurred. Hence, for these altitudes a breakdown of frost resistance can be excluded. These findings contradict the hypothesis of Rehfuess and Bosch (1986), who state that the sudden drop in temperature on 31 December 1978/1 January 1979 is responsible for spruce decline in the high altitudes of the highlands. Also in the lower parts of the Bavarian Forest, the effects were weak because silver fir, which is much more sensitive to frost, shows no distinct decrease in ring widths in consequence to this event (Elling 1993).

18.3.2 Influence of Deposition

During recent decades, the Fichtelgebirge was highly polluted by S deposition (SO_2, H_2SO_3, H_2SO_4). Since the end of the 1980s the input decreased distinctly (Fig. 18.4). The deposition of N is, however, still high (Alewell et al. 2000; Matzner et al., this Vol.). Between 1993 and 2001, on average, a total N deposition of 22.2 kg N ha^{-1} $year^{-1}$ was detected (Matzner et al., this Vol.). Also the ozone load is high in the Fichtelgebirge and increases with altitude (Klemm, this Vol.).

Impacts of these pollutants on growth in connection with frost are discussed above. Investigations of trends in tree rings of spruce trees from the Fichtelgebirge repeatedly revealed quick reductions of increment starting between 1965 and 1970 and ending in the mid-1980s. This is especially obvious on the western exposed mountain Schneeberg (Elling 1990). Investigations, using similar methods as at Coulissenhieb, of two sites at equal altitudes but differing aspects [one site with a western aspect (Sternseherin) and one site with an eastern aspect (Köhlerbrunnen)] found that, compared to the Coulissenhieb site, trees at both sites showed a more pronounced growth depression (Fig. 18.2). At the western exposed site, however, it is distinctly more severe and started earlier. At the Sternseherin site, growth reduction began in the 1960s (Fig. 18.2a) and at Köhlerbrunnen later in the 1970s (Fig. 18.2b). During the period of maximum growth suppression on the western and eastern slopes, seven and three missing rings (20 sampled trees each) were detected respectively. Differences in the degree of damage between the windward and leeward exposed sites are confirmed by several studies in the Fichtelgebirge and the Bavarian Forest (Elling 1990; Freilinger and Weber 1995; Picard et al. 1999). There is strong evidence that these differences are caused by the deposition of acid compounds, which is higher on windward-facing slopes (western) than at leeward-facing (eastern) sites (Katzensteiner and Glatzel 1997). This is in line with the findings that neither the growth reduction during the 1970s nor the recovery in the 1980s sufficiently can be explained by climatic influences.

The network of bio-indication for spruce of the Bayerisches Landesamt für Umweltschutz provided continuous data of S contents (micrograms per gram of dry matter) in 6-month-old spruce needles in Bavaria during the years 1977–1995 on a 16- to 16-km grid. These data show high temporal intercorrelations and strong relationships with the SO_2-emission data (Elling and Pfaffelmoser 1997). From these data the temporal and regional differences in S deposition were derived (Fig. 18.4).

To prove whether long-term growth variations of Norway spruce are affected by S deposition, the tree-ring data (stand mean curves) from the Fichtelgebirge sites were correlated with the SO_2-emission data from Germany (Table 18.2). The correlations give evidence that the long-term growth variations of Norway spruce in the Fichtelgebirge, especially on west-facing slopes, are affected by high and long-lasting S deposition. Decreasing growth and growth depressions in the 1970s and around 1980 cannot be explained by weather influences (see above) and stand dynamic processes.

These findings can be compared with the results of Wilson and Elling (2004) who investigated the temporal stability of climate–growth relationships of Norway spruce and silver fir chronologies over the last 500 years in the lower Bavarian forest region in southern Germany. They present compelling evidence that anomalous trends observed in the tree-ring data in the second half of the twentieth century are predominantly related to local SO_2 emissions from power plants and refineries.

Recovery of growth in the Fichtelgebirge during recent decades follows the strong decrease in SO_2 emission after 1986. Besides other influences (e.g. increased growing space after thinning, N input, increased atmospheric CO_2 concentration and climatic warming), an improved air quality after the mid-

Table 18.2. Correlations between time course of tree-ring widths and SO_2 emission data. For correlation stand, means of raw ring data were used and calculated for maximum possible time span. SO_2-emission data are strongly correlated with deposition data according to Elling and Pfaffelmoser (1997). *ns* Not significant; ** significant with $p < 0.01$; *** significant with $p < 0.001$

Site	SO_2 emission West Germany (sum of year, 1852–1995)	SO_2 emission Germany total (sum of year, 1970–1999)
Coulissenhieb	–0.40*** (*n*=123, 1873–1995)	–0.21 ns (*n*=28, 1970–1997)
Sternseherin	–0.80*** (*n*=131, 1860–1990)	–0.80*** (*n*=21, 1970–1990)
Köhlerbrunnen	–0.84*** (*n*=121, 1870–1990)	–0.64** (*n*=21, 1970–1990)
Turm	–0.68*** (*n*=144, 1852–1995)	–0.20 ns (*n*=28, 1970–1997)

1980s should be taken into account when analysing increasing growth trends in recent decades. Recovery effects must be considered also, if strong increasing growth trends or sudden step growth rises occur (Dittmar 2002).

These results suggest that the deposition of S compounds plays an important role in the complex phenomenon of growth depression and decline of spruce. Impacts of high N deposition are discussed below by comparing dendroecological results of studies in the Fichtelgebirge and the Erzgebirge.

Spruce trees at the high altitude site of Fichtelberg/Turm show a stronger growth reduction between 1970 and 1980 than trees at the lower elevated Coulissenhieb site (Figs. 18.1 and 18.3a). Further, the number of trees with missing rings is also higher (five at Fichtelberg/Turm and one at Coulissenhieb, respectively; 20 sample trees each). In addition, the recovery of radial growth in the 1980s is noticeable at the Coulissenhieb site, whereas at the Turm site, growth conditions do not appear to have improved in the 1980s and 1990s.

In comparison to the site in the western Erzgebirge (Fig. 18.3b), the Turm site is similar with respect to not only growth patterns until 1970 but also growth depression and the number of missing rings. In the western Erzgebirge, however, the recovery of radial growth after 1980 is much more pronounced and is observed in almost all investigated trees.

The trees from the site in the eastern Erzgebirge, which survived the extensive decline on ridge sites, show an intense and long-lasting growth depression. Missing rings were noticed in 5 of the 20 sampled trees. The growth depression occurs between 1970 and 1990. There is no substantial increase of radial increment in the 1980s. After 1990, however, tree-ring widths on average are wider than before the growth depression. In contrast to the high altitude sites of the Fichtelgebirge, growth curves are also close together again (Fig. 18.3 c). This points to a striking improvement of growth conditions after 1990. In addition, the low age of the trees must be considered, which probably made recovery easier. No older spruce and no unlimed stands could be found, however, because of the intensive decline in the recent past.

Processes reducing growth and causing decline of Norway spruce in the Fichtelgebirge can be assigned to different symptom complexes or syndromes. For the development of mountainous yellowing, both natural conditions and anthropogenic pollution are responsible. On base-poor parent rocks and under high precipitation, acidic soils are a natural consequence. The deposition of S and N cause additional acidification and leaching (Katzensteiner and Glatzel 1997). Also the removal of biomass enhances loss of nutrients. Hence, the very limited supply of Mg is further reduced (Schulze et al. 1989; Kölling et al. 1997). Sulfur dioxide taken up via the stomata increases the demand for neutralizing cations (Slovik et al. 1992). Specific microorganisms additionally impede the uptake of Mg (Devêvre et al. 1993, 1996). In yellowed needles suffering from Mg deficiency, ozone leads to a quick decomposition of chlorophyll and, in this way, to a distinct discoloration (Siefermann-Harms et al. 1998).

In the eastern Erzgebirge, another decline syndrome has to be considered. There, spruce is damaged by high SO_2 pollution under more favorable soil conditions. However, this also cannot be considered as a monofactorial situation but represents a complex process, in which site conditions and climate impacts are involved. Wind and frost especially play an important role (Wienhaus et al. 1994). Neither by visible symptoms nor by needle analyses could nutrient deficiencies be shown (Pfanz and Beyschlag 1991). However, the life span of needles was as short as 1 year or even less (Pelz and Materna 1964). This described phenomenon decreased with the reduced SO_2 pollution.

For the Erzgebirge, a transition zone of both syndromes is described between Rochlitz and Schwarzenberg (Wienhaus et al. 1994). Westwards of this line, mountainous yellowing is observed and eastward of this line the damage of spruce by high SO_2 pollution is predominant. The investigated spruce stands confirm this picture. All sites in the Fichtelgebirge show symptoms of mountainous yellowing. The site in the eastern Erzgebirge (Altenberg/Schellerhau) had no noticeable symptoms of discoloration. The spruce site in the western Erzgebirge (Horní Blatná/Zlatý kopez) is located near the mentioned border line Rochlitz–Schwarzenberg (ca. 10 km west of the Fichtelberg mountains). Its symptoms could be described as transitional: 12 of 20 sampled trees show discoloration in older needles, which is typical of mountainous yellowing. No noticeable discoloration was found in the other trees.

Due to the reduction of SO_2 pollution starting in the Erzgebirge, especially after 1990, radial increment of all trees at the Altenberg/Schellerhau site increased (Fig. 18.3 c). This is also visible at the western Erzgebirge site (Fig. 18.3b), but less pronounced. The SO_2 pollution in the Erzgebirge was accompanied by high deposition of basic dust and moderate N input (Zimmermann et al. 1998). Its reduction obviously allowed a distinct growth recovery.

A noticeable different growth pattern, reflecting another deposition situation and history, is apparent at the two spruce sites in the Fichtelgebirge (Coulissenhieb and Turm, Figs. 18.1 and 18.3a, respectively). Especially at the latter site, numerous trees show no convincing recovery after growth depression around 1980. At the Coulissenhieb site most of the trees show increasing tree-ring widths in the 1980s, but after 1990 growth reductions are visible again. This can be explained by a continuous pollution of the soils. In spite of the strong reduction of SO_2 pollution (Fig. 18.4) the trees are further stressed by other factors. High N input as well as the mobilization of stored sulfate in the soils are still causing acidification. On the other hand, the deposition of Ca and Mg is distinctly reduced. Hence, the concentration of Ca and Mg in the soil, the molar Ca/Al and Mg/Al ratios and the Ca and Mg content in needles strongly decreased in recent years (Alewell et al. 2000). Under Mg deficiency, spruce is also very sensitive to ozone pollution (Siefermann-Harms et al. 1998). The interplay of these influences can explain the weak recovery potential of spruce in higher altitudes of the Fichtelgebirge.

18.4 Conclusions

- Short-term growth variations of Norway spruce at the Coulissenhieb site and at comparable high altitude sites in the Fichtelgebirge and the Erzgebirge are mainly influenced by temperature variations during the vegetation period.
- Temporal and regional relationships between long-term growth variations and deposition give evidence that the input of atmospheric pollutants plays an important role in growth reduction and decline processes as well as in growth recovery and release.
- Regional differences between growth patterns in the Fichtelgebirge and the Erzgebirge reflect divergent deposition situations and histories. In the eastern Erzgebirge a distinct recovery of radial increment is obvious after 1990. However, recovery in the Fichtelgebirge started earlier in the mid-1980s is often, especially at high altitude sites, only weak and not convincing. This can be explained by specific soil conditions and a still ongoing acidification.
- The presented interpretation of tree-ring data demonstrates the potential of dendroecological studies to analyse the growth, vitality and damage history of trees and stands. From temporal and regional (site to site) compar-

isons, valuable contributions are possible to resolve causes of increasing growth, on the one hand, and forest decline phenomena, on the other. Hence, to enable this and to improve the database, additional spruce stands should be dendroecologically investigated in the Fichtelgebirge. This is especially important since declining processes are still ongoing at several sites and increasing growth trends only appear as short and not durable transition phase.

Acknowledgements. The authors thank the forest administrations Weißenstadt, Fichtelberg, Altenberg and Horní Blatná for friendly support, for providing data and for permission to undertake field work. We are grateful to Martina Mund, Pedro Gerstberger and Egbert Matzner for their cooperation and assistance, Thomas Rötzer for water balance calculations and the German Weather Service for providing us with climatic data. We gratefully acknowledge the contributions made by students of the Weihenstephan University of Applied Sciences in sampling wood cores and measuring tree-ring widths.

References

Alewell C, Manderscheid B, Gerstberger P, Matzner E (2000) Effects of reduced atmospheric deposition on soil solution chemistry and elemental contents of spruce needles in NE-Bavaria, Germany. J Plant Nutr Soil Sci 163:509–516

Altherr E, Unfried P, Hradetzky J, Hradetzky V (1978) Statistische Rindenbeziehungen als Hilfsmittel zur Auformung und Aufmessung unentrindeten Stammholzes. IV. Fichte, Tanne, Douglasie und Sitka-Fichte. Mitt Forstl Versuchsanst Forschungsanst Baden-Württemberg 90

Aniol RW (1983) Tree-ring analyses using CATRAS. Dendrochronologia 1:45–53

Aniol RW (1987) A new device for computer-assisted measurement of tree-ring widths. Dendrochronologia 5:135–141

Badeau V, Becker M, Bert D, Dupouey J-L, Lebourgeois F, Pircard J-F(1996) Long-term growth trends of trees: ten years of dendrochronological studies in France. In: Spiecker H, Mielikäinen K, Köhland M, Skovsgaard JP (eds) Growth trends in European forests – studies from 12 countries. EFI Research Rep 5. Springer, Berlin Heidelberg New York, 372 pp

Bart P (1987) Kronenmerkmale und Jahrringbau von geschädigten Hochlagenfichten im Nationalpark Bayerischer Wald. Diploma Thesis, Fachbereich Forstwirtschaft, Fachhochschule Weihenstephan

Beck W (2001) Waldwachstum unter anhaltendem Fremdstoffeintrag – Ergebnisse aus waldwachstumskundlichen und dendroökologischen Untersuchungen. Beitr Forstw Landschaftsökol 35(4):192–201

Becker M, Bert G, Bouchon J, Dupouey J, Picard J, Ulrich E (1995) Long-term changes in forest productivity in north-eastern France: the dendroecological approach. In: Landmann G, Bonneau M (eds) Forest decline and atmospheric deposition effects in the French mountains. Springer, Berlin Heidelberg New York, pp 143–156

Bendel M, Preidt H (1997) Wachstumsverlauf, Symptome und mögliche Ursachen einer Schädigung der Fichte im tschechischen Teil des westlichen Erzgebirges im Zusammenwirken von klimatischen Faktoren und Immissionsbelastung. Diploma Thesis, Fachhochschule Weihenstephan

Binkley D, Högberg P (1997) Does atmospheric deposition of nitrogen threaten Swedish forests? For Ecol Manage 92:119–152

Bosch C, Rehfuess KE (1988) Über die Rolle von Frostereignissen bei den "neuartigen" Waldschäden. Forstwiss Centralbl 107:123–130

Bredemeier M, Blanck K, Dohrenbusch A, Lamersdorf N, Meyer AC, Murach D, Parth A, Xu YJ (1998) The Solling roof project – site characteristics, experiments and results. For Ecol Manage 101:281–293

Däßler H-G, Ranft H (1986) Untersuchungen zur komplexen Wirkung von Immissions- und Frosteinfluß auf Fichtenwald in Mittelgebirgslagen. Allg Forstztg 41:340–343

Devêvre O, Garbaye J, Perrin R (1993) Experimental evidence of a deleterious soil microflora associated with Norway spruce decline in France and Germany. Plant Soil 148:145–153

Devêvre O, Garbaye J, Botton B (1996) Release of complexing organic acids by rhizosphere fungi in Norway spruce yellowing in acidic soils. Mycol Res 100:1367–1374

Dittmar C (1999) Radialzuwachs der Rotbuche (*Fagus sylvatica* L.) auf unterschiedlich immissionsbelasteten Standorten in Europa. Bayreuther Bodenkund Ber 67

Dittmar C (2002) Bedeutung der Jahrringbreite für die Waldschadensforschung. Stuttg Geogr Stud 113:117–131

Dittmar C, Elling W (1999) Jahrringbreite von Fichte und Buche in Abhängigkeit von Witterung und Höhenlage. Forstwiss Centralbl 118:251–270

Ehlich W, Pfadenhauer K (1992) Jahrringbau von zwei Fichtenbeständen mit sehr unterschiedlicher Nährstoffversorgung in Nordostbayern. Diploma Thesis, Fachhochschule Weihenstephan

Ellenberg H (1996) Vegetation Mitteleuropas mit den Alpen. 5th edn. Ulmer, Stuttgart

Ellenberg H, Mayer R, Schauermann J (1986) Ökosystemforschung. Ergebnisse des Sollingprojekts 1966–1986. Ulmer, Stuttgart

Elling W (1990) Schädigungsverlauf und Schädigungsgrad von Hochlagen-Fichtenbeständen in Nordostbayern. Allg Forstztg 45:74–77

Elling W (1993) Immissionen im Ursachenkomplex von Tannenschädigung und Tannensterben. Allg Forstztg 48:87–95

Elling W, Pfaffelmoser K (1997) Auswertung der Schwefeldaten des flächendeckenden Bioindikatornetzes Fichte. Abschlussbericht an das Bayerische Landesamt für Umweltschutz, Freising

Elling W, Dittmar C (2003) Neuartige Zuwachsdepressionen bei Buchen. Allg Forstztg/ Der Wald 58(1):42–45

Ewald J, Reuther M, Nechwatal J, Lang K (2000) Monitoring von Schäden in Waldökosystemen des bayerischen Alpenraumes. Materialien 155. Bayerisches Staatsministrium für Landesentwicklung und Umweltfragen, München

Franz C (1989) Zuwachsreaktionen von Fichten auf die extreme Frostsituation des Winters 1955/56 und deren mögliche Ursachen. Diploma Thesis, Fachhochschule Weihenstephan

Freilinger R, Weber J (1995) Jahrringuntersuchungen und Erfassung von Kronenmerkmalen an Fichte (*Picea abies*) in einer Höhe von 1250 m ü.NN auf einem westexponierten Hang (Forstamt Bodenmais) und einem ostexponierten Hang (Forstamt Bayerisch Eisenstein der Fürstl. Hohenzollernschen Hofkammer). Diploma Thesis, Fachbereich Forstwirtschaft, Fachhochschule Weihenstephan

Hofmann G (1995) Vegetationswandel in nordostdeutschen Kiefernwaldungen durch atmosphärischen Eintrag von Stickstoffverbindungen. Der Wald Berlin 45(8):262–267

Holzmann M (1998) Untersuchungen zu Schädigungen und Wachstumsverlauf von Hochlagenfichten anhand von Kronenmerkmalen und Jahrringbau im Bereich des Forstamtes Fichtelberg/Ofr. Diploma Thesis, Fachbereich Forstwirtschaft, Fachhochschule Weihenstephan

Kandler O, Dover C, Ziegler P (1979) Kälteresistenz der Fichte. I. Steuerung von Kälteresistenz, Kohlehydrat- und Proteinstoffwechsel durch Photoperiode und Temperatur. Ber Dtsch Bot Ges 92:225–241

Kaps S, Stegmaier C (2001) Wachstumsverhalten von Tanne (*Abies alba* L. [Karst.]) und Fichte (*Picea abies* L. [Karst.]) auf warm trockenem Standort im Wuchsbezirk "Taubergrund mit westlicher Fränkischer Platte". Diploma Thesis, Fachbereich Forstwirtschaft, Fachhochschule Weihenstephan

Katzensteiner K, Glatzel G (1997) Causes of magnesium deficiency in forest ecosystems. In: Hüttl RF, Schaaf W (eds) Magnesium deficiency in forest ecosystems. Kluwer, Dordrecht, pp 227–251

Keller T (1978) Frostschäden als Folge einer "latenten" Immissionsschädigung. Staub-Reinhalt. Luft 38:24–26

Keller T (1981) Folgen einer winterlichen SO_2-Belastung für die Fichte. Gartenbauwissenschaft 46:170–178

Keller T (1989) Oberirdische Wirkung von Luftschadstoffen: Neuere Ergebnisse. In: Schmidt-Vogt H (ed) Die Fichte, vol II/2. Parey, Hamburg, pp 280–314

Kölling C, Pauli B, Häberle K-H, Rehfuess KE (1997) Magnesium deficiency in young Norway spruce (*Picea abies* (L.) Karst.) trees induced by NH_4NO_3 application. Plant Soil 195:283–291

Kraft G (1884) Beiträge zur Lehre von den Durchforstungen, Schlagstellungen und Lichtungshieben. Klindworth, Hannover

Lamersdorf NP, Beier C, Blanck K, Bredemeier M, Cummins T, Farrell EP, Kreutzer K, Rasmussen L, Ryan M, Weis W, Xu Y-J (1998) Effect of drought experiments using roof installations on acidification/nitrification of soils. For Ecol Manage 101:95–109

Lochner H, Schirbel G (1991) Jahrringanalysen und Kronenmerkmale von zwei Fichtenbeständen auf 800 m Meereshöhe im Forstamt Weißenstadt. Diploma Thesis, Fachbereich Forstwirtschaft, Fachhochschule Weihenstephan

Materna J (1987) Zusammenhang zwischen Schäden an Fichte und Kiefer und Temperaturstürzen in immissionsbelasteten Gebieten. GSF-Ber 10/87. GSF-Forschungszentrum für Umwelt und Gesundheit, Neuherberg, pp 265–268

Mund M, Kummetz E, Hein M, Bauer GA, Schulze E-D (2002) Growth and carbon stocks of a spruce forest chronosequence in central Europe. For Ecol Manage 171:275–296

Murach D (1991) Feinwurzelumsätze auf bodensauren Fichtenstandorten. Forstarchiv 62:12–17

Murach D, Parth A (1999) Feinwurzelwachstum von Fichten beim Dach-Projekt im Solling. Allg Forstztg/DerWald 54:58–60

Pelz E, Materna J (1964) Beiträge zum Problem der individuellen Rauchhärte der Fichte. Arch Forstwesen 13:177–210

Pfanz H, Beyschlag W (1991) Photosynthetic performance of Norway spruce in relation to the nutrient status of the needles. A study in the forests of the Ore Mountains. GSF-Ber 24/91. GSF-Forschungszentrum für Umwelt und Gesundheit, Neuherberg, pp 523–527

Picard B, Arbeiter C, Mederer M, Till M (1999) Jahrringbau von Fichten (*Picea abies*) im Bereich der Borkenkäfer-Massenvermehrung in den Hochlagen des Nationalparks Bayerischer Wald. Diploma Thesis, Fachhochschule Weihenstephan

Pretzsch H (1999) Waldwachstum im Wandel. Forstwiss Centralbl 118:228–250

Rehfuess KE, Bosch C (1986) Experimentelle Untersuchungen zur Erkrankung der Fichte (*Picea abies* [L.] Karst.) auf sauren Böden der Hochlagen: Arbeitshypothese und Versuchsplan. Forstwiss Centralbl 105:201–206

Röhle H (1995) Zum Wachstum der Fichte auf Hochleistungsstandorten in Südbayern. Mitteilungen aus der Staatsforstverwaltung Bayerns Heft 48, München

Rötzer T, Dittmar C, Elling W (2004) A model for site specific estimation of the available soil water content and the evapotranspiration in forest ecosystems. Journal of Environmental Hydrology, accepted

Schlecht A (1998) Kronenmerkmale und Jahrringbau herrschender Fichten (*Picea abies* [L.] Karst.) im Forstamt Weißenstadt. Diploma Thesis, Fachhochschule Weihenstephan

Schulze E-D, Lange OL, Oren R (1989) Forest decline and air pollution. A study of spruce (*Picea abies*) on acid soils. Ecological studies, vol 77. Springer, Berlin Heidelberg New York

Senser M (1990) Influence of soil substrate and ozone plus acid mist on the frost resistance of young Norway spruce. Environ Pollut 64:265–278

Siefermann-Harms D, Payer HD, Lütz C (1998) Die Wirkung von Ozon auf die Nadelvergilbung bei jungen, unter Mg-Mangel kultivierten Clonfichten. http://bwplus.fzk.de

Slovik S, Kaiser WM, Körner C, Kindermann G, Heber U (1992) Quantifizierung der physiologischen Kausalkette von SO_2-Immissionsschäden. Allg Forstztg/Der Wald 47:800–805/913–920

Spiecker H, Mielikäinen K, Köhland M, Skovsgaard JP (1996) Growth trends in European forests – studies from 12 countries. EFI Research Rep 5. Springer, Berlin Heidelberg New York

Triebenbacher C (2001) Untersuchungen zu Schädigung und Wachstumsverlauf von Fichten im Osterzgebirge am Forstamt Altenberg anhand von Kronenmerkmalen und Bohrspananalysen. Diploma Thesis, Fachbereich Wald und Forstwirtschaft, Fachhochschule Weihenstephan

United Nations Economic Commission for Europe (UN/ECE) and European Commission (EC) (2001) Forest condition in Europe, 2001. Executive Report. Federal Research Centre for Forestry and Forest Products (BFH), Hamburg, 28 pp

Utschig H (1989) Waldwachstumskundliche Untersuchungen im Zusammenhang mit Waldschäden – Auswertung der Zuwachstrendanalyseflächen des Lehrstuhles für Waldwachstumskunde für die Fichte (*Picea abies* (L.) Karst.) in Bayern. Forstl Forschungsber München 97

Wentzel KF (1956) Winterfrost 1956 und Rauchschäden. Allg Forstztg 11:541–543

Wienhaus O, Liebold E, Zimmermann F (1994) Beziehungen zwischen Standort, Klima und immissionsbedingten Waldschäden in den Fichtenbeständen der Mittelgebirge. Forst Holz 49:411–415

Wilson R, Elling W (2004) Temporal instability in tree-growth/climate response in the Lower Bavarian Forest region: implications for dendroclimatic reconstructions. Trees 18(1):19–28

Wissenschaftlicher Beirat der Bundesregierung Globale Umweltveränderungen (WBGU) (1998) Die Anrechnung biologischer Quellen und Senken im Kyoto-Protokoll: Fortschritt oder Rückschritt für den globalen Umweltschutz. Sondergutachten, Bremerhaven

Zimmermann F, Fiebig J, Wienhaus O (1998) Immissionen und Depositionen. Forst wissensch. Beitr Tharandt 4:39–49

IV Soil Response

19 Environmental Controls on Concentrations and Fluxes of Dissolved Organic Matter in the Forest Floor and in Soil Solution

K. KALBITZ, T. ZUBER, J.-H. PARK, and E. MATZNER

19.1 Introduction

Dissolved organic matter (DOM) in soils plays an important role in many ecosystem processes. For example, DOM affects the transport of nutrients and pollutants and the weathering of soil minerals (Temminghoff et al. 1997; Raulund-Ramussen et al. 1998; Kalbitz et al. 2000). It also has great implications for the below-ground C and N cycle, because the leaching loss of DOM can account for a substantial portion of the annual litterfall C and N (McDowell and Likens 1988; Qualls et al. 1991; Hongve et al. 2000).

Although the release of DOM has been researched extensively, it is still not clear whether DOM originates primarily from recent litter or from relatively stable organic matter in the lower parts of organic horizons (Kalbitz et al. 2000; McDowell 2003). Contrasting views exist also concerning the relative significance of biotic and abiotic controls on DOM release from the forest floor and soils (Kalbitz et al. 2000; Qualls 2000; Neff and Asner 2001). Some field studies have suggested that DOM production in the forest floor is largely controlled by microbial activity, based on higher DOM concentrations observed in warmer seasons (Cronan and Aiken 1985; McDowell and Likens 1988; Dai et al. 1996). This temperature dependency of DOM production was supported in many laboratory experiments (summarized by Andersson et al. 2000; Kalbiz et al. 2000). Also increasing dissolved organic carbon (DOC) concentrations in surface waters of Great Britain were attributed to increasing temperature (Freeman et al. 2001). However, the responses of microbial DOM production mechanisms to changes in the inputs of organic matter (e.g., increased litter inputs following CO_2 enrichment, decreased input of organic matter) are still poorly understood (King et al. 2001; Hagedorn et al. 2002).

Water fluxes are a further key driver of DOM release (Tipping et al. 1999; Kalbitz et al. 2000; Judd and Kling 2002). DOM production and leaching can be increased by increased water fluxes through (1) the positive effect of moisture on microbial activity (Falkengren-Grerup and Tyler 1993), (2) the rewet-

Ecological Studies, Vol. 172
E. Matzner (Ed.), Biogeochemistry of Forested Catchments in a Changing Environment

ting effect after dry periods (Christ and David 1996a; Lundquist et al. 1999) and (3) a desorption of potentially soluble organic materials from the adsorbed phase (Christ and David 1996b).

Changes in soil solution chemistry induced by changes in atmospheric deposition can also affect DOM dynamics (Kalbitz et al. 2000). Increasing pH, decreasing concentrations of sulfate, calcium and aluminum as observed at the coniferous site Coulissenhieb and in many European countries after declining atmospheric deposition (Matzner et al., this Vol., Chap. 14) should change the solubility, sorption and biodegradation of DOM. On the other hand, the already relatively low atmospheric deposition did not largely change at the deciduous site Steinkreuz. Therefore, the comparison of long-term trends in DOC and DON concentrations at both sites could reveal effects of a drastic decline in acidic deposition on DOM dynamics.

Conclusions about controls on DOM dynamics from laboratory experiments were often not supported in the field, highlighting the importance of field experiments to explore the relationship between DOM dynamics and environmental factors (Kalbitz et al. 2000). Clear-cutting of forest stands provides a unique opportunity to study the response of dynamic controls on DOM, as clear-cutting affects almost all factors that influence DOM dynamics (Kalbitz et al. 2004). After clear-cutting, temperature and water fluxes often increased (Londo et al. 1999), whereas the input of organic matter by throughfall and litterfall should decrease. The disadvantage of such an experiment is that no single factor can be evaluated for its impact on DOM dynamics because obtained data will be the result of the combined effect of all changed environmental factors. Therefore, in another approach, a field experiment was conducted to manipulate the input of litter and throughfall in order to elucidate the impact of resource availability and water fluxes on DOM dynamics.

In this chapter, we want to reveal and, if possible, to quantify controlling factors on DOM dynamics, including the disclosure of temporal trends in DOM concentrations. To achieve this goal we used long-term DOM data in throughfall, forest floor and soil solution of a coniferous and a deciduous site, data from the clear-cut study at the coniferous site and from the litter and throughfall manipulation experiment at the deciduous site. From the main problems addressed by this book we highlight (1) the regulation of the dynamics of DOM, (2) the role of DOC for the C turnover in the ecosystem and especially in the soil and (3) the role of dissolved organic nitrogen (DON) for the N turnover in the ecosystem.

19.2 Experimental Design and Sampling

19.2.1 Site Coulissenhieb

The coniferous site Coulissenhieb is described in detail by Gerstberger et al. (this Vol.). There are four experimental plots at this site. The control plots (1 and 4) are situated in the old Norway spruce stand. Plot 3 was cleared in spring 1999 and plot 2 was cleared in spring 2000. Bulk precipitation was collected in an open area near the site whereas throughfall was sampled biweekly using five collectors per plot. Soil solutions were gained monthly as biweekly bulk samples, using ceramic suction cups. Five replicates at 20 and 90 cm soil depth were established at each plot. Sampling and measurements started in 1992. Leachates from the different horizons of the forest floor were collected by tension plate lysimeters using three replicates per plot and horizon (Kalbitz et al. 2004). Measurements in the forest floor started in 1999 using a biweekly sampling period. The experimental clear-cutting was combined with installation of temperature probes in the litter (Oi), fermented (Oe) and humified (Oa) horizons recording temperature at 10-min intervals.

19.2.2 Site Steinkreuz

The deciduous site Steinkreuz is described in detail by Gerstberger et al. (this Vol.). Bulk precipitation was collected in an open area near the site. Throughfall was sampled using 12 collectors which were pooled to form four composite samples in order to reduce the number of analyses. Soil solutions were gained monthly as biweekly bulk samples, using ceramic tension lysimeters with seven replicates at 10 cm depth and seven replicates at 60 cm depth. Sampling and measurements started in autumn 1994. In addition to this setup, litter and throughfall were experimentally manipulated at 16 plots (4 m^2) with four replicates each (Park and Matzner 2003). We established (1) no litter (exclusion of above-ground litter inputs), (2) double litter (doubling aboveground litter inputs), (3) double throughfall (doubling throughfall inputs) and (4) control plots (Park and Matzner 2003). At these plots zero-tension lysimeters were installed beneath the Oi and the Oa horizons. The plots were established in spring 1999 and measurements started in autumn 1999 using biweekly samples. Due to technical difficulties, the reported results for the double throughfall treatment refer only to the periods from April to December 2000 and from April to June 2001.

19.3 Analysis, Calculations and Statistics

Concentrations of DOC were analyzed by persulfate oxidation (Foss Heraeus Liqui TOC) up to 1998. From 1999 all DOC analyses were done by high temperature combustion (Elementar, highTOC). In 1998, all analyses were performed using both devices. The obtained significant regressions (r^2 >0.70) for each stratum at each site between these two methods were used to convert all DOC data measured by the Liqui TOC to the highTOC. Total nitrogen, nitrate and ammonium were measured using standard methods (Kalbitz et al. 2004). DON was calculated as the difference between total N and dissolved inorganic N (DIN: NO_3-N plus NH_4-N).

The annual fluxes in bulk precipitation, throughfall and in forest floor leachates were calculated using water volumes and concentrations of each lysimeter and collector, respectively, on a biweekly basis. Due to technical difficulties, some forest floor leachates were missing. In order to avoid underestimation of annual DOC and DON fluxes and inconsistencies of flux calculation, water fluxes of all forest floor leachates including missing samples were calculated from throughfall fluxes assuming fluxes of 80, 80 and 72 occurring in the Oi, Oe and the Oa horizons of the site Coulissenhieb, respectively (Kalbitz et al. 2004), and of 80 and 75 % occurring in the Oi and Oa horizons of the site Steinkreuz, respectively (Solinger et al. 2001). Fluxes in the mineral soil were calculated using the median of the annual concentration which was calculated including all single measurements of each sampler and the calculated annual water flux by HYDRUS 2D (Matzner et al., this Vol., Chap. 20). Water fluxes could not be successfully modeled in 10-cm depths of the Steinkreuz catchment.

We compared mean concentrations between the sites and depths using mean values of each stratum which were calculated as annual medians. Linear regression analyses were performed to identify temporal trends in DOC and DON concentrations. For these analyses we used the annual medians. Furthermore, we performed these linear regressions also for each sampler (collector for bulk precipitation and throughfall, suction cups). Temporal trend analyses of DOC and DON fluxes were not performed because of the crucial effect of water fluxes which were highly variable between the years. Differences between the control and the various treatments were tested by the nonparametric Kruskal-Wallis rank-test and the Mann-Whitney U-test.

19.4 Long-Term DOC and DON Concentrations and Fluxes at a Coniferous and a Deciduous Site

19.4.1 DOC and DON Concentrations: Patterns and Trends

Mean DOC and DON concentrations increased significantly in the following order: bulk precipitation < throughfall < forest floor and decreased in the upper (10–20 cm) and deeper (60–90 cm) mineral soil relative to the forest floor (Table 19.1). These results are in agreement with most of the studies dealing with DOM dynamics in forested ecosystems as reviewed by Michalzik et al. (2001). The observed patterns of DOC and DON concentrations (Table 19.1) highlight the relatively low DOM concentrations in bulk precipitation which will be enriched with organic matter during the passage of the forest canopy. High DOC and DON concentrations in throughfall should be the result of leaching of living biomass and of biological processes occurring in the canopy, e.g., interactions between aphids and microorganisms in the canopy contribute significantly to DOC concentrations in throughfall during May and June at these two sites (Stadler and Michalzik, this Vol.).

Both sites had similar DOC and DON concentrations of ~2 mg C l^{-1} and 0.15 mg N l^{-1}, respectively, in bulk precipitation, whereas in throughfall DON concentrations were higher at the deciduous site, reaching 0.73 mg N l^{-1}. We did not find long-term temporal trends in the input of organic matter by bulk precipitation and throughfall despite a strong decline in the input of major cations and anions at the coniferous site.

Table 19.1. DOC and DON concentrations in bulk precipitation, throughfall, forest floor and in the mineral soil at sites Coulissenhieb and Steinkreuz (means and standard deviations of annual medians from 1996 to 2001)

Stratum	Coulissenhieb DOC (mg C l^{-1})	Steinkreuz DOC (mg C l^{-1})	Coulissenhieb DON (mg N l^{-1})	Steinkreuz DON (mg N l^{-1})
Bulk precipitation	1.7± 0.5	2.0± 0.8	0.14±0.06	0.18±0.07
Throughfall	12.2± 1.2	13.3± 2.0	0.48±0.13	0.73±0.14
Oi horizon[a]	46.4±16.1	27.4± 5.3	1.43±0.51	1.02±0.19
Oe horizon[a]	39.5±10.7		1.15±0.28	
Oa horizon[a]	53.4±18.9	50.2±15.3	1.34±0.37	1.64±0.59
10/20 cm[b]	25.2± 6.0	17.6± 2.7	0.36±0.15	0.44±0.14
60/90 cm[c]	4.3± 0.4	5.1± 1.2	0.14±0.03	0.18±0.04

[a] Means and standard deviations of annual medians (1999–2001) of each lysimeter
[b] Sampling depth: Coulissenhieb: 20 cm, Steinkreuz: 10 cm
[c] Sampling depth: Coulissenhieb: 90 cm, Steinkreuz: 60 cm

The main DOM source was the forest floor with in general largest DOC and DON concentrations in the Oa horizon (Table 19.1). Only at the coniferous site did largest DON concentrations occur in the Oi horizon with an average annual mean of 1.43 mg N l^{-1}. Spruce needles seem to be a stronger source for DOC and DON than beech and oak leaves as illustrated by higher concentrations in the Oi horizon of the coniferous site than at the deciduous site. However, beneath the Oa horizons, DOC and DON concentrations were similar in both forests. Reasons for the differences in the litter horizon can only be partially explained. In the laboratory, a higher release of soluble organic matter from leaves was reported (Kuiters 1993), whereas soil solutions from mixed and coniferous stands often contain significantly more DOC and DON than those from hardwood stands (summarized by Kalbitz et al. 2000). Magill and Aber (2000) found high and low DOC and DON release from needles and leaves, respectively. Therefore, a general conclusion that coniferous forest floors have larger DOM concentrations than deciduous ones does not seem to be justified.

In the mineral soil DOC and DON concentrations declined with depth probably as a result of sorption onto Fe and Al oxides/hydroxides and microbial decay (Kalbitz et al. 2000). In the upper mineral soil this decrease is less pronounced at the coniferous site because soil solution at this depth (20 cm) is affected by the organic carbon-rich Bh and Bs horizons. These relatively high DOC concentrations in 20 cm of the coniferous site (25.2 mg C l^{-1}) demonstrate the DOM source function of soil organic matter and the ongoing podsolization at this site. Furthermore, the highest amounts of pedogenic oxides which are known to be the most effective sorbents of DOM (Kaiser and Zech 2000) are stored in the upper 20 cm at the deciduous site, whereas these main sorbents are concentrated somewhat deeper at the coniferous site.

In the deep mineral soil, DOC and DON concentrations declined to about 10 % of the values in the Oa horizon with slightly higher values at the deciduous site. Obviously, a 30-cm-deeper percolation at the coniferous site in comparison to the deciduous one only resulted in a minor additional DOM retention. At both sites, a sufficient number of sorption sites are available down to a depth of 60 cm to retain all carbon and nitrogen that can be retained.

The annual medians of DOC concentrations in the upper mineral soil of the coniferous site decreased significantly over the last 10 years from 35 mg C l^{-1} in 1993 to 24 mg C l^{-1} in 2001 (r=–0.76; Fig. 19.1). This trend was also significant for 8 of the 20 suction cups. The annual medians and DON concentrations in most of the suction cups have not changed significantly over the last 10 years.

In the upper mineral soil of the deciduous site, the annual medians of DOC concentrations did not show any trend (Fig. 19.1), whereas DOC concentrations increased significantly in four of the seven suction cups from 1995–2001 with significant correlation coefficients up to r=0.73 (p <0.0001). Annual medians of DON concentrations of the deciduous site increased significantly

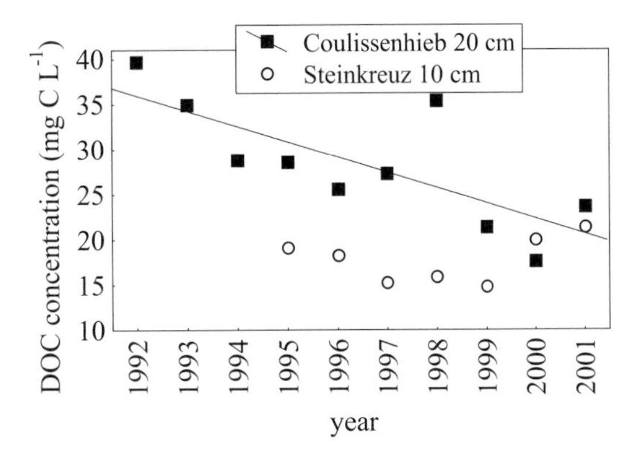

Fig. 19.1. Temporal trends of DOC concentrations (annual means) in the upper mineral soil of sites Coulissenhieb and Steinkreuz

from 1995 (0.36 mg N l^{-1}) to 2001 (0.57 mg N l^{-1}) (r=0.80). This trend could not be seen in the evaluation of each suction cup.

In summary, DOC concentrations decreased over the last 10 years in the upper mineral soil of the coniferous site whereas DOC and DON concentrations tended to increase at the deciduous site.

The opposite long-term dynamics of DOC concentrations between the coniferous and the deciduous site was even more evident in the deep mineral soil. Here, DOC concentrations decreased from 1992 (5.4 mg C l^{-1}) to 2001 (4.0 mg C l^{-1}) at the coniferous site. This trend was significant for all of the 20 suction cups and for the annual medians (r=–0.72) (Fig. 19.2) with no turnaround after 1998 as at 20 cm depth. In turn, annual medians of DOC concentrations increased in deep mineral soil of the deciduous site from 5.4 mg C l^{-1} in 1995 to 6.3 mg C l^{-1} in 2001 (Fig. 19.2). This trend was significant for four of the seven suction cups. DON concentrations did not change significantly during the study period at both sites.

Different changes in atmospheric deposition at our two sites should be responsible for the observed opposite temporal trends. At the deciduous site the already low deposition decreased only slightly in the last 10 years. In contrast, the coniferous site received more than 60 kg S ha^{-1} $year^{-1}$ in the late 1980s, which has declined to about 12 kg S ha^{-1} $year^{-1}$ nowadays (Matzner et al., this Vol., Chap. 14). Also the input of base cations decreased strongly at the coniferous site. These changes in atmospheric deposition were reflected in the soil with decreasing concentrations of calcium, aluminum and sulfur (Matzner et al., this Vol., Chap. 20). Moreover, the pH increased slightly in the soil solution. After these changes one might expect increasing DOC concentrations because of increasing solubility of organic matter (Kalbitz et al. 2000). However, we found an opposite trend which can only be explained by sulfur and aluminum effects. Sulfate and carbon compete for adsorption sites in the soil (Nodvin et al. 1986; Vance and David 1991; Kaiser and Zech 1996). The decreasing sulfate concentrations in the soil solution of the coniferous site could result in a larger sorption of DOC in the mineral soil resulting in decreasing DOC concentra-

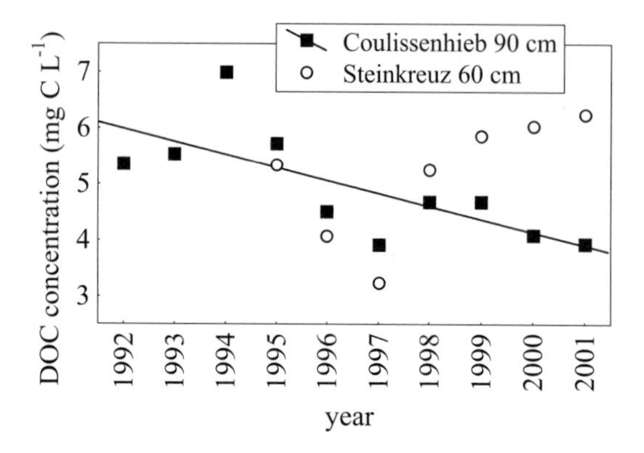

Fig. 19.2. Temporal trends of DOC concentrations (annual means) in the deeper mineral soil of sites Coulissenhieb and Steinkreuz

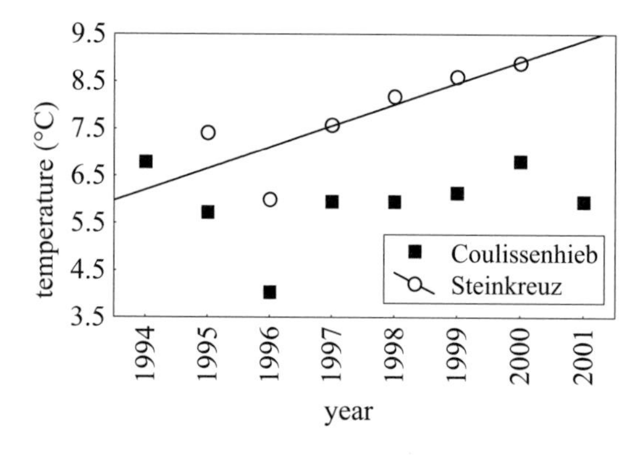

Fig. 19.3. Temporal trends of mean annual air temperature at sites Coulissenhieb and Steinkreuz

tions. Furthermore, the site Coulissenhieb lost 170 kg S in the last 9 years (Matzner et al., this Vol., Chap. 20) probably stored as sorbed sulfate, with the consequence of an increasing number of available sorption sites for DOM. However, our hypothesis is not supported by the findings of Kaiser and Zech (1997, 1998) who concluded that sulfate displaces DOC from adsorption sites only when present in concentrations >10 mmol l^{-1}, which never occurred at this site. A second explanation for decreasing DOC concentrations after decreasing atmospheric deposition might be an enhanced mineralization of DOM caused by decreasing Al concentrations in soil solution (Boudot et al. 1989; Schwesig et al. 2003).

Increasing DOC concentrations in the mineral soil of the deciduous site supported trends reported for British and Scandinavian surface waters (Skjelevale et al. 2000; Evans and Monteith 2001). These increasing DOC concentrations were attributed to be an effect of global warming (Freeman et al. 2001). Increasing mean annual air temperature at the deciduous site (Fig. 19.3; 1995–2000, r=0.81, p <0.05) seems to support this hypothesis whereas no such trend in temperature could be observed at the coniferous site.

19.4.2 Contribution of DOC and DON to C and N Fluxes in the Ecosystem

At both sites most of the C and N input to the soil was provided by litterfall. Another significant input of organic C and N to the forest floor was provided by throughfall. There were no differences in the input of DOC and DON by bulk precipitation and throughfall between the two sites. At the deciduous site Steinkreuz, DON fluxes in throughfall (~5 kg N ha^{-1} $year^{-1}$) were in the same range as those in forest floor leachates, whereas DOC fluxes were much larger in the forest floor than in throughfall (Table 19.2). This means that the forest canopy is a stronger source of DON than of DOC in comparison to the forest floor at the deciduous site Steinkreuz. Similar observations for deciduous sites were reported by Michalzik et al. (2001) and Solinger et al. (2001). Immobilization of DON in freshly fallen leaves might be one reason for this phenomenon (see below).

Larger water fluxes at the coniferous than at the deciduous site resulted also in larger DOC and DON fluxes in the forest floor (Table 19.2). DOC and DON fluxes in the Oa horizon were 16 % of C and N fluxes with above-ground litterfall at the coniferous site Coulissenhieb, whereas these percentages were 8 % for DOC and 11 % for DON at the deciduous site Steinkreuz. Past studies have shown that DOM release from forest floors ranges from 3–35 % of annual litterfall C and N (Yavitt and Fahey 1986; McDowell and Likens 1988; Qualls et al. 1991; Vance and David 1991; Hongve et al. 2000; Wagai and Sollins 2002). However, we have no information about litter input by roots at our sites. In a 40-year-old Norway spruce stand in the German Fichtelgebirge root litter contributes 46 % to root + needle litter (Godbold et al. 2003).

Table 19.2. DOC and DON fluxes in bulk precipitation, throughfall, forest floor and in the mineral soil at sites Coulissenhieb and Steinkreuz (means and standard deviations of annual fluxes from 1996 to 2001). *n.d.* Not determined because warter modeling was not successful

Stratum	Coulissenhieb DOC	Steinkreuz DOC	Coulissenhieb DON	Steinkreuz DON
	(kg C ha^{-1} $year^{-1}$)		(kg N ha^{-1} $year^{-1}$)	
Bulk precipitation	21± 6	23±10	1.6±0.2	1.7±0.7
Throughfall	109±19	92±25	4.1±0.9	4.8±0.8
Oi horizon	258±74	146±25	8.4±3.0	5.1±0.7
Oe horizon	234±69		6.7±2.3	
Oa horizon	282±81	203±45	7.4±2.6	6.7±1.6
10/20 cm[a]	161±71	n.d.	2.3±1.4	n.d.
60/90 cm[b]	22± 8	16± 7	0.7±0.3	0.5±0.2

[a] Sampling depth: Coulissenhieb: 20 cm, Steinkreuz: 10 cm
[b] Sampling depth: Coulissenhieb: 90 cm, Steinkreuz: 60 cm

The Oi and Oa horizons represented a net source of DOC and DON at both sites. At the coniferous site the Oi horizon was a larger net source of DOC and DON than the Oa horizon. In contrast, at the deciduous site, Oi and Oa horizons were similar net sources of DOC; however, the Oi horizon acted only as a minor net source of DON (Tables 19.3, 19.4). This small net release of DON from leaf litter could be attributed to DON immobilization during initial stages of decomposition (Aber and Melillo 1982; Staaf and Berg 1982; Berg and Cortina 1995). Higher deposition of inorganic N at the coniferous site Coulissenhieb in comparison to the deciduous site Steinkreuz (Matzner et al., this Vol., Chap. 14) might explain why such immobilization of DON did not occur at the site Coulissenhieb. Our explanation is supported by increasing DON concentrations with increasing addition of inorganic N (Currie et al. 1996; McDowell et al. 1998; Neff et al. 2000).

Although the highest DOC fluxes were measured beneath the Oa horizon of both sites, the largest net release occurred in the litter horizon of the coniferous site. This strong DOM source of litter becomes even more evident when taking into account the much smaller C and N stocks in the Oi than in the Oa horizon. Although a much higher proportion of forest floor C and N is stored in the Oi horizon of the deciduous site Steinkreuz (15 % in comparison to 6 % at the coniferous site), the Oi horizon of the coniferous site Coulissenhieb is a larger net DOM source than the Oi horizon of the deciduous site Steinkreuz. In turn, DOC and DON fluxes beneath the entire forest floor were only slightly higher at the coniferous than at the deciduous site despite much higher C and N stocks in the forest floor of the coniferous site. These results clearly demonstrate that C and N stocks cannot be used as an indicator of DOM fluxes in soils (Kalbitz et al. 2000; Michalzik et al. 2001). One and 0.5 % of the stored C and N in the Oa horizon, respectively, annually entered the mineral soil as DOC and DON at the coniferous site, whereas these percentages were almost twice as high at the deciduous site.

The relatively small net release of DOM from the Oa horizon does not mean that these large amounts of decomposed and humified organic material in the Oa horizon had no effect on DOC and DON release. It is more likely that a high proportion of DOM sampled underneath the Oa horizon also originated from this part of the forest floor as indicated by changing DOM properties (Kalbitz et al. 2004). The increases in specific UV absorbance, humification indices deduced from fluorescence spectra, and portions of the hydrophobic DOM fraction in the forest floor with depth (Kalbitz et al. 2004) indicated an enrichment of aromatic and complex structures and a selective decomposition of carbohydrates. Therefore, the largest DOC fluxes measured in the Oa horizon and increased aromaticity and complexity of DOM with increasing soil depth did suggest that DOM was mainly derived from decomposition, transformation and leaching of more decomposed organic matter.

DOC and DON fluxes decreased with depth in the mineral soil, resulting in much smaller fluxes in the deep mineral soil than in throughfall. The output

Table 19.3. Fluxes (kg ha[-1] year[-1]) and stocks (kg ha[-1]) of organic carbon and nitrogen at the site Coulissenhieb

	C stock	DOC flux	N stock	DON flux
Input fluxes				
Bulk precipitation		21		1.6
Throughfall		109		4.1
Litter	1,620[a]		41[a]	
Forest floor/soil				
Oi horizon	3,560	258	160	8.4
Net release of DOC/DON[b]		149		4.3
Oe horizon	25,410	234	1,260	6.7
Net release of DOC/DON[b]		−24		−1.7
Oa horizon	31,050	282	1,490	7.4
Net release of DOC/DON[b]		48		0.7
Mineral soil	110,040	22	6,720	0.7
Net release of DOC/DON[b,c] (retained amount in the mineral soil)		−260		−6.7

[a] C and N fluxes with litter: kg ha[-1] year[-1]
[b] Input flux from the overlaying horizon (stratum) was subtracted; negative values mean net retention
[c] Retained amount of C in the mineral soil is not equal to C storage. Mineralization has to be considered to estimate contribution of DOC to long-term C storage

Table 19.4. Fluxes (kg ha[-1] year[-1]) and stocks (kg ha[-1]) of organic carbon and nitrogen at the site Steinkreuz

	C stock	DOC flux	N stock	DON flux
Input fluxes				
Bulk precipitation		23		1.7
Throughfall		92		4.8
Litter	2,640[a]		65[a]	
Forest floor/soil				
Oi horizon	1,970	146	70	5.1
Net release of DOC/DON[b]		54		0.3
Oa horizon	12,820	203	730	6.7
Net release of DOC/DON[b]		57		1.6
Mineral soil	78,490	16	5,420	0.5
Net release of DOC/DON[b,c] (retained amount in the mineral soil)		−187		−6.2

[a] C and N fluxes with litter: kg ha[-1] year[-1]
[b] Input flux from the overlaying horizon (stratum) was subtracted; negative values mean net retention
[c] Retained amount of C in the mineral soil is not equal to C storage. Mineralization has to be considered to estimate contribution of DOC to long-term C storage

of DOC from the soil was in the same range as the input with bulk precipitation whereas the DON output from the soil was even lower than the input by bulk precipitation. That means all of the internal DOM fluxes provided by throughfall and the forest floor are retained in the mineral soil where it can be stored or mineralized. For atmospheric DON, the forests behaved as a strong sink. Moreover, DON plays only a minor role in N export from soils in N-saturated forests although large internal organic N fluxes occurred in the forest floor and the upper mineral soil. This is in contrast to pristine sites with low inputs of inorganic N where DON plays the major role in N export from soils (Hedin et al. 1995; Perakis and Hedin 2002).

The amount of retained organic C and N in the mineral soil was slightly higher at the coniferous site Coulissenhieb (e.g. 260 kg C ha^{-1} $year^{-1}$) than at the deciduous site Steinkreuz (e.g. 187 kg C ha^{-1} $year^{-1}$) (Tables 19.3, 19.4). This retained organic matter should be mainly adsorbed on pedogenic oxides and can contribute to the pool of stable organic matter in the mineral subsoil (Kaiser and Guggenberger 2000; Guggenberger and Kaiser 2003). It seems that the contribution of DOM to the stable pool of organic matter is rather small taking into account the stocks of organic matter in the mineral soil of 80–110 Mg ha^{-1}. However, within centuries and millenniums large proportions of organic matter in the mineral soil could be originated from DOM because this organic matter is up to several thousand years old (Kaiser et al. 2002) and Kaiser and Guggenberger (2000) revealed many similarities in organic matter composition comparing forest floor DOM and soil organic matter in the mineral subsoil. Furthermore, Kalbitz et al. (2003) showed that DOM originated from Oa horizons of forest floors could be very stable against microbial decay.

19.5 Environmental Controls on DOM Dynamics

In the previous section we showed that the forest floor is the main source of DOM at both sites. In this section, we want to evaluate the effects of the main controlling factors on DOM dynamics in the forest floor using the results of two field studies. At the coniferous site Coulissenhieb we investigated the effects of decreased inputs of organic matter by throughfall and litterfall after clear-cutting of a Norway spruce stand. Furthermore, effects of temperature and water fluxes were also addressed in this study because after clear-cutting increased temperatures and water fluxes can be expected. Double and No litter treatments of the manipulation experiment at the deciduous site Steinkreuz were used to address both effects of increasing and decreasing inputs of organic matter. The double throughfall treatment was used to evaluate the response of DOC and DON concentrations and fluxes.

19.5.1 Effects of Changing the Inputs of Organic Matter on the Dynamics of DOM

In this section, we want to test the hypothesis that decreasing inputs of organic matter will result in decreasing concentrations and fluxes of DOM in the forest floor, whereas increasing inputs should result in the opposite. Changing DOM concentrations and fluxes could be expected especially in the Oi horizon because of low storage of organic matter (Tables 19.3, 19.4). Large amounts of organic matter stored in the Oe and Oa horizons could mask effects of changed organic matter inputs on DOM.

At the coniferous site, needle litterfall C and N input were 1,600 kg C and 41 kg N ha^{-1} $year^{-1}$ before clear-cutting and can be assumed to be negligible after that (Berg and Gerstberger, this Vol.). However, input of litter by the upcoming ground vegetation cannot be excluded during our study and this litter input can be expected to increase in the next years following the clear-cutting. The input of organic matter by throughfall decreased from 107 kg DOC ha^{-1} $year^{-1}$ and 3.8 kg DON ha^{-1} $year^{-1}$ at the control plots to 25 kg DOC ha^{-1} $year^{-1}$ and 1.8 kg DON ha^{-1} $year^{-1}$ at the clear-cut plots (mean annual fluxes of 1999–2001) (Kalbitz et al. 2004).

After decreased input of organic matter, DOC and DON concentrations were significantly lower in the Oi horizon at the clear-cut plots during most seasons (Fig. 19.4). The combined effect of decreased concentrations and increased water fluxes (Kalbitz et al. 2004) resulted in unchanged fluxes of DOC and DON in the Oi horizon after clear-cutting (Fig. 19.4). It was surprising that the cessation of the litter input and the strong decline in throughfall DOM inputs (by 82 kg C and 2.0 kg N ha^{-1} $year^{-1}$) had no significant effect on the production of DOM in the Oi horizon. However, the concentrations of DOC and DON in the Oi horizon did decrease slightly. Therefore, resources limitation, as observed for soil respiration (Foster et al. 1980; McGill et al. 1986; Zak et al. 1994), affected DOC and DON production only slightly in the forest floor of the coniferous site.

Increasing DOM concentrations and fluxes after clear-cutting in the Oe and Oa horizons should not be seen as a result of decreasing input of organic matter but as the consequence of increasing temperature in the forest floor. These effects will be discussed in the next section dealing with temperature effects on DOM dynamics.

At the No litter plots of the deciduous site Steinkreuz, only small amounts of litter (approximately 13 % of those of the control) reached the forest floor during winter months, during which litter traps were not installed due to snowfall. Doubling of litter inputs increased the total amount of aboveground litter inputs by 76 % (Park and Matzner 2003). At the No litter plots, no distinct Oi litter was found in May 2001. At the Double Litter plots, total amounts of C and N in the Oi horizon were twice those of the control. No significant change in the C and N stocks of the Oe/Oa horizons was observed

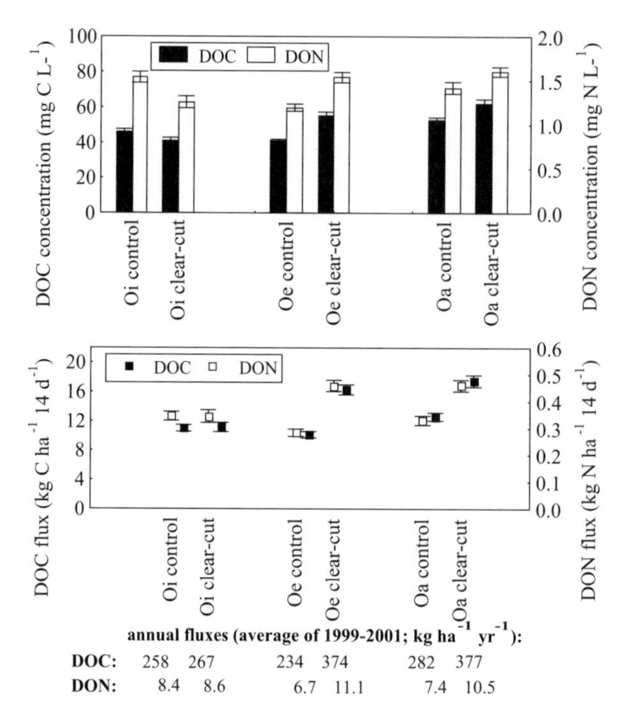

Fig. 19.4. Concentrations and fluxes of DOC and DON in the different horizons of forest floor at the coniferous site Coulissenhieb as affected by clear-cutting (means and standard errors)

among treatments. However, the C/N ratios of the whole forest floor significantly increased at the Double litter plots, reflecting increased recent litter as a consequence of litter addition.

Annual mean concentrations of DOC and DON significantly increased in the Oi and Oa horizons of the Double Litter plots compared with those of the control and No Litter plots (Table 19.5). The addition of above-ground litter inputs increased DOC release from the forest floor by 83 %, suggesting a sensitive response of forest floor DOM dynamics to an increase (76 %) in above-ground litter inputs (Park and Matzner 2003).

Besides concentrations and gross fluxes also the net releases of DOC and DON from the Oi horizons were significantly greater at the Double litter plots than at the control or No litter plots. The net release from the Oe/Oa horizons also tended to be greater at the Double litter plots.

The similar net DOM release observed for the Oi and Oe/Oa horizons at the control plots (Table 19.4), along with the positive response of DOM release at both horizons to litter addition (Table 19.5), suggest an equal importance of different forest floor horizons as a DOM source. In a laboratory leaching experiment with forest floor horizons collected from the study site (Park et al. 2002), each of three horizons was evaluated as making an almost equal contribution to the total release of DOC and DON. This is in conflict with some past studies (Qualls et al. 1991; Michalzik and Matzner

1999) that found no substantial contribution of the Oe/Oa horizons to DOC and DON fluxes.

Over the entire manipulation period, no marked changes in DOC and DON concentrations and fluxes were observed at the No Litter plots, with the only exception of decreased DOC concentrations and fluxes in Oi leachates (Table 19.5). The Oa leachates of the No Litter plots showed no significant change in concentrations and fluxes of DOC and DON throughout the observation period. However, the reduction in DOC release from the Oi horizons resulted in increased net releases of both DOC and DON from the Oe/Oa horizons after litterfall exclusion.

In order to examine the relationship between variations in DOM releases from the Oe/Oa horizons and changes in microbial activity following litter manipulations, rates of basal respiration and substrate-induced respiration (SIR) were measured for the samples collected in May 2001 (Park and Matzner 2003). Both respiration measurements showed only slightly higher values at the Double litter plots in comparison to the control, although DOM concentrations and fluxes increased remarkably (Park and Matzner 2003). It seems that the amount of available and water-soluble substrate is more important than the microbial activity for increased DOM release after doubling the litter input.

We found almost equal respiration rates at the control and No litter plots (Park and Matzner 2003). Since SIR can be used as a relative measure of metabolically active microbial biomass (Wardle and Parkinson 1990; Ross 1991; Wardle and Ghani 1995), the similar rates of SIR for the Oe/Oa samples from

Table 19.5. Concentrations and fluxes of DOC and DON in the forest floor as affected by experimentally manipulated above-ground litter and throughfall input at the deciduous site Steinkreuz

	DOC concentration[a] (mg C l^{-1})	DON concentration[a] (mg N l^{-1})	DOC flux[b] (kg C ha^{-1} $year^{-1}$)	DON flux[b] (kg N ha^{-1} $year^{-1}$)
Oi horizon				
Control	27.4± 5.3	1.02±0.19	146± 25	5.1±0.7
No Litter	20.6± 7.8	1.17±0.66	85± 14	4.5±0.9
Double Litter	48.3±13.4	1.65±0.30	246± 34	7.8±0.8
Double Throughfall	28.9± 7.9	1.15±0.34	183± 14	7.1±0.4
Oe/Oa horizon				
Control	50.2±15.3	1.64±0.59	203± 45	6.7±1.6
No Litter	55.7±30.7	2.04±1.15	204± 75	7.5±3.3
Double Litter	86.6±24.6	2.68±0.64	393± 94	12.0±2.6
Double Throughfall	55.9±19.8	1.98±0.61	347±100	12.0±1.0

[a] Means and standard deviations of annual medians of each lysimeter (1999–2001)
[b] Means and standard deviations of annual fluxes of each lysimeter (2000–2001)

the control and No Litter plots suggest that old forest floor horizons can provide resources for the subsistence of metabolically active microbes at least for 2 years, despite a substantial reduction in above-ground litter inputs. However, no significant changes in microbial activity following litterfall exclusion might indicate that the sustained or even slightly increased release of DOC and DON from the Oe/Oa horizons cannot be simply explained by enhanced microbial decomposition of old organic horizon materials. The higher contribution of more decomposed organic matter to DOM in the Oi and Oa horizon after litterfall exclusion should result in reduced microbial consumption of produced DOM because biodegradability of DOM decreased with increasing degree of decomposition of the material where DOM comes from (Kalbitz et al. 2003). This reduced DOM biodegradation should also contribute to the high DOM release after litterfall exclusion.

In summary, we can confirm our hypothesis that increasing inputs of organic matter will also result in increasing concentrations and fluxes of DOC and DON in all horizons of the forest floor. The combined effect of the increased amounts of microbial available and water-soluble substrate with increased microbial activity seems to be responsible for increasing DOM release.

We can also confirm our hypothesis that decreased input of easily decomposable organic matter will result in decreasing DOM concentrations and fluxes in the Oi horizon. Considering DOM fluxes in the entire forest floor, decreased input of organic matter will not result in decreasing DOM fluxes into the mineral soil in the first years. Large amounts of organic matter stored in the more decomposed parts of the forest floor and a higher microbial stability of released DOM compensate for the declined input of organic matter by litterfall. Therefore, the amount and composition of DOM percolating into the mineral soil should mainly depend on the large amounts of more decomposed organic matter in the Oe/Oa horizons.

19.5.2 Effects of Temperature on the Dynamics of DOM

We hypothesized a positive relationship between temperature and the production of DOM in the forest floor because of the dependence of DOM production from microbial activity. Clear-cutting of parts of the Norway spruce stand at Coulissenhieb increased mean annual temperatures by 1.5 °C in all horizons of the forest floor which should result in increasing soil microbial activity (Johnson et al. 1995; Londo et al. 1999).

Clear-cutting resulted in significantly increasing DOM concentrations and fluxes in the Oe and Oa horizons (Fig. 19.4). Because of increasing concentrations and water fluxes, clear-cutting led to higher fluxes while concentrations changed more moderately (Fig. 19.4). On average, DOC and DON concentrations were 36 and 28 % higher after clear-cutting and the respective average fluxes increased by 60 and 66 % in the Oe horizon. The increases in concentrations and fluxes were somewhat lower in the Oa horizon. DOC and DON concentrations were 18 and 13 % higher after clear-cutting, whereas the respective average fluxes were 34 and 42 % higher in the Oa horizon.

DOC and DON concentrations tended to be higher in the upper mineral soil (20 cm depth) at the clear-cut plots after increasing temperatures. Although the mean concentrations were significantly higher at the clear-cut than at the control plots, only in two suction cups could a significant trend of increasing DOC concentrations be proven after clear-cutting. In 90 cm depth, DOC and DON concentrations were not affected by increasing temperature after clear-cutting.

Increasing specific UV absorbance, increasing humification indices deduced from fluorescence spectra and decreasing $\delta^{13}C$ values of DOM indicated an increased contribution of aromatic and complex lignin-degradation products to forest floor DOM after clear-cutting (Kalbitz et al. 2004). Therefore, elevated microbial processing of more strongly decomposed organic matter seems to be the main reason for increasing DOM concentrations and fluxes in the forest floor after increasing temperature (Kalbitz et al. 2004). Also Dai et al. (2001) concluded that aromatic compounds contributed stronger to DOM after clear-cutting. These changes in DOM composition after increased temperature also indicated a larger microbial stability of DOM (Kalbitz et al. 2003).

The largest increase in DOC and DON concentrations and alterations of spectroscopic properties after clear-cutting were found in the Oe horizon (Kalbitz et al. 2004). The increase in DOC and DON concentrations by 30 % due to increased temperature by 1.5 °C shows that Q_{10} values for DOC may be higher than 2 and thus can exceed the values deduced from laboratory studies (Gödde et al. 1996; Freeman et al. 2001; Neff and Hooper 2002). We conclude that the Oe horizon is characterized by large amounts of organic matter with sufficiently high substrate value for microorganisms. This combination resulted in the largest impact of an increased microbial activity on DOM following clear-cutting. In contrast to the Oe horizon, the stock of organic matter and therefore the amount of potential DOM (Christ and David 1996b) in the Oi horizon is too small to maintain high DOC and DON concentrations after the clear-cutting for a longer period.

In the Oa horizon, DOC and DON concentrations and the values obtained from the spectroscopic measurements increased after clear-cutting to a smaller extent than in the Oe horizon (Kalbitz et al. 2004). The substrate value of organic matter in the Oa horizon should be lower than in the less decomposed Oe horizon. That might be the reason for smaller changes in DOM concentrations, fluxes and composition due to enhanced microbial activity in the Oa horizon in comparison to the Oe horizon.

We can confirm our hypothesis that an increased temperature after clear-cutting results in increasing DOC and DON concentrations and changed

DOM composition caused by increased microbial processing of more decomposed organic matter. Besides, the positive relationship between temperature and the annual medians of DOC concentration in the mineral soil of the deciduous site Steinkreuz is a further proof that increasing temperature should result in increasing DOM concentrations in soils.

19.5.3 Effects of Water Fluxes on the Dynamics of DOM

Increased water fluxes should result in increased DOM fluxes caused by enhanced production and leaching of DOM (see Introduction).

After clear-cutting, water fluxes with throughfall and seepage were on average 23 % higher because of missing interception and transpiration by the spruce trees (Kalbitz et al. 2004). Unchanged DOC and DON fluxes in the Oi horizon despite decreasing concentrations and increasing fluxes in the Oe and Oa horizons were a direct consequence of these elevated water fluxes. At the control plots, we did not find any relationship between water flux and DOC and DON concentrations. That means increased water fluxes will result in almost equivalently increased DOM fluxes (Michalzik and Matzner 1999), indicating that the pool of potential DOM (Christ and David 1996b) is not easily exhausted at high water fluxes. After clear-cutting, a weak negative correlation between concentrations and water fluxes existed, especially in the Oe horizon (DOC: r=–0.36, DON: r=–0.27; p <0.001) but also in the Oa horizon (DOC: r=–0.30, DON: r=–0.25; p <0.001). These correlations could indicate a kinetically restricted DOM leaching (Tipping 1998): The increase in the microbially mediated production of DOM following clear-cutting is not large enough in relation to the elevated water fluxes. This hypothesis is supported by the fact that the inverse correlations between DOC and DON concentrations with water fluxes did not occur in spring and summer, periods with a high microbial activity.

By experimentally increasing throughfall fluxes at the deciduous site Steinkreuz both DOC and DON concentrations decreased slightly in Oi leachates relative to those of the control, while no reduction in concentration was found for Oa leachates (Table 19.5; Park and Matzner 2003). Annual DOC and DON fluxes increased by 25 and 39 %, respectively, in the Oi horizon, and by 71 and 79 % beneath the Oe/Oa horizon (Table 19.5). These values were much higher than expected because double throughfall could only be applied from April to December 2000 and from April to June 2001. These findings support our assumption that the effect of increased water fluxes is not only based on simply enhanced leaching; also positive effects of additional water on microbial activity should be responsible for increased DOM fluxes after throughfall manipulation.

Net releases of DOC from each of the two forest floor horizons showed that additional amounts of DOC released by doubled throughfall fluxes were primarily derived from the Oe/Oa horizons (Park and Matzner 2003). In contrast to slight reductions in the net release from the Oi horizon (besides water, input of DOM with throughfall was also doubled), the Oe/Oa horizons released substantial amounts of additional DOC in response to increased water fluxes throughout the manipulation period. Considerable increases in net DOC release from the Oe/Oa horizons in response to increased water fluxes suggest that the majority of potential DOM exists in the lower forest floor horizons at this site. Our results also showed that the size of potential DOM pools in the upper forest floor (including fresh litter) is relatively small compared to that of the lower horizons.

We can confirm our hypothesis that increased water fluxes result in increased DOM fluxes. Whereas at the coniferous site Coulissenhieb a slight dilution effect was observed at high water fluxes, increased addition of throughfall resulted in increased DOM fluxes at the deciduous site Steinkreuz which were even higher than expected. Increased microbial activity due to increased moisture, increased addition of easily degradable organic matter by throughfall and increased leaching of potential soluble organic matter might be responsible for increased DOM fluxes.

19.6 Summary and Conclusions

The main DOM source at both sites was the forest floor. Comparing different sites, the amounts of organic matter stored in the forest floor and in the mineral soil cannot be used to predict DOM concentrations and fluxes. Reduced inputs of organic matter by litterfall and throughfall affected DOC and DON concentrations and fluxes only in the Oi horizon with small organic matter stocks. In deeper horizons of the forest floor (Oa) large stocks of organic matter prevent the decrease of DOM fluxes after reduced inputs of organic matter. Addition of litter resulted in increasing DOM fluxes in the forest floor. Increased temperature and water fluxes raised DOM fluxes in the forest floor, with the largest effects in deeper forest floor horizons. Accelerated microbial processing of organic matter should be responsible for increasing DOM concentrations after temperature rise. DOM entering the mineral soil is mainly derived from the large amounts of organic matter stored in the decomposed parts of the forest floor.

Predicting effects of the changing environment on DOM dynamics is rather difficult:

- Global warming might increase the role of DOM in the global C cycle because of increasing DOM fluxes and changes in DOM composition towards a larger microbial stability.
- Declining rates of acidic deposition result in decreasing DOC concentrations in the mineral soil. The release of stored sulfate from soil minerals

provides additional binding places for DOC. Furthermore, decreasing Al concentrations tend to increase the microbial decay of DOC.

- The net release of DON from the forest floor increases with increasing deposition of inorganic N which seems to decrease the microbial immobilization of DON in the forest floor.

Acknowledgements. We thank the members of the Central Analytical Department of BITÖK for support. This study was funded by the German Ministry of Education and Research (BMBF) under grant no. PT BEO 51-0339476 and the German Academic Exchange Service (DAAD).

References

Aber JD, Mellillo JM (1982) Nitrogen immobilization in decaying hardwood leaf litter as a function of initial nitrogen and lignin content. Can J Bot 60:2263–2269

Andersson S, Nilsson SI, Saetre P (2000) Leaching of dissolved organic carbon (DOC) and dissolved organic nitrogen (DON) in mor humus as affected by temperature and pH. Soil Biol Biochem 32:1–10

Berg B, Cortina J (1995) Nutrient dynamics in some decomposing leaf and needle litter types in a *Pinus sylvestris* forest. Scand J For Res 10:1–11

Boudot JP, Bel Hadj Brahim A, Steiman R, Seigle-Murandi F (1989) Biodegradation of synthetic organo-metallic complexes of iron and aluminium with selected metal to carbon ratios. Soil Biol Biochem 21:961–966

Christ MJ, David MB (1996a) Temperature and moisture effects on the production of dissolved organic carbon in a Spodosol. Soil Biol Biochem 28:1191–1199

Christ MJ, David MB (1996b) Dynamics of extractable organic carbon in Spodosol forest floors. Soil Biol Biochem 28:1171–1179

Cronan CS, Aiken GR (1985) Chemistry and transport of soluble humic substances in forested catchments of the Adirondack Park, New York. Geochim Cosmochim Acta 49:1697–1705

Currie WS, Aber JD, McDowell WH, Boone RD, Magill AH (1996) Vertical transport of dissolved organic C and N under long-term N amendments in pine and hardwood forests. Biogeochem 35:471–505

Dai KH, David MB, Vance GF (1996) Characterization of solid and dissolved carbon in a spruce-fir Spodosol. Biogeochem 35:339–365

Dai KH, Johnson CE, Driscoll CT (2001) Organic matter chemistry and dynamics in clear-cut and unmanaged hardwood forest ecosystems. Biogeochem 54:51–83

Evans CD, Monteith DT (2001) Chemical trends at lakes and streams in the UK Acid Waters Monitoring Network, 1988–2000: evidence for recent recovery at a national scale. Hydrol Earth Syst Sci 5:351–366

Falkengren-Grerup U, Tyler G (1993) The importance of soil acidity moisture exchangeable cation pools and organic matter solubility to the cationic composition of beech forest (*Fagus sylvatica* L) soil solution. Z Pflanzenernaehr Bodenkd 156:365–370

Foster NW, Beauchamp EG, Corke CT (1980) Microbial activity in a *Pinus banksiana* Lamb. forest floor amended with nitrogen and carbon. Can J Soil Sci 60:199–209

Freeman C, Evans CD, Monteith DT, Reynolds B, Fenner N (2001) Export of organic carbon from peat soils. Nature 412:785–785

Godbold DL, Fritz H-W, Jentschke G, Meesenburg H, Rademacher P (2003) Root turnover and root necromass accumulation of Norway spruce (*Picea abies*) are affected by soil acidity. Tree Physiol 23:915–921

Gödde M, David MB, Christ MJ, Kaupenjohann M, Vance GF (1996) Carbon mobilization from the forest floor under red spruce in the northeastern USA. Soil Biol Biochem 28:1181–1189

Guggenberger G, Kaiser K (2003) Dissolved organic matter in soil: challenging the paradigm of sorptive preservation. Geoderma 113:293–310

Hagedorn F, Blaser P, Siegwolf R (2002) Elevated atmospheric CO_2 and increased N deposition effects on dissolved organic carbon – clues from $\delta^{13}C$ signature. Soil Biol Biochem 34:355–366

Hedin LO, Armesto JJ, Johnson AH (1995) Patterns of nutrient loss from unpolluted, old-growth temperate forests. Evaluation of biogeochemical theory. Ecology 76:493–509

Hongve D, van Hees PAW, Lundström US (2000) Dissolved components in precipitation water percolated through forest litter. Eur J Soil Sci 51:667–677

Johnson CE, Driscoll CT, Fahey TJ, Siccama TG, Hughes JW (1995) Carbon dynamics following clear-cutting of a northern hardwood forest. In: McFee WW, Kelly JM (eds) Carbon forms and functions in forest soils. Soil Science Society of America, Madison, Wisconsin, pp 463–488

Judd KE, Kling GW (2002) Production and export of dissolved C in arctic tundra mesocosms: the roles of vegetation and water flow. Biogeochem 60:213–234

Kaiser K, Guggenberger G (2000) The role of DOM sorption to mineral surfaces in the preservation of organic matter in soils. Org Geochem 31:711–725

Kaiser K, Zech W (1996) Nitrate, sulfate, and biphosphate retention in acid forest soils affected by natural dissolved organic carbon. J Environ Qual 25:1325–1331

Kaiser K, Zech W (1997) Competitive sorption of dissolved organic matter fractions to soils and related mineral phases. Soil Sci Soc Am J 61:64–69

Kaiser K, Zech W (1998) Soil dissolved organic matter sorption as influenced by organic and sesquioxide coatings and sorbed sulfate. Soil Sci Soc Am J 62:129–136

Kaiser K, Zech W (2000) Dissolved organic matter sorption by mineral constituents of subsoil clay fractions. J Plant Nutr Soil Sci 163:531–535

Kaiser K, Eusterhues K, Rumpel C, Guggenberger G, Kögel-Knabner I (2002) Stabilization of organic matter by soil minerals – investigations of density and particle-size fractions from two acid forest soils. J Plant Nutr Soil Sci 165:451–459

Kalbitz K, Solinger S, Park J-H, Michalzik B, Matzner E (2000) Controls on the dynamics of dissolved organic matter in soils: a review. Soil Sci 165:277–304

Kalbitz K, Schmerwitz J, Schwesig D, Matzner E (2003) Biodegradation of soil-derived dissolved organic matter as related to its properties. Geoderma 113:273–291

Kalbitz K, Glaser B, Bol R (2004) Clear-cutting of a Norway spruce stand – implications for controls on the dynamics of dissolved organic matter in the forest floor. Eur J Soil Sci (in press)

King JS, Pregitzer KS, Zak DR, Sober J, Isebrands JG, Dickson RE, Hendrey GR, Karnosky DF (2001) Fine-root biomass and fluxes of soil carbon in young stands of paper birch and trembling aspen as affected by elevated atmospheric CO_2 and tropospheric O_3. Oecologia 128:237–250

Kuiters AT (1993) Dissolved organic matter in forest soils: sources, complexing properties and action on herbaceous plants. Chem Ecol 8:171–184

Londo AJ, Messina MG, Schoenholtz SH (1999) Forest harvesting effects on soil temperature, moisture, and respiration in a bottomland hardwood forest. Soil Sci Soc Am J 63:637–644

Lundquist EJ, Jackson LE, Scow KM (1999) Wet–dry cycles affect dissolved organic carbon in two Californian agricultural soils. Soil Biol Biochem 31:1031–1038

Magill AH, Aber JD (2000) Dissolved organic carbon and nitrogen relationships in forest litter as affected by nitrogen deposition. Soil Biol Biochem 32:603–613
McDowell WH (2003) Dissolved organic matter in soils – future directions and unanswered questions. Geoderma 113:179–186
McDowell WH, Likens GE (1988) Origin, composition, and flux of dissolved organic carbon in the Hubbard Brook Valley. Ecol Monogr 58:177–195
McDowell WH, Currie WS, Aber JD, Yano Y (1998) Effects of chronic nitrogen amendments on production of dissolved organic carbon and nitrogen in forest soils. Water Air Soil Pollut 105:175–182
McGill WB, Cannon KR, Robertson JA, Cook FD (1986) Dynamics of soil microbial biomass and water-soluble organic C in Breton L after 50 years of cropping to two rotations. Can J Soil Sci 66:1–19
Michalzik B, Matzner E (1999) Dynamics of dissolved organic nitrogen and carbon in a central European Norway spruce ecosystem. Eur J Soil Sci 50:579–590
Michalzik B, Kalbitz K, Park J-H, Solinger S, Matzner E (2001) Fluxes and concentrations of dissolved organic carbon and nitrogen – a synthesis for temperate forests. Biogeochemistry 52:173–205
Neff JC, Asner GP (2001) Dissolved organic carbon in terrestrial ecosystems: synthesis and a model. Ecosystems 4:29–48
Neff JC, Hooper DU (2002) Vegetation and climate controls on potential CO_2, DOC and DON production in northern latitude soils. Global Change Biol 8:872–884
Neff JC, Hobbie SE, Vitousek PM (2000) Nutrient and mineralogical control on dissolved organic C, N and P fluxes and stoichiometry in Hawaiian soils. Biogeochemistry 51:283–302
Nodvin SC, Driscoll CT, Likens GE (1986) Simple partitioning of anions and dissolved organic carbon in a forest soil. Soil Sci 142:27–35
Park J-H, Matzner E (2003) Controls on the release of dissolved organic carbon and nitrogen from a deciduous forest floor investigated by manipulations of aboveground litter inputs and water flux. Biogeochemistry 66:265–286
Park J-H, Kalbitz K, Matzner E (2002) Resource control on the production of dissolved organic carbon and nitrogen in a deciduous forest floor. Soil Biol Biochem 34:813–822
Perakis SS, Hedin LO (2002) Nitrogen loss from unpolluted South American forests mainly via dissolved organic compounds. Nature 415:416–419
Qualls RG (2000) Comparison of the behavior of soluble organic and inorganic nutrients in forest soils. For Ecol Manage 138:29–50
Qualls RG, Haines BL, Swank WT (1991) Fluxes of dissolved organic nutrients and humic substances in a deciduous forest. Ecology 72:254–266
Raulund-Rasmussen K, Borggaard OK, Hansen HCB, Olsson M (1998) Effect of natural soil solutes on weathering rates of soil minerals. Eur J Soil Sci 49:397–406
Ross DJ (1991) Microbial biomass in a stored soil: a comparison of different estimation procedures. Soil Biol Biochem 23:1005–1007
Schwesig D, Kalbitz K, Matzner E (2003) Effects of aluminium on the mineralization of dissolved organic carbon derived from forest floors. Eur J Soil Sci 54:311–322
Skjelevale BL, Andersen T, Halvorsen GA, Raddum GG, Heegaard E, Stoddard JL, Wright RF (2000) The 12 year report: acidification of surface water in Europe and North America; trends, biological recovery and heavy metals. ICP Waters Rep 52:1–115
Solinger S, Kalbitz K, Matzner E (2001) Controls on the dynamics of dissolved organic carbon and nitrogen in a central European deciduous forest. Biogeochemistry 55:327–349
Staaf H, Berg B (1982) Accumulation and release of plant nutrients in decomposing Scots pine needle litter. Long-term decomposition in a Scots pine forest II. Can J Bot 60:1561–1568

Temminghoff EJM, van der Zee SEATM, de Haan FAM (1997) Copper mobility in a copper-contaminated sandy soil as affected by pH and solid and dissolved organic matter. Environ Sci Technol 31:1109–1115

Tipping E (1998) Modelling the properties and behavior of dissolved organic matter in soils. Mitt Dtsch Bodenkd Ges 87:237–252

Tipping E, Woof C, Rigg E, Harrison AF, Inneson P, Taylor K, Benham D, Poskitt J, Rowland AP, Bol R, Harkness DD (1999) Climatic influences on the leaching of dissolved organic matter from upland UK moorland soils, investigated by a field manipulation experiment. Environ Int 25:83–95

Vance GF, David MB (1991) Forest soil response to acid and salt additions of sulfate. III. Solubilization and composition of dissolved organic carbon. Soil Sci 151:297–305

Wagai R, Sollins P (2002) Biodegradation and regeneration of water-soluble carbon in a forest soil: leaching column study. Biol Fertil Soils 35:18–26

Wardle DA, Parkinson D (1990) Comparison of physiological techniques for estimating the response of the soil microbial biomass to soil moisture. Soil Biol Biochem 22:817–823

Wardle DA, Ghani A (1995) Why is the strength of relationships between pairs of methods for estimating soil microbial biomass often so variable? Soil Biol Biochem 27:821–828

Yavitt JB, Fahey TJ (1986) Litter decay and leaching from the forest floor in *Pinus contorta* (lodgepole pine) ecosystems. J Ecol 74:525–545

Zak DR, Tillman D, Parmenter RR, Rice CW, Fisher FM, Vose J, Milchunas D, Martin CW (1994) Plant production and soil micro-organisms in late successional ecosystems – a continental-scale study. Ecology 75:2333–2347

20 Response of Soil Solution Chemistry and Solute Fluxes to Changing Deposition Rates

E. Matzner, T. Zuber, and G. Lischeid

20.1 Introduction

Soil properties determine to a large extent the functioning of terrestrial ecosystems. Changes in soil solution chemistry and element fluxes with soil solution (seepage) are most important for ground and surface water quality, and in relation to the nutrient uptake by roots. Among other factors, the influences of deposition, tree species, soil properties and hydrology on soil solution chemistry and seepage fluxes in forest ecosystems have been addressed in numerous papers in the past. At our sites the changes in atmospheric deposition in the last two decades were significant (Matzner et al., this Vol., Chap. 14). The effects of these changes on soil solution chemistry and seepage fluxes will be addressed in this chapter.

Depletion of nutrient cations from forest soils and the release of potentially toxic Al ions from soil minerals as a result of H^+ buffering are major effects of acidic deposition on forest soils (Abrahamsen et al. 1994; Ulrich 1994). The release of Al ions is of special interest because of the potential toxicity of Al to tree roots and other biota (Godbold 1994). Depletion of available (exchangeable) nutrient cations in the soil results if losses by seepage and plant net uptake exceed the inputs from deposition and weathering. The input of Ca, Mg and K by weathering is crucial in this budget, but difficult to estimate under field conditions. The rate of weathering and nutrient cation release can be estimated through various methods, e.g. by mineral analysis (Fölster 1985; Likens et al. 1998; Olsson and Melkerud 2000), using Sr isotope ratios as tracer for Ca weathering (Dijkstra et al. 2003), by establishing cation budgets for the soil (Johnson et al. 1968; Matzner 1989; Frogner 1990) or by using simulation models (Sverdrup and Warfvinge 1993). The long-term measurements of seepage fluxes and deposition at our sites provide a valuable database to establish cation budgets and estimate weathering rates.

Under the conditions of decreasing SO_4 and H^+ depositions, the question of the recovery of acidified soils and waters is a matter of concern (Likens et al. 1996; Stoddard et al. 1999; Alewell et al. 2000; Armbruster et al. 2003). The time

Ecological Studies, Vol. 172
E. Matzner (Ed.), Biogeochemistry of Forested Catchments in a Changing Environment

needed for recovery of soil solution acidity and the processes involved seem to differ in relation to soil conditions, with soil S pools and N behaviour as the main driving variables. At our sites, the deposition rates of acidity have become lower than the estimated critical loads (Matzner et al., this Vol., Chapter 14), providing a good opportunity to study the recovery of soil conditions and element fluxes with soil solution.

The long-term shifts in ecosystem functioning as a consequence of N deposition are often summarized under the term 'N saturation'. According to Aber et al. (1989), N saturation is reached if the storing capacity of the ecosystem is exhausted and nitrate losses with seepage or runoff exceed the natural background. The mechanisms of N accumulation in forest soils, their kinetics and limitations are yet unresolved and predictions of the N cycle in response to changing deposition and environmental conditions are very uncertain. Recent analysis of long-term time series in a number of European catchments indicated no increasing trend of nitrate losses with runoff, despite chronically high N depositions (Wright et al. 2001). This points to the large time scales necessary until the retention capacity of the ecosystems for N is exceeded.

Here, we present long-term data on soil solution chemistry and seepage fluxes for the coniferous Coulissenhieb and the deciduous Steinkreuz site. Element budgets of the soil are calculated with special emphasis on the temporal development and on the location of sinks and sources for specific elements. We concentrate on H^+, nutrient cations, Al, S and N.

20.2 Development of Soil Solution Chemistry

At both sites, soil solution was collected by ceramic suction cups installed at 20 and 90 cm depth at the Coulissenhieb site and at 60 cm depth at the Steinkreuz site. At Coulissenhieb, the spatial heterogeneity of soil solution chemistry was investigated in detail and in total 20 ceramic suction cups were installed in 1992 in each depth, organized in four fields (Manderscheid and Göttlein 1995; Manderscheid and Matzner 1995; Gerstberger 2001). Samples were taken fortnightly from each suction cup starting in mid-1992 and were analysed for major cations and anions. At Steinkreuz, seven suction cups were selected for the long-term monitoring at 60 cm depth. The selection was based on a screening phase with a larger number (20) of suction cups in that depth such that the seven suction cups represented the average concentrations. Sampling started fortnightly in autumn 1994 and was reduced to a monthly interval from 1998 on.

After 0.45-μm filtration, the analysis of the soil solutions covered pH (glass electrode), Na, K, Ca, Mg, Al, Mn, Fe (ICP-AES), NH_4, NO_3, SO_4, PO_4 and Cl (ion chromatography).

The average annual concentrations given in Figs. 20.1, 20.2 and 20.3 represent annual median concentrations of all measurements in a specific depth. Trends were also statistically evaluated based on the single measurements (data not shown). Trends are only interpreted if the linear regressions of concentrations vs. time were significant on a <0.001 level.

There were a number of pronounced changes in soil solution chemistry observed at both sites. Most consistent was the declining trend of the *sulfate* concentrations (Fig. 20.1). At Coulissenhieb, the concentrations of sulfate were generally much lower at 20 cm depth as compared to 90 cm depth. At both depths, the decline is obvious and almost parallel. At the 90 cm depth, concentrations decrease from around 8 to 6 mg SO_4-S l^{-1} by the end of the observation and at 20 cm depth from around 6 to 3 SO_4-S l^{-1}. At Steinkreuz, the decrease of the sulfate concentrations was from about 4.5 to 3.0 mg SO_4-S l^{-1}.

The concentrations of *nitrate* in the soil solution of the Coulissenhieb site were similar at 20 and 90 cm depths (Fig. 20.1). At both depths the concentrations peaked in 1996 and 1997 and from then on decreased to a minimum in 2001. At Steinkreuz the concentration range was similar to Coulissenhieb and there was a peak of the nitrate concentrations at 60 cm depth in 1999 and 2000. Lowest concentrations were observed here in 1998. At both sites no clear trend can be observed. Given the low concentrations in 2001 and high con-

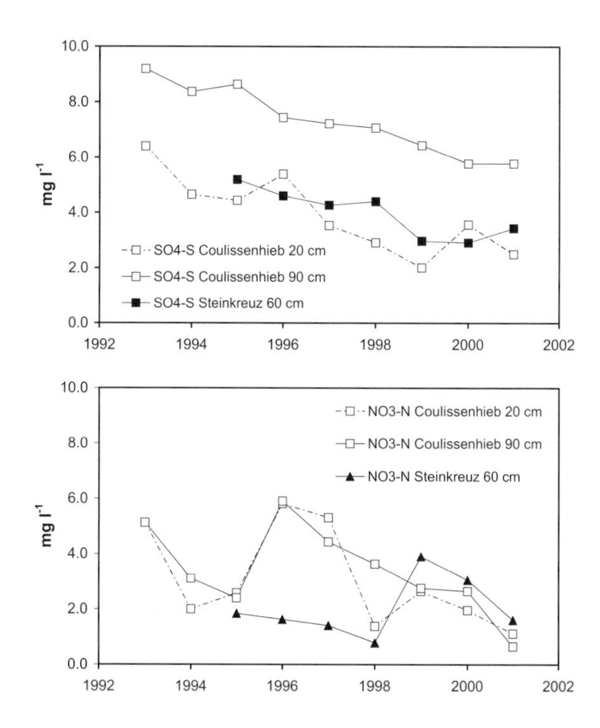

Fig. 20.1. Concentrations of sulfate and nitrate in soil solutions

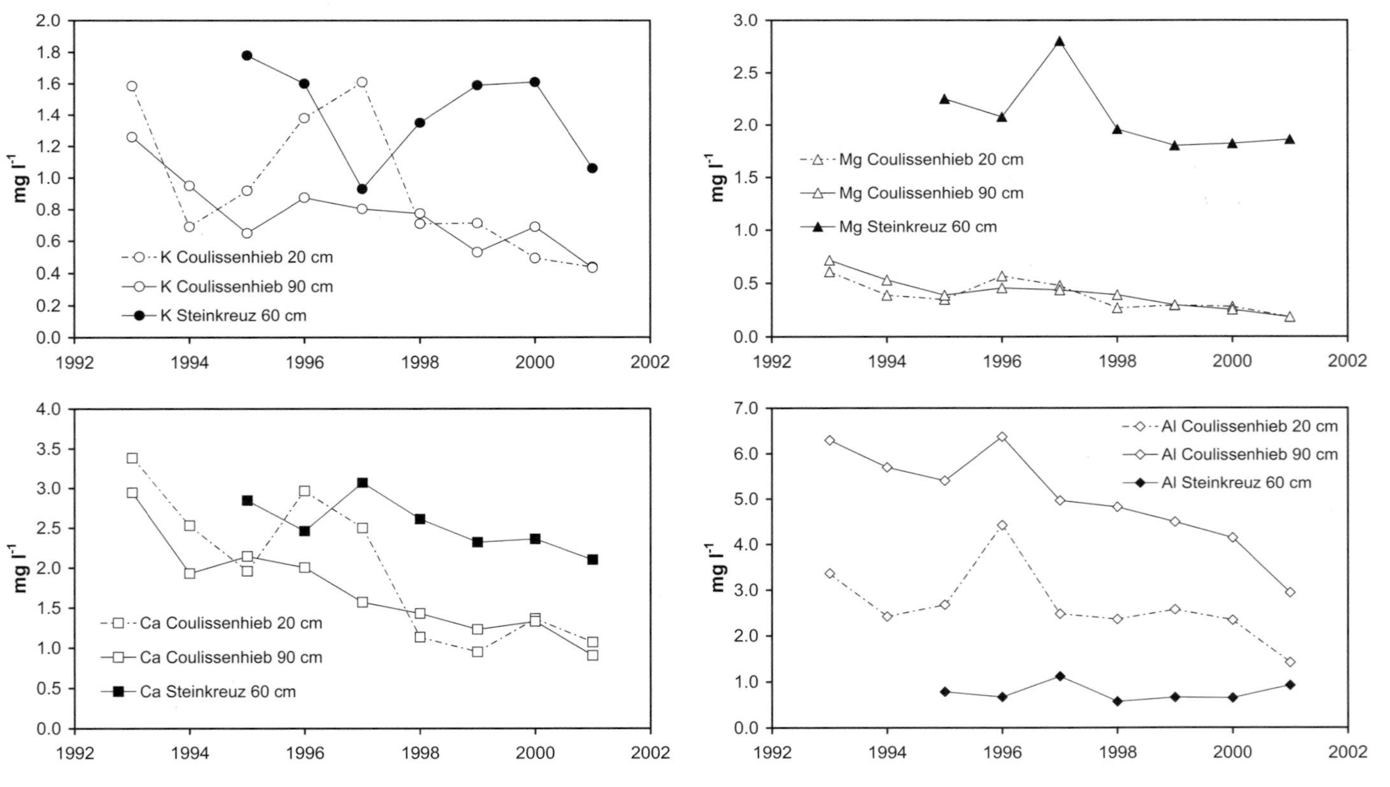

Fig. 20.2. Concentrations of K, Ca, Mg and Al in soil solutions

centrations in autumn1992 after the installation (not shown in Fig. 20.1) the statistical analysis of the single values indicated a slight decrease in nitrate concentrations at the Coulissenhieb site.

The NH_4 concentrations in soil solutions at both sites were always very low (<0.1 mg N l^{-1}, data not shown).

Similar to sulfate, the concentration of *Ca* in soil solutions decreased at the Coulissenhieb site at both depths by about 50 % and were in the range of 1.0–1.5 mg Ca l^{-1} by the end of the observation period (Fig. 20.2). The concentrations of Ca were similar at 20 and 90 cm depths in most years, with the exception of 1996 and 1997 where a peak in the concentrations was observed only at the 20 cm depth. The average Ca concentrations of the soil solution at the Steinkreuz site seem to decrease slightly from about 2.5 to 2.0 mg Mg l^{-1}, but the statistical analysis of the single measurement did not reveal a trend.

The concentrations of *Mg* at 20 and 90 cm depths at the Coulissenhieb site were surprisingly similar and in both cases decreased by about 50 % to 0.2 mg Mg l^{-1} in 2001 (Fig. 20.2). At Steinkreuz, the Mg concentrations of the soil solution were much higher compared to Coulissenhieb. When comparing the first 3 years with the last 3 years, a slight decrease in the concentrations is obvious, but based on the analysis of the single data this was statistically not significant.

In the case of *K*, the concentrations at Coulissenhieb tend to decrease at 90 cm depth as well, starting with values around 0.8 and reaching 0.4 mg l^{-1} in 2001 (Fig. 20.2). At 20 cm depth the concentrations were more dynamic and showed a pronounced peak in 1996 and 1997 similar to the Ca concentrations. The K concentrations of the soil solution at Steinkreuz had no trend.

The temporal patterns of *total Al* concentrations differed for the three time series (Fig. 20.2). The highest concentrations were found in general at 90 cm depth at the Coulissenhieb site. Here, there was also a clear decrease over time. The year 2001 had the lowest average Al concentration (2.9 mg l^{-1}), while at the beginning Al concentrations were between 5.4 and 6.4 mg l^{-1}. No clear trend was observed for the Al concentrations at 20 cm depth at Coulissenhieb. However, the lowest measured concentration was found in 2001 and the regression based on single values had a slight negative slope that was significant. At Steinkreuz, the concentrations of total Al in soil solutions were generally much lower than at Coulissenhieb (around 1 mg l^{-1}) and no trend was found.

There was also a slight increase in the *pH* of the soil solution at the 20 cm depth at the Coulissenhieb site (not shown). The pH increased here from about 3.7 to 3.9. At the 90 cm depth, the pH increased on average from about 4.1 to 4.2. At the Steinkreuz site, the pH of the soil solution at 60 cm depth was around 4.5 but without a trend.

The *Ca/Al ratio* is often used as an indicator of the risk of Al toxicity to tree roots (Cronan and Grigal 1995). Despite the decreasing deposition of sulfur and H^+ and the slightly increasing pH of soil solutions, the Ca/Al ratios of the

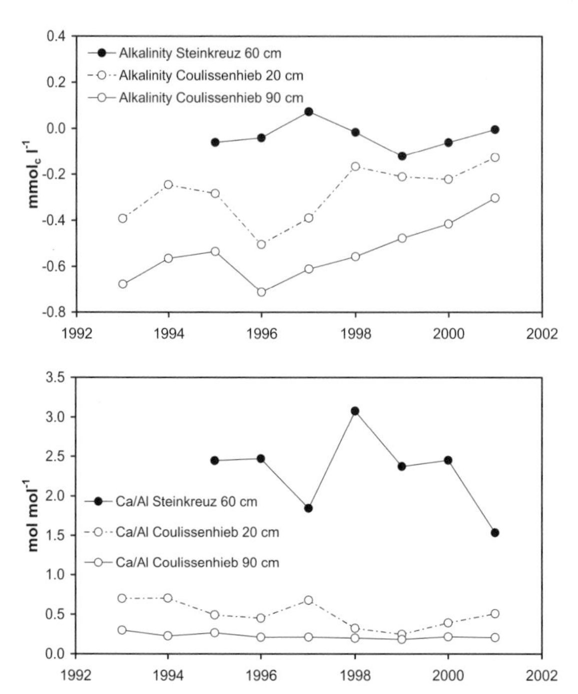

Fig. 20.3. Development of soil solution alkalinity and molar Ca/Al ratios

soil solution did not show an improvement (Fig. 20.3). On the contrary, there was a further decrease at the Coulissenhieb site when considering the 20 cm depth. Here, the Ca/Al ratios were generally low (<1.0) and the average of the first 3 years was 0.63, dropping in the last 3 years to 0.38. However, in the last 2 years the Ca/Al ratios at 20 cm depth increased following the minimum in 1999. At 90 cm depth, the Ca/Al ratio was even lower than at 20 cm depth and decreased from an average of 0.27 in the first 3 years to 0.20 in the last 3 years. In an earlier paper on soil solution data from Coulissenhieb (Alewell et al. 2000), the decrease in the Ca/Al ratio at 90 cm depth seemed to be greater, based on the fact that the ratios in 1992 were highest and the statistical treatment of the data was different. Here, we present annual average median concentrations only from 1993 on, while the paper of Alewell et al. (2000) presented arithmetic means for each single sampling date, starting in 1992.

At the Steinkreuz site, the average annual Ca/Al ratios of the soil solution were much higher (with a range of 1.4–1.9) than at the Coulissenhieb site. The lowest ratios were observed in the last 3 years, but the annual variation is high and no clear trend can be stated.

The *alkalinity* of the soil solution was calculated as (Na + K + Ca + Mg + NH_4) minus (Cl + sulfate + nitrate) and is given in Fig. 20.3. The alkalinity of the soil solution at 60 cm depth at the Steinkreuz site was around zero throughout the observation period and had no trend. The alkalinity of the soil solution at 20 cm depth at the Coulissenhieb site was negative throughout the whole observation period, with a minimum in 1996 and a constantly increasing trend afterwards. The highest alkalinity (–0.13 $mmol_c\ l^{-1}$) was observed in 2001. At 90 cm depth, the alkalinity of the soil solution was generally lower than at 20 cm, reaching also its minimum in 1996 and increasing from then on. The values in the last 3 years were the highest recorded and the maximum alkalinity occurred in 2001, similar to the 20 cm depth. The statistical analysis of the single values indicated a slight, but significant, positive trend of alkalinity both at 20 and 90 cm depth. Thus, a recovery of the soil solution from acidification can be stated.

20.3 Development of Element Fluxes with Soil Solution

We calculated daily seepage fluxes at different soil depths using the simulation model HYDRUS-2D (Simúnek et al. 1996). The model was calibrated to the site conditions and calculates seepage fluxes based on soil physical properties, precipitation and evapotranspiration data and root distribution. The model does not consider spatial heterogeneity of inputs and seepage within the site. In previous publications on seepage fluxes at the Coulissenhieb site, the spatial heterogeneity of throughfall and seepage fluxes was explicitly considered, and a different simulation tool was used (Manderscheid and Matzner 1995). We compared the different models calculating seepage rates at the Coulissenhieb site. Differences in water and element fluxes were minor if averages over more than 2 years are considered. In single years, differences of up to 20% might occur. In addition, we used the long-term Cl budget for the calibration of the HYDRUS-2D model, such that the soil input of Cl with throughfall matched the seepage losses at 60 and 90 cm depth. The influence of time resolution of the calculation was also tested. There was no significant difference in the results of seepage flux calculations between (1) fortnightly water fluxes times average fortnightly concentrations, with subsequent summation of the annual flux and (2) annual water flux times median annual concentrations. In order to avoid the interpolation of missing data needed for the short-term resolution, we used the latter calculation.

The hydrological conditions and the water fluxes with seepage differed strongly from year to year. The range of the annual seepage fluxes was 288–712 mm at Coulissenhieb (90 cm depth) and 204–444 mm at Steinkreuz (Tables 20.1 and 20.2). Since the variation in the annual element concentrations in soil solution was moderate, the variation in the water fluxes resulted in strong interannual variations of the element fluxes with seepage. This makes it difficult to identify trends. However, there were developments in the seepage fluxes especially at the *Coulissenhieb* site: the *Ca and Mg* fluxes, both at 20 cm and 90 cm depth, declined significantly (Table 20.1). At the 90 cm depth, the first 3-year averages of Ca and Mg fluxes were 12.5 and

Table 20.1. Coulissenhieb site element fluxes with seepage at 20 and 90 cm depths and budgets for the 20- and 90-cm soil depth. *TD* Total deposition

Year	H_2O (mm)	H (kg ha^{-1} year^{-1})	Na	K	Ca	Mg	Al	NO_3-N	SO_4-S	Cl
20 cm										
1993	511	1.41	9.3	8.1	17.3	3.1	17.2	26.2	32.7	11.8
1994	792	1.29	11.3	5.5	20.0	3.1	19.2	15.8	36.8	14.8
1995	835	1.84	12.8	7.7	16.4	2.9	22.4	21.5	37.0	13.2
1996	575	1.63	9.5	7.9	17.0	3.2	25.4	33.4	31.0	12.4
1997	402	1.12	5.1	6.5	10.1	1.9	10.0	21.3	14.2	7.5
1998	831	1.92	10.9	5.9	9.4	2.2	19.6	11.4	24.2	12.6
1999	715	1.14	9.1	5.1	6.8	2.1	18.4	18.9	14.4	10.8
2000	535	0.76	9.6	2.6	7.3	1.5	12.5	10.5	19.1	7.9
2001	698	1.08	10.6	3.0	7.5	1.3	9.9	7.8	17.4	9.7
Mean	655	1.4	9.8	5.8	12.4	2.4	17.2	18.5	25.2	11.2
90 cm										
1993	421	0.44	9.8	5.3	12.4	3.0	26.5	21.6	38.7	12.2
1994	661	0.37	11.5	6.3	12.8	3.5	37.7	20.5	55.3	14.4
1995	705	0.50	11.1	4.6	15.1	2.7	38.1	16.9	60.9	12.8
1996	483	0.48	8.6	4.2	9.7	2.2	30.8	28.5	35.9	11.3
1997	288	0.17	4.9	2.3	4.5	1.3	14.3	12.8	20.8	5.3
1998	712	0.46	12.0	5.5	10.2	2.8	34.3	25.9	50.3	13.5
1999	587	0.36	9.7	3.1	7.2	1.7	26.4	16.2	37.8	10.6
2000	430	0.22	7.1	3.0	5.7	1.1	17.8	11.4	24.8	6.6
2001	563	0.28	9.5	2.4	5.1	1.0	16.6	3.7	32.5	8.4
Mean	539	0.4	9.4	4.1	9.2	2.2	26.9	17.5	39.7	10.6
TD – 20 cm										
1993		0.39	–3.0	–4.6	–12.9	–2.8			–1.7	0.2
1994		–0.24	–3.4	–1.4	–15.7	–2.4			–10.9	0.0
1995		–0.42	–6.4	–3.9	–13.5	–2.5			–10.9	–0.7
1996		–0.58	–4.7	–4.8	–12.4	–2.7			0.9	–2.2
1997		–0.61	0.3	–3.1	–4.2	–1.2			3.1	2.5
1998		–0.96	–2.2	–0.8	–4.8	–1.5			–8.4	0.9
1999		–0.78	–2.9	2.0	–1.4	–1.4			–1.2	0.3
2000		–0.54	–4.1	1.8	–0.9	–0.7			–6.6	2.5
2001		–0.76	–4.8	0.8	–2.6	–0.5			–5.6	0.5
Mean		–0.5	–3.5	–1.6	–7.6	–1.8			–4.6	0.4
TD – 90 cm										
1993		1.36	–3.5	–1.8	–8.0	–2.7			–7.7	–0.2
1994		0.68	–3.5	–2.2	–8.4	–2.8			–29.4	0.5
1995		0.92	–4.8	–0.8	–12.2	–2.4			–34.8	–0.3
1996		0.58	–3.9	–1.1	–5.0	–1.7			–4.1	–1.1
1997		0.33	0.5	1.1	1.3	–0.6			–3.5	4.7
1998		0.49	–3.2	–0.4	–5.5	–2.1			–34.5	0.0
1999		0.00	–3.5	4.0	–1.8	–1.0			–24.6	0.6
2000		–0.01	–1.6	1.5	0.7	–0.3			–12.3	3.8
2001		0.03	–3.7	1.4	–0.2	–0.3			–20.7	1.8
Mean		0.5	–3.0	0.2	–4.4	–1.5			–19.0	1.1

Table 20.2. Steinkreuz site element fluxes with seepage at 60 cm depth and budgets for the 60-cm soil depth. *TD* Total deposition

Year	H_2O (mm)	H (kg ha^{-1} year^{-1})	Na	K	Ca	Mg	Al	NO_3-N	SO_4-S	Cl
60 cm										
1995	393	0.05	7.8	7.0	11.2	8.8	3.1	7.2	20.3	9.4
1996	308	0.08	5.9	4.9	7.6	6.4	2.1	5.0	14.2	6.9
1997	204	0.03	4.3	1.9	6.3	5.7	2.3	2.9	8.7	4.3
1998	314	0.11	5.4	4.2	8.2	6.2	1.8	2.5	13.8	9.7
1999	271	0.08	6.5	4.3	6.3	4.9	1.8	10.6	8.1	6.3
2000	224	0.06	5.2	3.6	5.3	4.1	1.5	6.8	6.5	5.7
2001	444	0.11	10.1	4.7	9.3	8.3	4.1	7.1	15.3	9.4
Mean	308	0.07	6.4	4.4	7.7	6.3	2.4	6.0	12.4	7.4
TD – 60 cm										
1995		0.19	–0.4	–3.2	–6.2	-8.1			-8.6	–0.3
1996		0.30	–2.8	–1.8	–4.5	-6.0			–2.9	0.3
1997		0.13	–1.2	1.2	–2.0	-5.0			–0.7	2.4
1998		0.43	–0.9	1.3	–2.4	-5.4			–6.5	–1.2
1999		0.06	–2.8	–0.3	–1.7	-4.2			–0.8	1.3
2000		0.00	–1.8	0.9	0.6	-3.4			0.0	1.2
2001		–0.07	–6.4	0.2	–2.8	-7.4			–8.5	–2.4
Mean		0.15	–2.3	–0.2	–2.7	-5.7			–4.0	0.2

3.2 kg ha^{-1} year^{-1}, respectively, while in the last 3 years, the average fluxes were only 6.0 and 1.3 kg ha^{-1} year^{-1}, equivalent to a reduction of 50 %. The decline in the Ca and Mg was similar at the 20 cm depth. Substantial declines in the fluxes of total Al and sulfate are also obvious. The *total Al* fluxes at 90 cm depth declined from 35 to 20 kg ha^{-1} year^{-1} and the *sulfate-S* fluxes from 51 to 32 kg ha^{-1} year^{-1} when comparing the averages of the first 3 years with the last 3 years (Table 20.1). Decreasing fluxes of Al and sulfate were also found at 20 cm depth, where the average fluxes of both elements were much lower compared to the 90 cm depth. The fluxes of K with seepage at the Coulissenhieb site also decreased at both depths with time. *K fluxes* were generally higher at the 20 cm depth as compared to the 90 cm depth.

Considering the development of the *nitrate* fluxes at Coulissenhieb, the situation is not as clear. At both depths, the nitrate fluxes were exceptionally low in 2000 and 2001 as a result of relatively low water fluxes in 2001 and very low concentrations in 2001. The nitrate fluxes at 20 and 90 cm depth were in a similar range. On average, the seepage flux of nitrate was 18.5 kg N ha^{-1} year^{-1} at 20 cm depth and 17.5 at 90 cm depth. We consider this difference minor, given the uncertainties in the determination of the seepage fluxes (Manderscheid and Matzner 1995) and the similar concentrations of nitrate in both depths. The nitrate fluxes with seepage at the Coulissenhieb site are very high in comparison with many other European forests (MacDonald et al. 2002) and the time series indicates that N saturation of this site has been reached for a long time.

The temporal development of the seepage fluxes at *Steinkreuz* did not reveal a trend for any of the elements considered (Table 20.2). The average fluxes of Na, K and Ca were in a similar range as compared to the 90 cm depth at the Coulissenhieb site. Fluxes of total Al, sulfate and nitrate were, on the contrary, much lower than at Coulissenhieb, while the Mg fluxes were substantially higher and similar to the Ca fluxes at the Steinkreuz site. The nitrate flux with seepage of 6.3 kg N ha^{-1} $year^{-1}$ represents a substantial N loss from the system. The nitrate losses were already observed from the beginning of the study and – similar to the Coulissenhieb site – the Steinkreuz site can be considered as N saturated according to Aber et al. (1989).

The sulfate concentrations had a negative trend at Steinkreuz, but did not cause a clear trend of the seepage fluxes, because the interannual variation was too high.

20.4 Element Budgets of the Soil

For those elements where input fluxes with total atmospheric deposition have been calculated (Chapter 14), budgets for the soil are established by subtracting the seepage fluxes from the total deposition (Tables 20.1 and 20.2). The resulting sink and source functions of the soil will be investigated with regard to their localization, trends and to driving processes.

20.4.1 Coulissenhieb

Na, K, Ca and Mg Budgets

For the elements Na, K, Ca and Mg a source function of the soil is evident when looking at the average budgets of the 20-cm soil compartment (Tables 20.1 and 20.2). Comparing the budget of the 90-cm soil layer with the 20-cm layer, a sink function of the deeper mineral soil results, with different intensity. In the case of K and Ca a strong sink is evident, while the sink for Na and Mg is only minor. In other words, the K and Ca fluxes with seepage at the 90-cm depth are less compared to those at 20 cm depth. We attribute this to the internal cycle of Ca and K and to the uptake of these cations by tree roots in the soil layers at >20 cm depth. The source function of the upper mineral soil is most likely due to the mineralization of the annual litterfall, followed by root uptake in the deeper soil. Another explanation of these patterns would be the storage of these elements in exchangeable form in deeper layers, but this seems very unlikely given the long time scales, the high Al saturation of the cation exchange capacity and the extremely low concentrations of exchangeable Ca (Gerstberger et al., this Vol.). As outlined by Gerstberger and Berg (this Vol.), the annual element input of K, Ca and Mg with litterfall at the Coulissenhieb site is in the order of 6, 15 and 1 kg ha^{-1} $year^{-1}$ and thus the mineralization of litter in the forest floor results in a substantial input of these cations to the mineral soil.

In the case of Ca and Mg, the source function of both soil layers in relation to total deposition had a clear trend. Highest net release of both elements was observed in the beginning of the measurements, while the net losses of Ca and Mg from the soil declined towards the end of the measurements. This dynamic provides a good opportunity to identify potential sources of the Ca and Mg net losses from the soil. The source function of the soil originates from losses of exchangeable pools and/or from the weathering of minerals. A decline of the source function that is observed at the Coulissenhieb may be caused by (1) the depletion of the exchangeable pools, (2) retardation of litter decomposition and subsequent accumulation of Ca and Mg in aggrading humus or (3) increased net uptake of the trees. The latter is considered unlikely, since the net uptake is rather small and should not increase dramatically at this mature site (Dittmar and Elling, this Vol.).

Considering the 90 cm depth, the net release of Ca and Mg from the soil in the last 3 years was only around 0.5 kg Ca ha^{-1} $year^{-1}$ and 0.5 kg Mg ha^{-1} $year^{-1}$. The weathering rates of minerals and the subsequent release of Ca and Mg are not likely to change in the time scale considered here. Thus, given a constant weathering input, the larger net release observed in the first years is likely due to losses from the exchangeable pool. The soil pools were investigated in 2000 (Gerstberger et al., this Vol.). The pools of exchangeable Ca and Mg were below the detection limit in mineral soil horizons deeper than 12 cm, indicating severe depletion of the mineral soil. Exchangeable Ca and Mg was only found in the forest floor and the A horizon.

When we accept the depletion of the exchangeable pools as the reason for the decreasing net release, and neglect a potential accumulation of Ca and Mg in the forest floor, then the observed release of Ca and Mg from the last years can be used to estimate inputs of these elements by weathering of minerals under field conditions. This needs assumptions on the net uptake and accumulation of Ca and Mg in the standing biomass. At a similar mature Norway spruce site in the Solling area, the net uptake into biomass of Ca and Mg was estimated at 5.0 and 1.1 kg ha^{-1} $year^{-1}$ respectively (Matzner 1988). Adding the very small net losses observed with seepage, the input rates by weathering of minerals in the Coulissenhieb soil up to 90 cm depth would be around 5.5 kg Ca ha^{-1} $year^{-1}$ and 1.6 kg Mg ha^{-1} $year^{-1}$.

The long-term K budget of the 90-cm soil layer is almost balanced, or as in the last 3 years, positive. The strong positive K budget in 1999 is a consequence of the exceptionally high total deposition rate calculated in Matzner et al. (this Vol., Chap. 14) and thus should interpreted with caution. If we take the K budget of the solute fluxes for balanced, the estimated K input by weathering would be equal to the net uptake by the stand, which might also be estimated from the Solling spruce site at 5 kg ha^{-1} $year^{-1}$ (Matzner 1988). However, the soil profile in 2000 contained in total about 200 kg K ha^{-1} in exchangeable form and a depletion of this pool cannot be excluded, resulting in an overestimation of K inputs from weathering.

In the case of Na, the budget of the soil is negative throughout the whole observation period and the average depletion calculated for the 90 cm depth amounts to 3 kg Na ha^{-1} $year^{-1}$. Net accumulation of Na in the biomass is very small (<1 kg ha^{-1} $year^{-1}$) and might be neglected (Matzner et al. 1982) as well as the exchangeable pools in the soil which were below detection limit in all mineral soil horizons. Thus, the rate of Na release by weathering is estimated at 3 kg ha^{-1} $year^{-1}$.

Summing up the estimated rates of cation inputs by mineral weathering for Na, Ca, Mg and K, a value of 0.6 $kmol_c$ ha^{-1} $year^{-1}$ results for the soil profile of 90 cm depth. The rates of cation release by weathering reported in the literature differ strongly according to the methods used and the site conditions. The rates observed for the Coulissenhieb site are in the same order of magnitude as those found for the Solling (Matzner 1989). Olsson and Melkerud (2000) estimated rates of Ca, Mg and K losses for post-glacial time scales in Scandinavian soils in the order of 6, 2 and 8 kg ha^{-1} $year^{-1}$, respectively, which is very similar to the rates estimated for Coulissenhieb. However, the Ca release from weathering was also estimated to be much lower: Likens et al. (1998) suggested 1 kg Ca ha^{-1} $year^{-1}$ based on mineral analysis for the Hubbard Brook catchment. Dijkstra et al. (2003), using Sr isotope ratios, estimated the Ca release by weathering to be <0.6 kg ha^{-1} $year^{-1}$ in Northern American forest soils.

N Budget

Establishing an *N budget* of the soil is difficult since the total deposition of N cannot be estimated and the throughfall fluxes will underestimate the actual N input. In most literature studies, the throughfall flux of mineral N is taken as the N input to the ecosystem and the soil in the absence of better estimates. Following this approach, the mineral N flux with throughfall at the Coulissenhieb site was 22 kg N ha^{-1} $year^{-1}$ on average without a trend and with relatively low interannual variation. On average 57 % of the mineral N in throughfall was represented by nitrate. The budget of the 90-cm soil compartment reveals a sink of N in the range of 5 kg N ha^{-1} $year^{-1}$. This is what can be estimated from the net N uptake into aggrading biomass of the mature spruce site. At the 20- and 90-cm depths, the N in solution is represented only by nitrate. Thus, the incoming NH_4 is taken up, immobilized or nitrified. Nitrate fluxes with seepage were similar at 20 and 90 cm depth. The uptake of nitrate in deeper soil layers seems to be low in comparison to K, where the flux decreased by about 30 % from the 20- to the 90-cm soil layer.

Taking the mineral N in throughfall as input, the N retention in the ecosystem is low. Losses of N via N_2O emissions seem to be negligible at this site (Drake et al., unpubl. data). The calculated N retention is highly variable with time mostly according to the hydrological conditions, but also to changes in soil solution nitrate. Especially the last 2 years showed a substantial retention of N in comparison to the years before. This points to the large time scales needed to evaluate the status of the N cycle in forest ecosystems.

S Budget

During the observation period, the soil of the Coulissenhieb site can be considered as a source of *sulfate*. The source of sulfate can be partly located in the topsoil up to 20 cm depth, losing 5 kg S ha^{-1} $year^{-1}$ on average (Table 20.1). There is no time trend visible in the source function of that layer, since the annual variation of the fluxes was too high. Considering the 90-cm layer, the sulfate release increased. The soil layer from 20–90 cm depth released on average about 14 kg ha^{-1} $year^{-1}$. Thus, 25 % of the overall S release was located in the 0- to 20-cm layer and 75 % in deeper soil layers. As in the 0- to 20-cm layer, the source function of the deeper soil for sulfate had no trend. On average the whole soil profile lost 19 kg S ha^{-1} $year^{-1}$. The sulfate released from the soil may originate from the desorption of inorganic sulfate and from the excess mineralization (net mineralization exceeds plant uptake) of organic S. The sulfate pools of the soil were determined in 2001. The pool of inorganic sulfate in the forest floor and 0–20 cm of the mineral soil is estimated at only 50 kg S ha^{-1} while the organic S pool up to that depth amounts to 900 kg ha^{-1}. The organic S pool in the deeper mineral soil is also substantial (400 kg ha^{-1} at 20–55 cm depth), and in a similar range to the inorganic sulfate pool in the 20- to 90-cm layer, which is estimated to be about 400 kg ha^{-1} (Manderscheid et al. 2000; Gerstberger et al., this Vol.). It seems very unlikely, that the long-lasting sulfate release from the upper soil (up to 20 cm depth), amounting to 5 kg S ha^{-1} $year^{-1}$, results from the depletion of the inorganic pool alone. The role of the organic S for the sulfate release from the soil is presently being investigated at the Coulissenhieb site in more detail (Prechtel 2004). Preliminary data indicate that net mineralization in the forest floor and upper mineral soil exceeds the estimated sulfate uptake of the roots by about 3.4 kg S ha^{-1} $year^{-1}$. Thus, a large part of the sulfate released from the upper soil layer seems to result from excess mineralization of organically bound S.

Al Budget

Aluminium is by far the dominant cation (about 65 % of the cation charge) in the soil solutions at the Coulissenhieb site. Since the input of Al from atmospheric deposition is negligible given the high Al fluxes with soil solution, the fluxes of Al at 20 and 90 cm depth can be taken as net release of Al from the soil. The Al release from the soil solid phase is a consequence of H^+ buffering, but the origin of Al in the soil solution is difficult to evaluate. Organically complexed Al as well as several inorganic binding forms of Al are involved (Mulder and Stein 1994; Berggren and Mulder 1995; Porebska and Mulder 1996; Matzner et al. 1998).

About two thirds of the total Al release in the whole soil profile was on average located in the 0- to 20-cm layer (Table 20.1). The fluxes of Al with soil solutions were often shown to be related to the inputs of acidity to the soil and to the fluxes of strong mineral acid anions like nitrate and sulfate (e.g. Dise et al. 2001). This is also shown by our data; the Al fluxes and the net release of Al from the soil decreased according to the declining fluxes of sulfate. The low seepage fluxes of nitrate in 2000 and 2001 caused a further decrease of the Al export in these years.

20.4.2 Steinkreuz

Na, K, Ca and Mg

With regard to the budgets of Na, K, Ca and Mg, most evident is the substantial loss of Mg from the soil profile at the Steinkreuz site. The difference between seepage and total deposition indicates an average loss of 5.8 kg Mg ha^{-1} $year^{-1}$ without a trend in the source function. The loss of Mg from the soil is enhanced by the net accumulation of Mg in the aggrading biomass amounting to 1–2 kg Mg ha^{-1} $year^{-1}$ (Feger 1997). The amount of exchangeable Mg in the soil profile in 2000 was substantially higher than at the Coulissenhieb site (Gerstberger et al. 2004), with about 35 kg Mg ha^{-1} in the A horizon and 225 kg ha^{-1} in the deeper B horizon (50–80 cm depth). The upper B horizons were totally depleted of Mg. Thus, the most important source of the Mg loss seems to be the exchangeable pool. The weathering input of Mg cannot be estimated since we cannot differentiate the weathering from losses of exchangeable Mg.

The long-term soil budgets of the Steinkreuz site for Na, K and Ca were similar to the Coulissenhieb soil. The Na budget was negative in each year, pointing to a net loss of 2.3 kg Na ha^{-1} $year^{-1}$. This is attributed to the weathering of Na containing feldspars, since the exchangeable Na pool is negligible (Gerstberger et al., this Vol.). The weathering rate and the resulting input of Na at Steinkreuz were similar to that of the Coulissenhieb soil.

The K budget of the soil was almost balanced when the solute fluxes are considered; the losses with seepage are similar to the estimated deposition rates. Only the net accumulation of K in the aggrading biomass leads to K losses from the soil. This export from the soil amounted to about 8 kg K ha^{-1} $year^{-1}$ in a beech forest in the Solling with similar soil conditions

(Matzner 1988). The exchangeable K pools of the soil at Steinkreuz were about 350 kg ha^{-1} and their depletion might explain part of the K losses from the mineral soil.

The Ca losses with seepage exceeded the estimated input by deposition on average by 2.8 kg Ca ha^{-1} $year^{-1}$. The annual variation of the Ca losses was high, and the Ca budget was even slightly positive in 2000. A net uptake of about 5 kg Ca ha^{-1} $year^{-1}$ in aggrading biomass can be assumed (Matzner 1988) to add to the total Ca losses of about 8 kg Ca ha^{-1} $year^{-1}$. Given the high amounts of exchangeable Ca, the loss can be explained by the depletion of exchangeable pools. Exchangeable Ca is mostly found in the A (230 kg ha^{-1}) and deeper B horizons (225 kg ha^{-1} at 50–80 cm depth), while the upper B horizons were totally Ca depleted (Gerstberger et al., this Vol.).

N Budget

With respect to the N budget, the uncertainties in the estimation of total deposition of N outlined above also hold for the beech site, but probably to a lesser extent compared to spruce because of the lower leaf area and the leafless winter period. The average annual mineral N flux in throughfall at the Steinkreuz site was 15 kg N ha^{-1} $year^{-1}$ and thus substantially lower compared to the Coulissenhieb site. Taking the throughfall flux as input, the N deposition at this site is in the range of values suggested as critical loads for N (see Chap. 14) which should avoid nitrate losses with seepage. However, the losses of nitrate at the 60-cm layer are substantial, amounting to about 6 kg N ha^{-1} $year^{-1}$ on average with relatively small variations over time. The calculated retention of N (9 kg N ha^{-1} $year^{-1}$) in the ecosystem is higher than that at the Coulissenhieb site. The N losses with seepage are quite high at Steinkreuz in relation to the estimated input of N. Taking the 60-cm layer as output from the ecosystem might be questionable, because the roots presumably reach deeper layers. The monitoring of soil water matrix potentials at 2 m depth indicated water uptake below the 60-cm layer (Lischeid, unpubl. data). Groundwater sampling at different depths showed that elevated nitrate concentrations occur at up to 3 m depth (Lischeid and Langusch 2004), thereby supporting the magnitude of nitrate losses from the ecosystem indicated by the seepage fluxes at the 60-cm layer. Nitrate concentrations in the shallow groundwater layer were in the upper range of those in the soil solution.

S Budget

Under the conditions of decreasing S deposition, the soil of the Steinkreuz site was a source of sulfate throughout the observation period. The average loss of sulfate from the 60-cm soil depth amounted to 4 kg S ha^{-1} $year^{-1}$. While the absolute loss is rather small, the sulfate flux with seepage was on average 50 % higher than the sulfate flux with throughfall. Data on the soil S pools are not available for the Steinkreuz site and no conclusion on the relative importance of sulfate desorption vs. mineralization of organic S can be made.

The *Al fluxes* and the net losses from the soil are much lower than at the Coulissenhieb site, but still 3.2 kg Al ha^{-1} $year^{-1}$ is transferred on average with the soil solution to deeper layers. No trend in the source function of the soil for Al is evident.

20.5 General Discussion

With respect to soil and water acidification and its reversal, the results from the Coulissenhieb site are most interesting, given the magnitude of the S and H^+ deposition and their reduction to levels suggested as critical loads. The strongly acidified soil with a high S storage is another specificum of this site. What were the effects of these changes on the soil conditions? First, we have to state that there are a number of positive developments that point to the reversal of soil solution acidity and to a decreased export of acidity into deeper layers and the groundwater. The alkalinity of the soil solution was in 2001 still negative, but the increasing trend was significant. The pH of the soil solutions also had a slight positive trend; however, the absolute values remained in the strongly acidified range. The Ca/Al ratios of the soil solution at 20 cm depth reached a minimum in 1999 and were slightly increasing in the following 2 years. The export of acidity (calculated as Al^{3+} + H^+) with seepage to deeper layers decreased in the observation period from 4.4 to 2.5 $kmol_c$ ha^{-1} $year^{-1}$ when comparing the average flux at 90 cm depth for the first 3 years with the last 3 years. This development is mainly caused by the declining sulfate fluxes with seepage. Together with the decreasing net export of Ca and Mg from the soil this indicates that the rate of soil acidification (defined as reduction of the acid neutralization capacity of the soil solid phase) has declined substantially.

These positive developments following the decreasing S and H^+ depositions are substantially delayed at the Coulissenhieb site by the release of formally stored S from the large inorganic and organic soil S pools. The net release of S from the soil profile amounted to 20 kg ha^{-1} $year^{-1}$ on average and exceeds the present S input by deposition by about 100 %. Since the desorption of sulfate and the S mineralization are both acidifying the soil solution, the recovery of the soil solution acidity is delayed. The behaviour of inorganic sulfate under conditions of decreasing S depositions is often described by desorption isotherms (for the Coulissenhieb site see Manderscheid et al. 2000). Using a simulation model, the release of sulfate from the soil solid phase can be quantified (e.g. the MAGIC model; Cosby et al. 2001; Prechtel et al. 2003) and predictions are made on the exhaustion of the inorganic S pool of the soil. Assuming a drastic drop in S deposition in 2000 and soil profile of 1 m depth, a MAGIC simulation predicted in the case of the Coulissenhieb site that the S fluxes with seepage will reach the input level after about 25 years (Matzner et al. 2001).

More difficult to describe is the reaction of the soil organic S pool to decreasing depositions. Including plant uptake and S mineralization in their model (PnET-BGC) reduced the discrepancy between measured and modelled SO_4 stream concentrations considerably (Gbondo-Tugbawa et al. 2002). The authors suggested that under conditions of decreasing S deposition, excess mineralization (net mineralization > plant uptake) becomes more important in controlling stream SO_4 losses than SO_4 desorption. At the Coulissenhieb site, the organic S pool in the soil exceeds the inorganic one, emphasizing the potential role of the organic pool as an S source through excess mineralization (Prechtel 2004). The kinetics of the sulfate excess mineralization, its dependence on S inputs and the size of the mobilizable pool need future consideration to improve predictions on the soil S behaviour under decreasing S inputs.

While the above positive trends indicate recovery from acidification, the development of the Ca and Mg concentrations in the soil solution is a *matter of concern*. The decline in concentrations was substantial, amounting to 50 % in the observation period and extremely low concentrations in 2001. The Ca/Al ratios in the soil solution, as an indicator of potential Al toxicity, only slightly increased in the last 2 years at 20 cm depth at Coulissenhieb but still decreased at 90 cm. The conditions for the trees in respect of Ca and Mg nutrition have further deteriorated in the last decade. This conclusion is confirmed by the temporal development of Ca and Mg contents in the needles at the Coulissenhieb site which decreased as well in the 1990s (Alewell et al. 2000). The reasons for the decreasing Ca and Mg concentrations seem to be, on the one hand, the total depletion of the exchangeable storage of these elements in the mineral soil. The soil inventory in 2000 found the exchangeable pools of both elements in soil horizons deeper than the A horizon to be below the detection limit. On the other hand, the decrease in the deposition of Ca and Mg, which has been documented for the 1980s and beginning of the 1990s (Matzner et al., this Vol., Chap. 14), seems to cause the decreasing soil solution concentrations. The effect of the decreasing deposition is buffered by the depletion of the soil exchangeable pools. Furthermore, the decrease in deposition and soil solution concentration has resulted in decreasing Ca and Mg uptake into the needles (Alewell et al. 2000). The trend observed in the soil solution concentrations from 1993–2001 thus reflects the long-term effects of the decreasing Ca and Mg depositions. The strong relation of Mg deposition to seepage output of Mg found here is supported by results from regional comparisons (Armbruster et al. 2002).

The input of Ca and Mg by weathering of minerals cannot mitigate the effects of decreasing Ca and Mg depositions. The rates are simply too low. The decreasing needle contents of Ca and Mg (Alewell et al. 2000) also indicate that the mycorrhiza induced weathering of minerals and the bypass of cation uptake from the soil solution cannot balance the depletion, as was suggested by van Breemen et al. (2000). Despite the Mg deficiency of the trees, substantial amounts of Mg are lost with seepage and are not taken up by the roots. Most of the seepage occurs in winter and spring with little root activity, which might explain the losses. However, the uptake of K and Ca from the 20- to 90-cm layer was substantial and the lack of Mg uptake in deeper layers seems to be related to the low Mg/Al ratios of the soil solution and to antagonistic effects of Al on Mg uptake (Godbold 1991).

The decreasing depositions of S and H^+ today have reached rates that are suggested as *critical loads* for forest ecosystems. The critical load concept is a steady state approach, assuming that deposition rates below the critical load will avoid detrimental effects on trees, soils and waters in the future. The concept does not include a time frame needed until these conditions are reached. As shown above, the mitigating effects of low deposition rates are clearly visible, but counteracted and delayed by the release of S from the soil, high nitrate losses and by the further depletion of Ca and Mg from the soil solid phase and the soil solution as a result of decreasing Ca and Mg depositions. Given the low inputs of Ca and Mg by weathering, it seems unlikely that the conditions for Ca and Mg supply of the trees will improve in the next decades without Ca and Mg being added to the soil as fertilizer. Liming of acidic forest soils has been postulated for decades as a measure to prevent further and mitigate existing acidification of forest soils (Ulrich et al. 1980). Despite atmospheric acid deposition reaching critical loads today, the addition of lime still seems to be necessary under conditions similar to those at the Coulissenhieb site in order to improve soil conditions and tree nutrition.

The situation at the *Steinkreuz* site in respect of soil acidification and its reversal is different from that at the Coulissenhieb site. The soil still has substantial amounts of exchangeable Ca and Mg, especially in the deeper soil layers, and the inputs of acidity have been lower. There was no trend of decreasing Ca and Mg concentrations in the soil solution, nor was any trend in Al or alkalinity observed. The situation seems to be rather stable at the moment. However, some concern arises from the large net export of Mg and Ca from the soil in the range of 7–8 kg ha^{-1} $year^{-1}$. In the case of Ca, this is mostly caused by the net uptake of the trees, while for Mg the seepage losses were most important. The latter result from the nitrate and sulfate losses with seepage accompanied by Mg as the dominant cation in solution. Sulfate is still the dominating anion in the soil solutions of the Steinkreuz site. Despite the low S deposition at this site, the Mg losses with seepage are substantial, partly because of the net release of sulfate from the soil. Since the net release of Ca and Mg is attributed to the depletion of exchangeable soil pools (assuming low rates in the sedimentary sandstone), this development seems critical if the losses continue for longer time periods.

With respect to the *fate of deposited N* and the phenomenon of *N saturation* the high losses of nitrate with seepage at the Coulissenhieb site indicate a small N retention capacity of the mature spruce site. The C/N ratio of the forest floor was introduced as an indicator for the N retention capacity of the ecosystem (Matzner and Grosholz 1997; Gundersen et al. 1998; MacDonald et al. 2002). If C/N ratios are less than 25, the input–output relation of N changes to a lower retention. The C/N ratio of the forest floor is about 23 at Coulissenhieb and 19 at Steinkreuz, indicating a high risk of N losses at both sites. The nitrate losses at Steinkreuz are to be expected, given the C/N ratio and the throughfall N flux, while those at Coulissenhieb are exceptionally high in comparison with other European forest ecosystems (MacDonald et al. 2002). The latter might be caused by the underestimation of N inputs by throughfall measurements at the Coulissenhieb site and/or by the low N demand of the mature spruce site.

The nitrate concentrations in soil solutions and fluxes at the Coulissenhieb site were low in 2001 compared to the years before. No clear answer can be given on why this occurred since the N deposition remained constant. As suggested in the synthesis (Matzner et al., this Vol., Chap. 25) increased growth rates in 2000 might have caused high N uptake. The long-term patterns of nitrate in soil solutions (Fig. 20.1) point to the need for long-term measurements to conclude on trends. In other long-term studies, periods of low and high nitrate in soil solutions had amplitudes of several years (Wesselink et al. 1994) and it would be too early to assume a long-lasting decrease of nitrate in soil solutions at the Coulissenhieb site.

20.6 Conclusions

- Under conditions similar to the Coulissenhieb site, recovery from acidification is found in soil solution data and solute fluxes in the soil, but the improvement is delayed by the release of sulfate, likely from both organic and inorganic S pools of the soil.
- The decreased deposition of Ca and Mg has led to a decline of the soil solution concentrations and solute fluxes of these nutrients. The rate of silicate weathering cannot compensate for the decreased deposition. Liming of strongly acidified forest soils is still recommended. These conclusions are valid for the conditions given at the Coulissenhieb site.
- At the Steinkreuz site, the loss of Mg indicates an ongoing depletion of the plant-available soil pools.
- Nitrate fluxes with seepage are substantial at both sites, indicating that the retention capacity of the ecosystem is exceeded by N deposition.
- The interannual variation of solute fluxes, especially those of N, is very high, emphasizing the need for long-term studies to evaluate the trends in element budgets.

Acknowledgements. We would like to thank the members of the Central Analytic Department of the Bayreuth Institute of Terrestrial Ecosystem Research (BITÖK) for the chemical analysis of numerous water samples and their help in the field sampling. Thanks are also due to Uwe Hell, Andreas Kolb, Gerhard Müller and Roland Blasek for their help in the field sampling and site maintenance. This chapter reports data collected over a long time period. We thank the former members of BITÖK, Bernhard Manderscheid and Christine Alewell for their work in the field and data processing in the past. This research was supported by the Federal Ministry of Education and Research (PT BEO51-0339476 C+D).

References

Aber JD, Nadelhoffer KJ, Steuder P, Melillo JM (1989) Nitrogen saturation in northern forest ecosystems – hypotheses and implications. BioScience 39:378–386

Abrahamsen G, Stuanes AO, Tveite B (1994) Long-term experiments with acid rain in Norwegian forest ecosystems. Ecological studies, vol 104. Springer, Berlin Heidelberg New York

Alewell C, Manderscheid B, Gerstberger P, Matzner E (2000) Effects of reduced atmospheric deposition on soil solution chemistry and elemental contents of spruce needles in NE-Bavaria, Germany. J Plant Nutr Soil Sci 163:509–516

Armbruster M, MacDonald J, Dise NB, Matzner E (2002) Throughfall and output fluxes of Mg in European forest ecosystems: a regional assessment. For Ecol Manage 164: 137–147

Armbruster M, Abiy M, Feger K-H (2003) The biogeochemistry of two forested catchments in the Black Forest and the eastern Ore Mountains (Germany). Biogeochem 65:341–368

Berggren D, Mulder J (1995) The role of organic matter in controlling Al solubility in acidic mineral soil horizons. Geochim Cosmochim Acta 59:4167–4180

Cosby BJ, Ferrier RC, Jenkins A, Wright RF (2001) Modelling the effects of acid deposition: refinement, adjustments and inclusion of nitrogen dynamics in the MAGIC model. Hydrol Earth Syst Sci 5:499–517

Cronan CS, Grigal DF (1995) Use of calcium/aluminium ratios as indicators of stress in forest ecosystems. J Environ Qual 24:209–226

Dise NB, Matzner E, Armbruster M, MacDonald J (2001) Aluminium output fluxes from forest ecosystems in Europe: a regional assessment. J Environ Qual 30:1747–1755

Dijkstra FA, van Breemen N, Jongmans AG, Davies GR, Likens GE (2003) Calcium weathering in forested soils and the effect of different tree species. Biogeochem 62:253–275

Feger KH (1997) Biogeochemistry of magnesium in forest ecosystems. In: Hüttl RF, Schaaf W (eds) Magnesium deficiency in forest ecosystems. Kluwer, Dordrecht, pp 67–99

Fölster H (1985) Proton consumption rates in Holocene and present-day weathering of acid forest soils. In: Drever JI (ed) The chemistry of weathering. Reidel, Dordrecht, pp 197–209

Frogner T (1990) The effect of acid deposition on cation fluxes in artificially acidified catchments in western Norway. Geochim Cosmochim Acta 54:769–780

Gbondo-Tugbawa SS, Driscoll CT, Mitchell MJ, Aber JD, Likens GE (2002) A model to simulate the response of a northern hardwood forest ecosystem to changes in S deposition. Ecol Appl 12:8–23

Gerstberger P (2001) Waldökosystemforschung in Nordbayern: Die BITÖK-Untersuchungsflächen im Fichtelgebirge und Steinkreuz. Bayreuther Forum Ökol 901–186

Godbold DL (1991) Aluminium decreases root growth and calcium and magnesium uptake in *Picea abies* seedlings. In: Wright RJ, Baligar VC, Murrmann RP (eds) Plant–soil interactions at low pH. Kluwer, Dordrecht, pp 747–753

Godbold DL (1994) Aluminium and heavy metal stress: from the rhizosphere to the whole plant. In: Godbold DL, Hüttermann A (eds) Effects of acid rain on forest processes. Wiley-Liss, New York, pp 231–264

Gundersen P, Callesen I, de Vries W (1998) Nitrate leaching in forest ecosystems is related to forest floor C/N ratios. Environ Pollut 102:403–407

Johnson NN, Likens GE, Borman FH, Pierce RS (1968) Rate of chemical weathering of silicate minerals in New Hampshire. Geochim Cosmochim Acta 32:531–545

Likens DE, Driscoll CT, Buso DC (1996) Long term effects of acid rain: response and recovery of a forested ecosystem. Science 272:244–246

Likens GE, Driscoll CT, Buso DC, Siccama TG, Johnson CE, Lovett GM, Fahey TJ, Reiners WA, Ryan DR, Martin CW, Bailey SW (1998) The biogeochemistry of calcium at Hubbard Brook. Biogeochem 41:89–173

Lischeid G, Langusch J (2004) Do the data tell the same story? Comparing the INCA model with a neural network analysis. Hydrol Earth Syst Sci (in press)

MacDonald JA, Dise NB, Matzner E, Armbruster M, Gundersen P, Forsius M (2002) Nitrogen input together with ecosystem nitrogen enrichment predict nitrate leaching from European forests. Global Change Biol 8:1028–1033

Manderscheid B, Göttlein A (1995) Wassereinzugsgebiet 'Lehstenbach' das BITÖK-Untersuchungsgebiet am Waldstein (Fichtelgebirge, NO-Bayern). Bayreuther Forum Ökol 18:1–84

Manderscheid B, Matzner E (1995) Spatial and temporal variability of soil solution chemistry and ion fluxes through the soil in a mature Norway spruce (*Picea abies* (L.) Karst) stand. Biogeochemistry 30:99–114

Manderscheid B, Schweisser T, Lischeid G, Alewell C, Matzner E (2000) Sulfate pools in the weathered substrata of a forested catchment. Soil Sci Soc Am J 64:1078–1082

Matzner E (1988) Der Stoffumsatz zweier Waldökosysteme im Solling. Ber Forschungszentrums Waldökosysteme/Waldsterben Univ Göttingen Reihe A40:1–217

Matzner E (1989) Acidic precipitation: case study: Solling, West Germany. In: Adriano DC, Havas M (eds) Advances in environmental science: acid precipitation, vol 1. Springer, Berlin Heidelberg New York, pp 39–83

Matzner E, Grosholz C (1997) Beziehung zwischen NO_3-Austrägen, C/N-Verhältnissen der Auflage und N-Einträgen in Fichtenwald (*Picea abies* Karst.) – Ökosystemen Mitteleuropas. Forstwiss Centralbl 116:39–44

Matzner E, Khanna PK, Meiwes KJ, Lindheim M, Prenzel J, Ulrich B (1982) Elementflüsse in Waldökosystemen im Solling – Datendokumentation. Göttinger Bodenkde Ber 71:1–216

Matzner E, Pijpers M, Holland W, Manderscheid B (1998) Aluminum in soil solutions of forest soils: influence of water flow and soil Al-pools. Soil Sci Soc Am J 62:445–454

Matzner E, Alewell C, Bittersohl J, Lischeid G, Kammerer G, Manderscheid B, Matschonat G, Moritz K, Tenhunen JD, Totsche K (2001) Biogeochemistry of a spruce forest catchment of the Fichtelgebirge in response to changing atmospheric deposition. In: Tenhunen JD, Lenz R, Hantschel R (eds) Ecosystem approaches to landscape management in central Europe. Ecological studies, vol 147. Springer, Berlin Heidelberg New York, pp 463–504

Mulder J, Stein A (1994) The solubility of aluminum in acidic forest soils: long-term changes due to acid deposition. Geochim Cosmochim Acta 58:85–94

Olsson MT, Melkerud PA (2000)Weathering in three podsolized pedons on glacial deposits in northern Sweden and central Finland. Geoderma 94:149–161

Porebska G, Mulder J (1996) The chemistry of aluminum in strongly acidified sandy soils in Poland. Eur J Soil Sci 47:81–87

Prechtel A (2004) Release of stored sulfate from acid forest soils in Europe under decreasing sulphur depositions. Bayreuther Forum Ökologie (in press)

Prechtel A, Armbruster M, Matzner E (2003) Modelling sulphate stream concentrations in the Black Forest catchments Schluchsee and Villingen. Hydrol Earth Syst Sci 7:552–560

Simúnek J, Sejna M, van Genuchten T (1996) HYDRUS-2D simulating water flow and solute transport in two-dimensional variably saturated media. International Ground Water Modeling Center, Golden, Colorado

Stoddard JL, Jeffries DS, Lükewille A, Clair TA, Dillon PJ, Driscoll CT, Forsius M, Johannessen M, Kahl JS, Kellogg JH, Kemp A, Mannio J, Monteith DT, Murdoch PS, Patrick S, Rebsdorf A, Skjelkvale BL, Stainton MP, Traaen T, van Dam H, Webster KE, Wieting J, Wilander A (1999) Regional trends in aquatic recovery from acidification in North America and Europe. Nature 401:575–578

Sverdrup H, Warfvinge P (1993) Calculating field weathering rates using a mechanistic geochemical model PROFILE. Appl Geochem 8:273–283

Ulrich B (1994) Nutrient and acid-base budget of central European forest ecosystems. In: Godbold DL, Hüttermann A (eds) Effects of acid rain on forest processes. Wiley-Liss, New York, pp 1–50

Ulrich B, Mayer R, Khanna PK (1980) Chemical changes due to acid precipitation in a loess-derived soil in central Europe. Soil Sci 130:193–199

Van Breemen N, Finlay R, Lundström U, Jongmans AG, Giesler R, Olsson M (2000) Mycorrhizal weathering: a true case of mineral plant nutrition? Biogeochemistry 49:53–67

Wesselink LG, Mulder J, Matzner E (1994) Modelling seasonal and long-term dynamics of anions in an acid forest soil, Solling, Germany. Geoderma 64:21–39

Wright RF, Alewell C, Cullen JM, Evans CD, Marchetto A, Moldan F, Prechtel A, Rogara M (2001) Trends in nitrogen deposition and leaching in acid sensitive streams in Europe. Hydrol Earth Syst Sci 5:299–310

21 Sequestration Rates for C and N in Soil Organic Matter at Four N-Polluted Temperate Forest Stands

B. Berg

21.1 Introduction

The quantification of terrestrial sources and sinks for carbon dioxide and nitrogen-based greenhouse gases is one of the most important tasks facing environmental scientists today. Central to this is the determination of mechanisms for carbon (C) and nitrogen (N) sequestration in the soil organic matter (SOM). The buildup rate of an SOM layer depends not only on the amount of litter fall and the quality of the decomposing plant litter but also on the completeness of its decomposition. A clear problem is that the accumulation of SOM is a slow process that normally spans generations of scientists, thereby causing continuity problems in studying its buildup as well as the mechanisms controlling it.

Numerous attempts have been made to use mathematical models to describe organic matter decomposition and thus SOM accumulation rates. Among the earliest and most widely used is the simple exponential model for zero-order kinetics, first proposed by Jenny et al. (1949). Critics of this model have focused on the inhomogeneous nature of litter and that the various processes involved in litter mass loss require more complex models as suggested, e.g., by Anderson (1973). Field studies using litter bags have confirmed that the rate of decay, which instantaneously is high, decreases over time (e.g. Fogel and Cromack 1977; Meentemeyer 1978), which is not accounted for in the simple zero-order kinetics model. Developments have resulted in two- and three-factorial models, in which each factor describes the degradation rate of a group of compounds, e.g. 'fast' and 'slow' fractions (Bunell et al. 1977; Couteaux et al. 1998).

If the decay rate approaches zero before all of the organic matter has been decomposed, the accumulated mass loss will asymptotically approach a limit value for decomposition. We define the term 'limit value' as the accumulated mass of decomposing litter that has been lost (in percent) at the time when the decomposition rate is zero. It can be estimated as (Berg and Ekbohm 1991):

Ecological Studies, Vol. 172
E. Matzner (Ed.), Biogeochemistry of Forested Catchments in a Changing Environment

$$\text{m.l.} = \text{m}\left(1 - e^{-kt/m}\right) \tag{1}$$

where m.l. is litter accumulated mass loss, t is time in days, m is the asymptotic level which the accumulated mass loss will ultimately reach and k is the decomposition rate at the beginning of the decay (the derivative of the function at t=0). The recalcitrant 'remaining fraction' (litter-to-humus or LH factor; Berg et al. 2001) is simply calculated as:

$$\text{Remaining fraction} = \left(100 - \text{limit value}\right)/100 \tag{2}$$

Asymptotic models for mass loss during decay have been proposed by several scientists. Such models have the advantage that the recalcitrant or extremely slowly decomposing part may be quantified. Using data for litter incubated in the field, Berg and Johansson (1998) calculated ca. 130 limit values for several litter types, incubated in litter bags in mainly undisturbed and unpolluted forests (e.g. Berg 2000). At the limit-value stage the litter remains contain lignin and lignified holocellulose (Berg et al. 1984), the lignin in part modified. Limit values have been significantly related to the three nutrients N, Mn and Ca (Berg 2000) which influence both the activity and composition of the lignolytic microbial community (Eriksson et al. 1990; Hatakka 2001). Decomposing litter will reach the limit value after different time periods. For example, among litter species incubated in a Scots pine (*Pinus sylvestris* L.) forest, leaves of grey alder (*Alnus incana* L.) reached the limit value within ca. 3 years, whereas for Scots pine needle litter it may take 10–14 years to reach within, say, 1 % of the limit value. For unpolluted systems, limit values have thus been related to the litter species in a first step, species being determining for nutrient levels, e.g. for N (cf. Berg 2000). Globally, the limit value has been related negatively to initial litter N concentrations and positively to those of Mn and Ca (e.g. Berg and McClaugherty 2003). Within species relationships between limit values and nutrients, N and Mn, for example, can be very specific (e.g. Berg et al. 1996). With the use of a database of 149 limit values, equations for calculations of limit values may be set up.

The remaining fraction for a particular forest litter type multiplied by the foliar litter fall (e.g. in kilograms per hectare per year) allows the accumulation rate of SOM to be estimated on, e.g., a yearly timescale. Given an estimate for the accumulated litter fall for a particular forest since the time it colonized a new area, the total accumulation of SOM over that time can also be calculated. Finally, it is possible to estimate the total accumulation of selected nutrients in SOM, such as C, N, P or trace metals, if the concentration of that component in the remaining fraction is known. Such an accumulation may be validated by comparing the estimate with the actually measured amounts in SOM.

In decomposing litter, the N concentration increases linearly with litter mass loss, the relationships being highly significant and with R^2 values nor-

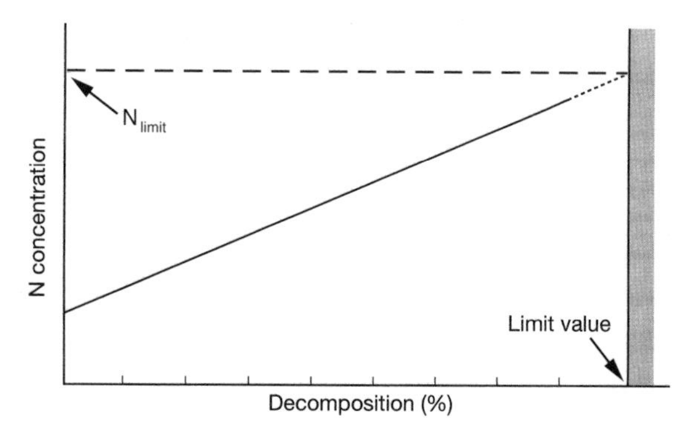

Fig. 21.1. Increasing N concentration in decomposing litter is linear to litter mass loss and N concentration at the limit value (N_{limit}) may be estimated by linear regression. The R^2 value for the relationship is normally above 0.9. (Berg et al. 1999)

mally well above 0.9 (Berg et al. 1999; Fig. 21.1). Different types of litter show different linear relationships, with the slopes depending on such factors as litter species and initial litter N concentration (Berg et al. 1999). The increase continues until decomposition reaches a stage – at the limit value – at which it is extremely slow (cf. Couteaux et al. 1998) or appears to cease completely (Berg et al. 1999).

By extrapolation of the linear relationship to the calculated limit value, the N concentration at the limit value (N_{limit}) is estimated:

$$N_{limit} = k * m + N_{init} \tag{3}$$

where k, the slope of the line, is a unique coefficient for a given set of litter decomposition data but still repeatable within a litter species decomposing under similar environmental conditions (Berg et al. 1999), and N_{init} is the initial litter N concentration (cf. Fig. 21.1). This calculation step was also used to estimate the concentration of N in SOM over 8 species and 48 studies (Berg et al. 1999).

Berg and Dise (2004) define the capacity of a given litter type to sequester N (N_{capac}; in milligrams per gram) as the amount of N remaining when the limit value is reached after decomposing an initial mass of 1 g of litter:

$$N_{capac}\left(mg\ g^{-1}\right) = N_{limit}\left(100 - m\right)/100 \tag{4}$$

Using the above methods and several data sets from decomposing litter from a good number of boreal forests, Berg and Dise (2004) calculated the expected amount of N to be sequestered in the SOM accumulated for four different periods. The calculated values were compared with the measured amounts of N in SOM sequestered over the same four periods, ranging from 120 to 2,984 years. This validation could explain more than 85% of the N sequestration for three of the four stands.

We have not considered root litter in the present evaluation for two main reasons: (1) because we have no information on root production in the stand

and (2) because how far decomposition of, e.g. fine roots goes is not clear. Visual observations indicate that fine roots in undisturbed environments may decompose completely.

The aim of this chapter is to present conceptual models for C and N sequestration in SOM in temperate, N-polluted forest stands. The model was validated for a Norway spruce and a European beech stand and then applied to two other stands, one with spruce and one with beech/oak. Both models have been developed and validated in previous studies (Berg et al. 1999, 2001; Berg and Dise 2004). The term 'soil organic matter' in this paper is used as defined by Waksman (1936) and Stevenson (1982).

21.2 Sequestration of Carbon in SOM Under Growing Stands of Norway Spruce and European Beech – A Model Validation

We validated the model for carbon sequestration as defined by Eqs. (1) and (2) by applying it to the Solling forest systems in which systematic measurements of amounts of carbon in SOM had been carried out for over 30 years. The used site (Solling) has plots with monocultures of European beech (*Fagus sylvatica* L.) and Norway spruce [*Picea abies* (L.) Karst.]. It is located in a nature reserve about 60 km NW of the city of Göttingen (51°31′N; 09°34′E), in central Germany and at an altitude of about 500 m. In 1995 the stand with European beech was 148 years old and that with Norway spruce 111 years. The average annual precipitation is 1,090 mm and annual average temperature 6.4 °C. Further descriptions are given by Ellenberg et al. (1986).

Measured data for accumulation of C in SOM was taken from Meiwes et al. (2002; Fig. 21.2). In the Norway spruce stand the amount of C increased from 25 tons ha^{-1} in 1968 to 56 tons ha^{-1} in 2001. In the stand with European beech the amount of C increased from ca. 15 tons ha^{-1} in 1965 to ca. 26 tons ha^{-1} in 2001. The measured data had a clear variation and to obtain an average accumulation or sequestration rate I regressed the measured accumulated amounts against time and obtained an average rate for the measured increase. For this I use the abbreviation AMSR (average measured sequestration rate) hereafter.

Average annual foliar litter fall was 3,284 (n=8) and 2,948 (n=7) kg ha^{-1} for the Norway spruce and European beech stands, respectively (Table 21.1). Norway spruce needle litter had an average N concentration of 10.8 mg g^{-1}, an Mn concentration of 1.2 mg g^{-1} and a Ca concentration of 4.0 mg g^{-1}. The foliar litter of European beech had an N concentration of 12.8 mg g^{-1}, an Mn concentration of 1.8 mg g^{-1} and a Ca concentration of 5.5 mg g^{-1} (Table 21.1) (Matzner et al. 1982).

At none of the sites were litter decomposition studies carried out that could be used for direct calculations of limit values; these were therefore modelled

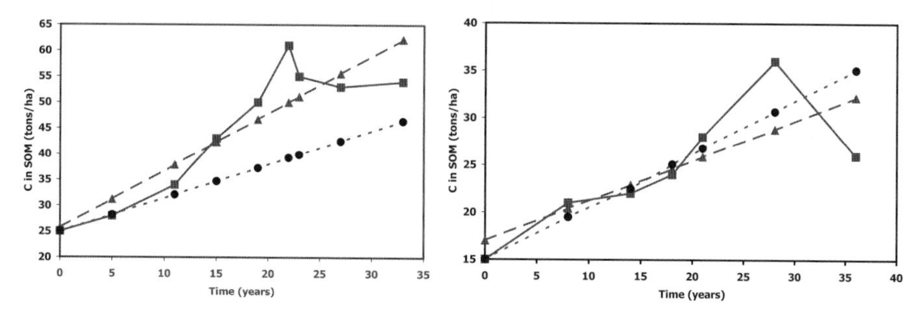

Fig. 21.2. Increase in SOM-C in the forest floor of a Norway spruce stand over 33 years (**a**) and that of a European beech stand over 36 years (**b**) in the Solling project. *Squares* and *solid line* indicate measured SOM-C data (taken from Meesenburg et al. 1999 and Meiwes et al. 2002); *triangles* and *dashed line* linear regressions of measured data as well as those modelled (*dotted line* and *circles*)

using available data for initial litter concentrations of N, Mn and Ca which have causal relationships to lignin degradation and are related to the limit values (cf. Berg et al. 1996). Using a database (Berg and Johansson 1998) with 149 calculated limit values, we developed two equations by relating limit values for Norway spruce to initial litter concentrations of N, Mn and Ca (Eq. 5).

For Norway spruce litter I had a set of 25 limit values for decomposition that did not have a normal distribution when compared to the initial N concentration for the corresponding litter. Therefore I created a set of average values based on litter N concentrations, using steps of 0.5 mg g^{-1} as an interval and obtained 11 values of which two were based on n=1 and were excluded since they were clear outliers. Thus an equation was created using the nine average limit values and average litter initial concentrations of N (N_{init}), Ca (Ca_{init}) and Mn (Mn_{init}):

$$\text{Limit value}_{spruce} = 88.1 - 2.99N_{init} + 0.98Mn_{init} + 0.54Ca_{init}$$
$$R^2 = 0.853;\ p < 0.01;\ n = 9. \qquad (5)$$

The range of litter nutrient concentrations in the original data set used to generate Eq. (5) were: N 4.2–13.6 mg g^{-1}, Mn 0.3–3.91 mg g^{-1} and Ca 2.7–31.7 mg g^{-1}. When applying Eq. (5) I used local litter nutrient concentrations (Table 21.1) to estimate the limit value for Norway spruce needle litter at site Solling and obtained a value of 60.5 %.

For European beech I used available data for deciduous leaf litter and related limit values to the litter chemical composition (Eq. 6). I used a data set with in all 26 sets of limit values for deciduous litter and initial litter concentrations for N_{init}, Ca_{init} and Mn_{init}. The values in that data set did not have a normal distribution. The range of litter nutrient concentrations in the data

Table 21.1. Data on average litter fall, its average concentrations of N, Mn and Ca, calculated limit values and N concentration at the limit value as well as concentrations of C and N in SOM

Site	Average foliar litter fall ($kg\ ha^{-1}\ year^{-1}$)	Concentrations in litter N ($mg\ g^{-1}$)	Mn (%)	Ca ($mg\ g^{-1}$)	Limit value	N conc. at limit value	Humus C/N	Humus org N	Reference
Norway spruce									
Solling	3,284	10.8	1.2	4.0	60.5	18.5	18.3	27.3	Matzner et al. (1982)
Coulissenhieb	2,141	12.1	0.3	6.1	55.4	19.1	20.7–22.6	20.9	Alewell (2001); Berg and Gerstberger (2004)
European beech									
Solling	2,948	12.8	1.8	5.5	61.9	24.4	25.5	19.6	Matzner et al. (1982)
Mixed deciduous stand at Steinkreuz									
European beech	3,118	13.3	1.8	7.0	63.9	25.4	17.8–18.9		Berg and Gerstberger (2004)
Mixed oak[a]	730	12.8	2.2	8.1	67.6	25.5	17.8–18.9		Berg and Gerstberger (2004)

[a] European oak and sessile oak

used to generate an equation were: N 5.6–16.8 mg g^{-1}, Mn 0.05–4.7 mg g^{-1} and Ca 5.6–21.2 mg g^{-1}. From this basic data set I had excluded all green litter. It turned out that for deciduous litter the green leaves formed a distinct group that deviated somewhat from the pattern of the normal litter. Using the same procedure as for Eq. (5) (above) I obtained an equation for n=13:

$$\text{Limit value} = 45.2 - 0.04\,\text{N}_{\text{init}} + 5.06\text{Mn}_{\text{init}} + 1.45\text{Ca}_{\text{init}}$$
$$R^2 = 0.423;\ p < 0.05;\ n = 13. \quad (6)$$

Using Eq. (6) and local leaf litter nutrient concentrations at the Solling site (Table 21.1) I estimated the limit value for European beech leaf litter and obtained a value of 61.9 %. The calculated limit values (Eqs. 5 and 6) gave remaining fractions of 0.39 and 0.38, respectively, for spruce and beech (Eq. 2). With the annual average litter fall of 3,284 and 2,948 kg ha^{-1} for spruce and beech, respectively, (Matzner et al. 1982; Table 21.1) and an assumed litter C concentration of 50 %, the annual increase for SOM-C became 649 kg ha^{-1} for spruce and 562 kg for beech (cf. Fig. 21.2). The *modelled sequestration rate* is abbreviated to MOD-SR.

I used the MOD-SR to recalculate the sequestration over 33 and 36 years for spruce and beech, respectively. I also compared the AMSR and MOD-SR for carbon (Fig. 21.2) and obtained for spruce the amount of 62 tons accumulated C using the AMSR and 46 tons using the MOD-SR, which means that the model equation could explain close to 76 % of the increase.

For European beech I obtained an accumulation rate of 32.2 tons of C using the relationship based on measured values (AMSR) and 35.1 tons using the model. The MOD-SR would thus overestimate the sequestration rate by 9 %.

In the above estimates we have used foliar litter only, whereas the SOM buildup may be supported also by recalcitrant remains of woody litter and of cones and nuts. Normally, woody litter, for example, should have low nutrient levels, especially for N, which will result in different kinetics, and decomposition patterns that are not predictable (e.g. Berg and McClaugherty 2003). In a more N-enriched stand the N concentration in wood could be higher and high enough to hamper the decomposition, and the decomposing litter may contribute to the SOM buildup. In comparison, in the unpolluted stands in northern Scandinavia used by Berg et al. (2001) for SOM budget calculations, better model fits were reached when the budgets were set up using foliar litter only.

21.3 Sequestration Rates of Carbon in SOM at the Sites Coulissenhieb and Steinkreuz

At the site Coulissenhieb the average concentrations of N, Mn and Ca in foliar litter fall were 12.1, 0.26 and 6.06 mg g^{-1}, respectively (Berg and Gerstberger, this Vol.), calculated as averages of four to five values. At the site Steinkreuz the average beech leaf litter concentrations of N, Mn and Ca were 13.3, 1.8 and 7.0 mg g^{-1}, respectively, and for the leaf litter fraction of the two oak species combined they were 12.8, 2.2 and 8.1, respectively. The average values were based on six samplings of litter fall.

No litter decomposition studies had been carried out at either of the sites that could be used for direct calculation of limit values, and this was therefore modelled using Eq. (5) for the site Coulissenhieb and Eq. (6) for the site Steinkreuz and available data for initial litter concentrations of N, Mn and Ca. The resulting limit values were 55.4 % for Norway spruce litter, 63.9 % for beech leaf litter and 67.6 % for oak leaves (Table 21.1). The ranges of litter nutrient concentrations in the original data set used to generate Eqs. (5) and (6) covered the ranges determined for both sites.

The average annual foliar litter fall at the site Coulissenhieb was 2,141 kg ha^{-1}. Using the modelled limit value, I estimated an annual SOM accumulation of 955 kg corresponding to 477 kg C ha^{-1}. At the site Steinkreuz the annual average foliar litter fall was 3,118 kg ha^{-1} for European beech and 730 kg ha^{-1} for the two oak species sessile oak (*Quercus petrea* Mattuschka Lieblein) and European oak (*Quercus robur* L.) (Table 21.1). With the modelled limit values (Eq. 6) for beech and oak leaf litter I estimated an annual SOM accumulation of 1,126 kg from the beech leaf litter and 236 kg from oak leaf litter. The total annual SOM accumulation from foliar litter was thus 1,362 kg ha^{-1}, corresponding to 681 kg C ha^{-1} $year^{-1}$.

I have assumed a linear increase in SOM-C like that found by Berg et al. (2001) in a study covering almost 3,000 years. In the present comparison with several measurements of SOM-C we cannot exclude a large variation among samplings, which may be interpreted as, e.g., a maximum amount or a plateau (cf. Fig. 21.2). Over a period of more than 30 years single trees may die and be taken out of the stand, which would decrease the litter input rate and thus also decrease the growth rate of the SOM layer.

The present approach with limit values may be compared with a completely different approach taken by Persson et al. (2000). Respiring Coulissenhieb samples at the laboratory, they estimated the annual respiration from the L, F and H layers in the field. We may make a bold comparison with the total litter fall at this site. Thus, Persson et al. (2000) estimated an annual soil heterotrophic respiration loss of 1,012 kg ha^{-1}. When we compare that amount with the measured annual C inflow in total litter fall which is ca. 1,623 kg we obtain an accumulation of 610 kg C ha^{-1} $year^{-1}$ or a mineralization of 62.4 %.

This number may be compared with the limit value of 55.4 % for Norway spruce needle litter. Considering the difference in methods, this may be a coincidence, yet it may be valuable to consider alternative methods to determine the stable SOM fraction formed from decomposing litter.

21.4 Sequestration of Nitrogen in SOM Under Growing Stands of Norway Spruce and European Beech at the Site Solling – A Model Validation

The monocultural forests of Norway spruce and European beech at the site Solling have long-term recordings of SOM-N accumulation in the forest floor. In the Norway spruce stand the measured amount increased over 33 years from 950–2,050 kg ha^{-1}. In the European beech stand the amount of N bound in SOM increased from ca. 800–1,200 kg ha^{-1} over 36 years (Fig. 21.3; Meiwes et al. 2002).

The average annual input of N with foliar litter fall was 35.2 kg ha^{-1} for the spruce stand and 36.3 kg ha^{-1} for that with beech (Table 21.2).

We modelled the increases in stored N using foliar litter fall, the estimated limit values, N concentration at the limit value using Eq. (3) and the capacity of the litter to store N (N_{capac}) using Eq. (4). To calculate N_{limit} we used a *k* value of 0.127 for spruce litter and 0.188 for beech, the values taken as averages from spruce and deciduous litter, respectively, from a paper by Berg et al. (1999). We thus obtained SOM-N accumulation rates of 31.4 and 27.4 kg N ha^{-1} $year^{-1}$ for the Norway spruce and the European beech stands, respectively.

To facilitate a comparison of the measured and the modelled sequestered amounts, we regressed them against time, compared the linear relationships

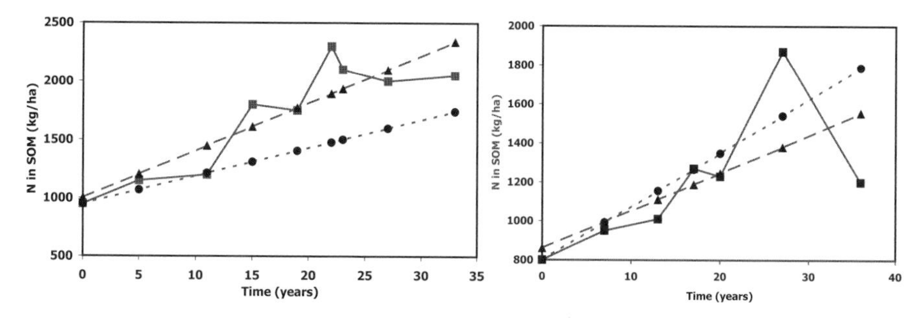

Fig. 21.3. Increase in SOM-N in the forest floor of Norway spruce over 33 years (**a**) and European beech over 36 years (**b**) in the Solling project. *Squares* and *solid line* indicate measured SOM-N data (taken from Meesenburg et al. 1999 and Meiwes et al. 2002); *triangles* and *dashed line* linear regression of measured data as well as those modelled (*dotted line* and *circles*)

Table 21.2. Measured values for annual N inflows and outflows and modelled values for N sequestration

Site	N input to the soil ($kg\ ha^{-1}\ year^{-1}$)		N sequestration		N loss with seepage water	Reference
	Deposition	Foliar litter	Measured	Modelled	Measured	
Norway spruce						
Solling	45	35.2	40.6	31.3	22.7	Marques (1999)
Coulissenhieb	40–25	25.9	–	18.2	22.1	Matzner et al. (this Vol.)
European beech						
Solling	35	36.3	19.3	27.4	5.2	Matzner (1988)
Mixed deciduous stand at						
Steinkreuz	22–17					Matzner et al. (this Vol.)
European beech		41.5		28.5		
Mixed oak		9.3		6.0		
Combined		50.8		34.5		

and obtained for the Norway spruce stand an average measured increase (AMSR) for nitrogen of 40.6 kg ha^{-1} $year^{-1}$, and in the European beech stand the measured amount of N bound in SOM increased over 36 years at a rate of 19.3 kg ha^{-1} $year^{-1}$ (Fig. 21.3). Thus the ratio of modelled to measured was 0.70 for the spruce stand and 1.42 for the beech stand.

21.5 How Much Can the Model Explain of the N Sequestration? – A Validation

The ratio between modelled (MOD-SR) and measured (AMSR) rates for N sequestration was ca. 70 % for the Norway spruce stand and 142 % for that of European beech, respectively (cf. Fig. 21.3). In their synthesis on SOM-N accumulation over 2,984, 2,081, 1,106 and 120 years, Berg and Dise (2004) compared the accumulated amounts of N – 761, 460, 163 and 18.8 g m^{-2}, respectively (Wardle et al. 1997) – with those modelled using the present approach. The 2,984-year-old stand was modelled to have stored 677 g m^{-2} which means 89 % of the 761 g actually measured in the field. The stands with an average age of 2,081 years had a modelled accumulation of 453.2 g m^{-2} which means that the model estimated ca. 99 % of what actually had accumulated (460 g). Finally, the stands aged on average 1,106 years were estimated to accumulate 213.2 g m^{-2} which was 131 % of the amount actually accumulated (163 g). For the 120-year-old monoculture of Scots pine they obtained 21.3 g m^{-2}; when measuring, this modelled amount was 133 % of that actually measured. Thus, the deviations of the modelled amounts at both the N-polluted Solling stands appear to be of the same magnitude as for the completely unpolluted ones.

The spruce stand at Solling accumulated N at a rate approximately twice that for the beech stand, although the N input with foliar litter to the forest floor was at least as high in the beech stand as in that with spruce. Litter species may have different capacities (as defined by Eq. 4) to store N as observed by Berg and Dise (2004). For unpolluted systems they found a two to three times higher capacity for beech leaves to store N as compared to spruce needles. Berg and Dise (2004) investigated and related the capacity to store N for six litter species and related the capacity to the litters' initial N concentrations. In the present case the initial N levels in litter were similar (Table 21.1), and the suggestion by Berg and Dise (2004) that a modified N level may be more important than species appears to hold in this case. In these N-polluted Solling stands the N_{capac} values were similar, with 27.4 and 24.0 mg N g^{-1} stored in recalcitrant remains from beech and spruce, respectively.

The difference between the amount of N in foliar litter input in the Norway spruce stand (35.2 kg) and that actually accumulating in SOM (40.6 kg) was 5.4 kg N ha^{-1} $year^{-1}$ and may be due to storage caused by N input by woody litter. This should be regarded as a speculation only based on the following discussion: without the repressing effect of N on the lignin degradation, wood may degrade very quickly when attacked by white rot or very slowly when attacked by brown rot, a pattern that is not yet predictable (Berg and McClaugherty 2003). At N levels high enough to repress the decomposition (cf. Eriksson et al. 1990) the degrading wood components may decompose towards a limit value, thus contributing to the SOM buildup.

For the beech stand at Solling the difference between the N input in foliar litter (36.3 kg) and the measured value for N sequestration (19.3 kg) was 16 kg, an amount available for root uptake or for outflow processes from the system.

21.6 Sequestration Rates of Nitrogen in SOM in a Norway Spruce Forest at the Sites Coulissenhieb and Steinkreuz

The site Coulissenhieb had an annual inflow of N in foliar litter fall of 25.9 kg ha^{-1} (Table 21.2). Using the limit values for spruce needle litter (55.4 %) and Eqs. (3) and (4), the calculated storage rate of N (MOD-SR) with origin in shed foliar litter was 18.2 kg ha^{-1}. Thus, ca. 7.7 kg ha^{-1} $year^{-1}$ would be available for root uptake from the decay of foliar litter.

At the site Steinkreuz the annual inflow of N in foliar litter fall was 50.8 kg ha^{-1} (Table 21.2). Using the limit values for beech leaf litter (65.2 %) and for oak (69.0 %) the modelled sequestration rate for N (MOD-SR) in SOM originating from shed foliar litter was 34.5 kg ha^{-1} (Table 21.2), thus leaving 16.3 kg ha^{-1} $year^{-1}$ for root uptake and losses from the system.

21.7 The Missing C and N Fractions – Can They Be Explained?

At the Solling spruce stand the annual C input in foliar litter fall to the forest floor was 1,642 kg ha^{-1}. The average measured accumulation rate of 1,106 kg ha^{-1} $year^{-1}$ means that the difference of 536 kg should have decomposed. The model predicts 649 kg C as annual storage rate based on needle litter only. The difference of 457 kg between measured and predicted is thus unaccounted for.

We may speculate that as for other forests, the foliar litter fall is ca. 70 % of the total (e.g. Mälkönen 1974; Berg et al. 1993). Based on that assumption the total amount of C in litter fall for this stand including twigs and cones would be ca. 2,350 kg C ha^{-1}, of which the C in twigs and cones fall alone would be ca. 710 kg ha^{-1}. We may speculate about what fraction of these components would be stabilized. Assuming that the fraction is of a similar magnitude as that for the needles the model may explain ca. 930 of the 1,106 kg annual sequestration. One reason to speculate about the woody litter is that in an N-polluted stand the N levels also in the woody material would increase and we may expect a behaviour similar to that of the needles, namely a hampering of the degradation rate leading to such a low rate that a limit value can be estimated.

For the European beech stand at Solling the measured C accumulation rate of 422 kg ha^{-1} means that of the 1,474 kg annual C input, ca. 1,050 kg should decompose. It may be difficult to develop a further discussion considering the large variation in measured SOM values for the beech stand (Fig. 21.2). Still, we may argue that for deciduous foliar litter very few limit values have been estimated based on mass-loss data, and for the calculation made in this chapter we combined information covering different deciduous species. For Scots pine and Norway spruce Berg et al. (1996) and Berg and Johansson (1998) found different relationships vs litter nutrients, resulting in different equations, and we cannot exclude that there may be a source of considerable error connected to the estimate of limit values for the decomposing litter, maybe in the order of 50 %.

When comparing the sequestration rate of N for the Norway spruce stand at Solling we may see that whereas the annual N input to the system in foliar litter is 35.2 kg, the measured storage rate is somewhat higher (40.6 kg), suggesting that woody litter with raised N concentrations may be the reason. For the estimate of N sequestration rate in the beech stand the above-mentioned uncertainty applies. Further, the variation in SOM-N determined over time was so uncertain that I did not obtain a significant linear relationship (Fig. 21.3).

21.8 How Stable Are the Long-Term Stored C and N Fractions in SOM?

The fact that the amount of N stored in SOM increased with time indicates a certain stability over time of the compound holding N. The long-term predictability based on the limit-value concept (Berg and Dise 2004) further supports this. There is, however, no proof that the N (and C) remaining in the litter when the limit value is reached is in a form that is completely stable to biological degradation. Thus, Couteaux et al. (1998) estimated three different fractions of C in SOM with very different degradation rates. The major, recalcitrant fraction, estimated to make up over 90 % of the SOM mass, had a decomposition rate of less than 0.0001 % day^{-1} which indicates a long-term stability.

The stability of stored C and N in SOM is in part dependent on the composition and activity of the microbial community and factors ruling them. A given amount of SOM that has accumulated for, e.g. a century, may be decomposed in a relatively short time if the limiting conditions for the microbial community change and possibly nutrient stress for the trees opens a mechanism for a high fungal activity (e.g. Hintikka and Näykki 1967). The other extreme is a more or less complete stability, a property that so far has been less easy to demonstrate. Still, Berg et al. (2001) and Berg and Dise (2004) reconstructed the amounts of C and N stored in mor humus with missing fractions of between 11 and 2 % for C, indicating that the C may have a long-term stability. For a period of 2,984 years the modelled N could explain 89 % of the measured sequestration (Berg and Dise 2004). These soil systems were located under growing forest stands.

Wardle et al. (1997) concluded that the oldest SOM-N in the stands studied by them (2,984 years) was biologically less available than that of the younger ones. That conclusion was based on experiments on the availability of N to plants. They also found that N concentrations in SOM (range ca. 1.0–1.5 %) were negatively related to the age of the system, which may be interpreted as there being a certain turnover of C but that N had been kept in the system which appeared to be N limited. For that very same system Berg and Dise (2004) used data from N_2 fixation studies and found that the amounts stored should be explained by the N_2 fixation rates for boreal coniferous systems, which supports an N limitation. We may speculate that although there may be a low turnover within the system, it appears that N is stored better than C. This conclusion appears valid for an undisturbed and unpolluted system, though.

21.9 Concluding Remarks

The present study encompassed four temperate N-polluted forests in central and southern Germany; the Solling site with monocultures of European beech and Norway spruce, the Coulissenhieb site with Norway spruce and the Steinkreuz site with a mix of mainly European beech and sessile oak. All sites had data on litter fall and the Solling site also had long-term measurements of accumulated SOM, including C and N determinations, allowing a validation.

In the stands of beech and spruce at Solling the average annual sequestration rates of C in SOM were measured to be 0.422 and 1.106 tons ha^{-1}, respectively. For N the rates were 19.3 and 40.6 kg ha^{-1} $year^{-1}$, respectively, in the beech and spruce stands. The model predicted 109 and 76 % of the measured C sequestration rate for the beech and spruce stands, respectively, and for N the model predicted sequestration rates of 142 and 70 %, respectively, of the measured rates.

The sequestration rates for the N-polluted stands Coulissenhieb and Steinkreuz were 0.477 and 0.679 tons ha^{-1} $year^{-1}$ for C, respectively, and 18.2 and 34.5 kg ha^{-1} $year^{-1}$ for N, respectively. The reconstruction of SOM buildup in a system limited in N has given us a tool to estimate the natural accumulation of N and C and thus the sequestration of potential greenhouse gases. Such a storage mechanism may be valid also under at least a certain pollution pressure; thus we may have a tool to estimate minimum amounts sequestered in N-polluted systems. The different capacities of different tree species to retain N in SOM may be positively related to the N status of the system, with the species being connected to systems with a higher N status having higher storage capacity.

Whereas it appears that the validations of C- and N-sequestration rates for boreal forests give acceptable results, the available data for deciduous forests have not yet allowed us to make an acceptable validation.

Acknowledgements. The study was carried out within the framework of the European Union project CN-ter (contract number QLK5-2001-00596).

References

Alewell C (2001) Predicting reversibility of acidification: the European sulfur story. Water Air Soil Pollut 130:1271–1276

Anderson J (1973) The breakdown and decomposition of sweet chestnut (*Castanea sativa* Mill.) and beech (*Fagus sylvatica* L.) leaf litter in two deciduous woodland soils. I. Breakdown, leaching and decomposition. Oecologia 12:251–274

Berg B (2000) Litter decomposition and organic matter turnover in northern forest soils. For Ecol Manage 133:12–22

Berg B, Ekbohm G (1991) Litter mass-loss rates and decomposition patterns in some needle and leaf litter types. Long-term decomposition in a Scots pine forest VII. Can J Bot 69:1449–1456

Berg B, Johansson MB (1998) Maximum limit for foliar litter decomposition – a synthesis of data from forest systems, part 1. In: Berg B (ed) A maximum limit for foliar litter decomposition – a synthesis of data from forest systems. Reports from the Departments of Forest Ecology and Forest Soils, vol 77, Swedish University of Agricultural Sciences, Uppsala, p 158

Berg B, Dise N (2004) Validating a new model for N sequestration in forest soil organic matter? Water Air Soil Pollut Focus (in press)

Berg B, McClaugherty C (2003) Plant litter. Decomposition. Humus formation. Carbon sequestration. Springer, Berlin Heidelberg New York

Berg B, Ekbohm G, McClaugherty CA (1984) Lignin and holocellulose relations during long-term decomposition of some forest litters. Long-term decomposition in a Scots pine forest IV. Can J Bot 62:2540–2550

Berg B, Berg M, Flower-Ellis JGK, Gallardo A, Johansson M, Lundmark JE, Madeira M (1993) Amounts of litterfall in some European coniferous forests. In: Breymeyer A (ed) Proc Scope Seminar, Conf Pap 18, Geography of Carbon Budget Processes in Terrestrial Ecosystems, Szymbark, 17–23 Aug 1991, pp 123–146

Berg B, Ekbohm G, Johansson MB, McClaugherty C, Rutigliano F, Virzo De Santo A (1996) Some foliar litter types have a maximum limit for decomposition – a synthesis of data from forest systems. Can J Bot 74:659–672

Berg B, Laskowski R, Virzo De Santo A (1999) Estimated N concentration in humus as based on initial N concentration in foliar litter – a synthesis. Can J Bot 77:1712–1722

Berg B, McClaugherty C, Virzo de Santo A, Johnson D (2001) Humus buildup in boreal forests – effects of litter fall and its N concentration. Can J For Res 31:988–998

Bunell F, Tait DEN, Flanagan PW, van Cleve K (1977) Microbial respiration and substrate weight loss. I. A general model of the influence of abiotic variables. Soil Biol Biochem 9:33–40

Couteaux MM, McTiernan K, Berg B, Szuberla D, Dardennes P (1998) Chemical composition and carbon mineralization potential of Scots pine needles at different stages of decomposition. Soil Biol Biochem 30:583–595

Ellenberg H, Mayer R, Schauermann J (1986) Ökosystemforschung: Ergebnisse des Sollingprojekts 1966–1986. Ulmer, Stuttgart

Eriksson KE, Blanchette RA, Ander P (1990) Microbial and enzymatic degradation of wood and wood components. Springer series in wood science. Springer, Berlin Heidelberg New York

Fogel R, Cromack K (1977) Effect of habitat and substrate quality on Douglas fir litter decomposition in western Oregon. Can J Bot 55:1632–1640

Hatakka A (2001) Biodegradation of lignin. In: Hofman M, Stein A (eds) Biopolymers, vol 1. Lignin, humic substances and coal. Wiley, Weinheim, pp 129–180

Hintikka V, Näykki O (1967) Notes on the effects of the fungus *Hydnellum ferrugineum* on forest soil and vegetation. Commun Inst Forest Fenn 62:1–22

Jenny H, Gessel SP, Bingham FS (1949) Comparative studies of decomposition rates of organic matter in temperate and tropical regions. Soil Sci 68:419–432

Mälkönen E (1974) Annual primary production and nutrient cycle in some Scots pine stands. Commun Inst Forest Fenn 84:5

Marques MC (1999) Eintrag von luftgetragenen partikelbebundenen Spurenstoffen in Wälder durch trockene Deposition. Wissenschaftlicher Verlag, Berlin

Matzner E (1988) Der Stoffumsatz zweier Waldökosysteme im Solling. Ber Forschungszentrums Waldökosysteme/Waldsterben A40:1–217

Matzner E, Khanna PK, Meiwes KJ, Lindheim J, Prenzel J, Ulrich B (1982) Elementflüsse in Waldökosystemen im Solling – Datendokumentation. Göttinger Bodenkd Ber 71: 1–267

Meesenburg H, Meiwes KJ, Bartens H (1999) Veränderung der Elementvorräte im Boden von Buchen- und Fichtenökosystemen in Solling. Ber Freiburger Forst Forsch 7:109–114

Meentemeyer V (1978) Macroclimate and lignin control of litter decomposition rates. Ecology 59:465–472

Meiwes KJ, Meesenburg H, Bartens H, Rademacher P, Khanna PK (2002) Accumulation of humus in the litter layer of forest stands at Solling. Possible causes and significance for the nutrient cycling. Forst Holz 13–14:428–433 (in German, English summary)

Persson T, Karlsson PS, Seyferth U, Sjöberg RM, Rudebeck A (2000) Carbon mineralization in European forest soils. In: Schulze ED (ed) Carbon and nitrogen cycling in European forest ecosystems. Ecological studies, vol 142. Springer, Berlin Heidelberg New York, pp 257–275

Stevenson FJ (1982) Humus chemistry. Genesis, composition, reactions. Wiley, New York

Waksman SA (1936) Humus. Williams and Wilkins, Baltimore

Wardle DA, Zachrisson O, Hörnberg G, Gallet C (1997) The influence of Island area on ecosystem properties. Science 277:1296–1299

22 Riparian Zones in a Forested Catchment: Hot Spots for Microbial Reductive Processes

K. KÜSEL and C. ALEWELL

22.1 Introduction

Anthropogenic deposition of protons, sulfate and nitrate has resulted in acidification of the groundwater in forested catchments (Hamm 1995). Nitrate and sulfate are leached from the upland aerated soils into lower situated soils and waterlogged fens with each rainfall. Under anoxic conditions in soils, nitrate and sulfate can be reduced by microorganisms and alkalinity is generated (Sexstone et al. 1984; Tiedje et al. 1984). Thus, depending on the soil characteristics, climatic parameters, and the composition of the soil microbiota, wetland soils may act as long-term sinks for deposited protons, sulfate, and nitrate. There is a principal difference between the retention of nitrate and sulfate in wetlands with respect to the sink function. The microbial reduction of nitrate leads to gaseous products, like or N_2O or N_2, which are lost to the atmosphere. Thus, the generation of alkalinity is permanent, and wetlands act as long-term sinks. In contrast, the reduction of sulfate leads to the formation of sulfide, which can be either fixed in the presence of iron as iron sulfides or as organic reduced sulfur, or it can be oxidized back to sulfate under changing redox conditions (Wieder and Lang 1988; Fig. 22.1). In the latter case, wetlands would not act as a long-term sink for sulfate.

In contrast to fresh-water or marine sediments, which have been studied in detail relative to the phylogenetic diversity of anaerobic bacteria and their anaerobic activities at oxic-anoxic gradients (e.g. Jørgensen 1978; Whiticar et al. 1986; Capone and Kiene 1988; Llobet-Brossa et al. 1998), little information is available from peatlands (Nedwell and Watson 1995; Wieder et al. 1996). However, wetlands are known to be significant sources of the greenhouse gas methane (Williams and Crawford 1984).

In this chapter, anaerobic microbial activities and geochemical parameters of different wetland sites of the Lehstenbach catchment are described to evaluate whether these wetlands act as long-term sinks for nitrate and sulfate. The field sites evaluated differ with respect to water saturation, vegetation, and the availability of iron.

Ecological Studies, Vol. 172
E. Matzner (Ed.), Biogeochemistry of Forested Catchments in a Changing Environment

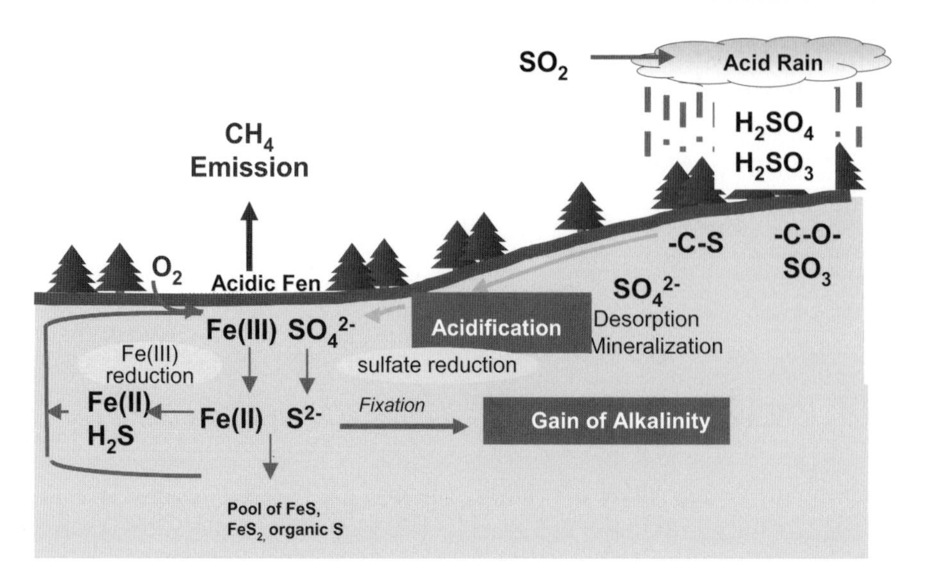

Fig. 22.1. Hypothetical model illustrating turnover of sulfur and iron in a forested catchment subjected to atmospheric deposition of sulfur

Varying methods including stable sulfur isotopes, ^{35}S radiolabeling studies, geochemical analyses, incubation experiments, and molecular biological studies are used to address the sink function of wetlands. It is hypothesized that the dissimilatory reduction of nitrate and sulfate in the wetlands contributes to the retention of nitrogen and sulfur and to the production of alkalinity to a significant extent in the Lehstenbach catchment.

22.2 Site Description of Wetland Sites

Approximately 30% of the Lehstenbach catchment is covered by wetland areas (Gerstberger et al., this Vol.). Some lower situated soils are only water saturated if the groundwater level increases during autumn. In these intermittent seeps, represented by the site Köhlerloh, the groundwater level reaches the soil surface during autumn, winter, and spring, whereas it is about 0.5–1 m below the surface during summer. After water saturation, depth profiles of the soil solution (0–40 cm) can be obtained with dialyses chambers. Köhlerloh is framed by a creek and a drainage trench to allow planting of spruce. Thus, drainage of the site is relatively efficient. Open areas on the site Köhlerloh are vegetated with *Sphagnum* mosses or are partly covered with a dense layer of *Vaccinium myrtilus*. Soils at Köhlerloh were classified as Dystric Gleysols.

Peatlands are defined as any area with a waterlogged, predominantly organic substrate of at least 30–40 cm thickness (Glaser 1987). In the Lehstenbach catchment, peatlands are dominated by minerotrophic fens that are in contact with nutrient-rich groundwater and runoff from uplands. The representative site Schlöppnerbrunnen I is a fen alternately covered with patches of *Sphagnum* mosses and with spruce stocking. The soil is classified as Fibric Histosol. The site Schlöppnerbrunnen II is also a fen completely overgrown by *Molinia caerula* grasses. The Fibric Histosols at this site are underlain by a Gleysol in 70–90 cm depth. The soil pH of both sites approximated 3.8 and 4.4, respectively. In the soil solution, the pH varied between 4 and 5.5 at Schlöppnerbrunnen I and between 4.5 and 6 at Schlöppnerbrunnen II. In contrast to Schlöppnerbrunnen I, the fen at Schlöppnerbrunnen II is enriched both with dithionite-extractable pedogenic iron oxides (Fe_d) and oxalate-extractable poorly crystallized iron oxides (Fe_o) (Table 22.1). Weakly crystallized Fe(III) oxides are the favored reducible forms of Fe(III) for microbial reduction (Munch and Ottow 1980). Schlöppnerbrunnen II receives anoxic Fe(II)-rich groundwater from intermittent seeps and fens located northeast in the Lehstenbach catchment. In the area of Schlöppnerbrunnen II, the groundwater effluences, and Fe(II) is oxidized in the presence of O_2 and precipitates as Fe(III) oxide in upper peat layers near the surface. Thus, considerable amounts of Fe(III) are available as alternative electron acceptor for the oxidation of organic matter at Schlöppnerbrunnen II.

Upland soils in the catchment are described in detail in a previous chapter (Gerstberger et al., this Vol.). For comparison, some data of the upland sites Coulissenhieb, Weidenbrunnen, and Gemös are used in this chapter.

Table 22.1. Geochemical parameters of characteristic field sites of the Lehstenbach catchment. Presented are the averages from duplicate soil samples. Fe_d Pedogenic Fe-oxides; Fe_o poorly crystallized Fe-oxides, hydroxides, and associated gels; *n.a.* data not available

Field site	Vegetation	Depth (cm)	Dry weight (%)	pH ($CaCl_2$)	C_{org} (%)	N_{total} (%)	Fe_d ($g\,kg^{-1}$)	Fe_o ($g\,kg^{-1}$)
Köhlerloh	*Sphagnum* mosses	0–10	21.61	3.08	37.2	1.98	2.11	1.97
		10–20	15.65	3.18	42.9	2.44	3.63	3.02
		20–30	12.47	3.25	34.7	1.92	2.66	2.25
Schlöppner-brunnen I	*Sphagnum* mosses, some spruce stockings	0–10	10.47	3.82	42.4	1.81	0.45	0.49
		10–20	9.40	3.87	44.0	1.88	n.a.	n.a.
		20–30	8.20	4.08	39.4	2.07	0.28	0.44
Schloppner-brunnen II	*Molinia* grasses	0–10	6.78	4.38	38.0	1.87	18.88	14.2
		10–20	6.03	4.11	40.0	1.51	9.78	8.72
		20–30	10.01	4.06	34.3	1.60	3.38	2.73

22.3 Sequential Microbial Reductive Processes in Soils: Theoretical Considerations and the Reality in the Lehstenbach Catchment

22.3.1 Theoretical Considerations

Microorganisms using other electron acceptors than O_2 for the oxidation of organic matter are regulated mainly by the availability of O_2, nitrate, Fe(III), sulfate, and CO_2 (Tiedje et al. 1984). In general, increasing water saturation of soils leading to anaerobiosis stimulates the activity of obligate anaerobes, like sulfate reducers or methanogens. However, sulfate-reducing prokaryotes have also been found to occur in oxidized and even oxic sediment layers (Canfield and DesMarais 1991). A temporary presence of O_2 in soils could be favorable for sulfate reducers, because sulfide can be oxidized to sulfate in oxic soil zones which renew the sulfate pool in habitats with sulfate-limiting conditions. However, the activity of the sulfate reducers and the quantitative role of competition between sulfate reduction and other respiratory processes in habitats with fluctuating redox conditions are poorly understood.

When O_2 is depleted in soils, terminal electron acceptors other than O_2 are utilized for the oxidation of organic matter (Tiedje et al. 1984; Smith and Arah 1986). The concentration of O_2 in soils is affected by O_2-driven respiration and the diffusion of atmospheric O_2 (Currie 1961). The diffusivity of O_2 decreases with increasing water content and decreasing porosity of the soil. In aerated soils, anaerobic processes are temporarily stimulated by rainfalls due to the transient presence of higher amounts of water (Parkin and Tiedje 1984; Sexstone et al. 1984). In anoxic soils, heavy rainfalls can increase the concentration of O_2 by mixing the soil water with aerated rain water. After drying or decreasing O_2-dependent respiratory activity, O_2 can easily diffuse back, and microbial products from fermentation or anaerobic respiration can be oxidized again. Permanent anoxic conditions in aerated soils seem to be restricted to microzones within soil aggregates or organic matter (van der Lee et al. 1999).

Flooding of soils initiates sequential reductive processes. The change from aerobic to anaerobic metabolism occurs at O_2 concentrations of less than 1% (Paul and Clark 1996). The anaerobic degradation of organic matter occurs via a complex network of trophic links that collectively terminates in the reduction of the most favorable and available electron acceptor (McInerney and Bryant 1981). In theory, the sequence in which electron acceptors are utilized is determined by the redox potentials of the half-cell reactions (Paul and Clark 1996). Thus, aerobic respiration (E_0'=+0.81 V) should be followed sequentially by the reduction of nitrate to N_2 by denitrifiers (E_0'=+0.75 V), the reduction of Fe(III) oxide to Fe(II) by iron reducers (E_0'=–0.1 V; see following clarification), the reduction of sulfate to sulfide by sulfate reducers (E_0'=–0.22 V), and the reduction of CO_2 to methane by methanogens

(E_0'=–0.25 V). The Fe(III)/Fe(II) half-cell covers a broad range of redox potentials (E_0'), from +0.1 V for ferrihydrite to –0.3 V for goethite, hematite, and magnetite (Straub et al. 2001). Most of the Fe(III) present in soils is in the form of insoluble oxides. Since the availability of Fe(III) oxides for microbial reduction decreases with increasing crystallinity and decreasing surface area (Munch and Ottow 1980; Roden and Zachara 1996), many soils are readily depleted of microbiologically reducible Fe(III). Dissimilatory sulfate reduction is one of the most important organic carbon mineralization processes in anoxic aquatic environments especially in marine sediments that have high concentrations of sulfate (Jørgensen 1978). Since terrestrial ecosystems have low sulfate concentrations, methanogenesis becomes the dominant electron-accepting process for the oxidation of organic matter in most waterlogged soils (Mayer and Conrad 1990; Peters and Conrad 1996). However, soils are very heterogeneous, and the presence of terminal electron acceptors can vary spatially and temporally. Thus, other anaerobic processes might be favored under certain in situ conditions.

22.3.2 Sequential Microbial Reductive Processes in Intermittent Seeps and Fens of the Lehstenbach Catchment

At Köhlerloh, depth profiles of the soil solution obtained after water saturation in autumn show typical geochemical gradients of electron acceptors that indicate the sequential utilization of nitrate, Fe(III), and sulfate with increasing soil depth. When nitrate is completely consumed at a depth of 13 cm, the concentrations of Fe(II) [i.e. the product of microbial Fe(III) reduction] increase, and the concentrations of sulfate decrease (Fig. 22.2). During late autumn and winter, nitrate is consumed completely, and the reduction of CO_2 becomes the main electron-accepting process for the anaerobic oxidation of organic matter. After the snowmelt, however, depth profiles of the soil solution both at intermittent seeps and fens show relatively high concentrations of nitrate (60–220 µM) throughout the soil profile, negligible concentrations of Fe(II) (5–40 µM) and NH_4^+ (0–10 µM), and equal concentrations of sulfate which approximate 120 µM. Apparently, mixing of the soil solution with O_2-rich surface water after the snowmelt and reduced microbial respiratory activity at low temperatures lead to oxidized conditions which counteracts nitrate-, Fe(III)-, and sulfate-consuming processes. After the snowmelt, average concentrations of nitrate and sulfate in soil solutions collected from terrestrial sites approximate 300 and 240 µM, respectively. Gradients of nitrate, Fe(II), and sulfate concentrations appear again in the soil solution at the intermittent seep Köhlerloh during spring (data not shown), indicating the onset of anaerobic microbial activities.

During early summer, nitrate is completely consumed in the soil solution, and concentrations of sulfate decrease from 90–40 µM with increasing soil

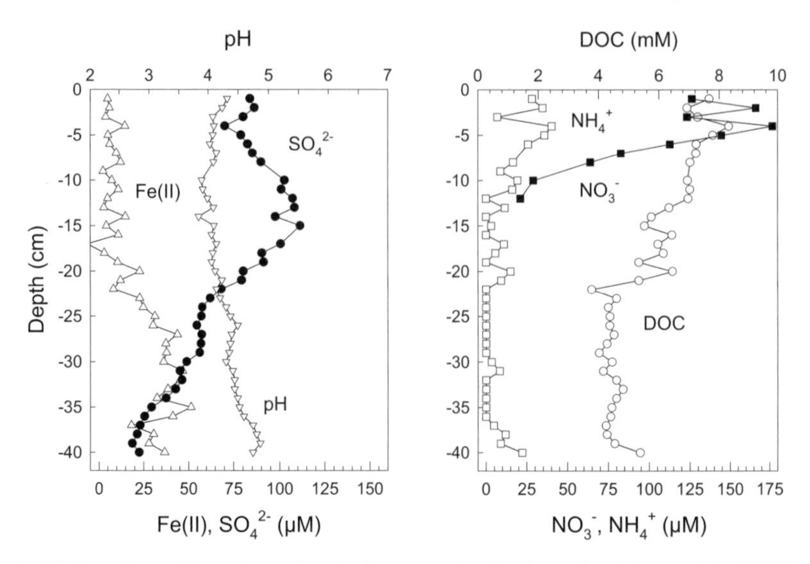

Fig. 22.2. Depth profiles of geochemical parameters in the soil solution of site Köhlerloh (November 2001). The soil solution was sampled with a dialysis chamber which consists of 40 1-cm cells covered with a membrane of 0.2-μm pore diameter. The water-filled chamber remained 2 weeks in the water-saturated soil for equilibration with the soil solution prior to sampling

Table 22.2. Nitrate concentrations in 90 cm soil depth at the wetland sites. Mean ± standard deviation are given for the time period April to September 2002

Site	*n*	NO_3^- (μM)	NH_4^+ (μM)
Schlöppnerbrunnen I	30	6.5±11.3	17.8±23.2
Schlöppnerbrunnen II	28	0	1.7± 1.7
Köhlerloh	29	0.8± 3.2	9.4±14.4

depth. In general, upper soil layers (0–40 cm) are not water saturated during late summer and autumn. Anoxic incubation experiments with soil samples obtained during summer demonstrate nitrate- and Fe(III)-reducing activities but negligible sulfate-reducing and methanogenic activities (data not shown). Thus, sulfate-reducing and methanogenic activities appear to be restricted to periods of water saturation during autumn and winter.

In the water-saturated fens of the Lehstenbach catchment, concentrations of nitrate are negligible during the year both in the upper 40 cm of the soil (<10 μM) and in the deeper soil solution (Table 22.2). Only after the snowmelt, nitrate can reach concentrations up to 300 μM in the upper 10 cm. Thus, the microbial reduction of nitrate appears to be an ongoing process in intermittent seeps and fens except during frozen and thawing periods.

22.4 Wetlands of the Lehstenbach Catchment as a Sink for Anthropogenic Nitrate

The adsorption capacity of nitrate in acidic forest soils is very low (Johnson et al. 1986). In declining forest sites, the amount of atmospheric nitrate recovered in spring water increases with increasing amounts of nitrate output, and surface flow can influence leaching of atmospheric nitrate especially in peaty waterlogged soils (Grennfelt and Hultberg 1986; Durka 1994). On the other hand, the main sink functions for nitrate in soils are plant uptake and/or microbial nitrate reduction to N_2O or N_2 (i.e. denitrification). While plant uptake of nitrogen might be assumed to be in equilibrium with decomposition and mineralization processes in a fully grown mature forest, denitrification can be an important sink for nitrate. The process of denitrification (Eq. 1) results in the production of alkalinity (consumption of one proton for each denitrified nitrate ion). If nitrate is fully reduced, N_2 will be emitted from the soils (Eq. 1). However, in acidic soils, intermediate products, especially N_2O, can also be produced (Eq. 2).

$$C_6H_{12}O_6 + 4NO_3^- + 4H^+ + O_2 \rightarrow 6CO_2 + 8H_2O + 2N_2 \qquad (1)$$

$$NO_3^- \rightarrow NO_2^- \rightarrow NO \rightarrow N_2O \rightarrow N_2 \qquad (2)$$

Since N_2O and N_2 will be released to the atmosphere, denitrification can be considered as a permanent sink for nitrate and acidity.

In 2002, all investigated wetland sites in the Lehstenbach catchment had very low to zero nitrate concentrations in the soil solutions and in the groundwater (Table 22.2, Fig. 22.3; for groundwater concentrations see Lischeid et al., this Vol.). Ammonium concentrations of wetland sites were as low as those of upland sites during the summer of 2002 in 90 cm depth. In the upper 40 cm, the concentrations of ammonium in the soil solutions collected from

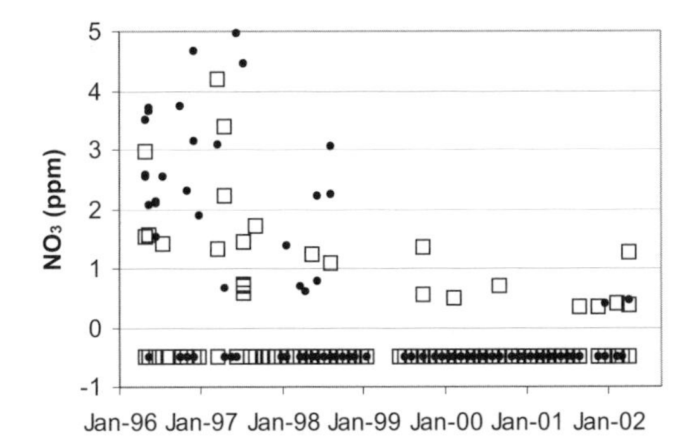

Fig. 22.3. Long-term nitrate concentrations at the intermittent seep Köhlerloh. Values below zero at –0.5 ppm indicate that samples were below the detection limit (0.25 ppm). *Squares* nitrate in 50 cm depth; *circles* nitrate in 100 cm depth

Schlöppnerbrunnen I and II vary between 20 and 300 μM during the year. The latter is due to the inhibition of nitrification under anoxic conditions.

A rough estimation of the nitrate sink can be done by considering the size of the wetland area. Since previous studies showed a relatively even distribution of the deposition of nitrate in the catchment (Manderscheid and Matzner 1995), we can assume that now roughly one third of the deposited nitrate is retained by denitrification. Unfortunately, it is not known how much nitrate is leached from upland soils and reduced in the wetlands.

22.5 Wetlands of the Lehstenbach Catchment as a Sink for Anthropogenic Sulfate

When discussing the retention of deposited sulfate in forest soils, the differentiation between adsorption of inorganic sulfate and the microbial reduction of sulfate is of major importance for the understanding and prediction of soil and water acidification. It has been shown that the adsorption of inorganic sulfate is highly reversible, which delays the recovery from soil and water acidification by decades (Mitchell et al. 1992; Alewell et al. 1997; Alewell 1998; Manderscheid et al. 2000). In soils of central Europe, the microbial reduction of sulfate can become an important mechanism to retain sulfur (Novak et al. 1994; Alewell and Giesemann 1996; Alewell and Gehre 1999), and rates of sulfate reduction might even exceed S inputs in North American soils (Spratt and Morgan 1990; Wieder et al. 1990). Furthermore, the reduction of sulfate is an important sink for protons (for overview see Giblin and Wieder 1992; Lindemann 1997). Vice versa, reoxidation of reduced inorganic sulfur leads to acidification (Giblin and Wieder 1992).

The retention of deposited sulfate due to sulfate reduction is reversible if sulfide is not fixed as inorganic or organic reduced S. Under short-term conditions, fixation of sulfur as inorganic reduced sulfur seems to play a major role in peatlands (Wieder and Lang 1988). At Schlöppnerbrunnen I, the amount of total inorganic reduced sulfur (TRIS) compounds, which comprise FeS_2, FeS, and S^0, approximates 0.05 μmol g (fresh weight soil)$^{-1}$, and acid volatile sulfur (AVS) concentrations, which comprise only FeS, are negligible. At Schlöppnerbrunnen II, TIRS concentrations increase from 0.3–0.5 μmol g (fresh weight soil)$^{-1}$ with increasing soil depth, and AVS concentrations approximate only 0.05 μmol g (fresh weight soil)$^{-1}$. At both sites, oxidation of iron sulfides could be demonstrated in situ with perforated tubes filled with an FeS-containing agar that showed a color change from black to yellow after a long period of rain (Paul 2003).

Under long-term anoxic conditions, inorganic reduced sulfur might be transformed into organic sulfur compounds (Brown 1985, 1986). In freshwater systems, organic reduced sulfur is the dominant long-term end-product of dissimilatory sulfate reduction (Giblin and Wieder 1992). The concentrations of total sulfur in the upper layers of Schlöppnerbrunnen I and II range from 3–10 g kg (dry weight soil)$^{-1}$ (Alewell and Novak 2001; Paul 2003). Even though the organic sulfur pools in central European systems seem to be relatively stable, the dynamics of biogenic sulfur cycling and its effect on the reversibility of acidification under changing environmental conditions are highly unpredictable.

A common method of measuring dissimilatory sulfate reduction is the radiolabeling method, in which tracer quantities of $^{35}SO_4^{2-}$ are injected into samples and the formation of reduced ^{35}S is measured (Ivanov 1956; Jørgensen 1978). The disadvantage of this method is the restriction to laboratory studies and to short time experiments. The analysis of stable isotopes of sulfur (^{34}S) is a useful tool for studying new aspects of the biogeochemical cycles of this element (Durka et al. 1999; Mitchell et al. 1999; Novak et al. 1999). Determination of $\delta^{34}S$ in the soil solution gives a fingerprint of the currently dominating process. The strong discrimination of the stable isotope ^{34}S during dissimilatory sulfate reduction results in an enrichment of the remaining soil solution sulfate and a depletion of ^{34}S in the final product (the organic sulfur). Determinations of $\delta^{34}S$ values in spring waters indicated that dissimilatory sulfate reduction is an ongoing process in springs from the Fichtelgebirge with sulfate concentrations less than 150 μM (Durka et al. 1999).

The investigated sites within the Lehstenbach catchment show different patterns in their stable isotope data (Alewell and Gehre 1999). Köhlerloh seems to be a 'hot spot' for the reduction of sulfate, because sulfate in the soil solution and groundwater is clearly enriched in ^{34}S compared to throughfall values (Fig. 22.4). In addition, ^{35}S radiolabeling studies indicate low sulfate reduction rates under laboratory conditions at 4 °C (Kaiser 1996). In contrast, the turnover of sulfur in the upland sites Coulissenhieb and Gemös is more influenced by sulfate desorption and/or S mineralization. However, soil samples of the Coulissenhieb show high potential rates of sulfate reduction at 20 °C (Alewell and Novak 2001). $\delta^{34}S$ values indicate for the site Weidenbrun-

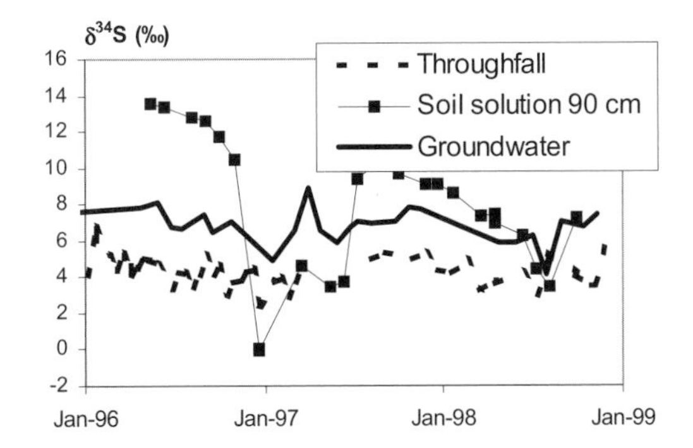

Fig. 22.4. $\delta^{34}S$ values in canopy throughfall, soil solution and groundwater at Köhlerloh. (Data from Alewell and Gehre 1999)

nen that the reduction of sulfate occurs in deeper soil layers (>1 m depth) as well as in horizontal groundwater flow from reduction zones (data not shown).

In order to obtain a long-term perspective on the dominant S turnover process, a second technique, the determination of $\delta^{34}S$ in the soil solid phase, is suitable. Several authors have shown that in aerated soils $\delta^{34}S$ values in organic soil sulfur increased with depth (Rolland et al. 1991; Giesemann et al. 1992; Gebauer et al. 1994; Mayer et al. 1995; Novak et al. 1996). Aging of organic matter includes ongoing mineralization. Because of the preferential mineralization and release of ^{32}S, the remaining organic sulfur will become heavier with increasing age (and thus increasing depth) in upland soils (Novak et al. 1996). Since organic soil sulfur is generally the dominant sulfur species in forest soils, a smooth increase of $\delta^{34}S$ ratios of total sulfur from the topsoil to mineral soil has been shown for a variety of forest upland soils throughout central Europe (Novak et al. 1996; Novak 1998). In fresh-water wetlands and peat soils, however, significantly lower $\delta^{34}S$ values have been determined (Novak et al. 1994, 1996; Morgan 1995).

$\delta^{34}S$ values of total soil sulfur at the upland site Coulissenhieb and the intermittent seep Köhlerloh (Fig. 22.5) indicated that dissimilatory sulfate reduction is not the primary control of S cycling at these sites. Even though we did not find a clear depth-dependent increase in $\delta^{34}S$ of total soil sulfur as was hypothesized for upland soils, $\delta^{34}S$ values in the mineral soil (Coulissenhieb) and deeper layer of the H horizon (Köhlerloh) are significantly higher than the top layer. Higher $\delta^{34}S$ values in the deeper layers are most likely due to

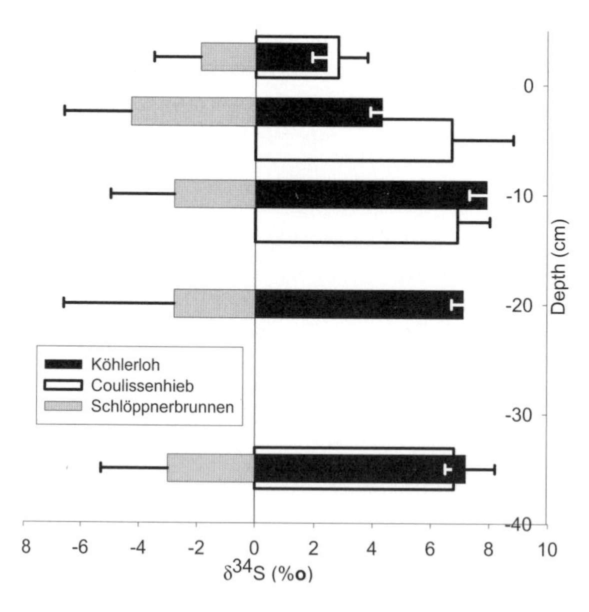

Fig. 22.5. Depth gradient of $\delta^{34}S$ values in total sulfur at the investigated sites. (Data from Alewell and Novak 2001)

preferential release of ^{32}S during mineralization and aging of the organic material. Thus, temporary water saturation alone or the genesis of histic H horizons at the site Köhlerloh are not a reliable indicator for long-term sulfate retention due to dissimilatory sulfate reduction. Thus, intermittent seeps seem to be only temporarily involved in the retention of reduced inorganic sulfur compounds. A prediction of the dynamics and a quantification of sink functions for sulfate in wetlands of the Lehstenbach catchment is difficult, if not impossible.

22.6 Ecophysiology and Phylogenetic Diversity of Sulfate-Reducing Prokaryotes in Fens of the Lehstenbach Catchment

Enumeration studies done with soil from Schlöppnerbrunnen I demonstrate that numbers of methanogens [10^6–10^7 cells g (fresh weight soil)$^{-1}$] cultured at pH 5 are only 10 times higher than those of sulfate reducers cultured at pH 5 (Bischof et al. 2003). Malate but not H_2 or ethanol can stimulate the consumption of sulfate. In contrast, supplemented acetate, formate, and lactate are inhibitory for the consumption of sulfate. Under sulfate-limiting conditions, sulfate reducers may act as secondary fermenters in synthrophic association with methanogens.

In contrast to the well-studied sulfate-reducing communities in marine sediments (Llobet-Brossa et al. 1998; Sahm et al. 1999; Ravenschlag et al. 2000, 2001), little is known about the phylogenetic diversity of sulfate-reducing prokaryotes in peatlands. In soil samples obtained from Schlöppnerbrunnen I and Schlöppnerbrunnen II, 16S rRNA gene-based oligonucleotide microarray analyses reveal the occurrence of bacteria affiliated with the deltaproteobacterial genera *Syntrophobacter* and *Desulfomonile* (Loy et al., submitted). Detailed sequence analysis of dissimilatory (bi)sulfite reductase genes (*dsrAB*) further reveal eleven distinct operational taxonomic units (OTUs) thereby reconfirming the presence of *Syntrophobacter wolinii*- and *Desulfomonile*-related species. Five of these eleven OTUs are of deltaproteobacterial origin (two OTUs most closely related to *Desulfobacca acetoxidans*), whereas six deeply branching OTUs are apparently not affiliated to any *dsrAB* amino acid sequence from yet recognized SRPs (Loy et al., submitted). Hybridization of DNA extracted from growth-positive MPN dilution series with the 16S rRNA gene-based microarray for SRPs indicate the presence of *Desulfobulbaceae*, *Desulfovibrionaceae*, *Syntrophobacter*, *Desulfovirga*, and *Desulforhabdus* in the enrichments. Amplification of dsrAB genes from the enrichments followed by cloning and sequencing yield a clone related to *Desulforhopalus singaporensis* (Bischof et al. 2003). Elucidating the ecological function of these unknown SRPs in acidic fens will require further study.

22.7 Competing Anaerobic Microbial Processes in Fens of the Lehstenbach Catchment

22.7.1 Sulfate Reduction Versus Methanogenesis

In general, northern peatlands are considered to be significant sources of atmospheric methane (Moore et al. 1998). Fens of the Lehstenbach catchment emit methane with rates that approximate 0.02–15 mM methane m^{-2} day^{-1} (Horn et al. 2003). Methane concentrations in the soil solution increase with increasing depth, and maximum concentrations (3.5 mM) are located in deeper peat layers in the central regions of the fens, whereas marginal regions (transition to upland soils) seem to have lower methane-producing activities with maximum concentrations of methane in the soil solution that approximate only 60 μM (Paul 2003). In contrast to rice paddy soils and pH-neutral wetlands, in which the methane production is attributed to acetoclastic methanogens (Whiticar et al. 1986; Capone and Kiene 1988), H_2 appears to be the dominant precursor for methanogenesis in slightly acidic peatlands (Williams and Crawford 1984; Landsdown et al. 1992). In fens of the Lehstenbach catchment, supplemented H_2 stimulates methanogenesis but not supplemented acetate. Phylogenetic members of the Methanobacteriaceae, Methanomicrobiales, and Methanosarcinaceae are apparently involved in the formation of methane (Horn et al. 2003).

Rates of methane formation can decrease if sulfate is temporarily present as alternative electron acceptor (Nedwell and Watson 1995). The marked depletion of ^{34}S in soil sulfur (Fig. 22.5) as well as potential sulfate reduction rates in soils of Schlöppnerbrunnen I measured with ^{35}S radiolabeling studies pointed to dissimilatory sulfate reduction as an important metabolism at this site (Alewell and Novak 2001). In anoxic microcosms, supplemental sulfate (300 μM) is rapidly reduced by the microbial community present at Schlöppnerbrunnen I and Schlöppnerbrunnen II (Loy et al., submitted) indicating a rapid turnover of sulfate in situ if sulfate is available. In the presence of supplemented sulfate, methanogenesis is inhibited up to 60% compared to unsupplemented controls (data not shown) which confirms that sulfate-reducing prokaryotes can outcompete methanogens in peatlands (Yavitt et al. 1987; Blodau et al. 2002). In fens of the Lehstenbach catchment, the concentration of sulfate in the soil solution varies over the year. Maximum sulfate concentrations of 180 μM and the lack of a sulfate gradient with increasing soil depth at Schlöppnerbrunnen I and II (Fig. 22.6, and data not shown) are indicative of negligible sulfate-reducing activities after the snowmelt. During late summer, the concentration of sulfate decrease to levels of 10–20 μM that approximate limit concentration sufficient for dissimilatory sulfate reduction (Lovley and Klug 1983). Thus, low concentrations of sulfate seem to be seasonally a limiting factor for the activity of sulfate-reducing prokaryotes in

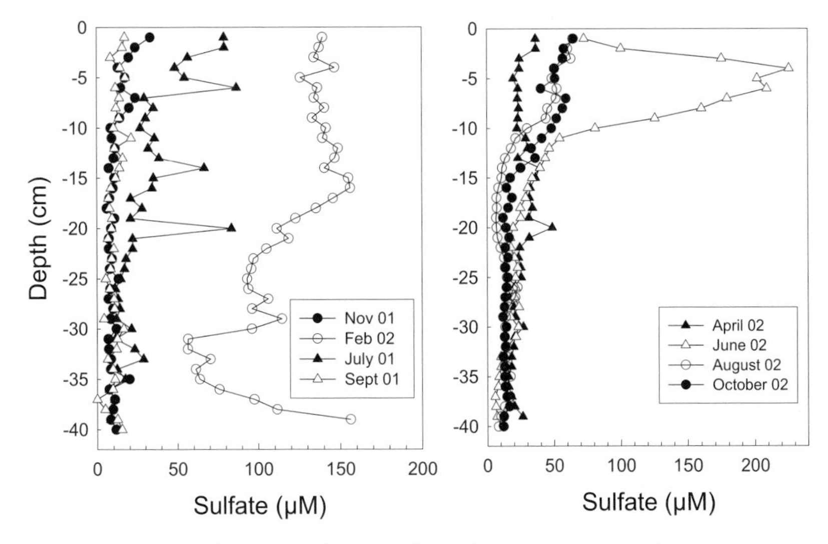

Fig. 22.6. Seasonal variations of depth profiles of sulfate concentrations in the soil solution of site Schlöppnerbrunnen II

Schlöppnerbrunnen I and II. Occasionally, higher concentrations of sulfate are detectable in upper peat layers (Fig. 22.6, see June 2002). Apparently, this increase is not due to atmospheric deposition of sulfate, but to the oxidation of reduced sulfur compounds in the presence of O_2 after mixing with oxygenated rain water after a heavy rainfall.

22.7.2 Sulfate Reduction Versus Reduction of Fe(III)

At Schlöppnerbrunnen I, concentrations of Fe(II) in the soil solution vary between <20 and 60 μM over the year, which corresponds to the low amounts of Fe(III) oxides present in this fen (Table 22.1) that are available for microbial reduction. Similar concentrations of dissolved Fe(II) are reported from some oligotrophic peatlands that receive most of their iron from atmospheric deposition (Blodau et al. 2002). Thus, it seems unlikely that dissimilatory Fe(III) reduction has a strong potential to inhibit the reduction of sulfate or the production of methane at this site.

At Schlöppnerbrunnen II, average concentrations of dissolved Fe(II) approximate 100–500 μM, and maximum concentrations can reach 6 mM (Küsel et al. 2002). In general, the concentrations of Fe(II) in the soil solution increase with increasing soil depth (Fig. 22.7), indicating the input of Fe(II)-rich groundwater. Since Fe(II) is rapidly oxidized in the presence of O_2 both chemically and biologically (Sobolev and Roden 2002), the presence of Fe(II) near the soil surface at Schlöppnerbrunnen II (Fig. 22.7) indicates that O_2 does

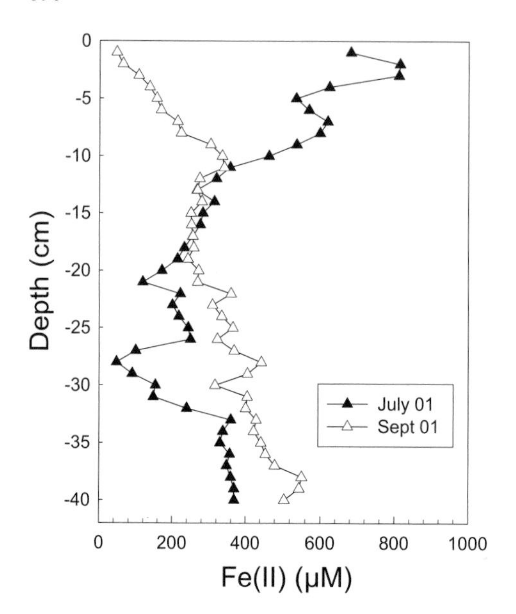

Fig. 22.7. Variations of depth profiles of Fe(II) concentrations in the soil solution of site Schlöppnerbrunnen II

not penetrate deep into the peat during most times of the year. Anoxic incubation experiments with peat samples from the surface layer obtained after the snowmelt demonstrate high Fe(III)-reducing but negligible sulfate-reducing or methanogenic activities (data not shown). Rates of Fe(III) reduction approximate 2.6 µmol Fe(II) g (fresh weight soil)$^{-1}$. When the Fe(III)-reducing activities decrease in microcosms, sulfate is consumed followed by the onset of methane formation (data not shown). After heavy rainfalls, the concentrations of dissolved Fe(II) decrease in the upper peat layers (Fig. 22.7; data from September 2001), which suggests an oxidation of Fe(II) near the surface. This spatially restricted cycling of iron is supportive for Fe(III)-reducing activities in the surface layers, but not for deeper soil layers. Thus, Fe(III) will be only temporarily and spatially an alternative electron acceptor for the anaerobic oxidation of organic matter at Schlöppnerbrunnen II.

22.8 Conclusions

In the investigated wetlands of the Lehstenbach catchment, nitrate and sulfate are reduced by microorganisms under anoxic conditions and alkalinity is generated. In intermittent seeps and fens, the reduction of nitrate is an ongoing process during the year. Since the end products of denitrification are released to the atmosphere, these wetlands can be considered as a permanent sink for nitrate and acidity. A rough estimation is that one third of the deposited nitrate in the catchment is reduced in the wetlands.

Estimation of the long-term buffering capacity of the deposited sulfate in the catchment is more complex, because a part of the reduced products derived from microbial sulfate reduction appears not to be fixed permanently in the wetlands. During periods of water saturation, intermittent seeps like the site Köhlerloh are hot spots for sulfate reduction. However, reduced sulfur is obviously not retained but reoxidized during dry periods in summer. In the fens Schlöppnerbrunnen I and II, high sulfate-reducing capacities are present throughout the year. However, the sulfate-reducing activity appears to be limited by the low availability of sulfate during late summer and autumn. Sequence analysis of dissimilatory (bi)sulfite reductase genes reveal the presence of *Syntrophobacter wolinii-*, *Desulfomonile-*, and *Desulfobacca*-related species and other yet recognized sulfate-reducing prokaryotes. Reduced sulfur seems to be temporarily fixed as inorganic reduced sulfur. Consecutively, a part of this pool is subjected to transformation into organic reduced sulfur. In upper peat layers, a part of the reduced sulfur is subjected to aerobic oxidation due to the input of oxygenated rainwater. Thus, the small pool of sulfate can be regenerated in upper peat layers. Oxygen can even penetrate deep in wetlands during the snowmelt due to the high runoff. Up till now, it remains unclear how much of the deposited sulfate cycles through reduced sulfur pools and is long-dated and stored, and how much is reoxidized and leaves the fens of the Lehstenbach catchment. Thus, despite the ongoing sulfate reduction in the wetlands, we cannot quantify the amount of reduced sulfur stored in the wetlands. Due to the low concentrations of sulfate available for microbial reduction, fens of the Lehstenbach catchment emit the greenhouse gas methane and, thus, contribute to global change.

Acknowledgments. The authors thank Gunnar Lischeid and Kerstin Bischof for helpful discussions, Sonja Trenz and Daria Schulz for technical assistance, and Harold L. Drake and Egbert Matzner for their support of this study. This project was financed by the German Ministry for Education and Science (BMBF, PT BEO 51-0339476 C, D).

References

Alewell C (1998) Investigating sulfate sorption and desorption of acid forest soils with special consideration of soil structure. Z Pflanzenernaehr Bodenkd 161:73–80

Alewell C, Gehre M (1999) Spatial and temporal heterogeneity of $\delta^{34}S$ values in a forested catchment as an indicator for the non conservative nature of the SO_4^{2-} ion. Biogeochemistry 47:319–333

Alewell C, Giesemann A (1996) Sulfate reduction in a forested catchment as indicated by $\delta^{34}S$ of soil solutions and runoff. Isot Environ Health Stud 32:203–210

Alewell C, Novak M (2001) Spotting zones of dissimilatory sulfate reduction in a forested catchment: the ^{34}S-^{35}S approach. Environ Pollut 112:369–377

Alewell C, Bredemeier M, Matzner E, Blanck K (1997) Soil solution response to experimentally reduced acid deposition in a forest ecosystem. J Environ Qual 26:658–665
Bischof K, Loy A, Wagner M, Drake HL, Küsel K (2003) Characterization of sulfate-reducing prokaryotes in a CH_4-emitting acidic fen. Abstract 4. In: Proc Gemeinsamer Kongress der DGHM, ÖGHMP und VAAM, BioSpectrum, Abstr PH053, Berlin, 103 pp
Blodau C, Roehm CL, Moore TR (2002) Iron, sulfur, and dissolved carbon dynamics in a northern peatland. Arch Hydrobiol 154:561–583
Brown KA (1985) Sulphur distribution and metabolism in waterlogged peat. Soil Biol Biochem 17:39–45
Brown KA (1986) Formation of organic sulphur in anaerobic peat. Soil Biol Biochem 18:131–140
Canfield DE (1989) Reactive iron in marine sediments. Geochim Cosmochim Acta 53:619–632
Canfield DE, DesMarais DJ (1991) Aerobic sulphate reduction in microbial mats. Science 251:1471–1473
Capone DG, Kiene RP (1988) Comparison of microbial dynamics in marine and freshwater sediments: contrasts in anaerobic carbon metabolism. Limnol Oceanogr 4:725–749
Cord-Ruwisch R, Seitz H-J, Conrad R (1988) The capacity of hydrogenotrophic anaerobic bacteria to compete for traces of hydrogen depends on the redox potential of the terminal electron acceptor. Arch Microbiol 149:350–357
Currie JA (1961) Gaseous diffusion in the aeration of aggregated soils. Soil Sci 92:40–45
Durka W (1994) Isotopenchemie des Nitrat, Nitrataustrag, Wasserchemie und Vegetation von Waldquellen im Fichtelgebirge (NO-Bayern). Bayreuther Forum Ökol 11:1–197
Durka W, Giesemann A, Schulze ED, Jäger HJ (1999) Stable sulfur isotopes in forest spring waters from the Fichtelgebirge (Germany). Isot Environ Health Stud 35:237–249
Gebauer G, Giesemann A, Schulze E-D, Jäger H-J (1994) Isotope ratios and concentrations of sulfur and nitrogen in needles and soils of *Picea abies* stands as influenced by atmospheric deposition of sulfur and nitrogen compounds. Plant Soil 164:267–281
Giblin AE, Wieder RK (1992) Sulphur cycling in marine and freshwater wetlands. In: Howarth RW, Stewart JWB, Ivanov MV (eds) Sulphur cycling on the continents: wetlands, terrestrial ecosystems, and associated water bodies. Scope 48. Wiley, Chichester, pp 85–123
Giesemann A, Jäger HJ, Feger KH (1992) Evaluation of the sulfur budget of a spruce ecosystem in the Black Forest by means of stable-isotope analysis. In: Hendry MJ, Krouse HR (eds) Proc Worksh Sulfur Transformations in Soil Ecosystems, 5–7 Nov, Saskatoon, Saskatchewan, Canada
Glaser PH (1987) The ecology of patterned boreal peatlands of northern Minnesota: a community profile. Biol Rep 85. US Fish and Wildlife Service, Washington, DC
Grennfelt P, Hultberg H (1986) Effects of nitrogen deposition on the acidification of terrestrial and aquatic ecosystems. Water Air Soil Pollut 30:945–963
Hamm A (1995) Saure Niederschläge und ihre Folgen für die Gewässer. Informationsberichte des Bayerischen Landesamts für Wasserwirtschaft 3/95, Munich, pp 37–44
Höpner T (1981) Design and use of a diffusion sampler for interstitial water from fine grained sample. Environ Tech Lett 2:187–196
Horn MA, Matthies C, Küsel K, Schramm A, Drake HL (2003) Hydrogenotrophic methanogenesis by moderately acid-tolerant methanogens of a methane-emitting acidic peat. Appl Environ Microbiol 69:74–83
Ivanov MV (1956) Isotopes in the determination of the sulfate-reduction rate in Lake Belovod. Mikrobiologia 25:305–309

Johnson DW, Cole DW, van Miegroet H, Horng FW (1986) Factors affecting anion movement and retention in four forest soils. Soil Sci Soc Am J 50:776–783

Jørgensen BB (1978) A comparison of methods for quantification of bacterial sulfate reduction in coastal marine sediments. I. Measurement with radiotracer techniques. Geomicrobiology J 1:11–27

Kaiser M (1996) Bestimmung von Sulfat-Reduktionsraten mit der $^{35}SO_4^{2-}$-Methode in säurebelasteten Quelleinzugsgebieten des Frankenwaldes und des Fichtelgebirges (Nordostbayern). Diploma Thesis, University of Bayreuth, Germany

Küsel K, Wagner C, Drake HL (1999) Enumeration and metabolic product profiles of the anaerobic microflora in the mineral soil and litter of a beech forest. FEMS Microbiol Ecol 29:91–103

Küsel K, Trenz S, Alewell C, Drake HL (2002) Effect of biogeochemical parameters on the reduction of sulfate and Fe(III) in slightly acidic bogs. In: Proc Annu Meetg American Society of Microbiology, Abstr N-125, Salt Lake City, 327 pp

Landsdown JM, Quay, PD, King SL (1992) CH_4 production via CO_2 reduction in a temperate bog: a source of ^{13}C-depleted CH_4. Geochim Cosmochim Acta 56:3493–3503

Lindemann J (1997) Quantifizierung biogeochemischer Eisen- und Sulfat-Umsetzungen in einem Quellmoor und deren Beitrag zur Säureneutralisierung in einem Einzugsgebiet des Frankenwaldes. Bayreuther Forum Ökol 51:1–271

Llobet-Brossa E, Rossello-Mora R, Amann R (1998) Microbial community composition of Wadden Sea sediments as revealed by fluorescence in situ hybridization. Appl Environ Microbiol 64:2691–2696

Lovley DR, Klug MJ (1983) Sulfate reducers can outcompete methanogens at freshwater sulfate concentrations. Appl Environ Microbiol 45:187–192

Lovley DR, Phillips EJP (1986) Availability of Fe(III) for microbial reduction in bottom sediments of the freshwater tidal Potomac River. Appl Environ Microbiol 52:751–757

Manderscheid B, Matzner E (1995) Spatial and temporal variability of soil solution chemistry and seepage water ion fluxes in a mature Norway spruce (*Picea abies* (L.) Karst.) stand. Biogeochemistry 30:99–114

Manderscheid B, Schweisser T, Lischeid G, Alewell C, Matzner E (2000) Sulfate pools in the weathered substrata of a forested catchment. Soil Sci Soc Am J 64:1078–1082

Mayer HP, Conrad R (1990) Factors influencing the population of methanogenic bacteria and the initiation of methane production upon flooding of paddy soil. FEMS Microbiol Ecol 73:103–112

Mayer B, Feger KH, Giesemann A, Jäger H-J (1995) Interpretation of sulfur cycling in two catchments in the Black Forest (Germany) using stable sulfur and oxygen isotope data. Biogeochemistry 30:31–58

McInerney MJ, Bryant MP (1981) Basic principles of basic conversions in anaerobic digestion and methanogenesis. In: Sofer SS, Zaborsky OR (eds) Biomass. Conversion processes for energy and fuels. Plenum Press, New York, pp 277–296

Mitchell MJ, Burke MK, Shepard JP (1992) Seasonal and spatial patterns of S, Ca and N dynamics of a northern hardwood ecosystem. Biogeochemistry 17:165–189

Mitchell MJ, Krouse RH, Mayer B, Stam AC, Zhang YM (1999) Use of stable isotopes in evaluating biogeochemistry of forest ecosystems. In: Kendall C, McDonnell J (eds) Isotope tracers in catchment hydrology. Elsevier, Amsterdam, pp 498–518

Moore TR, Roulet NT, Waddington JAM (1998) Uncertainty in predicting the effect of climatic change on the carbon cycle in Canadian peatlands. Clim Change 40:229–245

Morgan MD (1995) Modeling excess sulfur deposition on wetland soils using stable sulfur isotopes. Water Air Soil Pollut 79:299–307

Munch JC, Ottow JCG (1980) Preferential reduction of amorphous to crystalline iron oxides by bacterial activity. Soil Sci 129:15–21

Nedwell D, Watson A (1995) CH_4 production, oxidation and emissions in a UK ombrotrophic peat bog: Influence of SO_4^{2-} from acid rain. Soil Biol Biochem 27:893–903

Novak M (1998) $\delta^{34}S$ dynamics in the system bedrock–soil–runoff–atmosphere: results from the GEOMON network of small catchments, Czech Republic. In: Proc 9th Int Symp on Water–Rock Interaction, WRI-9, Taupo, New Zealand, 30 March–3 April

Novak M, Wieder RK, Schell WR (1994) Sulfur during early diagenesis in *Sphagnum* peat: insights from $\delta^{34}S$ ratio profiles in ^{210}Pb-dated peat cores. Limnol Oceanogr 39:1172–1185

Novak M, Botrell SH, Fottova D, Buzek F, Groscheova H, Zak K (1996) Sulfur isotope signals in forest soils of central Europe along an air pollution gradient. Environ Sci Tech 12:3473–3476

Novak M, Kirchner JW, Groscheova H, Havel M, Cerny J, Krejci R, Buzek F (1999) Sulfur isotope dynamics in two central European catchments affected by high atmospheric deposition of SO_x. Geochim Cosmochim Acta 64:367–383

Parkin TB, Tiedje JM (1984) Application of a soil core method to measure field denitrification rates. Soil Biol Biochem 16:323–330

Paul EA, Clark FE (1996) Soil microbiology and biochemistry, 2nd edn. Academic Press, San Diego, pp 12–33

Paul S (2003) Heterogenität der Sulfatreduktion im Einzugsgebiet des Lehstenbachs, Fichtelgebirge. Diploma Thesis, University of Bayreuth, Germany

Peters V, Conrad R (1996) Sequential reduction processes and initiation of CH_4 production upon flooding oxic upland soils. Soil Biol Biochem 28:371–382

Ravenschlag K, Sahm K, Knoblauch C, Jorgensen BB, Amann R (2000) Community structure, cellular rRNA content, and activity of sulfate-reducing bacteria in marine Arctic sediments. Appl Environ Microbiol 66:3592–3602

Ravenschlag K, Sahm K, Amann R (2001) Quantitative molecular analysis of the microbial community in marine Arctic sediments. Appl Environ Microbiol 67:387–395

Rolland W, Giesemann A, Feger KH, Jäger HJ (1991) Use of stable S isotopes in the assessment of S turnover in experimental forested catchments in the Black Forest, southwest Federal Republic of Germany. In: Proc Int Symp on The Use of Stable Isotopes in Plant Nutrition, Soil Fertility and Environmental Studies, Vienna, 1–5 Oct 1990, IAEA-SM-313/51, International Atomic Energy Agency, Vienna, Austria, pp 593–598

Roden ER, Wetzel RG (1996) Organic carbon oxidation and suppression of methane production by microbial Fe(III) oxide reduction in vegetated and unvegetated freshwater wetland sediments. Limnol Oceanogr 41:1733–1748

Roden ER, Zachara JM (1996) Microbial reduction of crystalline Fe(III) oxides: role of surface area and potential for cell growth. Environ Sci Tech 30:1618–1628

Sahm K, Knoblauch C, Pernthaler J, Amann R (1999) Phylogenetic affiliation and quantification of psychrophilic sulfate-reducing isolates in marine Arctic sediments. Appl Environ Microbiol 65:3976–3981

Sansone FJ, Martens CS (1982) Volatile fatty acid cycling in organic-rich marine sediments. Geochim Cosmochim Acta 46:1575–1589

Sexstone AJ, Parkin TB, Tiedje JM (1984) Temporal response of soil denitrification rates to rainfall and irrigation. Soil Sci Soc Am J 49:99–103

Smith KA, Arah JRM (1986) Anaerobic micro-environments in soil and the occurrence of anaerobic bacteria. In: Jensen V, Kjöller A, Sorensen LH (eds) Proc Microbial Communities in Soil FEMS Symp, no 33, Elsevier, London, pp 247–261

Sobolev D, Roden EE (2002) Evidence for rapid microscale bacteria redox cycling of iron in circumneutral environments. Ant Leeuwenhoek 81:587–597

Spratt HG, Morgan MD (1990) Sulfur cycling in a cedar dominated freshwater wetland. Limnol Oceanogr 35:1586–1593

Straub KL, Benz M, Schink B (2001) Iron metabolism in anoxic environments at near neutral pH.. FEMS Microbiol Ecol 34:181–186

Tiedje JM, Sexstone AJ, Parkin TB, Revsbech NP, Shelton DR (1984) Anaerobic processes in soil. Plant Soil 76:197–212

Van der Lee GEM, de Winder B, Bouten W, Tietema A (1999) Anoxic microsites in Douglas fir litter. Soil Biol Biochem 31:1295–1301

Whiticar MJ, Faber E, Schoell M (1986) Biogenic methane formation in marine and freshwater environments: CO_2 reduction vs. acetate fermentation – isotope evidence. Geochim Cosmochim Acta 50:693–707

Wieder RK, Lang GE (1988) Cycling of inorganic and organic sulphur in peat from Big Run Bog, West Virginia. Biogeochemistry 5:221–242

Wieder RK, Yavitt JB, Lang GE (1990) Methane production and sulfate reduction in two Appalachian peatlands. Biogeochemistry 10:81–104

Wieder RK, Novak M, Rodrigues D (1996) Sample drying, total sulfur and stable sulfur isotopic ratio determination in freshwater wetland peat. Soil Sci Soc Am J 60:949–952

Williams RT, Crawford RL (1984) Methane production in Minnesota peatlands. Appl Environ Microbiol 47:1266–1271

Yavitt JB, Lang GE, Wieder RK (1987) Control of carbon mineralization to CH_4 and CO_2 in anaerobic, *Sphagnum* derived peat from Big Run Bog, West Virginia. Biogeochemistry 4:141–157

V Catchment Response

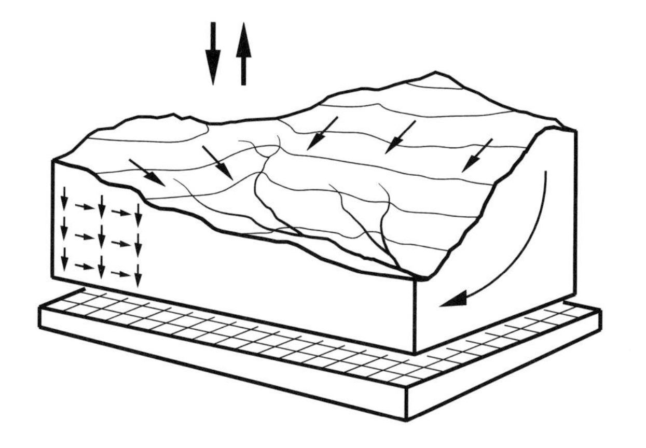

23 Dynamics of Runoff and Runoff Chemistry at the Lehstenbach and Steinkreuz Catchment

G. Lischeid, H. Lange, K. Moritz, and H. Büttcher

23.1 Introduction

Many functional biogeochemical aspects of terrestrial ecosystems are intimately linked to water transport processes. Most of the transport of subsurface matter occurs in the liquid phase. In addition, water fluxes exert a major control on, e.g., plant and microbial activity, which in turn has a major impact on matter fluxes and solute turnover in the subsoil.

In ecosystem science, many processes have been studied at the plot scale or in the laboratory. However, the problem arises of how to link a variety of single processes, facing the fact that a wealth of non-linear, scale-dependent relationships and feedback-loops were detected (cf. Schulze 1989). Thus, identification of the dominant processes directly at the scale of observation is increasingly recommended in addition to 'up-scaling' knowledge gained at smaller scales (Young and Beven 1994; Seyfried and Wilcox 1995; Blöschl 2001; Murray 2002).

The studies described in this chapter follow that approach. Strong emphasis is placed on quantitative descriptions of observed patterns in time and space in order to identify driving 'laws' (Dooge 1986), to relate the observed dynamics to processes, and to find a quantitative description (mathematical model) of these dynamics that can be used to extrapolate in time and space.

Sampling the catchment outlet generally is assumed to be a convenient way to infer information about a variety of biogeochemical processes at the catchment scale as it provides a spatial and temporal integral of the predominating catchment output fluxes for a number of chemical compounds of interest. Moreover, the short-term dynamics and long-term trends of the hydrograph and of solute concentrations in the catchment runoff can provide information about the predominating processes at the catchment scale and can be used to refine conceptual and mathematical models. Additional measurements inside the catchment, e.g., of soil solution, groundwater, and stream water at different sites, are used to relate the findings to within-catchment processes and thus to further constrain hypotheses and models. On the other hand, a recent

Ecological Studies, Vol. 172
E. Matzner (Ed.), Biogeochemistry of Forested Catchments in a Changing Environment

objection to this approach (Sivapalan 2003) seems to indicate that it is rather a new generic class of catchment-scale hydrological models that is required, since information from runoff chemistry and hydrology is limited.

It is clear that a continuum of time scales exists, which we conceptualize into 'short-term' and 'long-term' aspects. Following the common approach, this classification is driven by the temporal scale of the measurements, with a temporal resolution between minutes and hours in our studies. Small multiples of this basic time scale will be referred to as 'short-term' in the following text. The term 'long-term' is used to span approximately the total length of measurements, i.e., 1987–2002 at Lehstenbach and 1995–2002 at Steinkreuz. This is close to the time scale of the demanded risk assessment. On this time scale, changes in external (atmospheric) conditions became clearly visible.

The investigation of the dynamics of runoff and runoff chemistry in the two catchments consisted of three steps:

First, there was a focus on the short-term dynamics mainly in the catchment runoff. To that end, a variety of modern, non-linear methods were applied, including measures for information and complexity and artificial neural networks. These methods are described in more detail in the following section.

Second, spatial patterns were investigated. Here, not only was the spatial distribution of mean parameter values, e.g., solute concentration, analysed, but also spatial patterns of different types of short-term dynamics. Indeed, in many cases different sampling sites within the same catchment differed substantially with respect to short-term dynamics.

Third, long-term trends were studied. Again, this type of trend analysis explicitly took into account short-term dynamics and spatial patterns of that dynamics. Our experience is that the common (linear) trend analysis approach reveals only a rough impression of ongoing changes. In contrast, the short-term dynamics might be harder to identify, but then this gives a much more clear picture of the changing functioning of the system and allows for a more precise risk assessment. The SO_4 dynamics of the Lehstenbach catchment described below provides an excellent example of how short-term dynamics, spatial patterns and long-term trends are interrelated and linking results from these different aspects could be used for a sound risk assessment.

Prior to analysing short-term patterns it should be clear where and how these patterns are generated. For example, Hooper et al. (1998) and Hooper (2001) conclude from their measurements that the stream water solute concentration is determined by riparian zone processes rather than hillslope processes although hillslopes comprise the largest fraction of the catchment area. This is especially true for the short-term dynamics, but has to be taken into account for the long-term trends as well. As a consequence, the runoff generation process was investigated at both sites to serve as a necessary prerequisite to interpret the short-term dynamics of solute concentration and output fluxes.

23.2 Statistical Analysis

In this section, a comprehensive overview of the analysis methods employed is provided for the reader's convenience. It is not intended as a self-contained description; for more details to every method the reader is referred to the literature cited. In each case, we will discuss the domain of applicability with respect to short-term or long-term dynamics. An overview of the analysis techniques and their relevance in ecosystem research is provided by Lange (2004).

23.2.1 Auto- and Cross-Correlation

Given data sets of unknown spatiotemporal structure, it is our experience that a good starting point for a quick overview is the usage of standard techniques. The cross-correlation function is one example. It is a linear method, assumes stationarity of both partners, and is basically restricted to evenly spaced time series with a common temporal resolution and a sufficiently overlapping measurement period. Among these conditions, stationarity is the least critical one if the maximal time lag chosen is small enough (Honerkamp 1993). At any rate, it should be smaller than one quarter of the common observation period.

Cross-correlation analysis is simple enough to immediately apply an analytical significance test against the zero hypothesis (no linear correlation). One should be aware, though, that the presence of significant correlations might be no evidence of an actual dependence of one variable on another, and that its absence does not exclude non-linear correlations (cf. the section on recurrence plots).

23.2.2 Power Spectrum

Strongly connected to the autocorrelation function, the power spectrum has almost the same requirements as the data, but is a very different representation of the same information in Fourier space. Commonly investigated features comprise peaks (dominant frequencies), a possible scaling (indicated by power-law behaviour over an extended region) and cut-offs, e.g., in the high-frequency regime (pointing to a finite relaxation time) or in the low-frequency regime (vanishing autocorrelation at large lags). As the power spectrum covers the whole frequency range of the given record, it is a linear measure of temporal structures for short-term dynamics as well as long-term aspects. Depending on the quality and in particular the length of the data set, one could also investigate temporally local methods which take possible instationarities into account, the windowed Fourier transform and the wavelets. Their reliable interpretation, however, places high demands on the required number of data points.

23.2.3 Measures of Information and Complexity

The short-term dynamics of environmental time series can be characterized and classified by methods from information theory and statistical physics, in particular the symbolic dynamics (Lind and Marcus 1995). They provide us with different quantitative concepts of randomness, redundancy, information and complexity. Some of these measures rely on the details of the value distributions only (entropy-related quantities), while others include transition probabilities and non-linear correlation structure as well. All these measures strongly depend on the temporal resolution of the data which is intuitively appealing: given a type of environmental data (runoff rates, stream chemical concentrations in this context), they show low complexity at a coarse sampling rate (data close to noise), a low complexity as well at a very high sampling rate (redundant measurements) and a maximum somewhere in between. The information content of the time series is a monotonically increasing but non-linear function of randomness (Wackerbauer et al. 1994).

Among the measures available, we have selected two that seem to be suitable for our data sets and have a degree of robustness for given variable types as well (Lange 1999). The information content quantifier is the Mean Information Gain, or MIG for short (Wackerbauer et al. 1994), and the complexity quantifier is the Fluctuation Complexity or FC (Bates and Shephard 1993). These two quantities have the desired features, as can be demonstrated, e.g., for binary Bernoulli sequences (Lange 1999). Whereas MIG is non-linearly proportional to randomness, being more sensitive to structural changes in the region of low randomness, FC exhibits a maximum and vanishes for constant as well as completely random sequences. Thus, FC is close to our intuition of what a 'true' complexity measure should be.

23.2.4 Hurst Analysis

The phenomenon of persistence (extension of periods with systematic deviations from the long-term mean), first considered in the context of Nile floodings (Hurst 1951) and often present in hydrological data sets, is quantified by calculating the Hurst exponent H using the rescaled range or R/S statistics (Mandelbrot and Wallis 1969). This is a prototypical example of a long-term non-linear measure.

If Hurst scaling is found, the expected H values are $0 \leq H \leq 1$; $H \leq 0.5$ means antipersistent behaviour, which is seldom observed in experimental data, $H=0.5$ is ordinary Brownian motion and $H \geq 0.5$ is fractional Brownian motion with increasing persistence strength as H approaches 1.

23.2.5 Recurrence Plots and Recurrence Quantification Analysis

The theory of dynamic systems provides the embedding technique (Grassberger et al. 1991), in which consecutive values of individual time series are collected in vectors of a user-defined dimensionality. Euclidean distances in this vector space for all pairs of vectors are visualized in the recurrence plot of the system (Eckmann et al. 1987). Possible instationarities, trends, extreme periods, periodicities and many other (non-linear) features of the time series are exhibited by this technique (Zbilut et al. 1998a). Algebraic properties of this recurrence matrix are the basis for several derived quantities like the local recurrence rate, the degree of determinism, the Shannon entropy of line segments or an approximation to the highest Ljapunov exponent present in the system, summarized as recurrence quantification analysis or RQA (Zbilut et al. 1998a). The method does not presuppose stationarity or large amounts of data and is not very restrictive when data points are not equidistant. The method is, however, susceptible to the presence of noise in the data, rendering the interpretation more difficult but still possible when the parameters of the method are chosen carefully (Thiel et al. 2002).

A powerful extension of the RQA is cross-recurrence quantification (Zbilut et al. 1998b). After proper normalization, one can compare two different time series for the same observation period and elucidate the temporal development of their relationship in a non-linear fashion (Marwan and Kurths 2003). A particular strength of cross-recurrence plots is model testing, since weaknesses and strengths of model reconstructions are displayed on a local basis.

23.2.6 Scaling Behaviour and Multifractal Properties

Many geophysical phenomena exhibit self-similarity on various time scales, or lack intrinsic time constants that exert control on event size, especially in the extreme regime. However, the property of simple scaling, expressed as a power law with a single exponent α, is the exception rather than the rule. A theoretical framework to cope with the more general multiscaling, where the scaling exponent is a function of scale, is provided by the method of multifractals (Lovejoy and Schertzer 1990). The analytical approach of universal multifractals predicts the behaviour of higher moments of the time series, which can be tested against the observed data (Tessier et al. 1996). In addition, it provides a simple quantification in terms of the multifractal scaling exponent α of the extent and intensity of long-range correlations, by giving an estimate on the deepness of multifractality, from monofractal (α=0) to log-normal (α=2) multifractal (Salvadori et al. 2001).

23.2.7 Artificial Neural Networks

In this study artificial neural networks were applied to perform non-linear multivariate regressions, accounting for time lag effects. The objective was to identify driving variables of the observed dynamics in a statistical sense, to assess the minimum model complexity that is necessary to reproduce the observed dynamics, and to investigate how the short-term dynamics change in the long term.

The type of artificial neural networks used for this analysis is the multilayer perceptron-type network. It consists of a number of identical units, called nodes or neurons, that are arranged in subsequent layers. In the first layer, one neuron is required for every independent or input variable. Correspondingly, every neuron in the last or output layer corresponds to every dependent or output variable. Additional, so-called hidden layers are located between the input and the output layer, consisting of an arbitrary number of nodes. As a rule of thumb, the number of nodes in the hidden layers required for the neural network increases with the complexity of the relationships that shall be mapped. Here, only one hidden layer was used. The minimum number of nodes in this layer was between 3 and 5, determined by a trial and error approach or by a more systematic analysis called pruning.

Data processing by the neural network is performed by inputting the values of the corresponding single input variables. Every node of this layer passes the data value to every node of the subsequent layer. Here, all incoming values are summed up, transformed by an activity function and passed to the next layer of nodes. A logistic function was used as activity function that was identical for every node. However, for every link between nodes of subsequent layers there exists a factor called weight, which is used to multiply with the data value passed to the next layer. This weight is specific for every link and is adapted in an iterative way, called learning to optimize mapping of the input data onto the output data.

The learning algorithm used in this study is the resilient propagation (Riedmiller and Braun 1993). In addition, to prevent getting stuck in a local minimum of the error function, the weight matrix that yielded the best performance so far was altered randomly within a given range when the network error ceased to decrease. The training followed the common cross-validation approach: the data set was subdivided into a training, a validation and a test data set. The software used was the Stuttgart Neural Network Simulator (Zell et al. 1995) which is freely available for scientific purposes.

Further technical details, and examples of how they were applied to time series analysis, are given in Lischeid (2001a), Lischeid and Uhlenbrook (2002) and Lischeid and Langusch (2004).

23.3 Discharge

The Lehstenbach hydrograph has been monitored since 1987. The stream is perennial; even during long dry periods, runoff was never less than 0.25 mm day^{-1} due to groundwater discharge. In contrast, the Steinkreuz runoff usually runs dry during the growing season (Fig. 23.1).

In both catchments, the catchment runoff usually reacts within minutes or hours to rainstorms, concerning both discharge and solute concentration. Thus, understanding the process of runoff generation is a necessary prerequisite for understanding the short-term dynamics of solute concentration. In spite of substantial differences between bedrock material, stratification of the underground and differing fractions of wetlands, the process of runoff generation is very similar in both catchments.

The power spectrum of the Lehstenbach hydrograph (Fig. 23.2) demonstrates the property of scaling nicely. The exponent within the scaling region is representative for runoff series in general (Pandey et al. 1998) and close to the well-known '1/f noise', implying that the series lacks any preferred scale and possesses arbitrary long and intense correlations. It exhibits strongly intermittent behaviour and a particular long-term instationarity, which has been revealed by RQA. The annual cycle is present but not especially strong, despite ubiquitous snowmelt. Removing it leads to scaling behaviour over three orders of magnitude. In contrast to the autocorrelation analysis, the power spectrum is able to identify long-term memory (>1 year) effects that

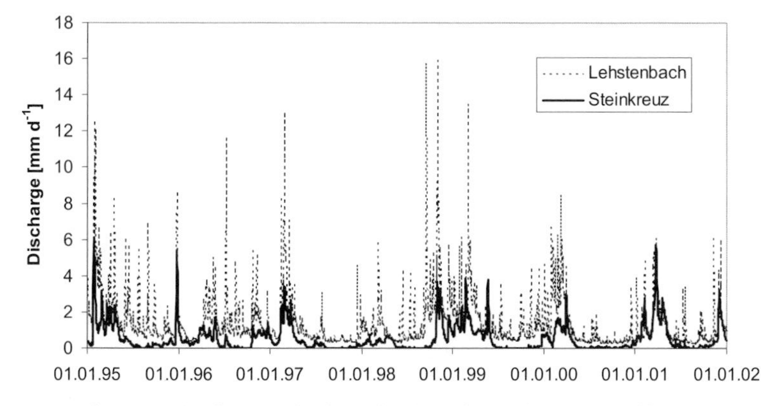

Fig. 23.1. Daily mean discharge of Lehstenbach and Steinkreuz runoff 1995–2002

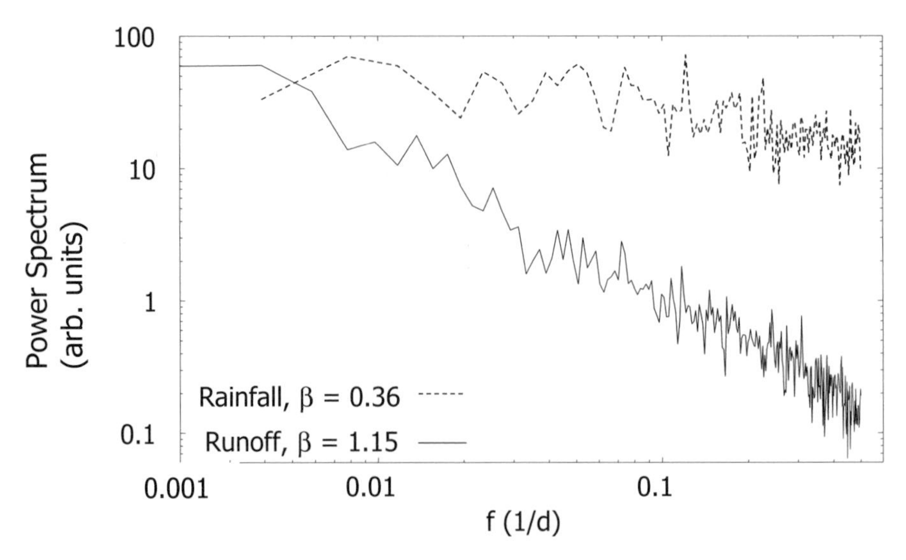

Fig. 23.2. Power spectra of bulk precipitation (*dotted line*) and runoff (*solid line*) at Lehstenbach. Although noisy, the difference in scaling exponents is obvious. Annual 'peak' in runoff [at a frequency (*f*) of 0.0027 day^{-1}] is rather a plateau and comparatively weak. Values for the exponents are in the typical range for these two variables

are masked by the seasonality of the signal. For example, during the early 1990s, discharge was rather low in some subsequent years. This was partly due to low amounts of precipitation. The rainfall (bulk precipitation) as measured at Lehstenbach has no visible annual cycle but has a non-trivial scaling part; however, the exponent (0.36) is much closer to the uncorrelated white noise case. Thus, part of the low discharge in the early 1990s has to be ascribed to an additional effect, that is, the depletion of the groundwater storage in preceding years.

The Steinkreuz hydrograph is strongly seasonal, and within the limits set by the finite measurement period (8 years) multiannual structures are indicated also (not shown). On a short-term basis, an autoregressive modelling with a set of discrete delays is successful, indicating the presence of distinct components of runoff generation.

The persistence, that is the strength of the long-term memory of the Lehstenbach data sets, is investigated using Hurst statistics (Fig. 23.3). Again, over roughly three orders of magnitude, we see strong indications of pronounced persistence for the Lehstenbach runoff, whereas it is much weaker for precipitation. Nevertheless, within errors the latter is also clearly different from the persistence–absent case. The Hurst exponent for runoff is within but at the upper end of observed H values in a survey of German and US rivers (Kamps 2001), and comparable to that of the Nile River. The theoretical (fractional Gaussian noise-based) relation $H=(1+\beta)/2$ is not confirmed within errors.

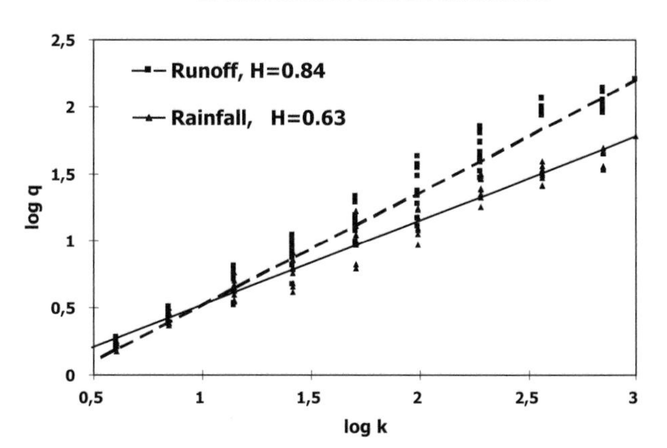

Fig. 23.3. Persistence of runoff and rainfall at Lehstenbach. Here, q is the Mandelbrot–Wallis R/S statistics (Mandelbrot and Wallis 1969), k is the time scale in days, and the least squares fit is applied to the ensemble mean of the individual determinations

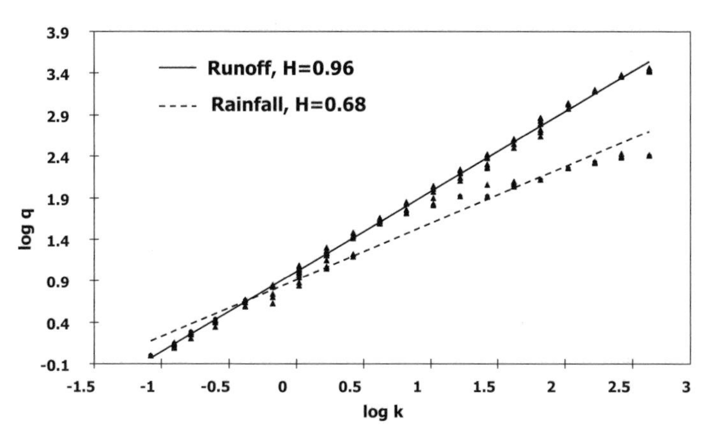

Fig. 23.4. Hurst analysis for the Steinkreuz catchment for 4 years. There is evidence of a saturation effect at high time scales, in particular for rainfall

In comparison, the Hurst analysis for the Steinkreuz catchment (Fig. 23.4) leads to a different conclusion. The best fitted exponent for runoff is actually extremely high, but this could be due to the short period investigated (4 years only). In addition, at the highest time scales, a deviation towards smaller values (saturation effect) occurs, which is particularly pronounced for the rainfall. This could originate in a finite memory effect, or non-Hurst behaviour. This could be partly explained by the fact that, in contrast to the Lehstenbach catchment, there is no extensive groundwater storage in the Steinkreuz catchment that could contribute to a multiannual memory effect of the catchment runoff.

In addition, further investigations (Seger 2002) give clear evidence that for the Lehstenbach catchment runoff, the scaling behaviour is not simple (characterized by a single scaling exponent) and fits into the multifractal framework (Pandey et al. 1998).

We next focus on information content and complexity of the hydrological variables which give hints on the 'optimal' time resolution of the measurements. Information content (MIG) and complexity (FC) were calculated, varying the temporal resolution via aggregation of consecutive values (Fig. 23.5), starting with daily values (lower left side of the graph) and ending with monthly (lower right side of the graph) aggregations (or equivalently means). The dependence of MIG and FC on the aggregation level is strong, but also characteristic for the respective variable: for runoff, MIG is increasing and FC as well, until a maximum is reached. The resolution at this maximum is site-dependent; it seems to be quite coarse for the Lehstenbach catchment, presumably reflecting contributions with long residence times (deep groundwater). In many cases, this maximum is reached in a few days only (Kamps 2001). For precipitation, MIG increases whereas FC decreases; here, for daily values we are already past the maximum, approaching more and more the random case (lower right corner in Fig. 23.5) at coarser resolutions. For this quantity, the maximal complexity is achieved when single rainfall events are captured by the data (requiring approximately hourly resolution). Finally, temperature has the very distinct property that above daily resolution, MIG does not vary at all when aggregating, whereas FC is decreasing. A step shift towards higher

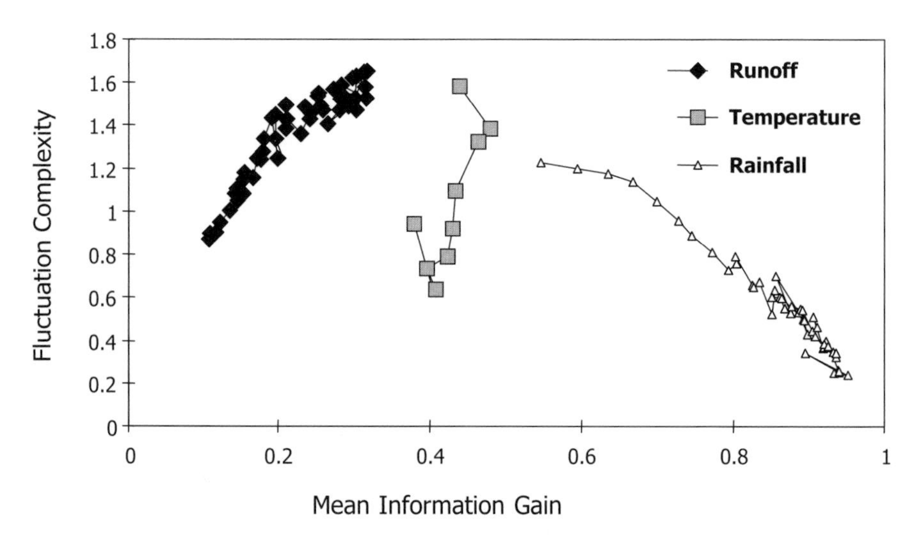

Fig. 23.5. Mean information gain (MIG) and fluctuation complexity (FC) for runoff, air temperature and rainfall at Lehstenbach. Temporal resolution is changed from daily (*lower left side of graph*) to monthly (*lower right side of graph*) values

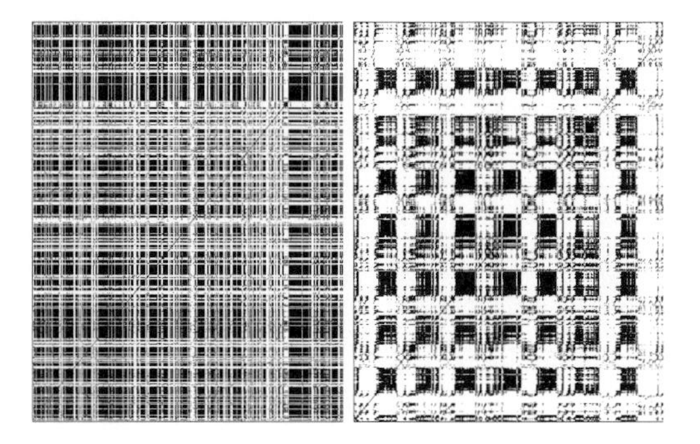

Fig. 23.6. Recurrence plots for precipitation (*left*) and runoff (*right*) at Lehstenbach. Eight years of data are shown (1991–1998)

values of MIG would presumably happen once the yearly cycle is encountered. This is, however, undeterminable for the method due to lack of data.

Mostly intriguing and even puzzling results stem from the RQA of Lehstenbach runoff. RQA is used here as a temporally local method to quantify the evolution of dynamic properties of the runoff; in other words, the details of instationary behaviour are elucidated. In Fig. 23.6, we show recurrence plots for 8 years of data for precipitation and runoff. Dark pixels indicate small distances and white lines indicate extreme (high) events. For precipitation, basically an irregular pattern of strong events and 'average' behaviour can be observed. The quantitative statistics (RQA, not shown) confirm the impression that this is nearly random behaviour, a typical property of rainfall (Haug 2001).

The recurrence plot for the runoff signal shows a chessboard-like pattern, pointing to a periodic process. This is of course the annual cycle, which, however, is weak for the whole measurement period. Looking at the plot, it is obvious that its strength is different from year to year, and that there are (wet) years when it is virtually absent. In other years (usually with low average flows), it is quite pronounced.

As an example of the RQA quantities, Fig. 23.7 presents the determinism of the runoff. This is a measure for the fraction of times when similar dynamics occurs in the time series; in other words, it gives the probability of using one portion of the series to successfully predict another portion with the same starting value, i.e., the determination of the time series by its initial value (hence the name). Its value (reported here as the fraction of the theoretical maximum) is dependent on the parameters of the method; however, its temporal course is a robust feature and thus open to interpretation.

As Fig. 23.7 shows, a particular long-term instationarity is present here. The determinism has a periodic (yearly) structure not reported before in the literature even for periodic processes. This is, however, plausibly attributable to the larger relative effect of noise in periods of low flow – the minima of determinism correspond to minima of runoff. In addition, it has a clear ten-

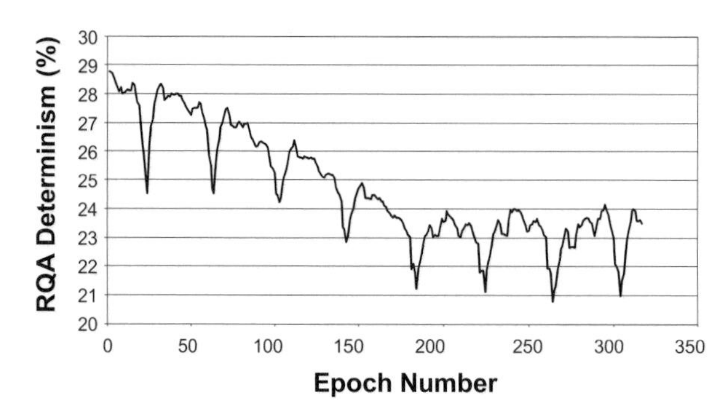

Fig. 23.7. Determinism as derived by recurrence quantification analysis (*RQA*) for Lehstenbach runoff. Values on x axis denote epoch number, where one epoch is 1 year long and the shift between adjacent epochs is 10 days

dency to decrease over recent years. This is related to an increasing noise level, whose origin, however, is unknown and not easily detected using windowed coefficients of variation for the runoff series (not shown). Two other RQA quantities, local recurrence rate and line length entropy, show similar trends over the measurement period (not shown). It is not obvious whether we observe a multiannual phenomenon of natural origin, like climatic variability, or the aging of a runoff weir or stream gauge, or other processes. The first one of these would have severe implications to long-term modelling of the hydrograph, as the performance of a hydrological model that works well for the first years of the measurement period is likely to decrease in quality in subsequent years.

Taken together, this ensemble of properties makes the Lehstenbach hydrograph unique among runoff streams from catchments of comparable size. An empirical separation of quick and slow components of the hydrograph signal seems nevertheless possible.

Mean residence time of the runoff water in both catchments and of groundwater at six different wells in the Lehstenbach catchment was determined based on ^{18}O data. The Lehstenbach data are given by Zahn (1995). Mean residence time for the runoff is given as 3.6 years, and differs between 2.1 and more than 4 years for six groundwater wells. For the Steinkreuz runoff, the analysis based on the approach by Burgman et al. (1987) yielded a mean residence time of about 1.3 years, with the 95 % confidence interval spanning the range between 0.6 and 2.7 years (Lischeid and Stichler, unpubl. data).

Lag times between the maximum rainfall intensity during single rainstorms and discharge at the catchment outlet were investigated for all 44 major rainstorms during the 1 June to 30 November 1998 period in the Lehstenbach catchment (Lischeid et al. 2002). Maximum time lag was 5 hours and minimum time lag was equal to the time resolution of discharge measurements, i.e., 1 h; in 64 % of cases, time lag was between 3 and 4 h. This time lag was not correlated with rainfall intensity, antecedent discharge or rate of increase of discharge.

Usually, the pressure signal propagates vertically with decreasing velocity through the soil. In most cases, the time lag between maximum precipitation intensity and maximum soil matrix potential or soil water content in 90 cm depth is about 2 h.

At Steinkreuz, the response of the catchment runoff to single convective rainstorms occasionally occurred within a few minutes, and the response of the soil matrix potential in 20, 35 and 90 cm depth within the time resolution of the sensors (1 h). Here, the spatial pattern of precipitation infiltration depends on canopy drip, root water uptake and stemflow. Due to the former, even at the end of the growing season, matrix potential at greater distance from the trunks was rarely less than –100 hPa. Groundwater recharge is observed during summer rainstorms close to the trunk of major beeches, where it is due to infiltration of stemflow (Chang and Matzner 2000). In general, vertical and lateral hydraulic gradients develop during the growing season according to the spatial pattern of throughfall and root water uptake which are highly variable in time, including occasional reversal of the gradients close to the beech trunks within hours.

There is strong evidence that neither infiltration excess overland flow nor lateral preferential flow in the unsaturated zone account for the quick response of the catchment runoff at Steinkreuz and Lehstenbach. Instead, the predominating runoff generation process is saturation excess overland flow in the riparian zone. In most cases, and with respect to most solutes, storm flow runoff water resembled that of pre-event topsoil water rather than that of precipitation. This was especially true for SO_4 and dissolved organic carbon (DOC) at Lehstenbach, and ^{18}O, SO_4, Ca and Si at Steinkreuz. Thus, preferential flow phenomena seem to be unimportant. However, it could be shown that this is in the first place due to rapid equilibration with the soil matrix in the Lehstenbach catchment. As a consequence, silica dynamics, for example, cannot be described as piston flow displacement due to its very slow kinetics (Lischeid et al. 2002).

Riparian wetlands comprise one third of the Lehstenbach catchment area, whereas wetlands do not exist at Steinkreuz. The role of biological redox processes in the Lehstenbach wetlands is described by Küsel and Alewell (this Vol.). The impact of denitrification and SO_4 reduction on the catchment budget and on stream water quality was investigated at the Köhlerloh site. Here, a first-order stream of about 100 m length is located within the riparian wetland. It is a perennial stream, where oxic groundwater discharges, which infiltrated at upslope sites. The discharging groundwater has to pass through the hypoxic zone of a few meters thickness underneath and along the stream. Soil solution in the vicinity of the stream is nearly completely depleted of nitrate, exhibits very low SO_4 and very high Fe and Mn concentrations, thus clearly indicating the role of redox processes. However, due to the short residence time of the discharging groundwater in the hypoxic zone, redox processes have only a minor effect on the stream water quality. Moreover, during rain-

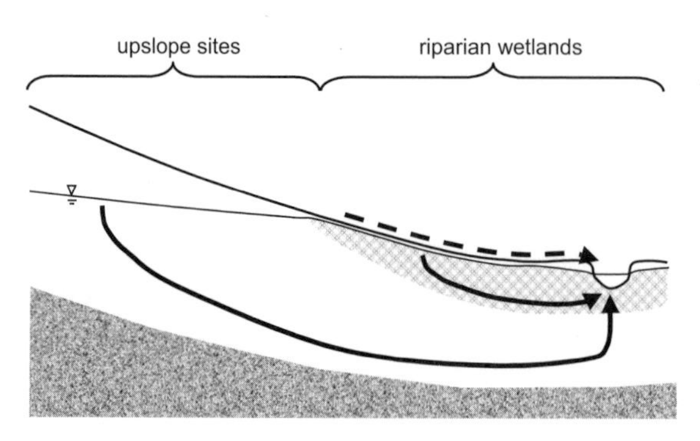

Fig. 23.8. Flow paths (*arrows*) and runoff generation processes in the Lehstenbach catchment. Anoxic part of aquifer is shown in *gray*. *Dashed arrow* indicates saturation excess overland flow during storm flow

storms surface runoff in the riparian zone bypasses the hypoxic zone, thus further reducing the beneficial effects of microbial reduction of nitrate and sulfate on stream water quality (Fig. 23.8; Lischeid et al. 2004). This seems to be the case in most of the wetland areas in the Lehstenbach catchment. The only exception seems to be the large wetland area Schlöppnerbrunnen II which is located close to the catchment outlet. Following the flow path of the groundwater that discharges close to this wetland reveals that roughly half of the groundwater infiltrated in wetland areas in the eastern part of the catchment. As a consequence, the groundwater well GW04, about 400 m upslope of the Schlöppnerbrunnen II wetland, is the only deep groundwater well with entirely anoxic water. Here, nitrate concentration rarely exceeded the detection limit, and very high Fe concentrations of about 10 mg l^{-1} were found all year-round. However, in spite of the clear evidence of redox processes found in the GW04 groundwater samples, SO_4 concentration was the highest of all groundwater wells at Lehstenbach. Solute concentration of the shallow groundwater, sampled at the Schlöppnerbrunnen II wetland in 1 and 2 m depth, was very similar to that of the GW04 well. Only the SO_4 concentration of the former was substantially less compared to the latter, indicating dissimilatory SO_4 reduction in the biologically active topsoil layer of the fen. However, the substantial increase of the ^{34}S values (Alewell, pers. comm.) of the sulfate provides strong evidence for repeated cycles of sulfate reduction and reoxidation, as observed at other sites (Dillon and LaZerte 1992).

23.4 Solute Concentration

At Lehstenbach, Ca and Na predominate among the cations, and SO_4 and NO_3 among the anions in stream water and groundwater. At Lehstenbach, SO_4 originates exclusively from atmospheric deposition. At Steinkreuz, Ca and Mg predominate both in the stream and in the groundwater. Sulfate is the prevailing anion in the shallow groundwater and in the stream during highflow, and HCO_3 in the deeper aquifer and during baseflow.

Both in the Lehstenbach and in the Steinkreuz, runoff concentration of all major solutes showed a marked short-term variability that was closely connected to the discharge dynamics. Moreover, these short-term dynamics clearly changed during the years of the study. Please note that this chapter focuses on solute concentration. The associated solute fluxes are presented in Chapter 24 (Lischeid et al., this Vol.).

23.4.1 Chloride

Chloride in the Lehstenbach catchment might serve as an illustrative example.

Roughly half of the annual Cl input in the catchment is due to road clearing in the western part of the catchment. As a consequence, most of the streams and groundwater wells downslope of the road exhibited high Cl concentration the whole year round. This holds for the western branch of the Lehstenbach stream as well.

The eastern tributary which has been sampled since 1987 at the confluence with the west tributary, about 10 m before the catchment runoff, is not contaminated by road salt. Here, Cl concentration decreased significantly (Spearman rank correlation, α <0.001), which is ascribed to decreasing atmospheric deposition. There is no corresponding trend for Na in that stream. In contrast, both Cl and Na concentration of the catchment runoff increased significantly (Spearman rank correlation, α <0.001) between 1987 and 2002 (Fig. 23.9).

The Cl time series of the catchment runoff exhibits remarkable short-term variance. A clear short-term decrease of Cl concentration was observed during discharge peaks, which were predominantly due to runoff generation in the riparian wetlands close to the catchment outlet. These wetlands were not contaminated by road salt. In contrast, it has to be assumed that most of the Cl

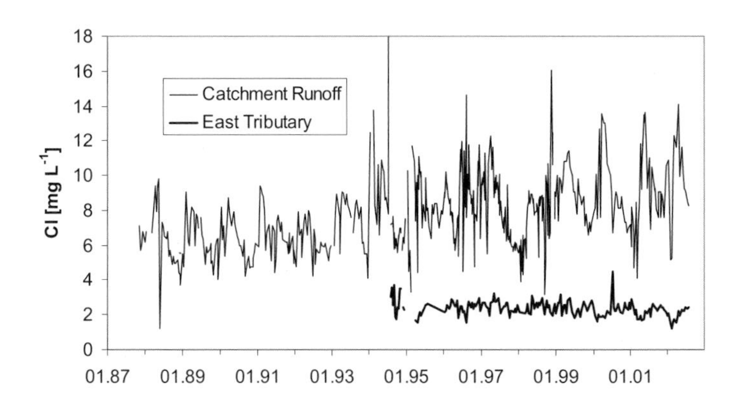

Fig. 23.9. Chloride concentration in the catchment runoff and east tributary of the Lehstenbach catchment

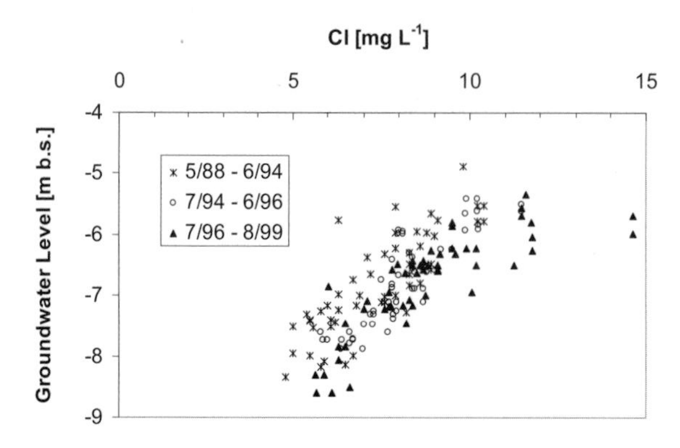

Fig. 23.10. Chloride concentration in catchment runoff during baseflow (<1 mm day^{-1}) versus groundwater level (well GW01) for different periods

is stored in the wetlands downslope of the road. Thus, the higher the groundwater level in the western upslope part of the catchment, the more these wetlands close to the road become hydraulically connected to the stream, and the higher the Cl concentration in the western branch of the Lehstenbach. As a consequence, there was a clear correlation between mean groundwater level and Cl concentration in the stream. Moreover, this regression exhibited a clear shift toward higher Cl concentration in recent years, indicating increasing Cl contamination (Fig. 23.10).

Due to a sequence of extraordinary dry years around 1990 (see above), the Lehstenbach groundwater level was fairly low in the first years of the observation period and increased in the early 1990s. This might explain part of the step-wise increase of Cl concentration in the catchment runoff (Fig. 23.9). However, Fig. 23.10 clearly indicates that the observed trend in the catchment runoff is not exclusively due to changed hydrological boundary conditions.

23.4.2 Sulfate

The turnover and residence time of sulfate was one of the major foci of BITÖK research in the Lehstenbach catchment. To elucidate the effort necessary to run a model, we performed a complexity analysis for both SO_4 and NO_3 from input (throughfall) and output (runoff), varying the temporal resolution by aggregating. At the highest resolution, all four variables have approximately the same information content and complexity, whereas at coarser resolution they fall into two different classes: whereas the input data reach the highest possible complexity values and are very similar to each other, the output variables keep lower complexity variables and show distinct behaviour (Fig. 23.11). A conclusion from that calculation is that the modelling task is easiest for the sulfate output, followed by the nitrate output.

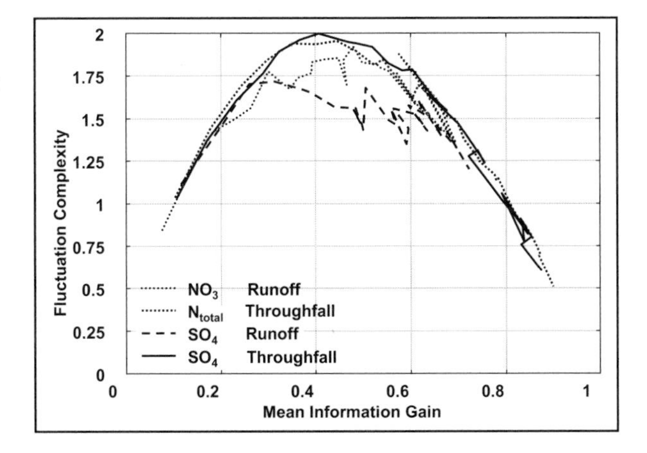

Fig. 23.11. Complexity analysis for sulfate and nitrate input and output at different aggregation levels

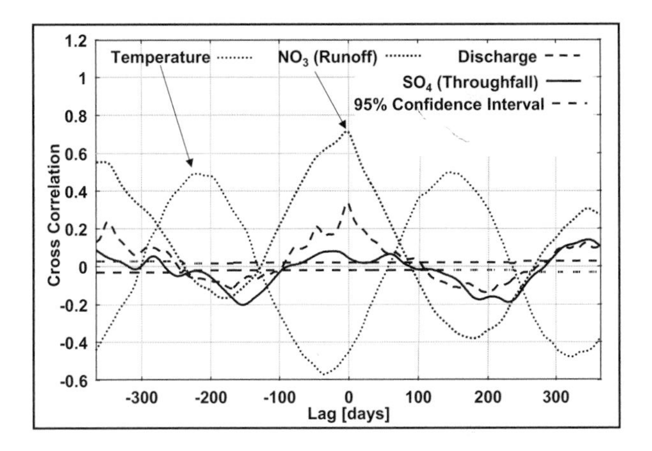

Fig. 23.12. Cross-correlation function of SO_4 concentrations in runoff compared to possible driving variables. Also included are 95% confidence intervals

As a next step towards an empirical model of the SO_4 dynamics in the catchment runoff, a cross-correlation analysis was performed for the SO_4 concentration in runoff with air temperature, discharge rate, SO_4 input (measured as throughfall) and NO_3 in runoff, as shown in Fig. 23.12. It is evident that there is a seasonality in the cross-correlation function as expected. The air temperature is strongly negative correlated with SO_4 at lag 0; however, the strongest (most negative) correlation occurs with the temperature about a month earlier. A model that includes the temperature of the previous month as input to the SO_4 signal should perform better than one with instantaneous values.

The positive correlation with NO_3 is instantaneous, as is the one with discharge rate, which is weaker but has a richer structure: the clear maximum at lag 0 and a number of smaller maxima aside point to different components of the runoff (fast and slow, shallow and groundwater-based), determining in

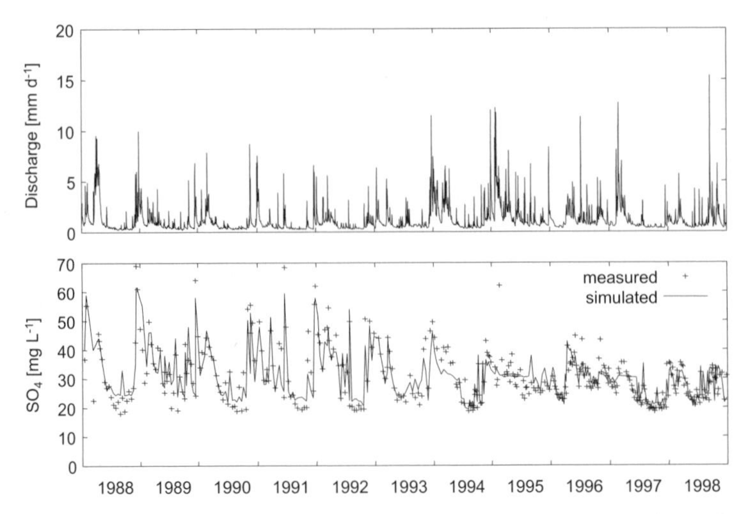

Fig. 23.13. Measured daily mean discharge (*above*) and measured and simulated SO_4 concentration (*below*) in catchment runoff. (Reprinted from Lischeid 2001a, copyright 2001, with permission from Elsevier)

part the SO_4 signal. Finally, the SO_4 input is only weakly correlated to the stream concentration. This means that the transforming process within the soil and groundwater is substantially altering the signal on its way through the system.

In a next step, the time series of SO_4 concentration in the Lehstenbach runoff were modelled using a multilayer perceptron network (Lischeid 2001a). It explained about 80 % of the short-term variance at daily time resolution (Fig. 23.13). Three out of a variety of candidate input variables were identified as driving variables: daily mean discharge at the sampling day, the antecedent runoff sum, i.e., the cumulative runoff since the last excess of a threshold value of 0.5 mm day^{-1} daily discharge, which separates baseflow from highflow conditions, and a sliding mean of throughfall concentration of SO_4. In contrast, for example, neither information about snowmelt nor air temperature data helped to increase the performance of the network. A dependency on temperature would have given hints on the role of biological processes.

The functional relationships revealed by the neural network are visualized in Fig. 23.14. The two planes span the range between the mean of the predicted value plus and minus the standard deviation, given by the results of 100 networks (see Lischeid 2001a for details). The larger the distance between the two planes at certain points, the larger the uncertainty of the neural network model.

It can be clearly seen that the short-term dependency of SO_4 concentration on discharge which was very pronounced at the end of the 1980s weakened

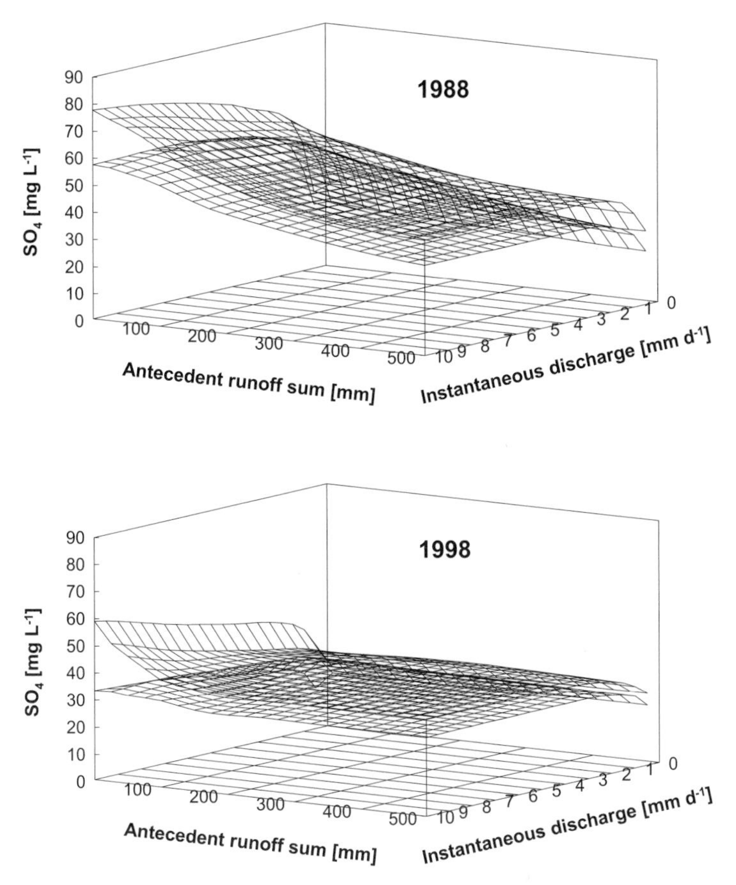

Fig. 23.14. SO_4 concentration in catchment runoff depending on instantaneous discharge and antecedent runoff sum predicted by the model. Mean SO_4 concentration in throughfall is 24 mg l^{-1}, corresponding to mean concentration measured in 1988 (*above*) and 4.8 mg l^{-1}, corresponding to mean concentration measured in 1998 (*below*), respectively. *Upper plane* Mean + standard deviation; *lower plane* mean – standard deviation (n=100) for every grid point. (Reprinted from Lischeid 2001a, copyright 2001, with permission from Elsevier)

during the 1990s. In contrast, there is no clear trend for baseflow conditions. These results confirm a simple perceptual model: due to the substantial decrease of SO_4 deposition in the catchment, SO_4 concentration of the topsoil solution decreased. This is confirmed by soil solution data (Matzner et al., this Vol., Chap. 20). The topsoil contributes to the catchment runoff only during storm flow (Lischeid et al. 2002). In contrast, deep groundwater discharge determines the Lehstenbach runoff during baseflow conditions. In general, groundwater SO_4 did not show a corresponding clear decrease as observed in the topsoil. To summarize, the neural network results indicate mixing of the

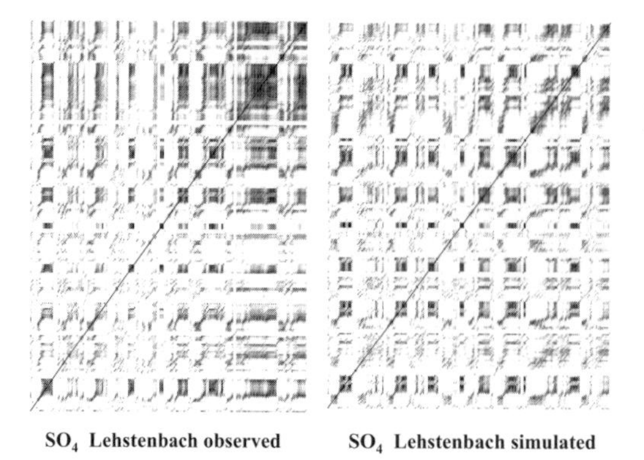

Fig. 23.15. Recurrence plots of sulfate in runoff as observed and as simulated using the neural network

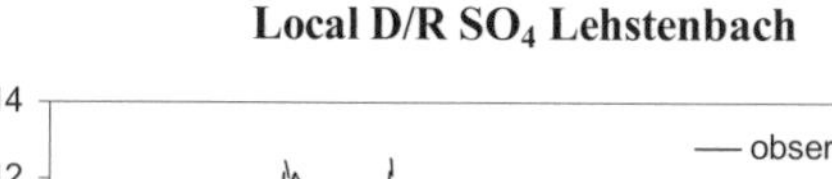

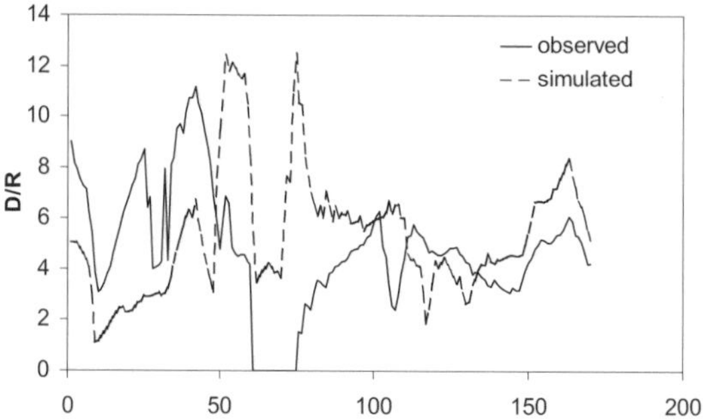

Fig. 23.16. Local determinism/recurrence (*D/R*) ratio as a function of epoch number for the measured and NN-simulated SO_4 signal in Lehstenbach runoff. Epoch length is 1 year and the shift between epochs is 10 days

deep groundwater component with constant rate and constant SO_4 concentration with the temporally varying soil water component, which clearly responded to the decreasing SO_4 deposition.

A detailed visualization of the quality of the neural network simulation is provided by a comparison of the corresponding recurrence plots in Fig. 23.15. The general agreement between the two time series is remarkably good, but the quality is not evenly spread throughout the simulation period. In particular, the varying strength of seasonality (extensions of the dark squares) is not reproduced by the simulation, and there is more variability at low concentrations than in the observation. Taking into account that recurrence plots are extremely sensitive to even minor deviations, for some years the agreement is excellent.

One of the RQA indicators discussed at the beginning of this chapter, the ratio of determinism to recurrence rate (D/R), is shown in Fig. 23.16 for comparison. The epochs in this plot are 1 year long and their offset is 10 days. It is clear that the later part of this simulation shows excellent agreement with the data; in the beginning, the simulation overestimates the degree of determinism of the measured series.

However, SO_4 concentration trends in the groundwater and at a number of upslope stream water sampling sites were puzzling. For example, in the deep groundwater wells GW04 and GW05 a highly significant and almost linear increase in SO_4 concentration was observed. Since the end of the 1990s, groundwater concentration at these sites clearly exceeded that of the catchment runoff. Correspondingly, SO_4 concentration was very high at the upslope spring Schlöppnerbrunnen, but did not show any clear trend before 1995, and a clear decrease thereafter (Fig. 23.17). Similar patterns were observed in the Lysina catchment, about 60 km east of the Lehstenbach catchment, which exhibits very similar geochemical properties (Büttcher 2001).

The observed differences of SO_4 dynamics at different sampling sites could not be explained by different geochemical properties of the soils and the aquifer. Instead, the hypothesis was developed that flow path length-depending SO_4 retardation due to sorption and desorption could explain the observed different dynamics at different sampling sites.

The SUNFLOW model was set up to test this hypothesis in a quantitative way. It simulates SO_4 transport by convection and sorption along one-dimensional flow paths. The longer the flow path, the more delayed and damped is the SO_4 breakthrough at the end of the flow path. The model assumes a clear stratification by age or flow path length, respectively, with the shortest flow

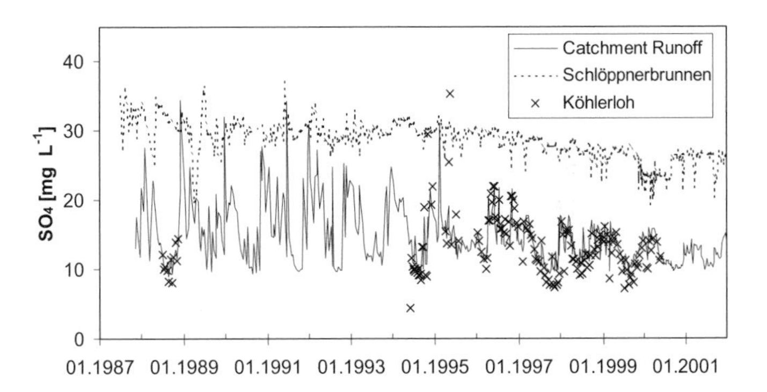

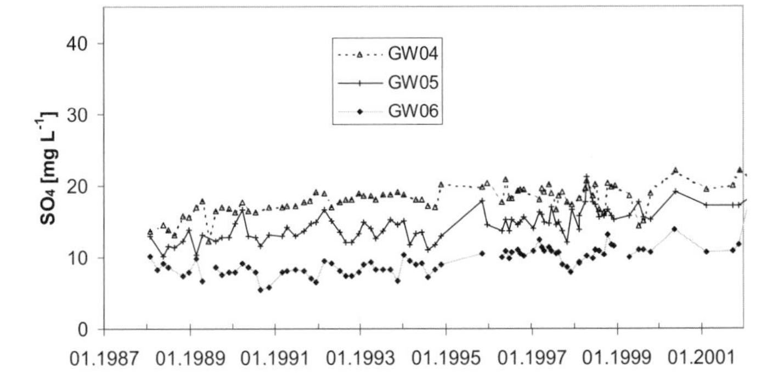

Fig. 23.17. Time series of SO_4 concentration in stream water (*above*) and groundwater (*below*) at various sampling sites in the Lehstenbach catchment

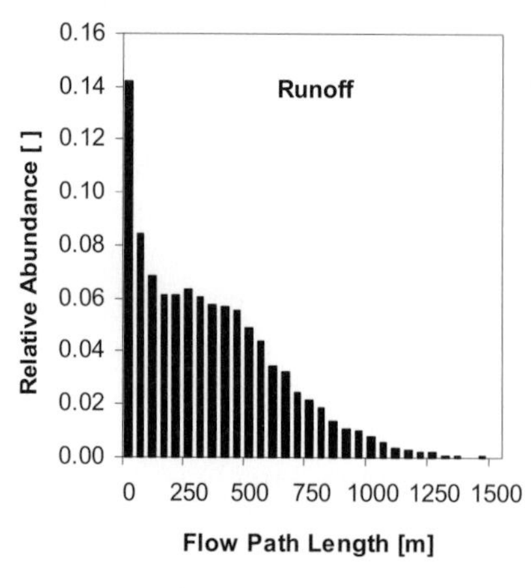

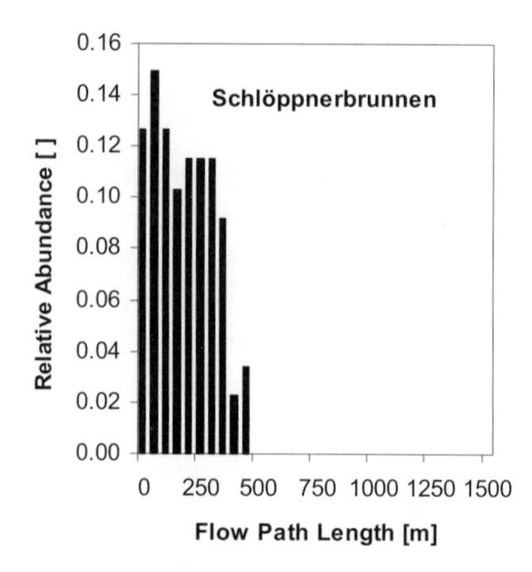

Fig. 23.18. Flow path length distribution for two stream water sampling sites in the Lehstenbach catchment

path, i.e., the shortest residence time, in the uppermost layer. Stream water and groundwater sampling sites are characterized by their flow path lengths distributions. Sulfate concentration at the sampling sites is calculated by integrating the contribution of single flow paths according to the flow path lengths distribution.

Flow path lengths at the Schlöppnerbrunnen upslope spring are fairly homogeneously distributed, with a maximum of about 500 m (Fig. 23.18). Thus, very short flow paths, i.e., the topsoil layer, comprise only a minor fraction of the spring runoff. According to the model, the decreasing SO_4 load of these layers was compensated for by an increase in the longer flow paths until the 1990s.

In contrast, the flow path lengths distribution of the catchment runoff is highly skewed with a clear maximum for the shortest flow path length. This fits with the observation that the hydrograph of the catchment runoff is much more flashy than that of the Schlöppnerbrunnen spring. In addition, the envelope given by the simulated contribution of the short and the long flow paths, respectively, roughly spans the range of observed short-term fluctuations (Fig. 23.19).

The model was applied without calibration by inverse modelling. Although parameterization of the deeper regolith is associated with substantial uncertainties, the model succeeded in explaining the differences of both the long-term and short-term SO_4 dynamics observed at different sites. The model depicts the observed decrease of SO_4 concentration in the topsoil, and the continuing increase in deeper groundwater layers. The longest flow paths, i.e., the deepest groundwater layers, deliver only small amounts of SO_4 to the stream.

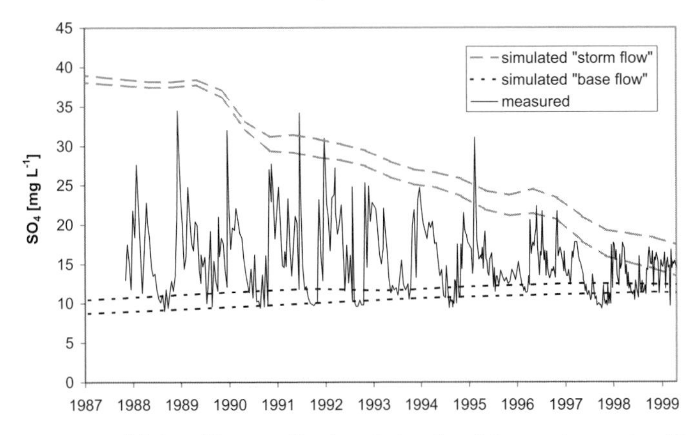

Fig. 23.19. Measured (biweekly sampling) and simulated (annual mean) sulfate concentration in Lehstenbach catchment runoff (parameterization 'slow'). Three shortest flow path length classes (<150 m length) are combined to yield 'storm flow', and the remaining flow paths to yield 'base flow'

In spite of its simplicity, there are still uncertainties with regard to the parameterization of the deeper aquifer where no drilling has been performed. Thus using that model for long-term prediction in a quantitative way should be avoided. However, it seems to be plausible that both the decrease in SO_4 concentration in the topsoil and the increase in the deeper groundwater will continue for years or even decades, although with decreasing rates. Actually, SO_4 concentration in the catchment runoff started even to decrease during discharge peaks in the late 1990s. In the Lysina catchment with similar geochemical properties, this trend became more clear: here, most of the SO_4 load of the stream is due to the deep groundwater contribution, thus shifting the episodes of maximal SO_4 contamination from highflow to baseflow periods. This is consistent with the SUNFLOW results (Büttcher 2001).

It is remarkable that the SUNFLOW model underestimates the total SO_4 pool of the Lehstenbach catchment (Büttcher 2001), although it does not account for additional long-term SO_4 sinks like incorporation into the biomass or finite SO_4 reduction. Sulfur isotope data clearly indicate that biological SO_4 reduction occurs (Alewell and Gehre 1999; Groscheová et al. 2000; Alewell and Novák 2001), but this seems to be a more transient and reversible sink.

Estimates of the total SO_4 pool stored in the upper 10-m layer of the Lehstenbach catchment that could be remobilized are in the range of about 50–100 times the annual output measured in 2000 (Manderscheid et al. 2000). Thus, although public concern about soil acidification substantially decreased during the last decade, acidification will continue and even increase to impair drinking water supply in the region.

Studying SO_4 dynamics at Steinkreuz has to take into account that often evaporites are found in the Lehrberg layers (Emmert 1985) which comprise the deepest layers of the catchment. However, long-term trends of SO_4 concentration in the Steinkreuz groundwater and stream water are consistent with the perceptual model of SO_4 dynamics developed for the Lehstenbach catchment.

The maximum of annual mean SO_4 concentration in throughfall was 1.96 mg l^{-1} in 1996, decreasing to 0.86 mg l^{-1} in 2001 (Matzner et al., this Vol., Chap. 14), which is less than 10 % of stream SO_4 concentration. A corresponding clear decrease of SO_4 concentration by one third was observed in soil solution, shallow groundwater and in the stream (Fig. 23.20). Sampling of the deep groundwater did not start before the end of 2000; a preliminary analysis seems to confirm a similar trend. Sulfate concentration in all of the components given in Fig. 23.20 clearly exceeded those of mean throughfall concentration. As has to be expected, SO_4 concentration was at a minimum in soil solution (60 and 90 cm depth). Please note that only during the first years of the study did soil solution SO_4 concentration occasionally exceed that of stream water (Fig. 23.20). Shallow groundwater SO_4 concentration was similar to that of the stream until the end of the 1990s, and fell below the latter in the last years of the study. Sulfate concentration of the upslope springs was remarkably similar to that of the catchment runoff throughout the whole observation period. As only the catchment runoff is likely to drain the deepest groundwater layer, it can be concluded that this deep layer is not an important SO_4 source.

In the last years of the study, spring and stream water exhibited the highest SO_4 concentration. It is assumed that this is due to high SO_4 loading of groundwater from intermediate depth, similarly as predicted for the Lehstenbach catchment by the SUNFLOW model. There is only one groundwater well of intermediate depth (4.5–5.5 m) in the Steinkreuz catchment that delivers water only after major rainstorms. Here, SO_4 concentration determined in

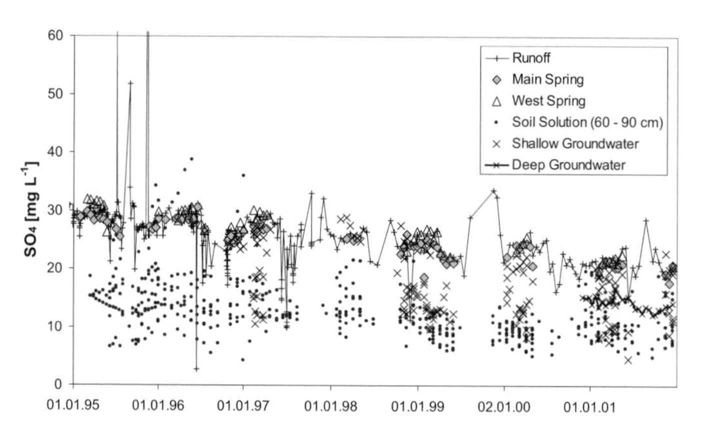

Fig. 23.20. Time series of SO_4 concentration in soil solution, shallow and deep groundwater, springs and runoff of the Steinkreuz catchment

four samples taken in 2001 was in the same range as that of stream water, confirming that assumption.

23.4.3 Nitrate

In both catchments, NO_3 is the almost exclusive form of inorganic nitrogen compounds in groundwater and stream water. Time series of NO_3 concentration in the Lehstenbach and Steinkreuz runoff exhibit marked short-term variance which is partly linked to that of discharge. These short-term dynamics was investigated by artificial neural networks. At both catchments, discharge and air temperature were identified as driving variables. Not only the values measured on the sampling day but also the mean values of the preceding 30-day period were required by the model (Lischeid et al. 1998; Lischeid and Langusch 2004). The neural network models explain 70 % of the variance at Steinkreuz and 40 % at Lehstenbach (Figs. 23.21 and 23.22, respectively).

The NO_3 concentration in the Lehstenbach catchment runoff decreased in the late 1980s in spite of constant atmospheric nitrogen deposition. Although the neural network model performance is rather low, it seems that the model can reproduce that trend (Fig. 23.22). This adds strength to the assumption that the observed trend is mainly due to interannual variations of climatolog-

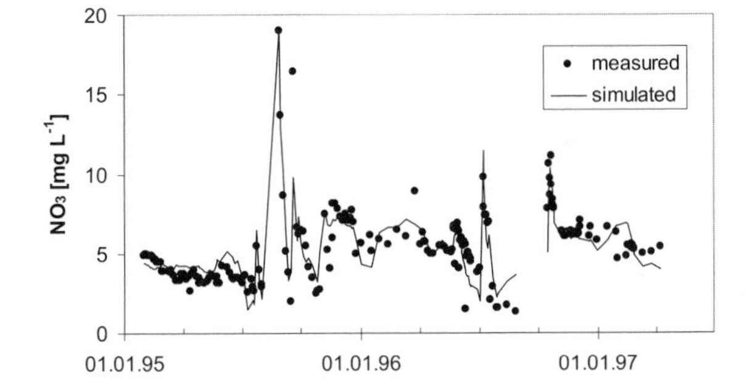

Fig. 23.21. Measured and simulated time series of NO_3 concentration in Steinkreuz runoff.

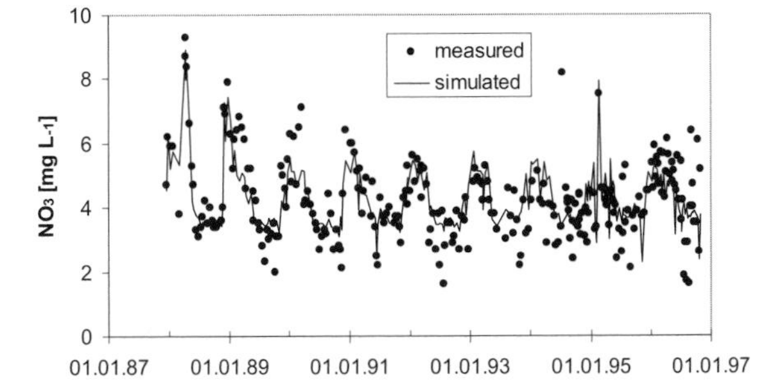

Fig. 23.22. Measured and simulated time series of NO_3 concentration in Lehstenbach runoff

ical and hydrological boundary conditions. Similar observations were made at various sites in central Europe and North America and are still poorly understood (Stoddard et al. 1999; Goodale et al. 2003).

As before with sulfate, we investigate the simulation quality, using cross-recurrence plots (Fig. 23.23). In this case, the last part of the simulation is somewhat worse than the others, and the reason is the loss of seasonality of the measured NO_3 concentrations. This is not covered by the model since it is temperature dependent and keeps seasonal throughout the simulation period.

Artificial neural networks that use the same input variables yield about the same performance for nitrate dynamics of the Lange Bramke runoff (Lischeid 2001b) in the Harz Mountains and the Lysina runoff (Lischeid et al. 2003) in the Czech Republic, about 60 km east of the Lehstenbach catchment. This is in accordance with the correlation analysis findings of, e.g., Sloan et al. (1994), Andersson and Lepistö (1998) and Arheimer and Lidén (2000), emphasizing a more general phenomenon.

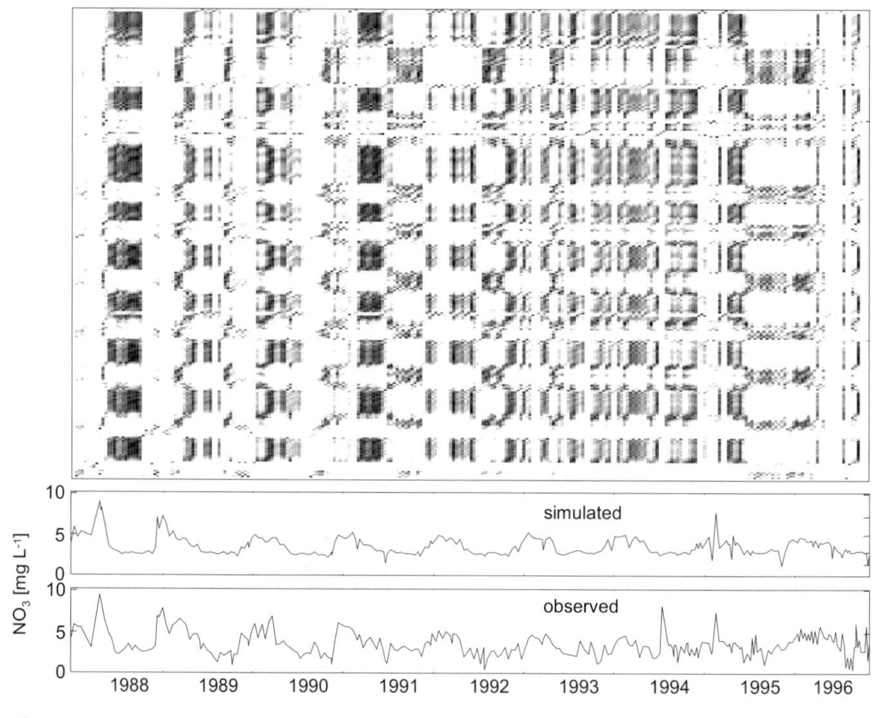

Fig. 23.23. Cross-recurrence plot of NO_3 as measured and simulated with the artificial neural network. The two time series are shown also: simulated (*above*) and measured (*below*)

Preceding analyses showed that increasing the time lag beyond 30 days does not improve the model. As the mean subsurface residence time of the discharging water in the Lehstenbach and Steinkreuz catchments is substantially longer, it must be concluded that the observed nitrate dynamics is predominantly generated close to the stream (cf. Fölster 2000). This is confirmed by the fact that none of the groundwater wells in either of the catchments exhibits any clear seasonality. Moreover, air temperature was identified as one of the key variables, pointing to the role of not only biota in the short-term nitrogen dynamics, but also processes close to the atmosphere, that is, the topsoil. In summary, nitrate dynamics in the runoff of these four catchments is unlikely to reflect hillslope processes, but may provide some information about processes in the riparian zone (Lischeid et al. 2004).

Combining these results with the outcome of various process studies in forested catchments (Creed et al. 1996; Sánchez Pérez et al. 1999; Fölster 2000; Mitchell 2001) suggests the following perceptual model: the higher the air and topsoil temperature, the higher the rates of decomposition of organic matter, nitrogen assimilation by microbes and plants and dissimilatory nitrate reduction. Thus the amount of nitrogen that can be mobilized and leached during the occasional periods of elevated seepage flux and groundwater recharge during rainstorms depends on the interplay of these processes.

Nitrogen input by throughfall exceeds the net uptake of the trees by far in both catchments. Denitrification in the riparian wetlands has to be taken into account as another important sink for nitrogen in the Lehstenbach catchment. At all of the four wetland sites where soil solution has been sampled since 1988, NO_3 concentration rarely exceeded the detection limit of 0.5 mg l^{-1} (Moritz et al. 1994; Küsel and Alewell, this Vol.). Thus, Küsel and Alewell (this Vol.) conclude that the fraction of nitrogen that is denitrified should be roughly equal to the fraction of wetland soils of about one third of the catchment area. This has been checked at the Köhlerloh site, where a perennial first-order stream of about 100 m length is entirely embedded within the wetland area. However, wetland soils comprise only about 10 % of this stream's catchment area. Thus, most of the groundwater that discharges into the stream infiltrated at well-aerated upland soils. As a consequence, oxygen saturation of less than 70 % was observed only in the upper 1- to 3-m layer in a groundwater well, about 5 m from the stream. Correspondingly, NO_3 concentration of the shallow groundwater was up to 50 % less than in the deeper groundwater. In contrast, NO_3 concentration only occasionally exceeded the detection limit of 0.5 mg l^{-1} in 0.5 and 1 m depth in the wetland soil (Küsel and Alewell, this Vol.). Ammonia and NO_2 concentrations are negligible in these samples.

One would expect that the stream water comprised a mixture of NO_3-depleted topsoil water and NO_3-rich deep groundwater. In fact, stream NO_3 concentration exceeded that of the (deep) groundwater at all sampling dates during 1999–2002 (Fig. 23.24). This is especially true for single discharge

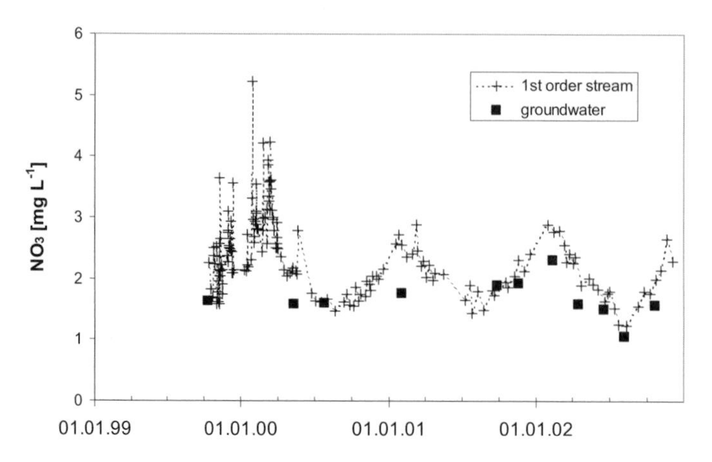

Fig. 23.24. NO_3 concentration in the first-order stream and in groundwater at the Köhlerloh site

peaks: NO_3 and DOC concentrations increase and Si concentration clearly decreases during storm flow (not shown). This points to an additional component, i.e., saturation overland flow, that bypasses the anoxic shallow groundwater zone (cf. Lischeid et al. 2002) and delivers considerable amounts of NO_3 to the stream (Fig. 23.8). The impact of this process on the catchment's nitrogen budget has not been precisely quantified up to now. However, it is concluded that the quantitative assessment given by Küsel and Alewell (this Vol.) overestimates the denitrification capacity of the catchment.

In contrast, denitrification in the Steinkreuz catchment seems to be restricted to clayey layers in the subsurface. Nitrogen concentration in all of the shallow groundwater wells, about 2–3 m below surface and roughly 100 m from the stream clearly exceeded that of the stream water sampled at two springs and at the catchment outlet (Fig. 23.25). The difference is ascribed to denitrification in the clayey layers of the aquifer.

The neural network analysis gave insight into the short-term nitrogen dynamics, but it cannot directly be applied for long-term risk assessment. However, the results achieved so far emphasize the role of biological processes, which in turn highly depend on short-term weather conditions. At Steinkreuz, short-term nitrate concentration peaks of up to 30 mg l^{-1} were observed in the stream, which is fairly high for a forested catchment. Shallow groundwater sampled underneath the rooting zone exhibited similar or even higher NO_3 concentrations after single heavy rainstorms, which were in the upper range of that observed in soil solution cups. This provides some evidence that the mobile fraction of the seepage water that feeds the aquifer exhibited systematically higher NO_3 concentrations than assessed by soil solution data. One reason for these pronounced NO_3 dynamics might be the fact that the area has belonged to the monastery of the nearby town of Ebrach since the beginning of the 12th century. The monastery did not allow local farmers to graze cattle in the forest or to remove litter from the forest, as was

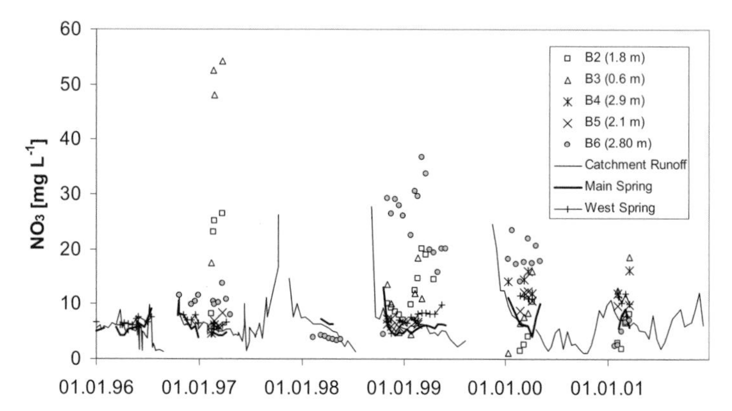

Fig. 23.25. NO_3 concentration in the stream (catchment outlet and two springs) and in the groundwater

done in most of the forested areas in central Europe during the Middle Ages. In addition, besides a small area of about 1.5 ha, historical maps do not indicate any agriculture in the catchment area. As a consequence, a nitrogen store of about 6,000 kg N ha^{-1} was able to develop in the topsoil (Lischeid and Langusch 2004), which might be easily available for microbial turnover. Moreover, the quality of that nitrogen pool, e.g., the C/N ratio, the fraction of easily decomposable compounds, etc., might contribute to an enhanced risk of being flushed into the groundwater or the stream.

Even if the NO_3 peak concentrations were extraordinary at Steinkreuz, the processes seem to be very similar at Lehstenbach as well as in other catchments: heavy rainstorms after prolonged warm periods mobilize substantial amounts of NO_3 from the forest floor, which leaches to the groundwater or into the stream. If the frequency of these events increases in the future, as predicted by climate change models, the NO_3 contamination of stream water and groundwater in forested areas is very likely to increase.

23.4.4 Dissolved Organic Carbon

In the runoff of the Lehstenbach catchment, concentration of dissolved organic carbon (DOC) is correlated with discharge in the short term (Spearman rank correlation, $\alpha < 0.01$). This is consistent with the perceptual model that most of the DOC is mobilized by saturation excess overland flow in the riparian zone (Lischeid et al. 2002). Besides, there is some evidence that the DOC dynamics of the Lehstenbach catchment runoff has changed since 1987. In spite of substantial scatter of the relationship between DOC concentration and discharge (Fig. 23.26), the offsets of the linear regressions differ significantly between the 1988–1992 and the 1998–2002 periods ($\alpha < 0.05$). Corresponding trends were observed at many other sites as well

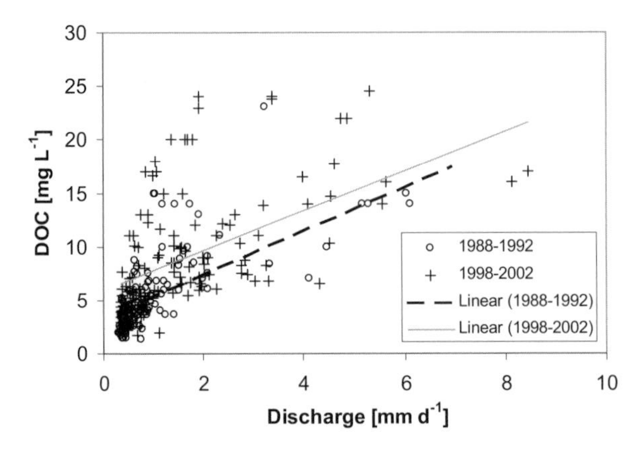

Fig. 23.26. Dissolved organic carbon (*DOC*) concentration versus discharge of Lehstenbach catchment runoff

and are ascribed to increasing temperature (Freeman et al. 2001). Thus, organic acids are likely to partly compensate for the pH effect of decreasing SO_4 concentration.

In the Steinkreuz runoff, concentration of DOC is inversely correlated with discharge in the short term (Spearman rank correlation, $\alpha < 0.01$). Similarly to NO_3, DOC concentration peaks were observed only during the first discharge peaks after extended dry, warm periods. Solute concentration, soil matrix potential and soil water content data confirm the assumption that these minor discharge peaks were generated in the riparian zone close to the stream. Here, DOC concentration of topsoil solution was rather high.

The zone of storm-flow generation extends with size of the discharge peaks. Thus, the higher the discharge, the lower the fraction of riparian zone water that contributes to stream flow, and the lower the DOC concentration. This is in contrast to the DOC dynamics at Lehstenbach, where discharge peaks are generated in the extended riparian zone wetlands.

23.4.5 Base Cations

Corresponding to the dominating anions, concentrations of base cations (Ca, Mg, Na, K) in the Steinkreuz and in the Lehstenbach catchment runoff vary substantially in the short term and are mostly closely related to discharge. Mass balances of the cations Ca, Mg and Na are positive in both catchments (Lischeid et al., this Vol.). Thus, it seems to be plausible that, according to the mobile anion concept (Nye and Greenland 1960; Seip 1980; Johnson et al. 1986), transport of these cations is driven by the high ionic strength of the infiltrating rainwater and thus can be regarded as a function of the long-term SO_4, N and Cl deposition. As the baseflow of the runoff of both catchments is dominated by deep groundwater, and that of storm flow by soil solution and precipitation (Lischeid 2001a; Lischeid et al. 2002), the short-term dependency of solute concentration on discharge gives valuable hints on the turnover of solutes in the catchments.

At Steinkreuz, Ca, Mg and Na concentration is inversely correlated with discharge (Spearman rank correlation, α <0.01). This is mostly due to the fact that concentration peaks are observed only during extended baseflow periods, pointing to the release of these cations in the deeper groundwater (cf. Lischeid et al., this Vol.). In contrast, low concentrations are not restricted to highflow periods. Moreover, there is no clear long-term trend for any of the base cations in the Steinkreuz runoff.

Analysis of the more comprehensive data set of the Lehstenbach catchment reveals different features. First, the longer time series allow us to investigate long-term shifts in the discharge–concentration relationship. Second, by comparing the dynamics of the catchment runoff and the east tributary, the impact of road salt contamination of the former can be assessed (Table 23.1).

It is remarkable that the positive correlation between solute concentration and discharge in the Lehstenbach catchment runoff substantially weakened for the anions SO_4 and NO_3 or even became inverse for the base cations Ca, Mg and K (Table 23.1). In contrast, there is no corresponding shift for Na and Cl. This points to two different paths of solute input into the catchment: First, atmospheric deposition of SO_4 and base cations, that have clearly decreased in recent decades (Matzner et al., this Vol., Chap. 14). As a consequence, not only did SO_4 concentration during discharge peaks decrease (see above), but also that of Ca and Mg. This is ascribed to the fact that since the end of the 1990s groundwater at a few meters depth exhibited higher SO_4 concentration and ionic strength compared to the topsoil solution (see above). Second, Na con-

Table 23.1. Correlation between solute concentration and discharge in the Lehstenbach catchment runoff and east tributary (Spearman rank correlation coefficient). Correlations are significant at the 0.01 and 0.05 (in *parentheses*) level of significance. *n.s.* Not significant

	Catchment runoff		East tributary
	1988–1992	1998–2002	1998–2002
Ca	0.647	−0.284	−0.318
Mg	0.723	(−0.185)	−0.260
K	n.s.	−0.307	−0.402
Na	−0.372	−0.283	−0.665
Cl	n.s.	n.s.	−0.369
SO_4	0.889	0.589	0.466
NO_3	0.366	n.s.	−0.498
DOC	0.778	0.676	0.565
Al	0.930	0.914	0.910

Table 23.2. Long-term trends of solute concentration in the Lehstenbach catchment runoff (1987–2002) and east tributary (1994–2002) (Spearman rank correlation coefficient for correlation with time). Correlations are significant at the 0.01 and 0.05 (in *parentheses*) level of significance. *n.s.* Not significant

	Catchment runoff 1987–2002	East tributary 1994–2002
Ca	−0.285	−0.356
Mg	−0.286	−0.348
K	−0.187	(−0.179)
Na	0.446	n.s.
Cl	0.453	−0.348
SO_4	−0.339	−0.272
NO_3	−0.164	−0.276
DOC	0.255	n.s.
Al	(−0.109)	(−0.164)

centration of the shallow groundwater in the wetlands downstream of the public road was mostly due to road salt contamination even in the first half of the monitoring period. Thus, precipitation and topsoil solution of uncontaminated wetlands close to the weir episodically decreased Na and Cl concentration during discharge peaks.

Long-term trends of solute concentration in the catchment runoff and the east tributary reveal a shift in the predominant source of contamination (Table 23.2). Significant trends are exclusively negative in the east tributary. The same holds true for most solutes in the Lehstenbach catchment runoff. In contrast, there is a clear increase in Cl and Na concentration, confirming the increasing contamination by road salt (see above). The identified increase of DOC concentration in the Lehstenbach catchment runoff, however, seems to be due in the first place to the fact that major discharge peaks with high DOC concentrations were much less frequent in the first dry years of the monitoring period (see above).

23.4.6 Protons and Alkalinity

Two of the key parameters for water quality are pH and alkalinity. Both are affected by a variety of processes and by different solutes. In contrast to Steinkreuz, both pH and alkalinity in the Lehstenbach runoff are very closely correlated with discharge (Fig. 23.27). Continuous measurements of pH in the Lehstenbach runoff at hourly intervals reveal that the increase in discharge and the decrease in pH occur simultaneously. This corresponds to higher DOC and SO_4 concentrations and lower base cation concentrations of

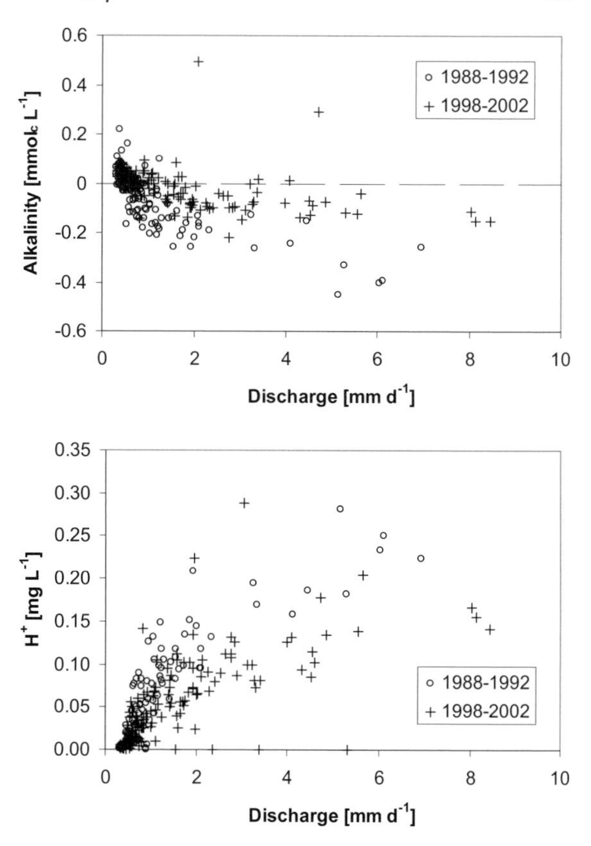

Fig. 23.27. Alkalinity (*above*) and H^+ concentration (*below*) in Lehstenbach catchment runoff depending on discharge

soil solutions and shallow groundwater which discharge into the stream only during storm flow. As these solutes have been subject to clear long-term trends since 1987, corresponding trends are to be expected for pH and alkalinity. Here, alkalinity was calculated as the sum of base cations Ca^{2+}, Mg^{2+}, Na^+ and K^+ minus that of SO_4^{2-}, NO_3^- and Cl^-, according to Reuss and Johnson (1986).

Taking into account that most of the short-term variance of theses parameters is explained by the hydrograph, in fact, clear trends can be identified (Fig. 23.27). Comparing the 1998–2002 and the 1988–1992 periods, storm-flow alkalinity increased by about 0.2 $mmol_c\ l^{-1}$. However, the minimum values of −0.1 $mmol_c\ l^{-1}$ were still in a critical range. Correspondingly, the H^+ concentration during storm flow was only about half of that during the period 1988–1992. However, this implies only a minor increase in terms of pH values, which increased from about 3.6–3.9 during major storms. Again, this is still in a critical range assumed to be harmful for many stream biota. Thus, signs of biological recovery from acidification are not anticipated yet (cf. Wright and Hauhs 1991).

23.5 Conclusions

Time series of discharge and the four-dimensional patterns of solute concentrations, i.e., the different dynamics at different groundwater and stream water sampling sites in the Steinkreuz and in the Lehstenbach catchment, were investigated using a variety of different approaches. Much of the variance in solute concentration could be traced back to two interacting sources of variance: firstly, the flow path length distribution proved to be a key variable to understanding and predicting spatial patterns. Secondly, a rather small number of episodic storm-flow events accounted both for the short-term dynamics of solute concentration and for a major fraction of the long-term solute output fluxes. The short-term dynamics of solute concentration clearly changed in the long-term for a variety of solutes in the streams and thus revealed more information about the 'changing functioning' than linear trend analyses.

Most of the variance in SO_4 concentration in both catchments was ascribed to the breakthrough of deposited SO_4 along subsoil flow paths of different lengths, being subject to convective transport and sorption to and desorption from the matrix. On the one hand, different sites differed with respect to the length distribution of the contributing flow paths. On the other hand, the short-term variance observed at various stream water sampling sites was traced back to the episodic contribution of the shortest flow paths during storm-flow peaks.

Corresponding to the long-term decrease of SO_4 loading of shallow groundwater and the streams during storm flow, H^+ concentration decreased and alkalinity increased in the Lehstenbach streams during storm flow. However, both parameters are still in a range that is assumed to be critical, and biological recovery is not anticipated yet.

In both catchments, the short-term NO_3 dynamics in the streams were generated in the riparian zone and did not reveal much information about hillslope processes. At Steinkreuz, there is some evidence of denitrification in clayey layers below the rooting zone. In contrast, denitrification in the Lehstenbach catchment occurred mainly in the riparian wetlands. However, during storm flow saturated area surface runoff bypassed the anoxic wetland zone. Thus, it is concluded that an estimate of denitrification in the catchment based on the area of wetlands will result in an overestimation.

At Lehstenbach, road salt input from a public road adds to the deterioration of groundwater and stream water quality. Its importance has increased continually since 1987, whereas that of atmospheric SO_4 deposition has decreased.

Atmospheric deposition and road salt input enhanced the ionic strength of stream water and groundwater in both catchments, and increased the leaching of the base cations Ca, Mg and Na in the Steinkreuz and the Lehstenbach catchments. As their concentrations depend on those of the predominating anions SO_4, NO_3 and Cl, which exhibit different long-term trends, understanding the dynamics of the latter provides the basis for better understanding the long-term development of base cation dynamics.

To conclude, public concern about acidifying deposition has decreased along with SO_4 deposition during the last two decades. In spite of this, both catchments are still considerably affected by ancient and recent deposition and presumably will continue to be so for decades. However, there is growing concern about the effect of climate change which might even intensify harmful long-term effects of anthropogenic deposition.

Acknowledgements. This chapter summarizes the results of numerous studies and of many discussions with a variety of colleagues to whom we are indebted. We gratefully acknowledge the contributions of Andreas Kolb, Uwe Hell, Uwe Wunderlich, Christine Alewell, Bernhard Manderscheid, Kirsten Küsel, Jochen Bittersohl and the central laboratory of BITÖK. Moreover, a variety of students assisted in data collecting, data analysis and modelling, namely Anette Hauck, Doris Haug, Daniel Kamps, Eva Plötscher, Andreas Poller, Tobias Rötting and Marcel Seger.

References

Alewell C, Gehre M (1999) Patterns of stable S isotopes in a forested catchment as indicators for biological S turnover. Biogeochemistry 47:319–333

Alewell C, Novák M (2001) Spotting zones of dissimilatory sulfate reduction in a forested catchment: the ^{34}S–^{35}S approach. Environ Pollut 112:369–377

Andersson L, Lepistö A (1998) Links between runoff generation, climate and nitrate-N leaching from forested catchments. Water Air Soil Pollut 105:227–237

Arheimer B, Lidén R (2000) Nitrogen and phosphorus concentrations from agricultural catchments – influence of spatial and temporal dynamics. J Hydrol 227:140–159

Bates JE, Shephard HK (1993) Measuring complexity using information fluctuation. Phys Lett A172:416–425

Blöschl G (2001) Scaling in hydrology. Hydrol Process 15:709–711

Burgman JO, Calles B, Westman F (1987) Conclusions from a ten year study of oxygen-18 in precipitation and runoff in Sweden. In: Proc IAEA Symp Isotope Techniques in Water Resources Development, 22 March, Vienna, pp 579–590

Büttcher H (2001) Random variability or reproducible spatial patterns? Investigating sulphate dynamics in forested catchments with a coupled transport sorption model. Diploma Thesis, University of Bayreuth

Chang S-C, Matzner E (2000) The effect of beech stemflow on spatial patterns of soil solution chemistry and seepage fluxes in a mixed beech/oak stand. Hydrol Process 14:135–144

Creed IF, Band LE, Foster NW, Morrison IK, Nicolson JA, Semkin RS, Jeffries DS (1996) Regulation of nitrate-N release from temperate forests: a test of the N flushing hypothesis. Water Resour Res 32:3337–3354

Dillon PJ, LaZerte BD (1992) Response of the Plastic Lake catchment, Ontario, to reduced sulphur deposition. Environ Pollut 77:211–217

Dooge JC (1986) Looking for hydrologic laws. Water Resour Res 22:46S–58S
Eckmann J-P, Kamphorst SO, Ruelle D (1987) Recurrence plots of dynamical systems. Europhys Lett 4:973–977
Emmert U (1985) Geologische Karte von Bayern 1:25.000. Erläuterungen zum Blatt Nr. 6128 Ebrach. Bayerisches Geologisches Landesamt, Munich
Fölster J (2000) The near-stream zone is a source of nitrogen in a Swedish forested catchment. J Environ Qual 29:883–893
Freeman C, Evans CD, Monteith DT, Reynolds B, Fenner N (2001) Export of organic carbon from peat soils. Nature 412:785
Goodale CL, Aber JD, Vitousek PM (2003) An unexpected nitrate decline in New Hampshire streams. Ecosystems 6:75–86
Grassberger P, Schreiber T, Schaffrath C (1991) Nonlinear time sequence analysis. Int J Bifurcation Chaos 1:521–547
Groscheová H, Novák M, Alewell C (2000) Changes in the $\delta^{34}S$ ratio of pore-water sulfate in incubated sphagnum peat. Wetlands 20:62–69
Haug D (2001) Wiederkehranalyse und Langzeitphänomene süddeutscher Abflusszeitreihen. Diploma Thesis, University of Bayreuth
Honerkamp J (1993) Stochastic dynamical systems: concepts, numerical methods, data analysis. Wiley, New York
Hooper RP (2001) Applying the scientific method to small catchment studies: a review of the Panola Mountain experience. Hydrol Processes 15:2039–2050
Hooper RP, Aulenbach BT, Burns DA, McDonnell J, Freer J, Kendall C, Beven K (1998) Riparian control of stream-water chemistry: implications for hydrochemical basin models. In: Proc HeadWater '98 Conf Hydrology, Water Resources and Ecology in Headwaters, Meran/Merano, Italy, April, IAHS-Publ 248, pp 451–458
Hurst HE (1951) Long-term storage capacity of reservoirs. Trans Am Soc Civil Eng 116:770–799
Johnson DW, Cole DW, van Miegroet H, Horng FW (1986) Factors affecting anion movement and retention in four forest soils. Soil Sci Soc Am J 50:776–783
Kamps D (2001) Querschnittsvergleich der Komplexität und des Informationsgehaltes hydrometeorologischer Zeitreihen. Diploma Thesis, University of Bayreuth
Lange H (1999) Time series analysis of ecosystem variables with complexity measures. Interjournal for complex systems 250, http://www.interjournal.org/cgi-bin/manuscript_abstract.cgi?7460
Lange H (2004) Time series in ecology. Nature Encycl Life Sci (in press)
Lind DA, Marcus B (1995) An introduction to symbolic dynamics and coding. Cambridge University Press, Cambridge
Lischeid G (2001a) Investigating short-term dynamics and long-term trends of SO_4 in the runoff of a forested catchment using artificial neural networks. J Hydrol 243:31–42
Lischeid G (2001b) Investigating trends of hydrochemical time series of small catchments by artificial neural networks. Phys Chem Earth Part B 26(1):15–18
Lischeid G, Langusch J (2004) Do the data tell the same story? Comparing the INCA model with a neural network analysis. Hydrol Earth Sys Sci (in press)
Lischeid G, Uhlenbrook S (2002) Checking a process-based catchment model by artificial neural networks. Hydrol Process 17:265–277
Lischeid G, Lange H, Hauhs M (1998) Neural network modelling of NO_3^- time series from small headwater catchments. In: Proc HeadWater '98 Conf Hydrology, Water Resources and Ecology in Headwaters, Meran/Merano, Italy, April, IAHS-Publ 248, pp 467–473
Lischeid G, Kolb A, Alewell C (2002) Apparent translatory flow in groundwater recharge and runoff generation. J Hydrol 265:195–211

Lischeid G, Büttcher H, Krám P, Hruška J (2003) Comparative analysis of hydrochemical time series of adjacent catchments by process-based and data-oriented modeling. In: Proc 3rd Int Conf on Water Resources and Environment Research (ICWRER), Dresden, 22–25 July 2002

Lischeid G, Kolb A, Alewell C, Paul S (2004) Impact of redox and transport processes in a riparian wetland on stream water quality. Hydrol Process (in press)

Lovejoy S, Schertzer D (1990) Multifractals, universality classes and satellite and radar measurements of cloud and rain fields. J Geophys Res 95:2021–2034

Mandelbrot BB, Wallis JR (1969) Robustness of the rescaled range R/S and the measurement of noncyclic long run statistical dependence. Water Resour Res 5:967–988

Manderscheid B, Schweisser T, Lischeid G, Alewell C, Matzner E (2000) Sulfate pools in the weathered substrata of a forested catchment. Soil Sci Soc Am J 64:1078–1082

Marwan N, Kurths J (2003) Nonlinear analysis of bivariate data with cross recurrence plots. Phys Lett A 302:299–307

Mitchell M (2001) Linkage of nitrate losses in watersheds to hydrological processes. Hydrol Process 15:3305–3307

Moritz K, Bittersohl J, Müller FX, Krebs M (1994) Auswirkungen des sauren Regens und des Waldsterbens auf das Grundwasser. Dokumentation der Methoden und Meßdaten des Entwicklungsvorhabens 1988–1992. Bayerisches Landesamt für Wasserwirtschaft, Munich, Materialien 40, pp 1–387

Murray AB (2002) Seeking explanation affects numerical modeling strategies. EOS Trans Am Geophys Un 83:418

Nye PH, Greenland DJ (1960) The soil under shifting cultivation, vol 51. Commonwealth Bureau of Soils Technical Communications, Commonwealth Agricultural Bureau, Farnham

Pandey G, Lovejoy S, Schertzer D (1998) Multifractal analysis of daily river flows including extremes for basins of five to two million square kilometers, one day to 75 years. J Hydrol 208:62–81

Reuss JO, Johnson DW (1986) Acid deposition and the acidification of soils and water. Ecological studies, vol 59. Springer, Berlin Heidelberg New York

Riedmiller M, Braun H (1993) A direct adaptive method for faster backpropagation learning: the Rprop algorithm. In: Proc IEEE Int Conf on Neural Networks (ICNN), San Francisco, pp 586–591

Salvadori G, Lovejoy S, Schertzer D (2001) Multifractal objective analysis: conditioning and interpolation. Stoch Environ Res Risk Assess 15:261–283

Sánchez Pérez JM, Trémolières M, Takatert N, Ackerer P, Eichhorn A, Maire G (1999) Quantification of nitrate removal by a flooded alluvial zone in the Ill floodplain (eastern France). Hydrobiologia 410:185–193

Schulze E-D (1989) Air pollution and forest decline in a spruce (*Picea abies*) forest. Science 244:776–783

Seger M (2002) Multifraktale Spektren ökosystemarer Zeitreihen. Diploma Thesis, University of Bayreuth

Seip HM (1980) Acidification of freshwater – sources and mechanisms. In: Drablos D, Tollan A (eds) Proc Int Conf Ecological Impact of Acid Precipitation, Sundefjord, Norway, pp 248–269

Seyfried MS, Wilcox BP (1995) Scale and the nature of spatial variability: field examples having implications for hydrologic modeling. Water Resour Res 31:173–184

Sivapalan M (2003) Process complexity at hillslope scale, process simplicity at the watershed scale: is there a connection? Hydrol Process 17:1037–1041

Sloan WT, Jenkins A, Eatherall A (1994) A simple model of stream nitrate concentration in forested and deforested catchments in mid-Wales. J Hydrol 158:61–78

Stoddard JL, Jeffries DS, Lükewille A, Clair TA, Dillon PJ, Driscoll CT, Forsius M, Johannessen M, Kahl JS, Kellogg JH, Kemp A, Mannio J, Monteith DT, Murdoch PS, Patrick

S, Rebsdorf A, Skjelkvale BL, Stainton MP, Traaen T, van Dam H, Webster KE, Wieting J, Wilander A (1999) Regional trends in aquatic recovery from acidification in North America and Europe. Nature 401:575–578

Tessier Y, Lovejoy S, Hubert P, Schertzer D, Pecknold S (1996) Multifractal analysis and modeling of rainfall and river flows and scaling, causal transfer functions. J Geophys Res 101:26427–26440

Thiel M, Romano MC, Kurths J, Meucci R, Allario E, Arecchi FT (2002) Influence of observational noise on the recurrence quantification analysis. Physica D 171:138–152

Wackerbauer R, Witt A, Atmanspacher H, Kurths J, Scheingraber H (1994) A comparative classification of complexity measures. Chaos Solitons Fractals 4:133–173

Wright RW, Hauhs M (1991) Reversibility of acidification: soils and surface waters. Proc R Soc Edinb 97B:169–191

Young PC, Beven KJ (1994) Data-based mechanistic modelling and the rainfall-flow non-linearity. Environmetrics 5:335–363

Zahn MT (1995) Transport von Säurebildnern im Untergrund und Bedeutung für die Grundwasserversauerung. In: Proc Int Symp Grundwasserversauerung durch Atmosphärische Deposition. Ursachen–Auswirkungen–Sanierungsstrategien, 26–28 Oct 1994, Bayreuth. Informationsber Bayerischen Landesamtes Wasserwirtschaft 3/95: 143- 151

Zbilut JP, Giuliani A, Webber CL (1998a) Detecting deterministic signals in exceptionally noisy environments using cross-recurrence quantification. Phys Lett A 246:122–128

Zbilut JP, Giuliani A, Webber Jr CL (1998b) Recurrence quantification analysis and principal components in the detection of short complex signals. Phys Lett A 237:131–135

Zell A, Mamier G, Vogt M, Mache N, Hübner R, Döring S, Herrmann K-U, Soyez T, Schmalzl M, Sommer T, Hatzigeorgiou A, Posselt D, Schreiner T, Kett B, Clemente G, Wieland J, Reczko M, Riedmiller M, Seemann M, Ritt M, DeCoster J, Biedermann J, Danz J, Wehrfritz C, Werner R, Berthold M, Orsier B (1995) Stuttgart neural network simulator user manual, version 4.1. Rep 6/95. Institute for Parallel and Distributed High Performance Systems, University of Stuttgart

24 Trends in the Input–Output Relations: The Catchment Budgets

G. Lischeid, C. Alewell, K. Moritz, and J. Bittersohl

24.1 Introduction

One important focus of biogeochemical ecosystem studies is that of identifying sinks and sources of various substances in the system. These might be either finite or transient sinks or sources. A quantification of element budgets within catchments is of crucial importance for the identification of critical ecosystem states such as nutrient loss, acidification and contamination of waters and deterioration of soils (Ulrich 1994). However, sinks might turn into sources and vice versa over time. Thus, long-term monitoring is important when interpreting element fluxes and budgets. The latter especially holds true because element budgets and fluxes are subject to high spatial and temporal variability, causing large errors in their estimation, as has been shown previously for the Lehstenbach catchment (Manderscheid and Matzner 1995; Manderscheid et al. 2000a; Alewell et al. 2004).

Earlier studies have shown that high N and SO_4^{2-} deposition has resulted in N and SO_4^{2-} accumulation and acidification of surface waters in the Lehstenbach catchment (Alewell et al. 2000a; Lischeid et al. 2000; Manderscheid et al. 2000a,b; Alewell 2001; Matzner et al. 2001). Furthermore, loss of base cations from forest soils can be accelerated by high anthropogenic deposition (Rhode et al. 1995; BML 1997; Yanai et al. 1999; Alewell 2001).

The aim of this chapter is (1) the quantification of output fluxes via the catchment runoff, (2) identification of sinks and sources of major elements in the catchments and (3) documentation of time series in element budgets in the Lehstenbach and the Steinkreuz catchments, with a comparison of the two catchments.

Ecological Studies, Vol. 172
E. Matzner (Ed.), Biogeochemistry of Forested Catchments in a Changing Environment

24.2 Determination of Catchment Budgets

Budgets are calculated by subtracting solute output via the catchment runoff from solute input by deposition. Input fluxes with bulk precipitation were calculated as concentration multiplied by precipitation amount. Here, the deposition data calculated by Matzner et al. (this Vol., Chap. 14) were used to establish budgets.

The difficulties associated with quantifying input fluxes via deposition are discussed by Klemm (this Vol.) and Matzner et al. (this Vol., Chap. 14). Among the solutes considered here, quantifying nitrogen deposition is subject to severe uncertainties as a considerable fraction of deposited nitrogen can be taken up in the tree canopies (Brumme et al. 1992; Lovett and Lindberg 1993; Boyce et al. 1996). Here, nitrogen input is quantified using throughfall data (canopy drip and stemflow). Thus, throughfall measurements are likely to underestimate the total nitrogen input.

In both catchments, nitrogen in deeper soil soution, groundwater and stream water is almost exclusively present as NO_3-N. In contrast, NH_4-N comprises roughly half of the total nitrogen in throughfall. In this chapter, the term DIN (dissolved inorganic nitrogen) summarizes NH_4-N and NO_3-N. NO_2-N was measured in most samples as well, but only occasionally exceeded the detection limit.

Quantifying output fluxes via the catchment runoff is less problematic. However, the following aspects should be considered.

First, determining output fluxes via the catchment runoff does not account for gaseous emissions or export from the catchment in the solid phase. The former is very likely to be negligible for most of the elements considered here. This does not hold, however, for water fluxes via evapotranspiration, carbon export via CO_2 and emissions of gaseous nitrogen or sulfur compounds, such as N_2, NH_3, N_2O and H_2S. As these fluxes were measured only during short periods and at small spatial scale, they comprise a part of the residuals of the budgets. The only exception concerns long-term measurements of CO_2 fluxes, which are reported by Rebmann et al. (this Vol.).

Export via the solid phase mainly consists of suspended sediments in the catchment runoff and biomass. The former is minimized by the V-notch weirs at the outlet of the Lehstenbach and Steinkreuz runoff, where most of the sediments are deposited before passing the weir. The only form of biomass export is for commercial use of the trunks of mature trees. However, this comprises only a minor portion of the total output fluxes for Mg, compared to that of the catchment runoff at Lehstenbach. In contrast, the net loss of K via harvest of the trees approximates the same order of magnitude as the output flux via the Lehstenbach catchment runoff. There was no clear-felling in the Steinkreuz catchment since the start of the measurements in 1995. Matzner et al. (this Vol., Chap. 20) relate nitrogen and organic carbon retention in the tree biomass to output fluxes via the catchments' runoffs.

Second, mass export in the liquid phase might partly bypass the weir. Overland flow is negligible. Mass export via groundwater across the catchment boundary cannot exactly be quantified in the two catchments. At Lehstenbach, it is likely to comprise only a negligible fraction of the total mass export. The catchment outlet is located in a small depression between two hills, where depth to the bedrock is only about 10 m, which is substantially less than in large parts of the catchments (Poller 2002). Although groundwater flow occurs in the fractures of the granite bedrock, the bedrock permeability is rather low.

In contrast, it is likely that the Steinkreuz runoff measured at the catchment outlet does not capture the full water export. First, the identification of the watershed boundaries based on topography is difficult, as the upper part of the catchment consists of a flat plane. Thus, the catchment area can only roughly be assessed based on topographic data.

Second, leakage of groundwater through the 30-m-thick clay layers at the base of the catchment cannot be excluded. Actually, the Steinkreuz runoff per unit area from 1995–1998 was only 81 and 84 %, respectively, of that of the Rauhe Ebrach and the Reiche Ebrach, the two main streams of the Steigerwald region. Assuming that there are no sinks or sources of Cl in the underground, the Cl budget for the 1995–2001 period suggests that the runoff measurement misses about 27 % of the total water output flux. However, both the Cl export and the annual runoff of the Steinkreuz vary considerably in time. In this chapter, the annual mean mass export of the Steinkreuz runoff is multiplied by a factor of 1.37, based on the Cl budget.

Third, solute concentration in the catchment runoff was determined based on grab samples. There are different methods of interpolating between the weekly or biweekly sampling dates. During a 2-year period, daily samples were taken at the Lehstenbach runoff (Alewell et al. 2004). These served as a reference to compare with different methods of calculating output fluxes based on the usual biweekly sampling intervals. These methods comprised:

1. Multiplying annual mean flux-weighted solute concentration by annual runoff.
2. Multiplying measured solute concentration with discharge of the respective 14-day interval.
3. Linear interpolation of solute concentration between subsequent sampling days, and multiplying by daily mean discharge.
4. Interpolating between subsequent biweekly samples using an autoregressive model, and multiplying interpolating daily concentration values by measured daily mean discharge.

The error of the 2-year mass flux of Cl, SO_4 and NO_3 was between 2 and 10 %, compared with daily samples (Alewell et al. 2004).

Following an alternative approach, non-linear regressions between solute concentration and discharge were used to interpolate solute concentration between the sampling dates. For some solutes, there was a clear long-term

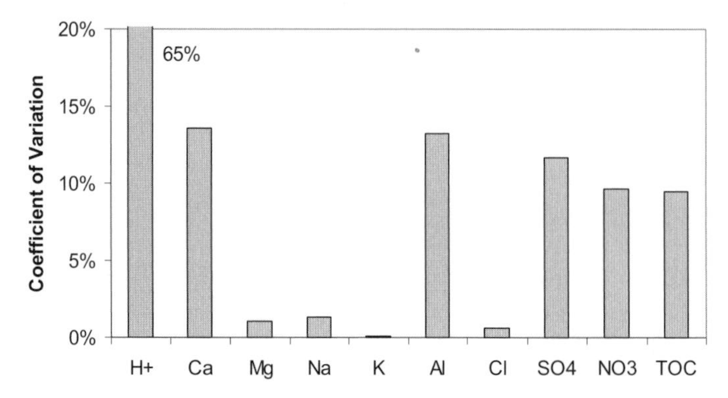

Fig. 24.1. Coefficient of variation (standard deviation over arithmetic mean) of annual mean solute export via the catchment runoff, calculated by four different methods for the 1988–2001 period

shift of the regression function that was accounted for by fitting different regression functions to the first and second half of the 1988–2001 period separately. Four different methods were compared:

1. Multiplying measured solute concentration with the corresponding 14-day mean discharge.
2. Interpolating solute concentration between sampling dates using the correlation between solute concentration and discharge determined in 1991.
3. Interpolating solute concentration between sampling dates using the correlation between solute concentration and discharge determined in 2002, accounting for two different periods, if necessary.
4. Multiplying annual mean flux-weighted solute concentration by annual runoff.

The coefficient of variation (standard deviation of the values of different methods divided by the arithmetic mean) of annual mean solute export via the catchment runoff, calculated by the four different methods for the 1988–2001 period, is given in Fig. 24.1. The results are nearly identical for Mg, Na, K and Cl. The coefficient of variation is between 9 and 14% for the remaining solutes except for H^+. Method 2 yielded about three times as high H^+ compared to the remaining methods. Besides, there was no clear bias of single methods with respect to different solutes.

These results illustrate some of the uncertainties associated with the calculation of output fluxes. Here, the focus is more on identifying major sinks and sources in the catchments rather than precisely quantifying their magnitude. Output fluxes via the catchment runoff were calculated for both catchments by multiplying flux-weighted annual mean concentration by annual water flux. For Steinkreuz, the respective time spans were calendar years, and for the Lehstenbach catchment they were hydrological years (November–October).

24.3 Output Fluxes Via Runoff

The Lehstenbach runoff is substantially affected by road salt application along the 1-km stretch of a public road in the western part of the catchment (Gerstberger et al., this Vol.). The Cl input via road salt was about the same order of magnitude as atmospheric deposition in the last decade (Lehmann 1999). To account for that effect, output fluxes calculated for the east tributary at the confluence close to the weir are given also. In either case, annual mean flux-weighted solute concentration is multiplied by the annual runoff measured at the catchment outlet.

The east tributary drains about 35 % of the catchment area that is not contaminated by road salt. There is no evidence that the eastern part of the catchment differs from the western part with respect to deposition, vegetation, soils, geochemistry or hydrology. Thus, the difference between the catchment runoff output fluxes and that of the east tributary are predominantly ascribed to the impact of road salt contamination.

The most pronounced differences in output fluxes of the Lehstenbach catchment runoff and the east tributary concern Cl and Na. Mean Cl concentration of the latter is only 24 %, and that of Na is 59 % compared to the former. In contrast, there is no difference with respect to NO_3 concentration, although the fraction of wetlands where denitrification has been proven (Küsel and Alewell, this Vol.) is about 47 % in the eastern part and only 35 % in the total Lehstenbach catchment.

Mean annual runoff of the Steinkreuz was about half compared to the Lehstenbach catchment, and annual mean DOC export was only 30 % in the 1995–2001 period (Fig. 24.2). The latter is most likely due to the fact that wet-

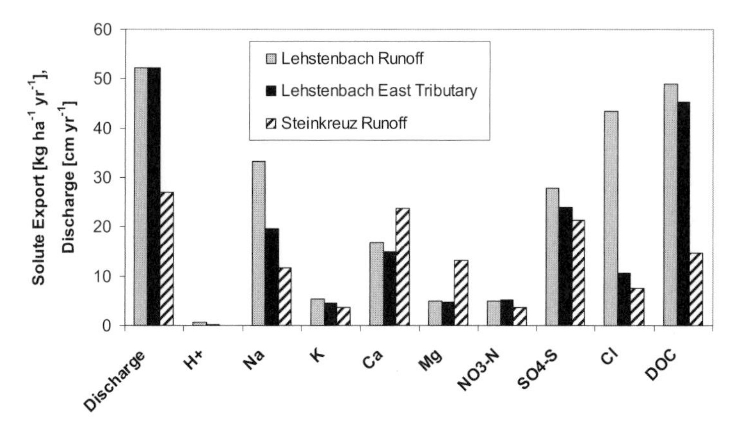

Fig. 24.2. Mean annual discharge and solute export via runoff in the Lehstenbach and Steinkreuz catchments 1995–2001

lands comprise about one third of the Lehstenbach catchment area, and do not exist in the Steinkreuz catchment. The reducing conditions in the wetland areas are likely to inhibit complete mineralization to CO_2 as well as immobilization and thus increase DOC export. Furthermore, soils at the Steinkreuz catchment are less acidified, and thus carbon might be recycled more efficiently. Export of protons of the Steinkreuz runoff was less than 0.01 kg ha^{-1} $year^{-1}$. In contrast, annual Mg export was more than twice, and Ca export 1.5 times that of the Lehstenbach, reflecting the different lithology of the sites. While the granite bedrock of the Lehstenbach is comparably poor in nutrient cations, soils are strongly acidified and depleted in nutrient cations with base saturations well below 10 % down to 2.5 m depth (Alewell et al. 2001). The Steinkreuz output fluxes of K, NO_3-N and SO_4-S were only slightly less compared to Lehstenbach.

Figures 24.3–24.5 show time series of annual output fluxes. The axes of Figs. 24.4 and 24.5 are equally scaled to facilitate comparison between the two.

At Lehstenbach, five subsequent exceptionally dry years were observed at the beginning of the observation period. In contrast, annual runoff exceeded 500 mm in 1994, 1995 and 1999. At Steinkreuz, measurements did not start before the end of 1994. Here 1995 and 2001 were the years with the highest and 1999 slightly lower annual runoff. Evaluating the trends of both catchments demonstrates clearly the importance of long-term measurements. Even with time series of a 13-year period as in the Lehstenbach catchment, the high temporal variability hampers identification of long-term trends or turns towards critical ecosystem states.

Much of the temporal variance of output fluxes is due to that of the annual catchment runoff, which clearly exceeds that of solute concentration at that time scale (Figs. 24.3–24.5). In the short term, however, several solutes are clearly correlated with discharge, e.g., SO_4, DOC and H^+, in the Lehstenbach runoff (Lischeid et al., this Vol.). Thus, annual output fluxes of these solutes

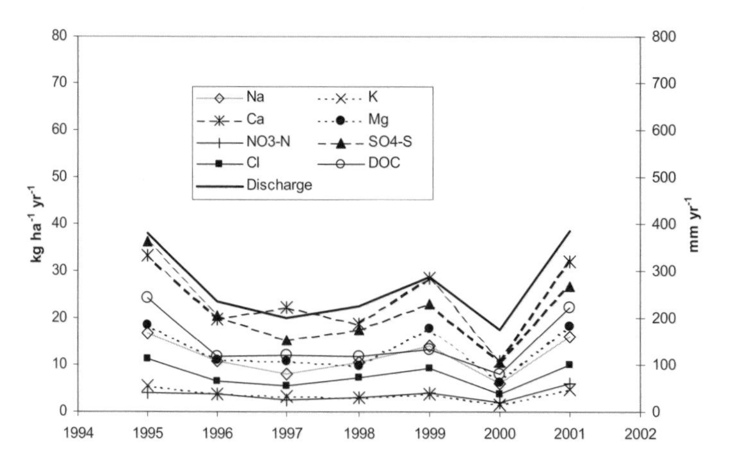

Fig. 24.3. Annual mass flux via the Steinkreuz catchment runoff

depend not only on the total annual runoff, but also on the frequency and magnitude of discharge peaks.

One striking example of that is given by the DOC mass fluxes (Figs. 24.4 and 24.5). On 14–15 September 1998, the second highest discharge peak since the start of the measurements was observed (Lischeid et al. 2002). Correspondingly, very high DOC concentrations were observed during the rising limb. By chance, the biweekly Lehstenbach east tributary grab sample was

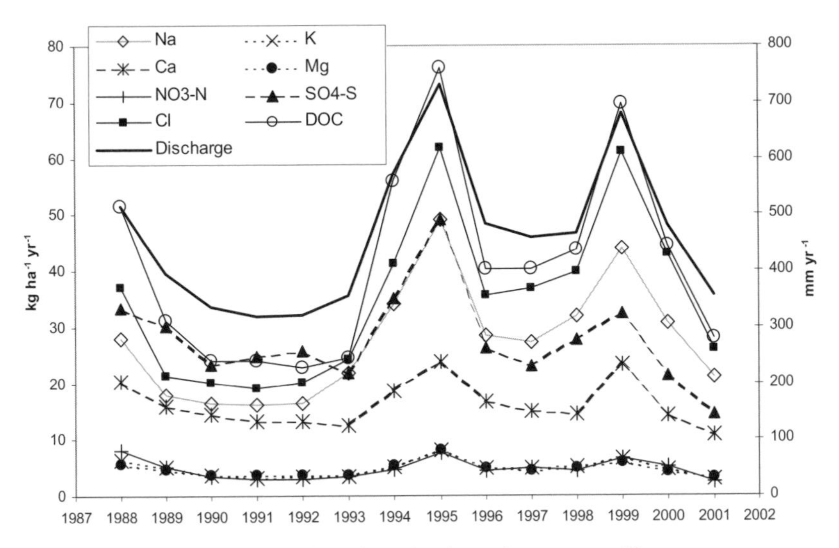

Fig. 24.4. Annual mass flux via the Lehstenbach catchment runoff

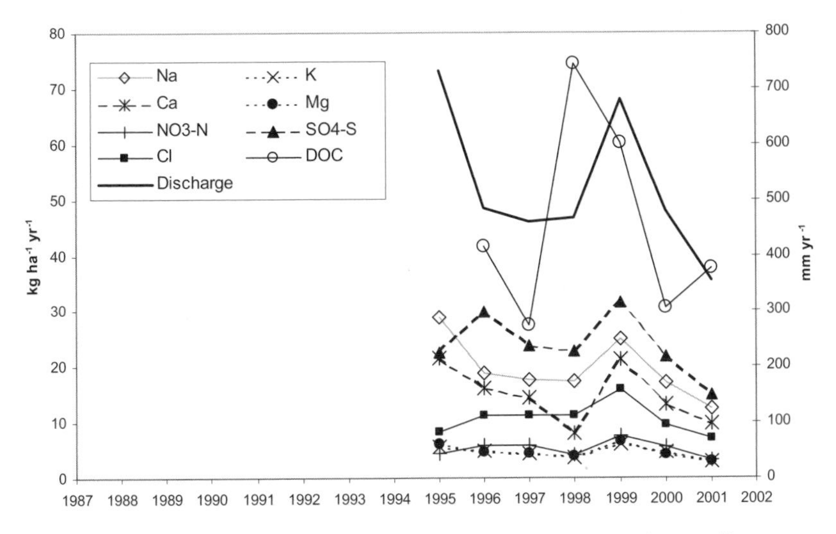

Fig. 24.5. Annual mass flux via the east tributary of the Lehstenbach runoff

taken during the rising limb with about 47 mg l^{-1} DOC. The corresponding grab sample of the Lehstenbach catchment runoff was taken 4 days later which yielded only about 22 mg l^{-1} DOC. This example illustrates one of the problems associated with identifying and interpreting long-term trends of solute concentration, especially for the Lehstenbach catchment (cf. Lischeid et al., this Vol.).

24.4 Mean Budgets

According to the mass budgets, the Steinkreuz catchment nearly completely buffered the proton input via deposition (Fig. 24.6). In contrast, it was clearly a source for Na, Ca, Mg and SO_4-S where output fluxes are between 2.5 and 20 times the deposition flux. The K budgets are nearly balanced. It has to be taken into account that the K export by tree harvesting of about 3 kg ha^{-1} $year^{-1}$ (cf. Fichter et al. 1998) is in the same order of magnitude compared to that by the Lehstenbach and Steinkreuz catchment runoff.

The nitrogen output via the catchment runoff was related to the throughfall flux. It is very likely that this is an underestimation of the nitrogen deposition, as canopy uptake of nitrogen was not considered. Correspondingly, the number given in Fig. 24.6 is likely to underestimate the nitrogen retention of the catchment which was assessed to be about three quarters of the nitrogen input.

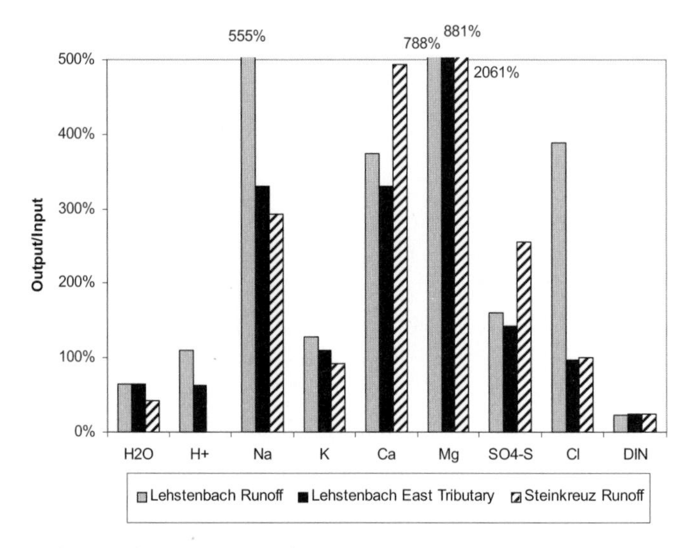

Fig. 24.6. Relation of annual output fluxes via the catchment runoff over total deposition or input via throughfall (DIN), respectively, in the Lehstenbach and Steinkreuz catchments 1995–2001

As stated above, the catchment runoff is impacted by road salt contamination in the western part of the catchment. Thus output fluxes were calculated using the same two different sets of solute concentration data as given above: solute concentration measured in the catchment runoff which was contaminated by road salt, and that of the uncontaminated east tributary before the confluence close to the weir. In either case, flux-weighted concentration data were multiplied by annual runoff sum measured at the catchment outlet.

As expected, the major differences concern Cl and Na output fluxes (Fig. 24.6). In contrast, K, Ca, Mg and SO_4-S output fluxes in the catchment runoff exceeded those of the east tributary by only 12–15%. This might be ascribed to the higher ionic strength of the road-salt-contaminated soils and streams. In contrast, the nitrogen budget seems not to have been impacted by road salt contamination (Fig. 24.6). Annual mean nitrogen retention of the Lehstenbach catchment was 16.9 kg ha^{-1} $year^{-1}$, and it was 11.8 kg ha^{-1} $year^{-1}$ in the Steinkreuz catchment, comprising three quarters of annual mean deposition in either catchment.

The mean budgets indicate a net release of Na, Ca, Mg and SO_4^{2-}-S of about 13.3, 10.1, 4.1 and 3.3 kg ha^{-1} $year^{-1}$ in the eastern part of the Lehstenbach catchment and 7.6, 18.6, 12.5 and 13.0 kg ha^{-1} $year^{-1}$ at Steinkreuz. The net export of the base cations Ca, Mg and Na in both catchments corresponds to the results of a variety of other studies (e.g., Houle et al. 1997; Reynolds et al. 2000; Nakagawa et al. 2001; Uyttendaele and Iroumé 2002).

In both the western and the eastern tributary, pH decreases by up to 2.5 units during single rainstorms. However, the amplitudes are slightly larger in the western tributary. As a consequence, the total H^+ export from the western tributary clearly exceeded that of the eastern tributary (Fig. 24.6).

The data indicate that most of the deposited nitrogen is retained in the catchments (Fig. 24.6). Annual mean net nitrogen storage in the spruce trees of the Lehstenbach catchment is about 4.5 kg ha^{-1} $year^{-1}$ (Mund 1996). This number closely corresponds, e.g., to annual mean net storage in the growing stock of 16 large river basins in the northeastern USA given by Goodale et al. (2002). However, if thinning is accounted for, it is about 10.5 kg ha^{-1} $year^{-1}$ in the Lehstenbach catchment (Bauer et al. 2000). In either case, that component comprises only a minor fraction of the deposited nitrogen at Lehstenbach and Steinkreuz. Another sink is denitrification in the wetlands that comprise about one third of the Lehstenbach catchment area. Thus, groundwater recharged in these areas is almost completely free of inorganic nitrogen. However, nitrogen-rich water bypasses the anoxic zones in the wetlands during storm flow. Thus, denitrification in the wetlands is assumed to comprise clearly less than one third of the total nitrogen deposition (Lischeid et al., this Vol.).

There are no wetlands in the Steinkreuz catchment. However, there is some evidence that denitrification in the clayey layers of the stratified aquifer is an important sink of deposited nitrogen (Lischeid et al., this Vol.).

The most important nitrogen sink in both catchments seems to be storage in the humus layer. According to the empirical model of Berg (this Vol.) the annual mean N sequestration in soil organic matter at Steinkreuz is about 34 kg ha^{-1} $year^{-1}$ which is about 1.7 times mean N input flux via throughfall (19.5 kg ha^{-1} $year^{-1}$), and roughly half of annual N flux via litter fall. In contrast, in the Lehstenbach catchment annual mean N sequestration in soil organic matter is only 18 kg ha^{-1} $year^{-1}$ (Berg, this Vol.).

In contrast to the Lehstenbach catchment natural sources of sulfur cannot be excluded for the Steinkreuz catchment. The underground of the Steinkreuz site consists of a vertical sequence of single aquifers which offers the opportunity to identify sinks and sources of different solutes below the rooting zone (Fig. 24.7). The two springs drain the upper layers of the underground, where the shallow groundwater wells are installed. The catchment runoff comprises all of the remaining components given in Fig. 24.7, in temporally differing fractions.

The concentrations of most solutes vary substantially in time (runoff, springs, shallow groundwater) and space (soil solution, shallow groundwater). Nevertheless, clear patterns emerge. The Ca and Mg concentration clearly increase with depth. Similarly, Na, K and Cl concentration of the deep groundwater exceeds that of the shallow groundwater and of the stream water by far. These findings are consistent with the fact that carbonates and evaporites are more abundant in the clayey Lehrberg layers at the base of the catchment compared to the overlying strata (Emmert 1985).

In contrast, SO_4 concentration increases from soil solution, shallow groundwater to stream water, but is considerably less in the deep groundwater. Although gypsum layers are reported in the Lehrberg layers (Emmert

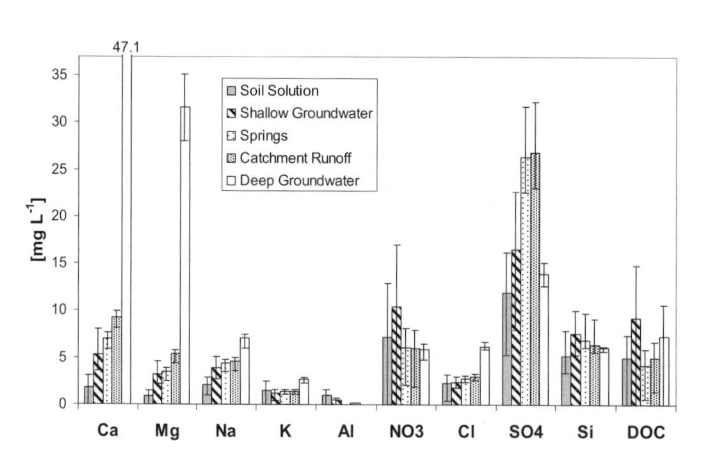

Fig. 24.7. Median, 10 % and 90 % percentiles of solute concentration in soil solution (60–90 cm depth), shallow groundwater (wells B02–B06, maximum depth 2.8 m), springs, catchment runoff and deep groundwater in the Steinkreuz catchment 1995–2001

1985), there is some additional evidence that a substantial fraction of the stream SO_4 is due to atmospheric deposition. Corresponding to the observed decrease of SO_4 deposition in the Steinkreuz catchment (Matzner et al., this Vol., Chap. 14), a clear decrease was observed in soil solution (Matzner et al., this Vol., Chap. 20), shallow groundwater and stream water in the 1995–2001 period (Lischeid et al., this Vol.). Thus, the depth profile of soil water and groundwater SO_4 concentration corresponds to that predicted by the SUNFLOW application for the Lehstenbach catchment (Lischeid et al., this Vol.). However, in contrast to the Lehstenbach catchment, the fraction of geogene SO_4 is very likely to increase with depth at the Steinkreuz site.

Although the median Si concentration tends to decrease from shallow groundwater to deep groundwater, this trend is hardly significant due to the great variability. Correspondingly, clear trends cannot be identified for DOC and NO_3. However, it is remarkable that the maximum NO_3 concentration values are found in the shallow groundwater, which is discussed in more detail in Lischeid et al. (this Vol.).

24.5 Changing Catchment Behaviour

Much of the temporal variance of the mass balances is due to that of the annual catchment runoff, which clearly exceeds that of throughfall water flux as well as the variance of the mean concentrations in throughfall and runoff for most solutes. Thus, annual runoff sum is given additionally in Figs. 24.8–24.12. Negative values indicate net release of elements from the catchment. Correspondingly, the right *y* axis is scaled upside down to allow direct comparison with annual changes of solute budgets.

Annual mass budgets were calculated for the hydrological years (November–October) 1988–2001, using LfW 01 deposition data for quantifying solute input (cf. Matzner et al., this Vol., Chap. 14). For DIN, LfW 01 throughfall data were used.

Mass budgets were negative during the whole period for Ca, Mg, Na and Cl (Fig. 24.8). As stated above, mass budgets of the latter are impacted by road salt input which could not be accounted for in a quantitative way. The only solute that exhibited positive values was DIN. Here, no trend can be detected. Similarly, there was no clear trend for the K budgets that fluctuated around zero (Fig. 24.9).

Budgets of H^+ and SO_4 changed from clearly positive to clearly negative values in the mid-1990s (Fig. 24.9). This is in accordance with the results of a variety of studies in Europe and North America (Stoddard et al. 1999; Alewell et al. 2001; Prechtel et al. 2001). In both catchments now the SO_4 output exceeds SO_4 deposition. In the Lehstenbach catchment, the observed net release of SO_4 can be clearly traced back to desorption of formerly deposited

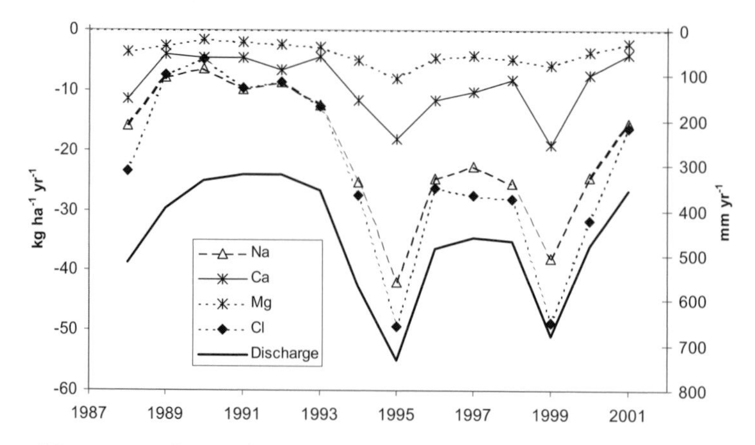

Fig. 24.8. Time series of annual budgets for the Lehstenbach catchment (input via deposition minus output flux via the catchment runoff)

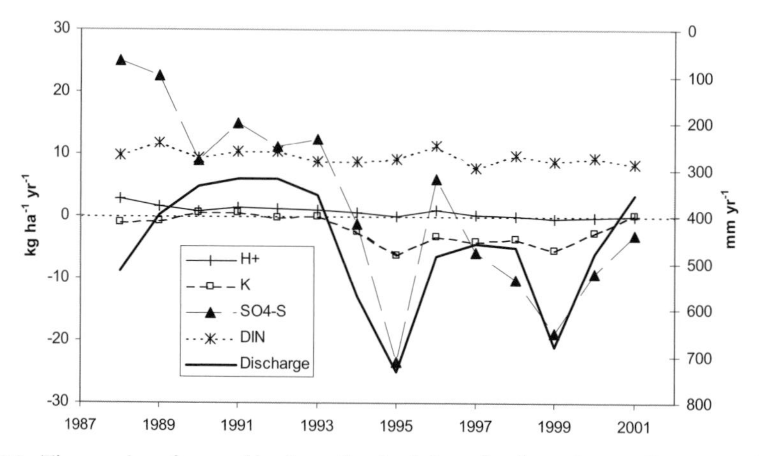

Fig. 24.9. Time series of annual budgets for the Lehstenbach catchment [input via deposition or throughfall (DIN) minus output flux via the catchment runoff]

sulfate in the deeper layers of the aquifer (Alewell 1998; Manderscheid et al. 2000a; Büttcher 2001).

To compensate for the effect of annual variations of runoff on the catchment's budget, a linear trend analysis was performed on the ratio of annual net budgets to annual runoff (Fig. 24.10). Highly significant trends were detected for Na, Cl and SO_4-S, for which an increasing net release was identified. This corresponds to the trend of Cl concentration in the stream (Lischeid et al., this Vol.) as well as to the results of the SUNFLOW model of decreasing SO_4 release from the topsoil and continuing increase of SO_4 concentration in the deeper groundwater (Lischeid et al., this Vol.).

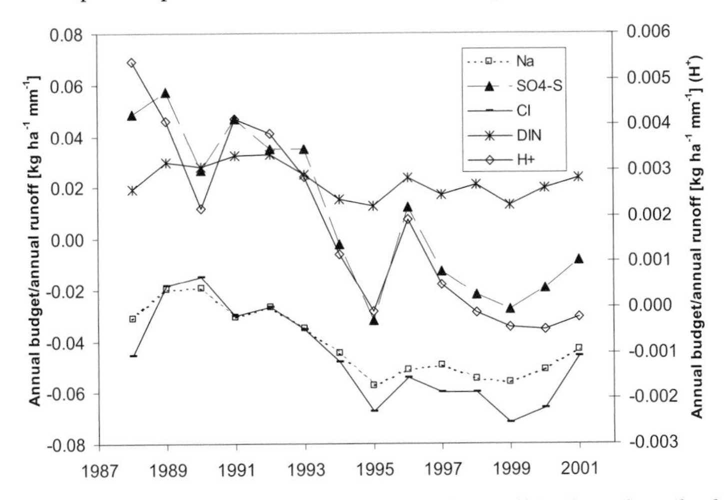

Fig. 24.10. Ratio of annual solute budgets to annual runoff in the Lehstenbach catchment

Alewell et al. (2001) identified an increase in the net release of the sum of 'base cations' (Na, K, Ca, Mg) in various forested catchments in the low mountain ranges in Germany including the Lehstenbach catchment. Part of that increase might be due to higher annual mean discharge in the second half of the monitoring period at Lehstenbach (Fig. 24.8). Relating annual mean budgets to discharge, however, confirms an increase since the late 1990s for the total Lehstenbach catchment as well as for the eastern part of the catchment, whereas the ratios tended to decrease before. In addition, the decreasing Ca and Mg content in the spruce needles in the Lehstenbach catchment clearly points to a crucial depletion of the pool of these nutrients in the topsoil (Alewell et al. 2000b). Obviously, the release of these cations by weathering cannot compensate for uptake by the plants and enhanced leaching.

Similar to Lehstenbach, the annual mass budgets of the Steinkreuz catchment highly depend on the amount of annual runoff. However, in spite of that variability, the balance was clearly negative for Ca, Mg, Na and SO_4-S throughout the 1995–2001 period (Fig. 24.11). In contrast, it fluctuated around zero for K and was strictly positive for inorganic nitrogen (Fig. 24.12). The data suggest a slightly decreasing nitrogen retention capacity of the Steinkreuz catchment.

So far, we have no clear indication for an imminent increase in nitrogen release and thus increasing risk of nitrogen saturation, as defined by Ågren and Bosatta (1988), Aber et al. (1989) or Stoddard (1994), in either of the two catchments. However, it has to be considered that the most important nitrogen sink is the buildup of organic mass in the forest floor (Berg, this Vol.) which might be susceptible to climatic change in the long term (Wright and Hauhs 1991). As stated above (Lischeid et al., this Vol.), substantial amounts of

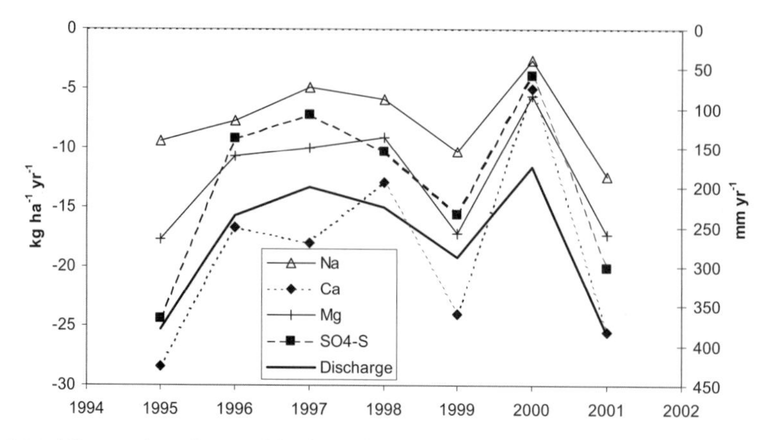

Fig. 24.11. Time series of annual budgets for the Steinkreuz catchment (input via deposition minus output flux via the catchment runoff)

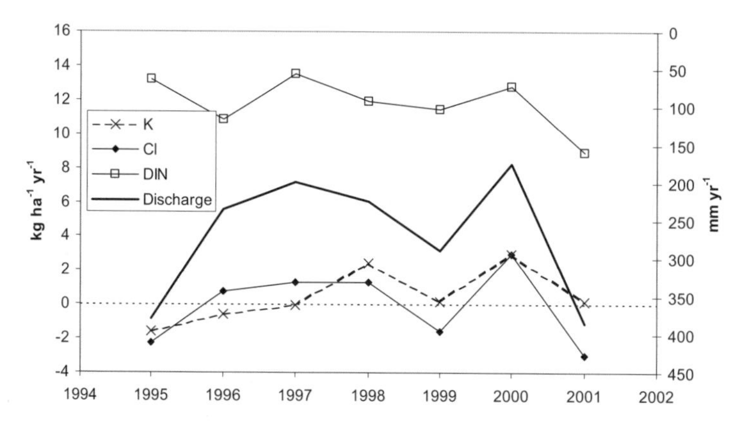

Fig. 24.12. Time series of annual budgets for the Steinkreuz catchment [input via deposition or throughfall (DIN) minus output flux via the catchment runoff]

nitrogen are released during rainstorms after dry, warm periods. If the frequency of these events increases due to climate change, it has to be considered that the nitrogen loading of groundwater and stream water is likely to increase as well. These processes are still not well understood, which became obvious when unexplained multiyear oscillations of NO_3 concentration were observed in the runoff of European and North American catchments (Driscoll et al. 1995; Stoddard et al. 1999; Goodale et al. 2003).

24.6 Conclusions: Long-Term Implications

In spite of highly differing bedrock geology and petrology, the budgets of both catchments exhibited similar features if the road-salt-contaminated part of the Lehstenbach catchment is disregarded. Both catchments acted as long-term sources of the base cations Ca, Mg and Na and of dissolved organic carbon. In the Steinkreuz catchment it could be shown that a large fraction of the base cations is released in the deeper layers of the aquifer.

The Lehstenbach catchment clearly shows a turn in the mid-1990s from retention to release of sulfate and protons. This is due to the release of formerly deposited SO_4 that has been desorbed from the matrix due to decreasing SO_4 concentration of the infiltrating rainwater. Now both the Lehstenbach and the Steinkreuz catchment clearly act as SO_4 sources.

In contrast, most of the deposited nitrogen does not show up in the catchments' runoffs. Here, incorporation in the building up of the humus layer and denitrification, either in the Lehstenbach wetlands or in the Steinkreuz clayey layers of the aquifer, seemed to act as the major sinks. So far it can only be speculated how stable the soil organic nitrogen pool is, facing changing climatic boundary conditions.

Both in the Lehstenbach and in the Steinkreuz catchment, the K annual mean input via deposition and output via the catchment runoff were nearly balanced. There is a clear net export of the base cations Ca, Mg and Na in both catchments. In the Steinkreuz catchment it could be shown that a large fraction of these base cations is released in the deeper layers of the aquifer.

The most pronounced difference between the two catchments concerned the protons which are almost entirely buffered in the Steinkreuz catchment, but are only to a minor degree retained in the Lehstenbach catchment. Again, this is traced back to the substantially lower pool of base cations that are released in the subsoil, due to different lithology.

Interannual variability of the budgets of most of the solutes are in the first place due to that of annual discharge, which in turn exhibited a clear trend at Lehstenbach, due to a sequence of exceptionally dry years in the first half of the monitoring period. This makes an analysis of the long-term trends difficult. In addition, shorter time series did not allow us to detect significant trends at the Steinkreuz catchment. At Lehstenbach, however, relating the annual budgets to annual discharge revealed a clearly decreasing net release of SO_4 and H^+ in the Lehstenbach catchment. On the other hand, Cl and Na export from the road-salt-contaminated part of the catchment has clearly increased since the late 1980s, pointing to a shift in contamination of the Lehstenbach groundwater and streams.

Acknowledgements. We would like to thank the members of the Central Analytic Department of the Bayreuth Institute of Terrestrial Ecosystem Research (BITÖK) for the chemical analysis of numerous water samples and for their help in the field sampling. Thanks are also due to Uwe Hell, Wolfgang Hofmann, Andreas Kolb, Georg Walther and Uwe Wunderlich for their help in the field sampling and site maintenance. This research was supported by the Federal Ministry of Education and Research (PT BEO51-0339476 C+D).

References

Aber J, Nadelhoffer KJ, Steudler P, Melillo JM (1989) Nitrogen saturation in northern forest ecosystems: hypotheses and implications. BioScience 39:378–386

Ågren GI, Bosatta E (1988) Nitrogen saturation of terrestrial ecosystems. Environ Pollut 54:185–197

Alewell C (1998) Investigating sulfate sorption and desorption of acid forest soils with special consideration of soil structure. J Plant Nutr Soil Sci 161:73–80

Alewell C (2001) Predicting reversibility of acidification: the European sulfur story. Water Air Soil Pollut 130:1271–1276

Alewell C, Armbruster M, Bittersohl J, Evans CD, Meesenburg H, Moritz K, Prechtel A (2001) Are there signs of acidification reversal after two decades of reduced acid input in the low mountain ranges of Germany? Hydrol Earth Syst Sci 5:367–378

Alewell C, Manderscheid B, Meesenburg H, Bittersohl J (2000a) Is acidification still an ecological threat? Nature 407:856–857

Alewell C, Manderscheid B, Gerstberger P, Matzner E (2000b) Effects of reduced atmospheric deposition on soil solution chemistry and elemental contents of spruce needles in NE-Bavaria, Germany. J Plant Nutr Soil Sci 163:509–516

Alewell C, Lischeid G, Hell U, Manderscheid B (2004) High temporal resolution of ion fluxes in semi-natural ecosystems – gain of information or waste of resources? Biogeochemistry (in press)

Bauer G, Persson H, Persson T, Mund M, Hein M, Kummetz E, Matteucci G, van Oene H, Scarascia-Mugnozza G, Schulze E-D (2000) Linking plant nutrition and ecosystem processes. In: Schulze E-D (ed) Carbon and nitrogen cycling in European forest ecosystems. Springer, Berlin Heidelberg New York, pp 63–98

BML (1997) Deutscher Waldbodenbericht 1996. Bundesministerium für Ernährung Landwirtschaft und Forsten, Bonn

Boyce RL, Friedland AJ, Chamberlain CP, Poulsen SR (1996) Direct canopy nitrogen uptake from ^{15}N-labeled wet deposition by mature red spruce. Can J For Res 26:1539–1547

Brumme R, Leimcke U, Matzner E (1992) Interception and uptake of NH_4 and NO_3 from wet deposition by above-ground parts of young beech (*Fagus silvatica* L.) trees. Plant Soil 142:272–279

Büttcher H (2001) Random variability or reproducible spatial patterns? Investigating sulphate dynamics in forested catchments with a coupled transport sorption model. Diploma Thesis, University of Bayreuth, Germany

Driscoll CT, Postek KM, Kretser W, Raynal DJ (1995) Long-term trends in the chemistry of precipitation and lake water in the Adirondack region of New York, USA. Water Air Soil Pollut 5:583–588

Emmert U (1985) Geologische Karte von Bayern 1:25.000. Erläuterungen zum Blatt Nr. 6128 Ebrach. Bayerisches Geologisches Landesamt, Munich

Fichter J, Dambrine E, Turpault M-P, Ranger J (1998) Base cation supply in spruce and beech ecosystems of the Strengbach catchment (Vosges Mountains, N-E France). Water Air Soil Pollut 104:125–148

Goodale CL, Lajtha K, Nadelhoffer KJ, Boyer EW, Jaworski NA (2002) Forest nitrogen sinks in large eastern US watersheds: estimates from forest inventory and an ecosystem model. Biogeochemistry 57/58:239–266

Goodale CL, Aber JD, Vitousek PM (2003) An unexpected nitrate decline in New Hampshire streams. Ecosystems 6:75–86

Houle D, Paquin R, Camiré C, Ouimet R, Duchesne L (1997) Response of the Lake Clair Watershed (Duchesnay, Quebec) to changes in precipitation chemistry (1988–1994). Can J For Res 27:1813–1821

Lehmann M (1999) Untersuchung der Belastungen von Rohwässern der BEW GmbH am Ochsenkopf (Fichtelgebirge). Diploma Thesis, University of Bayreuth, Germany

Lischeid G, Moritz K, Bittersohl J, Alewell C, Matzner E (2000) Sinks of anthropogenic nitrogen and sulphate in the Lehstenbach catchment (Fichtelgebirge): lessons learned concerning reversibility. Silva Gabreta 4:41–50

Lischeid G, Kolb A, Alewell C (2002) Apparent translatory flow in groundwater recharge and runoff generation. J Hydrol 265:195–211

Lovett GM, Lindberg SE (1993) Atmospheric deposition and canopy interactions of nitrogen in forests. Can J For Res 23:1603–1616

Manderscheid B, Matzner E (1995) Spatial and temporal variation of soil solution chemistry and ion fluxes through the soil in a mature Norway spruce (*Picea abies* (L.) Karst.) stand. Biogeochemistry 30:99–114

Manderscheid B, Schweisser T, Lischeid G, Alewell C, Matzner E (2000a) Sulfate pools in the weathered substrata of a forested catchment. Soil Sci Soc Am J 64:1078–1082

Manderscheid B, Jungnickel C, Alewell C (2000b) Spatial variability of sulfate isotherms in forest soils at different scales and its implications for the modeling of soil sulfate fluxes. Soil Sci 165:848–857

Matzner E, Alewell C, Bittersohl J, Lischeid G, Kammerer G, Manderscheid B, Matschonat G, Moritz K, Tenhunen JD, Totsche K (2001) Biogeochemistry of a spruce forest catchment of the Fichtelgebirge in response to changing atmospheric deposition. In: Tenhunen JD, Lenz R, Hantschel R (eds) Ecosystem approaches to landscape management in central Europe. Ecological studies, vol 147. Springer, Berlin Heidelberg New York, pp 463–504

Mund M (1996) Wachstum und oberirdische Biomasse von Fichtenbeständen (*Picea abies* (L.) Karst.) in einer Periode anthropogener Stickstoffeinträge. Diploma Thesis, University of Bayreuth, Germany

Nakagawa Y, Li CH, Iwatsubo G (2001) Element budgets in a forested watershed in southern China: estimation of a proton budget. Water Air Soil Pollut 130:715–720

Poller A (2002) Elektrotomographische Erkundung und geostatistische Modellierung von Aquifereigenschaften im Lehstenbach-Einzugsgebiet. Diploma thesis, University of Bayreuth, Germany

Prechtel A, Alewell C, Armbruster M, Bittersohl J, Cullen J, Evans CD, Helliwell R, Kopacek J, Marchetto A, Matzner E, Meesenburg H, Moldan F, Moritz K, Vesely J, Wright RF (2001) Response of sulphur dynamics in European catchments to decreasing sulphate deposition. Hydrol Earth Syst Sci 5:311–326

Reynolds B, Wood MJ, Truscott AM, Brittain SA, Williams DL (2000) Cycling of nutrient base cations in a twelve year old Sitka spruce plantation in upland mid-Wales. Hydrol Earth Syst Sci 4:311–321

Rhode H, Grenfelt P, Wisniewski J, Ågren C, Bengtsson G, Hultberg H, Johansson K, Kaupi P, Kucera V, Oskarson H, Pihl Karlsson G, Rasmussen L, Rosseland B, Schotte L, Sellden G, Thörnelöf E (1995) Acid reign '95? – conference summary statement from the 5th international conference on acidic deposition. Science and policy. Kluwer, Göteborg, pp 1–15

Stoddard JL (1994) Long-term changes in watershed retention of nitrogen: its causes and aquatic consequences. In: Baker LA (ed) Environmental chemistry of lakes and reservoirs. Adv Chem Ser 237:223–284

Stoddard JL, Jeffries DS, Lückewille A, Clair TA, Dillon PJ, Driscoll CT, Forsius M, Johannessen M, Kahl JS, Kellogg JH, Kemp A, Mannio J, Monteith DT, Murdoch PS, Patrick S, Rebsdorf A, Skjelkvåle BL, Steinton MP, Traan T, van Dam H, Webster KE, Wieting J, Wilander A (1999) Regional trends in aquatic recovery from acidification in North America and Europe. Nature 401:575–578

Ulrich B (1994) Nutrient and acid-base budget of central European forest ecosystems. In: Godbold DL, Hüttermann A (eds) Effects of acid rain on forest processes. Wiley-Liss, New York, pp 1–50

Uyttendaele GYP, Iroumé A (2002) The solute budget of a forest catchment and solute fluxes within a *Pinus radiata* and a secondary native forest site, southern Chile. Hydrol Process 16:2521–2536

Wright RW, Hauhs M (1991) Reversibility of acidification: soils and surface waters. Proc R Soc Edinb 97B:169–191

Yanai RD, Siccama TG, Arthur MA, Federer CA, Friedland AJ (1999) Accumulation and depletion of base cations in forest floors in the northeastern United States. Ecology 80:2774–2787

VI Synthesis

25 Biogeochemistry of Two Forested Catchments in a Changing Environment: A Synthesis

E. MATZNER, B. KÖSTNER, and G. LISCHEID

25.1 The Changing Environment

The environmental conditions that influence the functioning of our catchments have changed substantially in the last two decades. These changes comprise chemical and physical properties of the atmosphere and the deposition rates of substances from the atmosphere to the catchments.

Most striking in our catchments was the reduction of the SO_2 mixing ratios in the atmosphere (Klemm, this Vol.), coinciding with a decrease in the S and H^+ deposition (Matzner et al., this Vol., Chap. 14). This development has been observed in many countries in Europe and North America and is caused by emission control initiated in the 1980s. Figure 25.1 demonstrates the strong link between the development of the SO_2 mixing ratio and the average sulfate concentrations in throughfall of the two spruce sites in the Lehstenbach catchment. The concentrations decline almost parallel with a last peak in 1996 and reach a plateau of low concentrations from 1998 onwards. At the same time, the S deposition in the Lehstenbach catchment decreased from almost 60 to about 12 kg ha^{-1} $year^{-1}$. The parallel development of both parameters supports the use of throughfall measurements to estimate total S deposition in forest ecosystems.

The trend of the S depositions is also reflected in the S content of spruce needles, which decreased at the Coulissenhieb site (Alewell et al. 2000) and elsewhere (Dittmar and Elling, this Vol.).

Due to reductions in dust emissions, the deposition rates of Ca and Mg have also decreased in Europe and in North America, mainly in the 1980s and early 1990s (Hedin et al. 1994; Matzner and Meiwes 1994). Our data show that there was no further decrease from 1993 onwards, and that the deposition rates of Ca and Mg in both catchments have stabilized at a rather low rate.

Despite ongoing efforts to reduce N emissions into the atmosphere, no trend in the N fluxes with throughfall was observed in either catchment. The emissions of NO_x are assumed to have decreased in Germany (Klemm, this Vol.), but the uncertainties in calculating nationwide NO_x emissions are great.

Ecological Studies, Vol. 172
E. Matzner (Ed.), Biogeochemistry of Forested Catchments in a Changing Environment

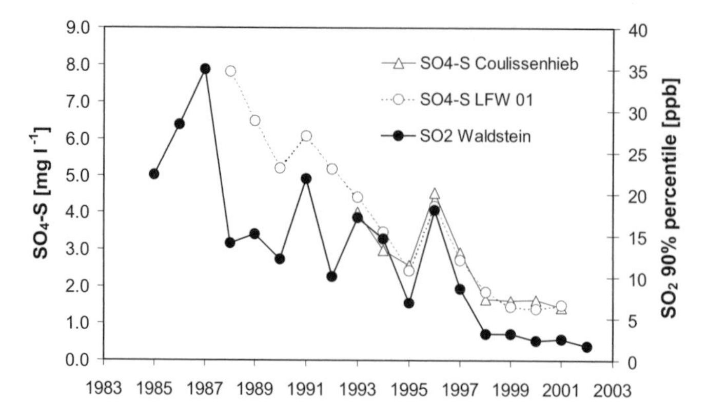

Fig. 25.1. Development of SO_2 mixing ratios and sulfate concentrations in throughfall at the Lehstenbach catchment

Moreover, NO_x emissions from ships may have been substantially underestimated so far (Corbett and Koehler 2003; Endresen et al. 2003).

Taking the throughfall fluxes as an indicator of N deposition, and considering other long-term studies in German forest ecosystems (Matzner and Meiwes 1994), the N inputs have remained unchanged at a high level for more than 30 years. Correspondingly, neither NO_x nor NH_3 mixing ratios at the Lehstenbach site have shown any clear trend since 1994 (Klemm, this Vol.). Thus, the development of the deposition rates of N does not coincide with the estimated emissions from industry and traffic, which was the case for S.

Our data emphasize the role of fog deposition in the high-altitude forests in the case of N. As outlined by Wrzesinsky et al. (this Vol.), the fog deposition at the Lehstenbach catchment accounted for only 4.5 % of the water input but for about 100 % of the bulk deposition of N over a 6-month measurement period. The absolute fog deposition of mineral N was about 8.7 kg ha^{-1} 6 $months^{-1}$. If we assume that fog deposition equals bulk deposition throughout the year at the Lehstenbach catchment (the latter amounting on average to 12.6 kg N ha^{-1} $year^{-1}$; Matzner et al., this Vol., Chap. 14), fog deposition + bulk deposition of mineral N would be in the range of 25 kg N ha^{-1} $year^{-1}$. This number still underestimates the total N deposition, since particle and gas deposition is not accounted for. In a similar Norway spruce stand in the German Solling area, Marques (1999) estimated the total deposition of N at 45 kg ha^{-1} $year^{-1}$, of which 7 kg ha^{-1} was due to fog deposition, but 23 kg ha^{-1} was due to dry deposition of particles and gases (mainly HNO_3 gas). The N pollution of the Solling site seems to be somewhat higher compared to the Lehstenbach, since the throughfall N fluxes there are about 30 kg N ha^{-1} $year^{-1}$, while those at Lehstenbach are around 22 kg ha^{-1}. Canopy uptake of N at the Solling site is estimated at 15 kg N ha^{-1} $year^{-1}$ (Marques 1999). Horn et al. (1989) calculated rates of total deposition of N for two forested sites in the Fichtelgebirge at 28–39 kg N ha^{-1} $year^{-1}$, of which 14–25 kg ha^{-1} would be canopy uptake and not measurable in throughfall.

Estimating the long-term total N deposition in our catchments thus remains difficult and only a range should be given: for the Lehstenbach site, a minimum of 25 kg N ha^{-1} $year^{-1}$ seems to be a conservative estimate of the long-term N deposition, while the maximum might be as high as 40 kg ha^{-1} $year^{-1}$. These rates result in an estimated N canopy uptake of 3–15 kg ha^{-1} $year^{-1}$ which is in the range of most other estimates for coniferous forests (Harrison et al. 2000). The conditions in the deciduous stand at Steinkreuz are not in favour of dry and occult (fog) deposition, because of the low elevation and the leafless winter period. Lacking any measurement on dry and occult deposition at Steinkreuz and considering experimental estimates of N canopy uptake by beech leaves (Brumme et al. 1992), the long-term total N deposition at Steinkreuz should be in the range of 17–22 kg ha^{-1} $year^{-1}$, resulting in a canopy uptake of 2–7 kg N ha^{-1} $year^{-1}$. The uncertainties in the rates of total deposition of N are thus substantial and must be taken into account when discussing the N budgets of the forest stands and catchments.

Besides the changes in element inputs by deposition, the Fichtelgebirge area was subjected to significant warming. The average increase in the annual temperature over the last four decades was 1.2, and 2 K in the winter months (Foken, this Vol.). So far, no significant changes have been detected in precipitation. The increase in temperature is also found in other regions of Germany and thus affects the Steinkreuz catchment as well. However, the exact degree of temperature change for the Steinkreuz catchment has not been assessed.

Overall, the environmental conditions of the two catchments have changed dramatically. Especially in the cold climate of the Fichtelgebirge, the increasing temperatures cause an extension of the vegetation period and might affect a number of processes, like tree growth, soil respiration, DOC release and gas exchange between vegetation and atmosphere as outlined below.

The decreasing inputs of acidity and S have reached a level that is considered as equal or below critical loads for forest ecosystems. At the same time, N inputs have remained at a high level, presumably for several decades and the nutrient cation input has decreased. The response of biogeochemical cycles, soil conditions and tree growth to these complex changes will thus be of special interest.

25.2 Recovery from Acidification of Soils and Waters

The acidification of soils and waters by the deposition of sulfuric and nitric acid is a well-documented phenomenon and causal relationships between deposition and effects are, to a large extent, well understood (van Breemen et al. 1983; Reuss and Johnson 1986; Sullivan 2000). Under conditions of decreasing S, Ca and Mg depositions, the recovery from past acidification is presently in focus (Likens et al. 1996; Jenkins et al. 2001).

The data presented in this volume on soil solution chemistry, throughfall fluxes, soil budgets (Matzner et al., this Vol., Chap. 20), catchment budgets (Lischeid et al., this Vol., Chap. 24) and runoff dynamics (Lischeid et al., this Vol., Chap. 23) have revealed a number of positive developments, indicating a recovery from acidification.

In summary, in the time frame of 1988–2001, the developments at the Coulissenhieb site and in the Lehstenbach catchment were:

- A substantial decrease in the acidity fluxes with seepage to deeper soil layers: the fluxes of acidity ($Al^{3+} + H^+$) at 90 cm soil depth and the acidification rate of the soil decreased by almost 50 %.
- An increase in soil solution alkalinity from 1996 onwards: from −0.5 to −0.15 $mmol_c\ l^{-1}$ in 20 cm soil depth and from −0.7 to −0.3 $mmol_c\ l^{-1}$ in 90 cm depth.
- A slight increase in the Ca/Al ratio of the soil solution in 20 cm depth from 1999 onwards.
- Decreasing net export of Ca and Mg from the soil.
- Decreasing leaching of cations from the canopy: especially in the case of Ca and Mg the calculated rates of canopy leaching decreased substantially as a result of the decreasing H^+ deposition and buffering on the leaves. The leaching of cations has to be compensated in the long run by increased uptake of cations by the roots, causing a specific H^+ load of the rhizosphere (Ulrich 1983). Thus, the decrease in leaching should have a beneficial effect on acidification parameters of the rhizosphere.

These positive developments indicated from the plot-scale soil solution studies are partly supported by the catchment budgets and by the developments in runoff chemistry. In the case of the Lehstenbach catchment, we observed:

- An increasing net release of SO_4 and H^+ in the Lehstenbach catchment, which is mainly caused by the reduction in input whereas output fluxes remain rather constant.
- A change in the S budget of the catchment from a sink to a source, indicating the depletion of sulfate pools in the soils and deeper aquifer.
- Decreasing SO_4 concentrations especially under high runoff conditions that coincide with the decreasing SO_4 concentrations in the upper soil. The upper soil layers are more important for runoff generation under high flow conditions.
- The extremely low alkalinity of runoff observed during storm flow increased by more than 0.2 $mmol_c\ l^{-1}$. The H^+ concentration in the Lehstenbach runoff during storm flow was only half by the end of the 1990s compared to the end of the 1980s. However, storm-flow pH values are still less than 4.

While there are clear signs of recovery of soil and waters, some of the results are still a matter of concern:

- The S release from inorganic and organic soil pools and from the aquifer delays the recovery. The Lehstenbach watershed has very high soil S pools and thus the mitigating effects of decreasing S deposition are delayed for several decades. In some groundwater wells the sulfate concentrations still increase, indicating that the S concentration peak is still on its way along the longest flow paths in the catchment. In the Steigerwald catchment, SO_4 concentration of the shallow groundwater decreased by 30 % from 1995–2001, whereas no trend is detectable in the deep groundwater. However, water acidification is of less concern at Steigerwald compared to Lehstenbach due to higher contents of both geogenic base cations and geogenic SO_4.
- The Ca and Mg concentrations in soil solution at the Coulissenhieb site decreased in the observation period and are extremely low today. The input of Ca and Mg to the mobile soil pools by weathering of minerals is very low and will not compensate for the past and present losses.
- The Ca and Mg concentrations in runoff at the Lehstenbach catchment also decreased, especially in periods of high flow. This coincides with the generally decreasing trends of Ca and Mg deposition and concentrations in the soil solutions.
- The concentrations of DOC in runoff of the Lehstenbach catchment tend to increase and partly compensate for the effect of decreasing SO_4 concentrations. Increase in surface-water DOC has been observed in other European catchments without a clear picture of what is causing the increase (Skjelevale et al. 2000; Evans and Monteith 2001; Freeman et al. 2001). Increasing temperature triggers DOM release from soils (Gödde et al. 1996; Freeman et al. 2001) and the amount and intensity of precipitation have also been shown to influence DOM production and transport positively (Michalzik and Matzner 1999; Tipping et al. 1999; Park and Matzner 2003). Furthermore, the increase in pH of precipitation associated with decreasing S deposition might be involved in the rise in DOM concentrations.
- There seems to be no long-term alkalinity production due to microbial SO_4 reduction in the wetland soils of the Lehstenbach catchment, followed by accumulation of organic and inorganic reduced S forms. The reduction of SO_4 is an ongoing process in the wetlands but is counteracted by the reoxidation in periods of higher oxygen supply, the latter being caused by changing hydrologic conditions (Küsel and Alewell, this Vol.).
- During the observation period, tree nutrition, indicated by the Ca and Mg contents of the needles, deteriorated (Alewell et al. 2000) and the trees have visible symptoms of Mg deficiency (yellowing of older needles).
- At the Steigerwald, the losses of Mg and Ca from the soil are substantial. The upper soil layers are largely depleted in exchangeable Ca and Mg compared to the deeper layers and it seems that the losses are now causing depletion of the deeper soil layers.

In summary, what is the relevance of these developments for ecosystem functions and their use? In the context of acidification and its reversal, the quality of ground- and surface water as well as the development of tree growth are of special importance.

The general development with respect to the acidification of soil solutions and runoff points to an improvement, but the aquifer of the Lehstenbach catchment is a large buffer preventing clear signals in runoff yet. This is similar to other catchments with deep aquifers (Alewell et al. 2001; Evans et al. 2001). Considering the sulfate in groundwater at Lehstenbach, maximum concentrations have not been reached in some wells. The level and frequency of extremely acidic runoff during high flow events have decreased, offering a first sign of a long-lasting recovery process in runoff. On the other hand, increasing DOC concentrations in runoff during high flow seem to partly compensate for the decreasing load of air-borne acidity in the topsoil. This should be followed with care, since increasing DOC concentrations in surface waters will have a strong impact on the functioning of fresh-water ecosystems.

In most forest ecosystems, the growth rate of the trees is assumed to be N limited under conditions of low N input (Tamm 1991). An increase in the supply of the limiting nutrient will stimulate tree growth as long as other nutrients and factors are not limiting. In many European countries, tree growth has increased in the past four to five decades (Spieker et al. 1996), and N inputs from deposition are likely to be the major cause for the increase. The tree ring and growth analyses at the Coulissenhieb site (Dittmar and Elling, this Vol.) confirm these general findings: the basal area increment at the site Coulissenhieb increased with time, especially up to about 1970. From 1975–1985 the basal area increment was significantly lower than the average from 1960–1997, indicating a growth depression. Following 1985 the basal area increment increased again, reaching its highest values around 1990. The basal area increment from 1985–1997 was about at the same level as compared to the 1960–1970 period. There was no obvious decline in basal area growth with stand age. The lack of an age trend in growth and unexpected high growth were also observed by Mund et al. (2002) when investigating growth rates in a chronosequence of Norway spruce sites in the Lehstenbach catchment.

As shown by tree ring analysis at the Coulissenhieb site, the annual growth was best correlated with temperature (Dittmar and Elling, this Vol.), pointing to temperature as the main growth-limiting factor at the Coulissenhieb site.

The reason for the growth depression from 1975–1985 might be seen in the culmination of SO_2 emissions and S (and acidity) deposition rates in that period. Growth reductions during the period of 1975–1985 were also observed in stands of the eastern Fichtelgebirge (Pretzsch 1996; Dittmar and Elling, this Vol.). The occurrence of forest decline phenomena in the Fichtelgebirge is also documented for this period (Schulze et al. 1989a). The SO_2

emissions and S depositions declined from 1985 onwards, which seems to improve the growth conditions again.

The high growth rates of the last decade are partly expected because of the general warming trend and the high N depositions. On the other hand, the depletion of the soil and soil solution of Ca and Mg continued in that period, leading to very low concentrations of these nutrient cations in soil solutions (Matzner et al., this Vol., Chap. 20). The obvious Mg deficiency of the trees and the high N/Mg ratios in needles (Alewell et al. 2000) still allow a high growth rate. One reason for that is the fact that the Mg deficiency is typical for older needles with low photosynthetic activity. Furthermore, the photosynthetic capacity of spruce needles was not correlated with their Mg content above a threshold of approx. 0.3 mg Mg g_{dw}^{-1} (Lange et al. 1989; Oren and Zimmermann 1989). Also needle shedding may enhance the light-use efficiency of remaining needles and trees (Beyschlag et al. 1994). The high N/Mg ratios are likely caused by the high N inputs in combination with poor soil conditions as revealed from fertilizer studies and long-term field observations (Flückiger and Braun 1998; Dusquesnay et al. 2000).

From growth analysis of Norway spruce stands in Norway, Nellemann and Thomsen (2001) reported that the growth-stimulating effect of N depositions turned into a growth depression after about two decades if N deposition was >15 kg N ha^{-1} $year^{-1}$. Such growth patterns of trees as a function of chronically high N depositions were also postulated in connection with the N-saturation hypothesis (Aber et al. 1989). However, these patterns cannot be confirmed for the Coulissenhieb site, subjected to an N deposition of >25 kg ha^{-1} $year^{-1}$ for several decades. The conditions found at the Coulissenhieb site are sustaining a high growth rate despite long-lasting N depositions and despite severe soil acidification and Mg deficiency. The reason for this might be specifically favourable conditions at this site, namely the high precipitation, the general increase in temperature, low transpiration rates and the lack of drought stress (Köstner et al. 2001). This does not imply that other tree physiological parameters, like frost hardiness, drought tolerance or pest resistance, were unaffected, but these were not investigated and seem to have little effect on the long-term growth rates at this site. High N depositions may lead to lower fine root biomass (George et al. 1999), similar to the effects of strong soil acidity (Godbold 1994; Matzner and Murach 1995). The lower fine root biomass in relation to shoot mass and leaf area may cause a higher susceptibility to drought stress under conditions of low precipitation and high transpiration. However, this chain of events is hard to demonstrate under field conditions and was questioned, e.g., for Swedish conditions by Binkley and Högberg (1997).

25.3 Fate of Deposited Nitrogen

The N pollution of temperate forest ecosystems and the various effects of N on forest ecosystems have been a matter of debate for more than two decades (e.g. Binkley and Högberg 1997; Bobbink et al. 2003). Our catchments have been subjected to relatively high and almost constant N depositions for several decades, thereby providing a good opportunity to study the long-term response to chronically high N depositions.

There was no significant trend in the runoff fluxes of mineral N in either catchment, indicating that the sinks of N at the catchment scale have a more or less constant strength. In Table 25.1, the N sinks are compiled and compared with the estimated N deposition, as the only source of N in the system. The fixation of N_2 by free-living organisms is not taken into account because of the lack of measurements. Estimates of non-symbiotic N_2 fixation in forests vary to a large degree: Limmer and Drake (1996) reported only small amounts of N_2 fixation (<1 kg ha^{-1} year^{-1}) in the soil of the Coulissenhieb spruce site, but Cleveland et al. (1999) estimated an average rate of non-symbiotic N_2 fixation in temperate forests of 6 kg N ha^{-1} year^{-1}. A large proportion of that was attributed to N_2 fixation in the phyllosphere, but this was not studied at our sites.

When interpreting the N budget in Table 25.1, one has to keep in mind the uncertainties associated with the measurements and estimates of the different fluxes. Those uncertainties are difficult to assess and sometimes can only be estimated rather than calculated with a certain statistical precision. For example, the estimated range of deposition is already quite large, especially in the case of the Coulissenhieb site. Unfortunately, those fluxes that can be easily measured (e.g. N export via the catchment runoff and sequestration in biomass) are often the smallest. In a rough general estimate, the single fluxes might be considered with an uncertainty of ±30 %.

The N fluxes with seepage clearly exceed those with runoff in both catchments. The difference is attributed to denitrification in the groundwater and riparian zones reaching a much higher rate (12.4 kg N ha^{-1} year^{-1}) at the Lehstenbach catchment as compared to Steinkreuz. At Lehstenbach, denitrification seems to be a substantial N export that exceeds the N sequestration by the growing stand and the export with runoff. The reason for the high rates of denitrification in the Lehstenbach catchment are seen in the riparian zones represented by the bogs and peats. As shown by Küsel and Alewell (this Vol.), these wetlands almost completely reduce NO_3, leading the authors to the conclusion that 30 % of the deposited N is denitrified, according to the area of wetlands in the Lehstenbach catchment.

Taking the estimated N deposition rate (line 1, Table 25.1), the denitrification would then be in the range of 8–13 kg N ha^{-1} year^{-1}. This independent estimate of denitrification at the catchment scale from the wetland studies is

Table 25.1. Average N fluxes and budgets of the sites and catchments (kg N ha^{-1} year^{-1}).

		Coulissenhieb site and Lehstenbach catchment	Steinkreuz site and catchment
1	Estimated range of total N deposition (see above)	25–40	17–22
2	Mineral N flux with throughfall (from) Matzner et al., this Vol., Chap. 14)	22.2	15.4
3	N loss with seepage at the site scale (from Matzner et al., this Vol., Chap. 20)	17.5	6.0
4	N loss with runoff from the catchment (from Lischeid et al., this Vol., Chap. 24)	5.1	3.6
5	Denitrification in groundwater and riparian zones (calculated as 3–4)	12.4	2.4
6*	Above- and below-ground biomass increment at the sites	10.5	14.0
7	Accumulation of N in the aggrading forest floor (from Berg, this Vol., Chap. 21)	18	34
8	$\sum$ N fate calculated as 4+5+6+7 or as 3+6+7	46	54

* *Coulissenhieb*: data from Bauer et al. (2000): calculated as total actual N pool of the stand divided by stand age. *Steinkreuz*: estimate from the 'Schacht' beech site in the Fichtelgebirge (Bauer et al. 2000) calculated as total actual N pool of the stand divided by stand age. At both sites the harvested biomass up to the present is taken into account according to Schober (1979). Harvested biomass is 75 % of the standing crop in the case of Norway spruce and 100 % in the case of beech

similar to the one calculated from the difference in runoff and seepage. Both approaches are supportive. However, part of the deposited N may bypass the anoxic wetland soils via near-surface runoff during storm flow, reducing the capacity of that sink. According to a rough estimate, about half of the annual runoff during the 1988–2001 period occurred during storm flow (Lischeid et al., this Vol., Chap. 23). In addition, NO_3 concentration in first-order streams that discharge from two of the investigated wetlands clearly exceeded the detection limit even during long dry summer periods, when NO_3 could not be detected in adjacent piezometers and soil suction cups. As a consequence, the denitrification in the wetlands might be overestimated if based only on the area of the wetlands. The denitrification in the Steigerwald catchment is located in the groundwater, since riparian zones are lacking. Depth gradients of groundwater NO_3 (Lischeid et al., this Vol., Chap. 23) support this conclusion. Denitrification losses are much less at Steinkreuz than those at the Lehstenbach catchment, but are approaching the N losses with runoff.

The N sequestration by the aggrading biomass is roughly estimated for the Coulissenhieb and Steinkreuz sites (see legend of Table 25.1) at 10.5 and 14 kg N ha^{-1} $year^{-1}$, respectively. In the case of the Lehstenbach catchment, a growth rate for the whole catchment is difficult to assess, since there is a significant variation in age classes and edaphic conditions in the catchment. At Steinkreuz, the site and the catchment have similar stand properties. The numbers indicate that a substantial amount of N is bound in the aggrading biomass.

In both catchments, the most important N sink seems to be the soil organic N pool. As estimated from the limit value concept of decomposition (Berg, this Vol.), the accumulation of N in the aggrading forest floor is 18 and 34 kg N ha^{-1} $year^{-1}$ for the Coulissenhieb and Steinkreuz site, respectively. The soil organic N pool was also identified as the major sink of deposited N in several other studies (e.g., Matzner 1989; Magill et al. 1997), but the mechanisms of the N sequestration are not fully understood. Besides the inhibition of decomposition in later stages of litter decay, biological and chemical immobilization of N are suggested as retention mechanisms (Johnson et al. 2000; Davidson et al. 2003).

The sum of sinks and outputs clearly exceeds the N fluxes with throughfall, questioning the use of throughfall data as total N inputs in temperate forest ecosystems. Overall, the estimated and measured N sinks and outputs at Coulissenhieb are in the upper range of the estimated total N deposition, but clearly exceed the N deposition at Steigerwald. There is substantial uncertainty in the calculated sum of N fates that results from the uncertainty in estimating each specific flux. Assuming a 30 % uncertainty for each of the N sinks and outputs and independence of those, the calculation of the error of the sum according to Gaussian error propagation yields a range of uncertainty of the Σ of N fates of about ± 9 kg N ha^{-1} $year^{-1}$ for Coulissenhieb and 19 kg N ha^{-1} $year^{-1}$ for Steinkreuz. Thus, the estimated N sinks and outputs do not necessarily exceed the range of estimated deposition rates at the Coulissenhieb site. The sum of the N fates would require an N input of about 46 kg N ha^{-1} $year^{-1}$ to balance the budget.

The sum of N sinks and output fluxes exceeds the estimated range of N deposition at the Steinkreuz site by far, even if the uncertainty of ± 19 kg N ha^{-1} $year^{-1}$ for the sum of fates is considered. Thus, there are indications from these data that unaccounted N inputs by deposition and canopy uptake or by non-symbiotic N fixation in soils and the phyllosphere might occur. Accepting the magnitude of the estimated N sinks at Steinkreuz and balancing the N budget would require an N deposition of 54 kg N ha^{-1} $year^{-1}$ or a non-symbiotic N fixation in the ecosystem of about 32 kg N ha^{-1} $year^{-1}$ which is five times the average rate estimated by Cleveland et al. (1999) for temperate ecosystems. The discrepancy of the N budget at Steigerwald is mostly due to the very high rate of N sequestration in the aggrading forest floor (for discussion of this see Berg, this Vol.). Because of the small database on litter decomposition and its limit values in deciduous stands, this number has to be considered with caution. However, the N budget of the Steigerwald sites calls for future studies on unaccounted N inputs as well as on long-term forest floor N pools in deciduous sites.

The long-term data from the two catchments and sites point out the strong interannual variation of the N fluxes with seepage and runoff. In the case of seepage fluxes, the interannual variation was by a factor of 4 at both sites. This emphasizes the need for long-term measurements to characterize the N status of terrestrial ecosystems. At the Coulissenhieb site, the NO_3 concentrations in seepage tended to decrease especially in 2001, despite the constant depositions. Furthermore, there was a slight decrease in the nitrate concentrations in runoff. The latter is related to hydrological conditions that were not in favour of NO_3 leaching in the last years of the study (Lischeid et al., this Vol., Chap. 23). One explanation for the unexpected decrease in NO_3 concentrations and fluxes with seepage at the Coulissenhieb site in 2000 and 2001 might be the exceptionally warm year 2000. In 2000 the average temperature was 1.8 K higher than the long-term mean and the estimated net C sequestration of the ecosystem was almost two times higher than the average (Rebmann et al., this Vol.) According to the tree ring analysis, temperature is a limiting factor under the site conditions of the Coulissenhieb site. A high growth rate in 2000 would fix a larger than average proportion of N in the aggrading biomass and thereby explain the low nitrate concentrations in soil solutions. On the other hand, increasing rainfall intensity in combination with increasing temperatures can increase the amount of N that is flushed from the forest floor into the groundwater and into the stream, as has been shown at the Steigerwald catchment (Lischeid et al., this Vol., Chap. 23). Due to a number of nonlinear effects, a quantitative assessment of these interactions is hardly possible (Goodale et al. 2003).

At both sites, the N fluxes with seepage exceeded those with runoff from the catchments, indicating significant denitrification in the groundwater (Steinkreuz) and riparian zones (Lehstenbach). Thus, if input-output relations are considered, it makes a difference if data from sites or catchments are considered. The large variation of the N outputs with seepage and runoff in relation to N deposition and the C/N ratio of the soil (e.g. MacDonald et al. 2002) might be partly explained by the joint use of data from sites and catchments in regional assessments.

In summary, our investigations of the N cycle have shown that:

- Nitrate losses with seepage are high at both sites.
- The sinks and losses of N are quite stable in both catchments, and there was no clear trend of increasing nitrate concentrations and fluxes with seepage and runoff despite constant N depositions.
- Denitrification in groundwater and riparian zones is a major export of N at the catchment scale.
- The N accumulation in the aggrading forest floor exceeds the N accumulation in the aggrading biomass.

- The N losses and sinks at the Coulissenhieb site are in the upper range of (or slightly exceed) the estimated N inputs by deposition.
- The N budget of the Steigerwald site remained unclosed. Future research on the role of N_2 fixation and on the long-term development of N pools in the forest floor is needed.

The catchments and sites are a sink for deposited N, despite the fact that N saturation (Aber et al. 1989) has been reached for many years. The strength of the N sinks is influenced by positive feedbacks to the N availability/deposition. This is clearly the case for the N accumulation in the aggrading forest floor, which is driven by increased litter fall amounts and high N contents in litter under high N depositions. The same is true for the N accumulation in the aggrading biomass which is also increased by the N-induced higher growth rates. Lastly the denitrification in the upper horizons of the riparian zones of the Lehstenbach catchment during dry periods is almost complete, indicating that the rate of denitrification is nitrate limited.

What would happen if the N deposition continues at the present rate? We conclude that the N sinks in both catchments are stable for a long time. As shown by Berg and Dise (2004), the N accumulation in aggrading forest floors can continue for decades to centuries if no disturbance of the ecosystem occurs. Whether the accumulated N in the soil organic matter poses a long-term risk, which could lead to nitrate losses with seepage, e.g. during the phase of stand regeneration, remains an open question. Our own measurements on the N dynamics in a clear-cut experiment in the Lehstenbach catchment do not support such a scenario: on the contrary, the nitrate fluxes in the forest floor percolates even decreased after clear-cut (Kalbitz, pers. comm.), supporting the conclusion that the risk of high nitrate losses in the regeneration phase is low.

While an increase in the N losses from the catchments under the present conditions seems unlikely, the calculated rates of denitrification and the measured NO_3 losses, especially at the stand scale, are not acceptable from the viewpoint of sustainable development of ecosystem functions. The present NO_3 losses with seepage and runoff are a substantial source of soil acidification and water pollution, and a significant part of the denitrification will result in the emission of N_2O under the low pH conditions of the riparian zones. Thus, a decrease in the N deposition is necessary. What would be the effect of decreasing N depositions? In roof experiments, with a drastic and instantaneous decrease of N inputs, the NO_3 in seepage and soil solutions decreased strongly as soon as 1–2 years after treatment (Bredemeier et al. 1998). The decrease in NO_3 availability was shown to be the first response to decreasing N inputs. Following these studies, we assume that a decrease in N depositions would, in the time scale of up to one decade, mainly reduce the denitrification rate and the losses of nitrate rather than influence the other sinks.

25.4 Carbon Sequestration

The C sequestration of terrestrial ecosystems has been investigated in numerous papers using very different approaches and scales and trying to quantify the role of forests as sinks of atmospheric CO_2 (e.g. Sedjo 1992; Dixon et al. 1994; Valentini et al. 2000; Valentini 2003). Here we compile the estimates from atmospheric CO_2 exchange (net ecosystem exchange, NEE) with growth measurements and estimates of soil C sequestration for the Coulissenhieb and the adjacent Weidenbrunnen site in the Lehstenbach watershed (Table 25.2). No data on NEE are available for the Steinkreuz catchment.

The data from Table 25.2 indicate a number of interesting points: first, it becomes obvious that the DOC export with seepage from the ecosystem can be neglected in the C budget, since the flux is less than the estimated error of the other fluxes. However, the soil internal DOC turnover might contribute significantly to the C sequestration of the soil. The DOC flux with soil solution from the forest floor into the mineral soil amounts to 282 kg C ha^{-1} $year^{-1}$ and the net retention in the mineral soil is 260 kg C ha^{-1} $year^{-1}$ (Kalbitz et al., this Vol.). The fate of the DOC in the mineral soil is not fully understood at the moment. An unknown part of the DOC will be mineralized to CO_2, but a significant part might also be stabilized for a long time by interaction with soil Al (Schwesig et al. 2003a,b) or by adsorption to the solid phase (Kaiser et al. 2002). Both processes will add to the large stable C pool of the mineral soil at this site.

By far the largest C sink in the ecosystem is the increment of the standing biomass of the trees, measured at 1,760 kg C ha^{-1} $year^{-1}$ for the Coulissenhieb site (Mund et al. 2002). As outlined above, the growth of the mature Coulissenhieb site is unexpectedly high, presumably as an effect of the N deposition and general warming tendency of the climate. Mund et al. (2002) also report

Table 25.2. Average C fluxes and budget for the Coulissenhieb/Weidenbrunnen site (kg C ha^{-1} $year^{-1}$)

1	Net ecosystem CO_2 sink measured at the tower of the Weidenbrunnen site (from Rebmann et al., this Vol.)	550
2	Above-ground increment of the Coulissenhieb site (stem biomass with bark) (Mund et al. 2002)	1,760
3	DOC export with seepage at 90 cm soil depth at the Coulissenhieb site (from Kalbitz et al., this Vol.)	22
4	DOC retained (mineralized or stored) in the mineral soil (from Kalbitz et al., this Vol.)	260
5	Estimated accumulation of C in the aggrading forest floor of the Coulissenhieb site (from Berg, this Vol.)	480

growth rates of younger Norway spruce sites in the Lehstenbach catchment, which all exceed the growth of the Coulissenhieb, reaching a maximum of 3,910 kg C ha^{-1} $year^{-1}$ at a 43-year-old site.

The estimates for the C accumulation in the aggrading forest floor, based on the limit value concept of litter decomposition (Berg, this Vol.), yielded an accumulation of 480 kg C ha^{-1} $year^{-1}$. The limit value concept was validated under different conditions, also using the long-term forest floor inventories of the Norway spruce Solling site. At the Solling, the average annual C accumulation in the forest floor under Norway spruce over a 30-year period was measured by periodic soil inventories at 800 kg C ha^{-1} $year^{-1}$ (Meesenburg et al. 1999; Berg, this Vol.). In a European transect study, Schulze et al. (2000) estimated the C sequestration of the soil using totally different methods to be on average 1,440 kg C ha^{-1} $year^{-1}$. These rates are higher than those estimated at the Coulissenhieb and Solling sites, but they also include the potential accumulation of root litter and DOC in the whole soil profile. At the Coulissenhieb site the heterotrophic soil respiration as estimated by Schulze et al. (2000) was, however, much less than the rates given by Subke et al. (this Vol.).

The total C sequestration of the ecosystem as derived from growth measurements and estimates of forest floor accumulation yields a number of 2,240 kg C ha^{-1} $year^{-1}$, which will further increase if we consider below-ground biomass accumulation, root litter accumulation (both unknown) and assume that part of the DOC transferred into the mineral soil will not be mineralized (maximum 260 kg C ha^{-1} $year^{-1}$).

There is a large discrepancy in the different estimates of C sequestration at the Lehstenbach watershed: the micrometeorological CO_2 exchange measurements revealed very low C uptake of the ecosystem with an average of 550 kg C ha^{-1} $year^{-1}$, which is only 24 % of the estimated C sequestration by growth and forest floor accumulation. The rate of C sequestration derived from the CO_2 budget is also low in comparison to other European forests: Valentini et al. (2000) gave a range of C sequestration based on CO_2 exchange measurements from <1,000 to about 6,000 kg C ha^{-1} $year^{-1}$, with most of the 27 sites having sequestration rates >1,000 kg C ha^{-1} $year^{-1}$. The CO_2 exchange measurements were conducted at the tower of the Weidenbrunnen site adjacent to the Coulissenhieb. The Weidenbrunnen site is a declining forest with a high mortality of the trees that might presently be a net C source. However, the footprint analysis (Rebmann et al., this Vol.) revealed that only <20 % of the observations can be attributed to the Weidenbrunnen site itself, while >80 % of the observations of the CO_2 exchange reflect the C budget of the forested sites outside the Weidenbrunnen site. Thus, the low rates of C sequestration revealed by the CO_2 exchange measurements cannot be explained by the potential C source in the Weidenbrunnen site. The reason for the discrepancy remains unclear. The highest confidence in the different estimates lies probably in the measured above-ground growth rates, which already exceeded the CO_2 exchange measurements by a factor of 3. The growth period investigated by Mund et al. (2002) was obviously favourable for growth at that site and might be slightly higher than the growth in the last years of the study (Dittmar and Elling, this Vol.). However, even when assuming that the growth rates of the last years are 30 % less than those found by Mund et al. (2002), the different approaches to measure C sequestration do not coincide. The numbers in Table 25.2 indicate that the different methodological approaches to calculate C sequestration of terrestrial ecosystems (especially micrometeorological, on the one hand, and growth assessment and soil inventory studies, on the other) need to be systematically compared for a larger number of sites in the future (cf. Curtis et al. 2002; Ehman et al. 2002).

In summary, our data showed that the Lehstenbach catchment is a sink for atmospheric CO_2, with the tree increment representing the largest part. Besides, the C sequestration in the forest floor following litter decomposition is a significant C sink. The estimates of ecosystem C sequestration based on CO_2 budgets and micrometeorological measurements did not match the estimates based on other approaches.

25.5 Water Fluxes Between Vegetation and Environment

Evaporation and transpiration from terrestrial ecosystems contribute on average 65 % to terrestrial precipitation. The importance of evapotranspiration (ET) is expected to increase with temperature rise intensifying the hydrological turnover at the global and ecosystem scale (Jackson et al. 2001). Plants as pathways of water from soil to atmosphere represent an important controlling factor by (1) root physiology, rooting depth and root biomass, (2) stomatal conductance at the leaf and canopy level, (3) genetic potentials and composition of species and (4) developmental stage, biomass and land cover. Plants also play an important role in the pathway from atmosphere to the soil by interception of water, and in conjunction with upper soil layers in preventing excessive runoff.

Within the framework of ecosystem studies in the Lehstenbach and Steinkreuz catchments, the contribution of vegetation to water vapour fluxes has been studied for several years. Most of the investigations in spruce forests of the Lehstenbach catchment are presented elsewhere (Wedler et al. 1996; Alsheimer et al. 1998; Köstner et al. 2001, 2002). It was found that seasonal and maximum canopy transpiration (E_c) decreased strongly with age, reaching only up to 1.6 mm day^{-1} at the 140-year-old site Coulissenhieb. The decrease was independent of leaf area index within a range from 5–8. Highest E_c was observed for small patches of spruce regrowth (10–20 years old) in more open areas, reaching up to 7 mm day^{-1} at the patch level (Köstner 2001). On summer days, forest floor and ground vegetation ET contributed up to 45 % of total ET at Coulissenhieb, whereas the contribution in 40-year-old stands reached 20 %.

As is typical for closed deciduous forests, the ground vegetation of the Steigerwald sites was only sparsely developed. Thus, contributions from the forest floor to total water vapour fluxes were almost confined to litter and soil evaporation. At the site Steinkreuz (cf. Köstner et al., this Vol.) soil evaporation was estimated gravimetrically by lysimeters (Schmidt, in prep.), reaching on average 0.38 mm day^{-1} before leaf unfolding and 0.58 mm day^{-1} on summer days (on average 28 % of total ET).

Comparisons of ET with other water flux components contribute to the validation of water fluxes at the stand and canopy level (Table 25.3). In a first approximation, ET may be compared with the difference between canopy throughfall and seepage at the stand level. This is exemplified for the site Coulissenhieb in the years 1994 and 1995. Significant tree sap flow could be measured between the middle of April and October. Canopy transpiration derived from scaled-up sap flow reached 121 and 110 mm $season^{-1}$ in 1994 and 1995, respectively. Taking into account an average contribution of 40 %

Table 25.3. Overview of water balance components derived at stand and catchment level at Lehstenbach and Steinkreuz (stand level: *Picea abies*, 140-year-old stand Coulissenhieb; *Fagus sylvatica*, *Quercus petraea*, 110-year-old stand Steinkreuz) (all fluxes in mm $year^{-1}$)

		Coulissenhieb site and Lehstenbach catchment		Steinkreuz site and catchment			
1	Year of study	1994	1995	1996	1998	1999	2000
2	Bulk precipitation	1,143	1,312	724	838	786	767
3	Catchment runoff	571	732	236	225	289	174
4	Total evaporation at catchment level	572	583	488	613	497	593
5	Throughfall	870	1015	576	615	589	581
6	Canopy interception	273	297	148	223	197	186
7	Seepage	661	705	308	314	271	224
8	Canopy transpiration	121	110	230		178	
9	Evapotranspiration of forest-floor and ground vegetation	82	75	64		50	
10	Evapotranspiration from catchment water balance	299	283	340	390	300	407
11	Evapotranspiration from site water balance	209	310	268	301	318	357
12	Scaled-up evapotranspiration at stand level	203	185	294		228	

from the forest floor during summer, ET sums up to 203 mm season^{-1} in 1994 and 185 mm season^{-1} in 1995. The respective differences between throughfall and seepage amounted to 209 and 310 mm year^{-1}, respectively. While estimates of ET agree in 1994, a discrepancy of 125 mm occurs in 1995. Monthly E_c was found to be best positively correlated with average vapour pressure deficit of the air, followed by radiation and air temperature. Seasonal averages of vapour pressure deficit, radiation and temperature were higher in 1994. Thus, from a physiological point of view and in accordance with observations, higher E_c should be expected in 1994. Further, exceptionally high rainfalls occurred in June 1995, reducing monthly E_c by more than half of that of 1994 and accounting for most of the differences between 1994 and 1995. Therefore, it has to be assumed that the greater difference between throughfall and seepage in 1995 was caused more by physically than physiologically controlled processes. Overall, it has to be stated that such comparisons based on measurements at different spatial integration are difficult. As discussed by Oren et al. (1998), comparisons of scaled-up sap flow and local soil water balance may be inappropriate due to spatial variation resulting in variances of up to 30 %.

It should be further noted that outside the period of active sap flow, leaf transpiration may occur on warm and sunny days even throughout winter months. Simulations of E_c by the adapted stand gas-exchange model STANDFLUX resulted in 4 % (8 mm) non-seasonal E_c (November–March) related to annual E_c (Falge 1997). This amount of water is obviously taken from internal storage. Tree water storage may be refilled by above-ground intake of intercepted water. Based on the specific conductance of bark and twigs (Katz et al. 1989), their absolute surface at the sites and the number of interception events after dry periods, it was estimated that the amount of water directly taken up in the tree crowns during the season could reach up to 15 % of E_c (determined by sap flow) or up to 8 % of interception (Köstner et al. 2002).

Interception represents another important part of the water balance. Within the years of study (1993–2001) at the site Coulissenhieb, annual interception (calculated from the difference between bulk precipitation and throughfall) reached on average 24 % (range 21–28 %) of precipitation which is somewhat less than the average of literature data from spruce forests reaching 30 % of precipitation (Köstner 2001). At the catchment level, ET results from precipitation, interception and runoff (see Table 25.3). Annual ET of the catchment water balance amounted on average (1993–2001) to 303 mm year^{-1} (28 % of bulk precipitation). The complete catchment water balance was modelled based on spatially predicted environmental variables using TOPMODEL and functional dependencies of E_c and ground vegetation and soil ET on vapour pressure deficit and on stand age (Ostendorf and Manderscheid 1997). For a simulation period of 5 years, the different flux components resulted in 176 mm year^{-1} for E_c, 91 mm year^{-1} for E_f, 223 mm year^{-1} for interception and 345 mm year^{-1} for runoff. Further, an unexplained residual of

59 mm $year^{-1}$ (6.6 % of precipitation) remained. Within this frame of uncertainty, ET derived from measurements at the plot level does not contradict ET derived from the catchment water balance.

In the Steinkreuz catchment, the difference between throughfall and seepage at the plot level amounted 318 mm $year^{-1}$ in 1999, while an ET of 300 mm $year^{-1}$ was estimated from the catchment water balance (Table 25.3). Assuming on average 28 % of ET from soil and litter, total ET based on tree sap flow and lysimeters sums up to 244 mm $season^{-1}$. Thus, ET from tree canopy and soil reached 76–81 % of indirect estimations. As discussed above, the discrepancy lays within the range of uncertainties due to spatial heterogeneity. Further, seepage in 60 cm soil depth may not have been deep enough to include the complete water extraction by roots because best correlation between sap flow and soil water content was found for 90 cm soil depth in the year 1999. Soil matrix potential in 200 cm depth in late summer clearly indicated root water uptake from that depth. However, there seems to be a general tendency for direct determinations of canopy water vapour fluxes (sap flow, eddy covariance) to be lower than indirect determinations (Köstner 2001).

Plant roots represent another important controlling factor of water transport from soil to atmosphere. Compared with the canopy–atmosphere exchange, relatively little information is available on the soil–root interface, especially on the role of woody roots. Woody roots may contribute to root water uptake (Kramer 1946; Braun 1988), but this has not been quantified for old-growth trees so far. Both axial flow and radial uptake of water by woody roots (diameter 0.2–4.6 cm) were studied in 140-year-old Norway spruce at the site Coulissenhieb and at the mixed beech–oak stand at the site Steinkreuz (Lindenmair et al., this Vol.). Compared to axial flux densities (up to 100 g cm^{-2} day^{-1}) in spruce, radial uptake rates reached 0.001–0.02 g cm^{-2} day^{-1}, which is in a similar order of magnitude to above-ground uptake through bark (0.005 g cm^{-2} MPa^{-1}; Katz et al. 1989). Compared to spruce, surface-related radial uptake rates in beech and oak roots were reduced by half. This is potentially related to the higher swelling capacity of cork in conifer roots (Braun 1988). On the basis of radial transport rates and root biomass of the site (Mund et al. 2002), it was estimated that water uptake by coarse roots contributed between 6 and 16 % to daily tree sap flow (Lindenmair et al., this Vol.). If this range holds over longer periods, it is comparable to the amount of above-ground tree water intake by interception. Also the function may be similar supplying the tissue water storage via short-distance transport. However, for a given diurnal change in root water content of 10 %, it would take several days to refill the tissue by radial transport only. The most important ecophysiological meaning seems to be additional local water supply of the tissue, which is especially important to bridge transient periods of drought or dormancy (cf. Waring and Running 1978; Braun 1988). More generally, the findings add to our current understanding of hydraulic transport and redistribution processes in the soil–root system (Dawson 1993; Burgess et al. 1998; Brooks et al. 2002).

Various types of models are used to simulate or scale-up water vapour fluxes in time and space. Emphasis may be laid on transport processes between vegetation and the atmospheric boundary layer being influenced by horizontal and vertical exchange (Berger et al., this Vol.). Such models represent vegetation in horizontally homogenous layers using physically based equations and empirical parameterizations (energy balance/BIGLEAF approach). Whereas gas-exchange models simulating both water vapour and CO_2 fluxes include physiological processes of light capture and photosynthesis of plants, a very detailed gas-exchange model was used to describe CO_2 and water vapour fluxes of leaf clouds in crown sectors (Fleck et al., this Vol.). In this model, gas exchange of crown sectors depends on spatially detailed crown microclimate, spatial distribution of plant organs and leaf nitrogen, an important determinant of photosynthetic electron transport and carboxylation capacities. However, leaves could contain even more N than useful for photosynthesis, potentially indicating N surplus. Further, there was no reduction in the complexity of light responses from the leaf to branch scale. Obviously, the response of individual leaves is dominated by the leaf microsite. Thus, structurally detailed gas-exchange models of tree crowns may in future be based on individual leaves rather than on crown sections or branches. Models describing crown internal processes will help to explain processes at the whole-canopy level. A stand-level gas exchange model has been applied to simulate effects of structural changes on stand CO_2 exchange and transpiration in mixed beech–oak stands (Köstner et al., this Vol.). Especially important was the different behaviour of CO_2 and water vapour fluxes along structural gradients. While a significant increase of canopy transpiration and conductance with increasing percentage of beech and increasing leaf area index was observed, the behaviour of net photosynthesis was less distinct due to increasing respiration at high leaf area index. Consequently, water-use efficiency (carbon flux/water flux) and growth are affected in the long term. Reductions in water-use efficiency were more pronounced in beech than in oak. For spruce sites in the Lehstenbach catchment, a decline in water-use efficiency with stand age was observed (Köstner et al. 2002). Thus, water-use efficiency may become an important selective factor at high temperatures and in dry conditions.

In summary, our investigations of water fluxes have shown:

- Agreement or at least no contradiction within the given range of variation when water fluxes scaled-up from tree sap flow and forest floor were compared with indirect estimations of evaporation from site and catchment water balance.
- A significant reduction of canopy transpiration with stand age in monospecific stands of *Picea abies* which is partly compensated at the whole-stand level by increasing evapotranspiration from forest floor and ground vegetation in old stands.

- Higher rates of transpiration and photosynthesis at the leaf, crown and forest canopy level of *Fagus sylvatica* compared with *Quercus petraea*. These are related to higher nitrogen-use efficiency of beech at the leaf level and reduced vitality of oak at the crown and stand level due to periodical defoliation by caterpillars.
- Higher rates and stronger increase of canopy transpiration with increasing leaf area index and stand basal area of beech compared to oak and spruce.
- That under high N input and absence of severe soil drought, beech alone is expected to perform better than mixed stands. With respect to competitiveness and ecological amplitude, *Fagus sylvatica* profits most from high N availability.
- That the contribution of woody roots to water absorption of trees is significant and comparable to above-ground water intake through bark. Special advantages seem to be local water supply and bridging of transient periods (drought, dormancy).
- That detailed modelling of crown parts related to structure, N content of leaves and species-specific photosynthetic capacities is possible. However, the complexity of structure and function may be better addressed at the leaf level within crowns than at the level of branches because variation in fine-scaled information was not principally reduced from leaf to branch level.

Overall, it can be stated that ecosystem water fluxes differed with species composition, structure and age, offering potential control to ecosystem management. With respect to ecosystem functioning under the observed and expected climatic change (air temperature rise, increased evaporative demand, decreasing soil water availability, high N input), there is an increasing need to view water fluxes in conjunction with carbon and nitrogen. There is still lack of knowledge of how water-use efficiency performs in general with stand development, how combinations of rising CO_2 concentrations, increasing C sequestration and drought will affect water-use efficiency in the long term, and how species-specific capacities will interact with increasing N input.

25.6 Atmospheric Controls on Processes and Fluxes

Atmospheric controls on the processes and fluxes in ecosystems are very diverse, ranging from diurnal dynamics of meteorological variables, the influence of extreme weather events, like frosts, hurricanes and precipitation events, to long-term patterns of temperature, humidity and precipitation. The selection of the critical parameters depends on the processes considered and the local conditions. Predicting the effects of future climate changes needs the understanding of these interactions.

In this chapter we concentrate on the findings related to precipitation and on temperature effects on soil processes and fluxes at the stand and catchment scale and on tree growth.

The influence of the *precipitation regime* on the concentrations and fluxes with runoff were very obvious. At the Lehstenbach catchment, the analysis of runoff-generating processes revealed that under conditions of high flow (precipitation) the runoff chemistry changes, which is attributed to a larger proportion of runoff originating from surface-near soil horizons (Lischeid et al., this Vol., Chap. 23). The evaluation of the long-term trends in runoff chemistry has to take these effects into account. A similar observation is made in the Steinkreuz catchment: with increasing precipitation: the upper groundwater layer becomes more important for the regulation of runoff chemistry. In both catchments the correlation between runoff amount and element fluxes is very obvious. There is very little dilution of element concentrations in runoff under high flow; on the contrary, for some elements concentrations increase with runoff amount. Thus, the precipitation regime finally determines to a large extent the runoff fluxes of elements and their annual variations.

The same is true for the soil fluxes of DOC from the forest floor into the mineral soil, which were strongly and positively related to the amount of precipitation (Kalbitz et al., this Vol.). The investigations of redox processes in the wetland soils of the Lehstenbach catchment also showed the influence of precipitation on the generation of oxygen-rich zones in the upper soil layers after high rates of infiltration (Küsel and Alewell, this Vol.). During heavy rainstorms a large fraction of the nitrogen of the throughfall bypasses the zone of denitrification (Lischeid et al., this Vol., Chap. 23). As a consequence, the catchment's nitrogen budget highly depends on the number and intensity of rainstorms.

Furthermore, we showed that the soil respiration ceased after periods of drought, even under the high precipitation regime and generally moist soil conditions of the Lehstenbach catchment (Subke et al., this Vol.).

So far, there are no indications for changing precipitation regimes at our catchments. However, our results demonstrate that changes in the precipitation amount, distribution and intensity will have a strong effect on the functioning of these ecosystems.

No significant soil drought effects on canopy transpiration have been observed during the years of study in spruce stands of the Lehstenbach catchment (Köstner et al. 2001). However, it can be expected that, comparable with dry conditions during the winter periods, spruce is able to cope with dry periods under reduced physiological activity. In the Steinkreuz catchment, a continuous decrease in soil water content during the season was observed (Köstner et al., this Vol.) which may have led to earlier leaf discoloration in beech while transpiration rates of oak were relatively increased at the end of the season. As discussed in several studies (see Köstner et al., this Vol.), the drought tolerance of beech may have been underestimated or changing environmental conditions (temperature, N input) may counteract drought effects on growth rate.

A number of contributions to this volume have explicitly addressed *temperature* effects on processes and fluxes. Soil temperature was shown to be the major driving factor for soil respiration at sites in the Lehstenbach catchment (Subke et al., this Vol.). In the temperature range of 5–15 °C the Q_{10} values of soil respiration were 2.4, similar to the estimated average for temperate ecosystems (Raich and Schlesinger 1992). The investigation on the temperature dependence of the DOC release from the forest floor (Kalbitz et al., this Vol.) resulted in a somewhat higher Q_{10} value of about 4, indicating that different soil processes need to be considered separately in their response to changing temperatures. Lastly, the tree ring analysis revealed a positive correlation of the annual growth rates with the temperature at the Coulissenhieb site. Growth under these conditions is largely limited by low temperatures (Dittmar and Elling, this Vol.).

The average temperatures have increased substantially in the Fichtelgebirge area by about 1.2 K (annual average) in the last four decades and by 2 K when considering only the winter months. The temperature increase seems to be (to an unknown part) responsible for the unexpected high tree growth at the Lehstenbach catchment. A long-term study at an old-growth Norway spruce site in the lower Erzgebirge (Tharandt forest) showed that an increase of 1 K in annual mean temperature was associated with an average extension of the vegetation period by 6.5 days (Niemand et al. 2003). Annual net ecosystem exchange of C determined by eddy-covariance technique at the Tharandt forest (Grünwald 2003) increased with length of the vegetation period by 5 g C m^{-2} day^{-1} on average. This indicates that positive correlations between growth or net photosynthesis and temperature on an annual basis are more related to a higher number of growing days than to higher daily air temperatures. This is especially important in the light of potential adaptation of plants to higher temperatures (Bolstad et al. 2003). The latter would mitigate the potential effect of warming, while the effect of the extension of the growing period would be unaffected.

At the leaf level, the relationship between temperature and net photosynthesis of spruce follows an optimum curve with maximum values at approx. 20 °C. On a daily basis, simulated stand net photosynthesis at Coulissenhieb reached maximum rates between 10 and 15 °C and declined on warmer days due to higher respiration rates (Falge 1997). In general, net photosynthesis as well as canopy transpiration (see Köstner et al., this Vol.) were best correlated with photosynthetic photon flux density and vapour pressure deficit of the air on an hourly and daily basis, while on a monthly basis, canopy transpiration also increased linearly with temperature, which was not the case for net photosynthesis.

The measurements of the element fluxes and concentrations date back only for about one decade, and a signature of the increasing temperature cannot easily be found in the field data with different, interfering environmental factors. Whether the increased temperatures over the last four decades have increased or decreased the C sequestration of the total ecosystems remains an open question. A general increase in the growth rates of the trees at the Coulissenhieb site is very likely at this site (see above). From the Q_{10} values of soil respiration at Coulissenhieb (Subke et al., this Vol.) it would be likely that the present rates are about 14 % higher than those prior to 1960 if the rates are not resource limited. Whether this has resulted in a net loss of C from the soil depends (1) on the relation between the respiration of soil organic matter to root respiration and (2) on the changes of above- and below-ground litter inputs over the last four decades. The long-term effect of increasing temperatures on the rate of soil respiration and the C pool of soils might be overestimated, because the reactive pool of soil organic matter is small and the response might be short-lived (Glardina and Ryan 2000; Schlesinger et al. 2000; Melillo et al. 2002). The potential decrease of soil organic matter storage would need to be subtracted from the growth-triggering effect of increasing temperatures in order to estimate the changes in the C sequestration of the total ecosystem. Such an analysis remains purely speculative given the available data.

25.7 Criteria of Sustainable Use

The term 'sustainable use' of forest ecosystems (in German '*Nachhaltigkeit*') was first defined by the German forester von Carlowitz (1713), based on the idea that wood removal from the forest should not exceed the growth rate. Besides the tree growth rate, the cycling of elements and water may be used as a criterion for sustainability of ecosystem use and development. These cycles should be managed to fulfil the ecosystem functions in the long run without polluting the atmosphere and hydrosphere. Furthermore, the protection of species and their functions is considered as a criterion for sustainability (Holling 1986; Chapin et al. 1996; Christensen et al. 1996).

Our results show that the cycling of elements in the catchments results in a pollution of the atmosphere and the hydrosphere and thus criteria of sustainability are violated. This is especially true for N. The rates of denitrification are substantial in both catchments and we assume that a significant proportion of the denitrification under these conditions results in the formation of N_2O and NO (Brumme et al. 1999; Papen and Butterbach-Bahl 1999), polluting the atmosphere. The NO_3 export from the sites with seepage to the groundwater and the NO_3 fluxes in runoff clearly indicate the N saturation of the systems and the NO_3 pollution of the hydrosphere. In addition, the fluxes of acidity with seepage are still substantial, despite significant reductions of S depositions, acidifying deeper soil layers and groundwater.

In order to achieve a sustainable development of forest ecosystems under the influence of air pollutant deposition, the 'critical load concept' has been introduced (Nilsson and Grennfelt 1988; van der Salm and de Vries 2001). Critical loads for the N deposition are estimated from empirical data (Bobbink et al. 2003) or by steady state mass balance (Schulze et al. 1989b; van der Salm and de Vries 2001). The latter is mainly based on estimates of the N accumulation in biomass and soil organic matter. For temperate forest ecosystems the critical loads for N deposition derived from both approaches are often given in the range of 10–20 kg N ha^{-1} $year^{-1}$, depending on the tree species and the effect considered.

Calculating the critical loads needs to quantify the actual rate of N deposition and N sinks (see above) and compare those with the estimated critical load. Our results showed that the N sink in the aggrading forest floor is higher than assumed in the steady state mass balance equations used today. This suggests a higher critical load of N depositions. On the other hand, we also showed that the estimates of N deposition based on throughfall measurements substantially underestimate the N deposition. Thus, the present approaches to calculate the critical load of N seem to underestimate both the sinks and the inputs of N. Adding the stable sinks of N in the biomass and the forest floor, according to the data from Table 25.1, and assuming that those will be 30 % lower at the critical load than under today's depositions, the critical load would amount to 20 kg N ha^{-1} $year^{-1}$ for the Coulissenhieb site. In relation to the estimated total N inputs as calculated from the sum of N fates in Table 25.1, our data suggest a reduction of the N depositions by about 50 % in the case of the Coulissenhieb site to match the critical load. We did not estimate critical loads for the Steinkreuz site, given the uncertainties in the N budget of this site.

The estimates of the C and N sequestration in the soil showed the accumulation of substantial amounts of C and N, which is considered beneficial from the viewpoint of atmospheric CO_2 and N pollution of the atmosphere and hydrosphere. We assume that these pools are bound rather stable under conditions of regular forest management and pose only a small risk to the long-term sustainable use of these ecosystems.

The deposition rates of acidity have reached a level that is in the range of the critical load assumed for such sites. In fact, there were clear signs of recovery of soil solution and runoff acidity. However, the calculation of the weathering rates for the whole soil profile of the Coulissenhieb site yielded a rate of only 0.6 $kmol_c$ ha^{-1} $year^{-1}$ (Matzner et al., this Vol., Chap. 20), which seems too low to substantially raise the Ca, Mg and K concentrations in the soil solution and at the exchange sites of the soil solid phase. The concentrations of Ca and Mg at the Coulissenhieb site have steadily decreased and are extremely low today. The Norway spruce sites in the Lehstenbach catchment generally show symptoms of Mg deficiency, in some cases followed by needle losses and high mortality. Our data furthermore showed the strong ongoing depletion of the exchangeable Ca and Mg pool of the soil at the Steigerwald site. We can only speculate on the long-term effect of these losses at Steigerwald, but changes in the tree nutrition and groundwater chemistry are likely if time scales of several decades are considered.

The need for liming of acidified forest soils to improve the Mg nutrition and to increase the buffer capacity of the soil has been postulated in the past under the influence of high acid deposition rates (Ulrich et al. 1980), but also was a matter of debate because of potential negative effects of soil organic matter pools and nitrate leaching (e.g., Kreutzer 1995). From our data, we conclude that a natural improvement of the Ca and Mg availability in the soil is unlikely. Thus, without adding these elements to the biogeochemical cycle of forest ecosystems on strongly acidified soils, an improvement of tree nutrition, especially of the Mg supply, seems rather unlikely. From the viewpoint of biogeochemical cycles, soil solution and groundwater acidity and from the viewpoint of the Mg supply to trees, moderate liming of sites like the Coulissenhieb with Mg-rich lime seems necessary. The need for liming cannot be supported from the growth development of the sites. The growth of the Coulissenhieb site was unexpectedly high and recovered after a depression from 1975–1985 (Mund et al. 2002; Dittmar and Elling, this Vol.). We do not know how growth at the Coulissenhieb site would have been under better Ca and Mg supply. Results from other liming trials indicate that tree growth may increase after liming (Tveite et al. 1991), but may also be more or less ineffective (Vejre et al. 2001). Under more N-limited conditions, like in Scandinavia, liming can even cause a decrease in tree growth (Derome and Pätilä 1989).

25.8 Overall Conclusions

The results presented in the different chapters of this volume have substantially improved our qualitative and quantitative understanding of water and element fluxes in forested catchments and their driving forces.

Generally speaking, our conclusions with respect to the *structure of the research program* are:

- The interannual variation of almost all fluxes considered is substantial, which calls for the need of long-term research programs to analyse the biogeochemistry of terrestrial ecosystems.
- The initiation of an interdisciplinary research program proved to be absolutely necessary to investigate the problems addressed.
- Different approaches to quantify the same fluxes, e.g., evapotranspiration or carbon budgets, partly yielded substantially different results which illustrates the need to combine methods.

Moreover, the long-term studies of the Lehstenbach and the Steinkreuz catchment revealed some *intrinsic properties* that seem to be typical for ecosystems but are often ignored:

- The effect of single events, and thus of non-linear effects, of climatic factors on the catchments' budgets was clearly demonstrated, e.g., for output fluxes via the catchment runoff.
- Shifts in the short-term dynamics or runoff concentrations rather than long-term trends might be better indicators of change. Discharge dynamics turned out to be most relevant for runoff fluxes.
- Multiannual structures and interannual long-term memory effects reduce the probability of using one portion of the series to successfully predict another portion changed between different periods.
- The hydrological flow paths in catchments change with the precipitation regime and need to be considered when interpreting and predicting runoff dynamics.
- The information content of time series is often too low for the calibration or validation of process-based models.
- Species- and structure-related quantifications of water fluxes emphasize the role of species and tree size and provide the basis for potential controls by forest management. Quantifications of water fluxes at different levels of integration show reasonable results, although uncertainties remain due to spatial heterogeneity.

Considering the *quantitative results*, important general conclusions are as follows:

- The degree of acidification of the soil solution is slowly improving, despite the reduction of Ca and Mg deposition and the release of formally stored organic and inorganic S from the soil.
- The N budget at the plot scale differed significantly from the catchment scale, most likely because of denitrification in groundwater and riparian zones. With regard to the fate of N, plot studies and catchment studies cannot easily be compared.
- The growth of the trees is unexpectedly high, resulting in a strong N sink in the ecosystem. Variation of the growth rates is assumed to explain part of the variation of annual N losses with seepage.
- The major C sink in the ecosystems is represented by the accumulation of biomass. Besides, the accumulation of soil organic matter driven by litter decomposition and dissolved organic carbon accumulation significantly adds to the C sequestration of the ecosystems.
- The present deposition of acidity seems to match the critical load for these ecosystems. For strongly acidified soils, liming is recommended to improve the Mg nutrition of the trees and the soil chemical conditions. Criteria for sustainable use are still violated mainly because of the high N depositions.

25.9 Outlook

While substantial progress has been made in the last decade by biogeochemical research at BITÖK and elsewhere, a number of problems remain for future research.

We see major gaps in the quantitative understanding of soil organic matter turnover and the related C and N sequestration of soils, especially in the case of broad-leaved stands. Establishing more accurate C and N budgets of soils under different boundary conditions is a challenging task for the future. The obvious discrepancies in the C budgets for whole ecosystems derived from different methods also call for a systematic assessment.

With more extreme shifts of boundary conditions (temperature, soil water, N), the species-specific potentials of adaptation will become a more important selective factor in forest communities. Increasing growth and C sequestration during the last 20 years have been governed by temperature rise, reduced acidity and increased N input. The water cycle, including its effects on cloud formation and radiation, may be more relevant for ecosystem functioning in future in many parts of central Europe. There is still a lack of knowledge on interrelationships between water, N and C at the species and ecosystem level, and how functions change with age and stand development. Especially, improved understanding of processes at the soil–root interface and appropriate implementation of processes in models will be critical to predict species and ecosystem responses and to evaluate management options.

In recent years, the role of biodiversity of primary and secondary producers for ecosystem functioning has been addressed in a few studies, mainly in grassland communities. With respect to forest ecosystems, large uncertainties still exist and the methodical problems are severe, given the long life span and development of trees. Promising developments, especially in the characterization of the immense biodiversity of soil microorganisms, have been made and call for future research on that issue also in forests.

Not only may changing climatic conditions in the future be seen based on average temperature and precipitation, but also the frequency of critical conditions may change. For example, the frequency of drying/wetting cycles of the soil, of extreme droughts or of freezing/thawing may be much more important for ecosystem functioning than the long-term shifts. This also holds for catastrophic events, like storms and extreme rainfalls. Investigating the effects of such extreme events should be a task of future ecosystem research.

Thus, a large number of challenging questions remain. Our hope is that biogeochemical studies in terrestrial ecosystems will find a prominent place in the future research programs and that the awareness of their benefits for the use of ecosystem services will increase.

References

Aber JD, Nadelhoffer KJ, Steuder P, Melillo JM (1989) Nitrogen saturation in northern forest ecosystems – hypotheses and implications. BioScience 39:378–386

Alewell C, Manderscheid B, Gerstberger P, Matzner E (2000) Effects of reduced atmospheric deposition on soil solution chemistry and elemental contents of spruce needles in NE-Bavaria, Germany. J Plant Nutr Soil Sci 163:509–516

Alewell C, Armbruster M, Bittersohl J, Evans CD, Meesenburg H, Moritz K, Prechtel A (2001) Are there signs of acidification reversal in freshwater of the low mountain ranges in Germany? Hydrol Earth Syst Sci 5:367–378

Alsheimer M, Köstner B, Falge E, Tenhunen JD (1998) Temporal and spatial variation in transpiration of Norway spruce stands within a forested catchment of the Fichtelgebirge, Germany. Ann Sci For 55:103–124

Bauer GA, Persson H, Persson T, Mund M, Hein M, Kummetz E, Matteucci G, van Oene H, Scarascia-Mugnozza G, Schulze E-D (2000) Linking plant nutrition to ecosystem processes. In: Schulze E-D (ed) Carbon and nitrogen cycling in European forest ecosystems. Ecological studies, vol 142. Springer, Berlin Heidelberg New York, pp 63–98

Berg B, Dise N (2004) Validating a new model for N sequestration in forest soil organic matter. Water Air Soil Pollut Focus (in press)

Beyschlag W, Ryel RJ, Dietsch Ch (1994) Shedding of older needle age classes does not necessarily reduce photosynthetic primary production of Norway spruce. Trees 9:51–59

Binkley D, Högberg P (1997) Does atmospheric deposition of nitrogen threaten Swedish forests? For Ecol Manage 92:119–152

Bobbink R, Ashmore M, Braun S, Flückiger W, van den Wyngaert IJJ (2003) Empirical nitrogen critical loads for natural and semi-natural ecosystems. Background document for the expert workshop on empirical critical loads for nitrogen on (semi)natural ecosystems. Swiss Agency for the Environment, Forests and Landscape, Berne

Bolstad PV, Reich P, Lee T (2003) Rapid temperature acclimation of leaf respiration rates in *Quercus alba* and *Quercus rubra*. Tree Physiol 23:969–976

Braun HJ (1988) Bau und Leben der Bäume, 2nd edn. Rombach, Freiburg, 295 pp

Bredemeier M, Blanck K, Xu Y-J, Tietema A, Boxman AW, Emmett B, Moldan F, Gundersen P, Schleppi P, Wrigth RF (1998) Input–output budgets at the NITREX sites. For Ecol Manage 101:57–64

Brooks JR, Meinzer FC, Coulombe R, Gregg J (2002) Hydraulic redistribution of soil water during summer drought in two contrasting Pacific Northwest forests. Tree Physiol 22:1107–1117

Brumme R, Leimcke U, Matzner E (1992) The uptake of NH_4 and NO_3 from wet deposition by above ground parts of young beech (*Fagus silvatica* L.) trees. Plant Soil 142:273–279

Brumme R, Borken W, Finke S (1999) Hierarchical control on nitrous oxide emission in forest ecosystems. Global Biogeochem Cyc 13:1137–1148

Burgess SSO, Adams MA, Turner NC, Beverly CR, Ong CK (1998) The redistribution of soil water by tree root systems. Oecologia 115:306–311

Chapin FG, Torn MS, Tateno M (1996) Principles of ecosystem sustainability. Am Nat 148:1016–1034

Christensen NL, Bartuska AM, Brown JH, Carpenter S, D'Antonio C, Francis R, Franklin JF, MacMahon JA, Noss RF, Parsons DJ, Peterson CH, Turner MG, Woodmansee RG (1996) The report of the Ecological Society of America Committee on the scientific bases for ecosystem management. Ecol Appl 6:665–691

Cleveland CC, Townsend AR, Schimel DS, Fisher H, Howarth RW, Hedin LO, Perakis SS, Latty EF, von Fischer JC, Elseroad A, Wasson MF (1999) Global patterns of terrestrial biological nitrogen (N_2) fixation in natural ecosystems. Global Biogeochem Cyc 13:623–645

Corbett JJ, Koehler HW (2003) Updated emissions from ocean shipping. J Geophys Res 108(D20):4650

Curtis PS, Hanson PJ, Bolstad P, Barford C, Randolph JC, Schmid HP, Wilson KB (2002) Biometric and eddy-covariance based estimates of annual carbon storage in five eastern North American deciduous forests. Agric For Meteorol 113:3–19

Davidson EA, Chorover J, Dail DB (2003) A mechanism of abiotic immobilization of nitrate in forest ecosystems: the ferrous wheel hypothesis. Global Change Biol 9:228–236

Dawson TE (1993) Hydraulic lift and water use by plants: implications for water balance, performance and plant–plant interactions. Oecologia 95:565–574

Derome J, Pätilä A (1989) The liming of forest soils in Finland. Meddelser fran Norsk Inst Skogforskning 42:147–155

Dixon RK, Brown S, Houghton RA, Solomon AM, Trexler MC, Wisniewski J (1994) Carbon pools and flux of global forest ecosystems. Science 263:185–190

Dusquesnay A, Dupouey JL, Clement A, Ulrich E, Le Tacon F (2000) Spatial and temporal variability of foliar mineral concentration in beech (*Fagus sylvatica*) stand in northeastern France. Tree Physiol 20:13–22

Ehman JL, Schmid HP, Grimmond CSB, Randolph JC, Hanson PJ, Wayson CA, Cropley FD (2002) An initial intercomparison of micrometeorological and ecological inventory estimates of carbon exchange in a mid-latitude deciduous forest. Global Change Biol 8:575–589

Endresen Ø, Sørgård E, Sundet JK, Dalsøren SB, Isaksen ISA, Berglen TF, Gravir G (2003) Emission from international sea transportation and environmental impact. J Geophys Res 108(D17):4560

Evans CD, Monteith DT (2001) Chemical trends at lakes and streams in the UK Acid Waters Monitoring Network, 1988–2000: evidence for recent recovery at a national scale. Hydrol Earth Syst Sci 5:351–366

Evans CD, Cullen JM, Alewell C, Kopácek J, Marchetto A, Moldan F, Prechtel A, Rogora M, Veselý J, Wright RF (2001) Recovery from acidification in European surface waters. Hydrol Earth Syst Sci 5:283–297

Falge E (1997) Die Bedeutung der Kronendachtranspiration von Fichtenbeständen (*Picea abies* (L.) Karst.) mit unterschiedlichen Modellierungsansätzen. Bayreuther Forum Ökol 48:1–221

Flückiger W, Braun S (1998) Nitrogen deposition in Swiss forests and its possible relevance for leaf nutrient status, parasite attacks and soil acidification. Environ Pollut 102:69–76

Freeman C, Evans CD, Monteith DT (2001) Export of organic carbon from peat soils. Nature 412:785

George E, Kircher S, Schwarz P, Tesar A, Seitz B (1999) Effect of varied soil nitrogen supply on growth and nutrient uptake of young Norway spruce plants grown in a shaded environment. J Plant Nutr Soil Sci 162:301–307

Glardina CP, Ryan MG (2000) Evidence that decomposition rates of organic carbon in mineral soil do not vary with temperature. Nature 404:858–861

Gödde M, David MB, Christ MJ, Kaupenjohann M, Vance GF (1996) Carbon mobilization from the forest floor under red spruce in the northeastern USA. Soil Biol Biochem 28:1181–1189

Godbold DL (1994) Aluminium and heavy metal stress: from the rhizosphere to the whole plant. In: Godbold DL, Hüttermann A (eds) Effects of acid rain on forest processes. Wiley-Liss, New York, pp 231–264
Goodale CL, Aber JD, Vitousek PM (2003) An unexpected nitrate decline in New Hampshire streams. Ecosystems 6:75–86
Grünwald T (2003) Langfristige Beobachtungen von Kohlendioxidflüssen mittels Eddy-Kovarianz-Technik über einem Altfichtenbestand im Tharandter Wald. Dissertation, Technische Universität Dresden
Harrison AF, Schulze ED, Gebauer G, Bruckner G (2000) Canopy uptake and utilization of atmospheric nitrogen. In: Schulze ED (ed) Carbon and nitrogen cycling in European forest ecosystems. Ecological studies, vol 142. Springer, Berlin Heidelberg New York, pp 171–188
Hedin LO, Granat L, Likens GE, Buishand TA, Galloway JN, Butler TJ, Rodhe H (1994) Steep declines in atmospheric base cations in regions of Europe and North America. Nature 367:351–354
Holling CS (1986) Resilience of ecosystems: local surprise and global change. In: Clark WC, Munn, RE (eds) Sustainable development and the biosphere, chap 10. Cambridge University Press, Cambridge, pp 292–317
Jackson RB, Carpenter SR, Clifford ND, McKnight DM, Naiman RJ, Postel SL, Running SW (2001) Water in a changing world. Ecol Appl 11(4):1027–1045
Horn R, Schulze ED, Hantschel R (1989) Nutrient balance and element cycling in healthy and declining Norway spruce stands. Ecological studies, vol 77. Springer, Berlin Heidelberg New York, pp 444–454
Jenkins A, Ferrier RC, Wright RF (2001) Assessment of recovery of European surface waters from acidification 1979 to 2000. Special issue. Hydrol Earth Syst Sci 5:273–542
Johnson DW, Cheng W, Burke IC (2000) Biotic and abiotic nitrogen retention in a variety of forest soils Soil Sci Soc Am J 64:1503–1514
Kaiser K, Eusterhues K, Rumpel C, Guggenberger G, Kögel-Knabner I (2002) Stabilization of organic matter by soil minerals – investigations of density and particle-size fractions from two acid forest soils. J Plant Nutr Soil Sci 165:451–459
Katz C, Oren R, Schulze E-D, Milburn JA (1989) Uptake of water and solutes through twigs of *Picea abies*. Trees 3:33–37
Köstner B (2001) Evaporation and transpiration from coniferous and broad-leaved forests in central Europe – relevance of patch-level studies for spatial scaling. Meteorol Atmos Phys 76:69–82
Köstner B, Tenhunen JD, Alsheimer M, Wedler M, Scharfenberg H-J, Zimmermann R, Falge E, Joss U (2001) Controls on evapotranspiration in a spruce forest catchment of the Fichtelgebirge. In: Tenhunen JD, Lenz R, Hantschel R (eds) Ecosystem approaches to landscape management in central Europe. Ecological studies, vol 147. Springer, Berlin Heidelberg New York, pp 377–415
Köstner B, Falge E, Tenhunen JD (2002) Age-related effects on leaf area/sapwood area relationships, canopy transpiration, and carbon gain of *Picea abies* stands in the Fichtelgebirge/Germany. Tree Physiol 22:567–574
Kramer PJ (1946) Absorption of water through suberized roots of trees. Plant Physiol 21:37–41
Kreutzer K (1995) Effects of forest liming on soil processes. Plant Soil 168/169:447–470
Lange OL, Weikert RM, Wedler M, Gebel J, Heber U (1989) Photosynthese und Nährstoffversorgung von Fichten aus einem Waldschadensgebiet auf basenarmen Untergrund. Allg Forst-Zeitschr (AFZ) 3:10
Likens DE, Driscoll CT, Buso DC (1996) Long term effects of acid rain: response and recovery of a forested ecosystem. Science 272:244–246

Limmer C, Drake HL (1996) Nonsymbiotic N_2 fixation by acidic and pH-neutral forest soils: aerobic and anaerobic differentials. Soil Biol Biochem 28:177–183

MacDonald JA, Dise NB, Matzner E, Armbruster M, Gundersen P, Forsius M (2002) Nitrogen input together with ecosystem nitrogen enrichment predict nitrate leaching from European forests. Global Change Biol 8:1028–1033

Magill AH, Aber JD, Hendricks JJ, Bowden RD, Melillo JM, Steudler PA (1997) Biogeochemical response of forest ecosystem to simulated chronic nitrogen deposition. Ecol Appl 7:402–415

Marques MC (1999) Eintrag von luftgetragenen partikelgebundenen Spurenstoffen in Wälder durch trockene Deposition. Wissenschafltícher Verlag, Berlin

Matzner E (1989) Acidic precipitation: case study: Solling, West Germany. In: Adriano DC, Havas M (eds) Advances in environmental science: acid precipitation, vol 1. Springer, Berlin Heidelberg New York, pp 39–83

Matzner E, Meiwes KJ (1994) Long-term development of element fluxes with bulk precipitation and throughfall in two German forests. J Environ Qual 23:162–166

Matzner E, Murach D (1995) Soil changes induced by air pollutant deposition and their implication for forests in central Europe. Water Air Soil Pollut 85:63–76

Meesenburg H, Meiwes KJ, Bartens H (1999) Veränderung der Elementvorräte im Boden von Buchen- und Fichtenökosystemen im Solling. Freiburger Forstl Forsch 7:109–114

Melillo JM, Steudler PA, Aber JD, Newkirk K, Lux H, Bowles FP, Catricala C, Magill A, Ahrens T, Morrisseau S (2002) Soil warming and carbon-cycle feedbacks to the climate system. Science 298:2173–2176

Michalzik B, Matzner E (1999) Dynamics of dissolved organic nitrogen and carbon in a central European Norway spruce ecosystem. Eur J Soil Sci 50:579–590

Mund M, Kummetz E, Hein M, Bauer GA, Schulze ED (2002) Growth and carbon stocks of a spruce forest chronosequence in central Europe. For Ecol Manage 171:275–296

Nellemann C, Thomsen MG (2001) Long-term changes in forest growth: potential effects of nitrogen deposition and acidification. Water Air Soil Pollut 128:197–205

Niemand C, Köstner B, Prasse H, Grünwald T, Bernhofer C (2003) Charakterisierung von Phänophasen und Vegetationsperiode im Tharandter Wald auf phänologischer und meteorologischer Datenbasis. Tharandter Klimaprotokolle 9:164–166

Nilsson J, Grenfelt P (1988) Critical loads for sulphur and nitrogen: report. Nordic Council of Ministers, Copenhagen

Oren R, Zimmermann R (1989) CO_2-assimilation and the carbon budget of healthy and declining Norway spruce stands. Ecological studies, vol 77. Springer, Berlin Heidelberg New York, pp 352–368

Oren R, Phillips N, Katul G, Ewers BE, Pataki DE (1998) Scaling xylem sap flux and soil water balance and calculation variance: a method for partitioning water flux in forests. Ann Sci For 55:191–216

Ostendorf B, Manderscheid B (1997) Seasonal modelling of catchment water balance: a two-level cascading modification of TOPMODEL to increase the realism of spatio-temporal processes. Hydrol Process 11:1231–1242

Papen H, Butterbach-Bahl K (1999) A 3-year continuous record of nitrogen trace gas fluxes from untreated and limed soil of a N-saturated spruce and beech forest ecosystem in Germany, 1. N_2O emission. Geophys Res 104:18487–18503

Park JH, Matzner E (2003) Controls on the release of dissolved organic carbon and nitrogen from a deciduous forest floor investigated by manipulations of aboveground litter inputs and water flux. Biogeochemistry 66:265–286

Pretzsch H (1996) Growth trends of forests in Germany. In: Spiecker H, Mielikäinen K, Köhl M, Skovsgaard JP (eds) Growth trends in European forests. Springer, Berlin Heidelberg New York, pp 107–131

Raich JW, Schlesinger WH (1992) The global carbon dioxide flux in soil respiration and its relationship to vegetation and climate. Tellus 44B:81–99

Reuss JO, Johnson DW (1986) Acid deposition and the acidification of soils and waters. Springer, Berlin Heidelberg New York

Schlesinger WH, Winkler JP, Megonigal JP (2000) Soils and the global carbon cycle. In: Wigley TML, Schimmel DS (eds) The carbon cycle. Cambridge University Press, Cambridge, pp 93–101

Schober R (1979) Ertragstafeln wichtiger Baumarten. Sauerländers Verlag, Frankfurt

Schulze ED, Lange OL Oren R (1989a) Forest decline and air pollution. A study of spruce (*Picea abies*) on acid soils. Ecological studies, vol 77. Springer, Berlin Heidelberg New York

Schulze E-D, de Vries W, Hauhs M, Rosen K, Rasmussen L, Tamm O, Nilsson J (1989b) Critical loads for nitrogen deposition on forest ecosystems. Water Air Soil Pollut 48:451–456

Schulze ED, Högberg P, van Oene H, Persson T, Harrison AF, Read D, Kjoller A, Matteucci G (2000) Interactions between the carbon and nitrogen cycle and the role of biodiversity: a synopsis of a study along a north–south transect through Europe. Ecological studies, vol 142. Springer, Berlin Heidelberg New York, pp 468–490

Schwesig D, Kalbitz K, Matzner E (2003a) Effects of aluminium on the mineralization of dissolved organic carbon derived from forest floors. Eur J Soil Sci 54:311–322

Schwesig D, Kalbitz K, Matzner E (2003b) Mineralization of dissolved organic carbon in mineral soil solution of two forest soils. J Plant Nutr Soil Sci 166:585–593

Sedjo RA (1992) Temperate forest ecosystems in the global carbon cycle. Ambio 21:274–277

Skjelevale BL, Andersen T, Halvorsen GA, Raddum GG, Heegaard E, Stoddard JL, Wright RF (2000) The 12 year report: acidification of surface water in Europe and North America; trends, biological recovery and heavy metals. ICP Waters Rep 52:1–115

Spieker H, Mielikäinen R, Köhl M, Skorgsgaard JP (1996) Growth trends in European forests. Springer, Berlin Heidelberg New York

Sullivan TJ (2000) Aquatic effects of acidic deposition. Lewis Publishers, New York, pp 1–373

Tamm CO (1991) Nitrogen in terrestrial ecosystems. Springer, Berlin Heidelberg New York

Tipping E, Wolf C, Rigg E, Harrison AF, Ineson P, Taylor K, Benham D, Roskitt J, Rowland AP, Bol R, Harkness DD (1999) Climatic influences on the leaching of dissolved organic matter from upland UK moorland soils investigated by a field manipulation experiment. Environ Int 25:83–95

Tveite B, Abrahamsen G, Stuanes AO (1991) Liming and wet acid deposition effects on tree growth and nutrition: experimental results. In: Zöttl HW, Hüttl RF (eds) Management of nutrition in forests under stress. Kluwer, Dordrecht, pp 409–422

Ulrich B (1983) A concept of forest ecosystems stability and of acid deposition as driving force for destabilization. In: Ulrich B, Pankrath J (eds) Effects of accumulation of air pollutants in forest ecosystems. Reidel, Dordrecht, pp 1–29

Ulrich B, Mayer R, Khanna PK (1980) Chemical changes due to acid precipitation in a loess-derived soil in central Europe. Soil Sci 130:193–199

Valentini R (ed) (2003) Fluxes of carbon, water and energy of European forests. Ecological studies, vol 164. Springer, Berlin Heidelberg New York

Valentini R, Matteucci G, Dolman AJ, Schulze ED, Rebmann C, Moors EJ, Granier A, Gross P, Jensen NO, Pilegaard K, Lindroth A, Grelle A, Bernhofer C, Grünwald T, Aubinet M, Ceulemans R, Kowalski AS, Vesala T, Rannik Ü, Berbigier P, Loustau D, Gudmundsson J, Thorgelrsson H, Ibrom A, Morgenstern K, Clement R, Moncrieff J, Montagnani L,

Minerbi S, Jarvis PG (2000) Respiration as the main determinant of carbon balance in European forests. Nature 404:861–865
Van Breemen N, Mulder J, Driscoll CT (1983) Acidification and alkalinization of soils. Plant Soil 75:283–308
Van der Salm C, de Vries W (2001) A review of the calculation procedure for critical acid loads for terrestrial ecosystems. Sci Total Environ 271:11–25
Vejre H, Ingerslev M, Raulund-Rasmussen K (2001) Fertilization of Danish forests: a review of experiments. Scand J For Res 16:502–513
Von Carlowitz HC (1713) Sylvicultura Oekonimica oder Haußwirthliche Nachricht und Naturgemäßige Anweisung zur wilden Baum-Zucht. Braun, Leipzig
Waring RH, Running SW (1978) Sapwood water storage: its contribution to transpiration and effect upon water conductance through stems of old growth Douglas fir. Plant Cell Environ 1:131–140
Wedler M, Köstner B, Tenhunen JD (1996) Understory contribution to stand total water loss at an old Norway spruce forest. Verh Ges Ökol 26:69–77

Subject Index

Ecological Studies
Volumes published since 1998

Volume 132
Vegetation of the Tropical Pacific Islands (1998)
D. Mueller-Dombois and F.R. Fosberg

Volume 133
Aquatic Humic Substances: Ecology and Biogeochemistry (1998)
D.O. Hessen and L.J. Tranvik (Eds.)

Volume 134
Oxidant Air Pollution Impacts in the Montane Forests of Southern California (1999)
P.R. Miller and J.R. McBride (Eds.)

Volume 135
Predation in Vertebrate Communities: The Białowieża Primeval Forest as a Case Study (1998)
B. Jędrzejewska and W. Jędrzejewski

Volume 136
Landscape Disturbance and Biodiversity in Mediterranean-Type Ecosystems (1998)
P.W. Rundel, G. Montenegro, and F.M. Jaksic (Eds.)

Volume 137
Ecology of Mediterranean Evergreen Oak Forests (1999)
F. Rodà et al. (Eds.)

Volume 138
Fire, Climate Change and Carbon Cycling in the North American Boreal Forest (2000)
E.S. Kasischke and B. Stocks (Eds.)

Volume 139
Responses of Northern U.S. Forests to Environmental Change (2000)
R. Mickler, R.A. Birdsey, and J. Hom (Eds.)

Volume 140
Rainforest Ecosystems of East Kalimantan: El Niño, Drought, Fire and Human Impacts (2000)
E. Guhardja et al. (Eds.)

Volume 141
Activity Patterns in Small Mammals: An Ecological Approach (2000)
S. Halle and N.C. Stenseth (Eds.)

Volume 142
Carbon and Nitrogen Cycling in European Forest Ecosystems (2000)
E.-D. Schulze (Ed.)

Volume 143
Global Climate Change and Human Impacts on Forest Ecosystems: Postglacial Development, Present Situation and Future Trends in Central Europe (2001)
J. Puhe and B. Ulrich

Volume 144
Coastal Marine Ecosystems of Latin America (2001)
U. Seeliger and B. Kjerfve (Eds.)

Volume 145
Ecology and Evolution of the Freshwater Mussels Unionoida (2001)
G. Bauer and K. Wächtler (Eds.)

Volume 146
Inselbergs: Biotic Diversity of Isolated Rock Outcrops in Tropical and Temperate Regions (2000)
S. Porembski and W. Barthlott (Eds.)

Volume 147
Ecosystem Approaches to Landscape Management in Central Europe (2001)
J.D. Tenhunen, R. Lenz, and R. Hantschel (Eds.)

Volume 148
A Systems Analysis of the Baltic Sea (2001)
F.V. Wulff, L.A. Rahm, and P. Larsson (Eds.)

Volume 149
Banded Vegetation Patterning in Arid and Semiarid Environments (2001)
D. Tongway and J. Seghieri (Eds.)

Volume 150
Biological Soil Crusts: Structure, Function, and Management (2001, 2003)
J. Belnap and O.L. Lange (Eds.)

Volume 151
Ecological Comparisons of Sedimentary Shores (2001)
K. Reise (Ed.)

Volume 152
Global Biodiversity in a Changing Environment: Scenarios for the 21st Century (2001)
F.S. Chapin, O. Sala, and E. Huber-Sannwald (Eds.)

Volume 153
UV Radiation and Arctic Ecosystems (2002)
D.O. Hessen (Ed.)

Volume 154
Geoecology of Antarctic Ice-Free Coastal Landscapes (2002)
L. Beyer and M. Bölter (Eds.)

Volume 155
Conserving Biological Diversity in East African Forests: A Study of the Eastern Arc Mountains (2002)
W.D. Newmark

Volume 156
Urban Air Pollution and Forests: Resources at Risk in the Mexico City Air Basin (2002)
M.E. Fenn, L. I. de Bauer, and T. Hernández-Tejeda (Eds.)

Volume 157
Mycorrhizal Ecology (2002, 2003)
M.G.A. van der Heijden and I.R. Sanders (Eds.)

Volume 158
Diversity and Interaction in a Temperate Forest Community: Ogawa Forest Reserve of Japan (2002)
T. Nakashizuka and Y. Matsumoto (Eds.)

Volume 159
Big-Leaf Mahogany: Genetic Resources, Ecology and Management (2003)
A. E. Lugo, J. C. Figueroa Colón, and M. Alayón (Eds.)

Volume 160
Fire and Climatic Change in Temperate Ecosystems of the Western Americas (2003)
T. T. Veblen et al. (Eds.)

Volume 161
Competition and Coexistence (2002)
U. Sommer and B. Worm (Eds.)

Volume 162
How Landscapes Change: Human Disturbance and Ecosystem Fragmentation in the Americas (2003)
G.A. Bradshaw and P.A. Marquet (Eds.)

Volume 163
Fluxes of Carbon, Water and Energy of European Forests (2003)
R. Valentini (Ed.)

Volume 164
Herbivory of Leaf-Cutting Ants: A Case Study on *Atta colombica* in the Tropical Rainforest of Panama (2003)
R. Wirth, H. Herz, R.J. Ryel, W. Beyschlag, B. Hölldobler

Volume 165
Population Viability in Plants: Conservation, Management, and Modeling of Rare Plants (2003)
C.A Brigham, M.W. Schwartz (Eds.)

Volume 166
North American Temperate Deciduous Forest Responses to Changing Precipitation Regimes (2003)
P. Hanson and S.D. Wullschleger (Eds.)

Volume 167
Alpine Biodiversity in Europe (2003)
L. Nagy, G. Grabherr, Ch. Körner, D. Thompson (Eds.)

Volume 168
Root Ecology (2003)
H. de Kroon and E.J.W. Visser (Eds.)

Volume 169
Fire in Tropical Savannas: The Kapalga Experiment (2003)
A.N. Andersen, G.D. Cook, and R.J. Williams (Eds.)

Volume 170
Molecular Ecotoxicology of Plants (2004)
H. Sandermann (Ed.)

Volume 171
Coastal Dunes: Ecology and Conservation (2004)
M.L. Martínez and N. Psuty (Eds.)

Volume 172
Biogeochemistry of Forested Catchments: in a Changing Environment A German Case Study (2004)
E. Matzner (Ed.)

Printing: Saladruck, Berlin
Binding: Stein+Lehmann, Berlin